Selected Works of A. N. Kolmogorov

Mathematics and Its Applications (*Soviet Series*)

Volume 25

Selected Works of A. N. Kolmogorov

Volume I
Mathematics and Mechanics

edited by

V. M. Tikhomirov

Translated from the Russian by V. M. Volosov

SPRINGER-SCIENCE+BUSINESS MEDIA, B.V.

Library of Congress Cataloging-in-Publication Data

Kolmogorov, A. N. (Andreĭ Nikolaevich), 1903-
[Selections. English. 1991]
Selected works of A.N. Kolmogorov / edited by V.M. Tikhomirov ; author, A.N. Kolmogorov.
p. cm. -- (Mathematics and its applications. Soviet series ; 25)
Includes index.
Contents: v. 1. Mathematics and mechanics
ISBN 978-94-010-5347-1 ISBN 978-94-011-3030-1 (eBook)
DOI 10.1007/978-94-011-3030-1
1. Mathematics. 2. Mechanics, Analytic. 3. Kolmogorov, A. N. (Andreĭ Nikolaevich), 1903- . I. Tikhomirov, V. M. (Vladimir Mikhaĭlovich), 1934- . II. Title. III. Series: Mathematics and its applications (Kluwer Academic Publishers). Soviet series ; 25.
QA3.K792513 1991
510--dc20 91-15705

ISBN 978-94-010-5347-1

Printed on acid-free paper

This is an annotated translation from the original work
Избранные труды
МАТЕМАТИКА И МЕХАНИКА
Published by Nauka, Moscow, © 1985

Contents

Series Editor's Preface v

Editor's Foreword xiii

Andrei Nikolaevich Kolmogorov. A brief biography xv

Preface to the English edition xix

1. A Fourier-Lebesgue series divergent almost everywhere 1
2. On the order of magnitude of the coefficients of Fourier-Lebesgue series 5
3. A remark on the study of the convergence of Fourier series 8
4. On convergence of Fourier series (in collaboration with G.A. Seliverstov) 10
5. Axiomatic definition of the integral 13
6. On the limits of generalization of the integral 15
   I. Introduction (15). II. Definition of $\mathcal{K}$ systems (19). III (21). IV (22). V (26). VI (27).
7. On the possibility of a general definition of derivative, integral and summation of divergent series 33
8. On conjugate harmonic functions and Fourier series 35
9. On the tertium non datur principle 40
   Introduction (40). I. Formal and intuitionistic points of view (40). II. Axioms of propositional logic (44). III. Particular propositional logic and its domain of applicability (51). IV. Pseudo-truth mathematics (55). V. Applications (60).
10. On convergence of Fourier series (in collaboration with G.A. Seliverstov) 69
11. A Fourier-Lebesgue series divergent of everywhere 74
12. On convergence of series of orthogonal functions (in collaboration with D.E. Men'shov) 76
13. On operations on sets 87
14. On the Denjoy integration process 95
15. On the topological group-theoretic foundation of geometry 97

16. Studies on the concept of integral 100
Chapter 1. Introduction 100
Chapter 2. The first integration theory (Countable partitions) . . . 105
§1. Definition and introductory remarks (105). §2. Definition and elementary properties of the integral (108). §3. Differential equivalence and a second definition of the integral (110). §4. A remark on infinite values of the integral (114). §5. Some special cases of integrability. Semi-additive functions and functions of bounded variation (116). §6. Some results on homogeneous functions (119). §7. Integrals of the type $\int_E f(x)\phi(dE)$ (122). §8. Examples of $\mathfrak{M}$-systems (126).
Chapter 3. The second integration theory (Finite partitions) . . . 128
§1. Introductory remarks (128). §2. Decomposable systems (129). §3. On the interrelation between the two methods of integration (132). §4. Examples (137).
Appendix 1. The integral as a many-valued function 138
Appendix 2. On the justification of differentiation on abstract sets . 140
17. On the notion of mean 144
18. On the compactness of sets of functions in the case of convergence in mean 147
19. On the interpretation of intuitionistic logic 151
20. On the foundation of projective geometry 159
21. On measure theory 161
Introduction (161). §1. Non-expanding mappings (163). §2. Maximum measure $\overline{\mu}^k(E)$ (164). §3. Minimum measure $\underline{\mu}^k(E)$ (167). §4. Some special properties of the measure functions $\overline{\mu}^k(E)$ and $\underline{\mu}^k(E)$ (168). §5. Cases when a linear measure is uniquely defined (170). §6. Measure theory in general metric spaces (172). §7. Uniqueness theorem for point sets in Euclidean spaces and its relation with surface measure theory (173). Appendix. New definition of measure of a linear set (178).
22. On points of discontinuity of functions of two variables (in collaboration with I.Ya. Verchenko) 181
23. On normability of a general topological linear space 183
24. Continuation of the study of points of discontinuity of functions of two variables (in collaboration with I.Ya. Verchenko) 187

25. On the convergence of series in orthogonal polynomials 190
26. Laplace transformation in linear spaces 194
27. On the order of magnitude of the remainder term in the Fourier series of differentiable functions 196
§1. (197). §2. Asymptotic estimation of $C_n^{(p)}$ for $p = 2i + 1$ ($i \geq 0$) (198). §3. Asymptotic estimation of $C_n^{(p)}$ for $p = 2i$ ($i \geq 1$) (200).
28. On the best approximation of functions of a given class 202
29. On duality in combinatorial topology 206
30. Homology rings of complexes and locally bicompact spaces 214
§1. The homology ring of a complex (214). §2. The homology ring of a locally bicompact space (219). §3. The case of two-sided manifolds (219).
31. Finite coverings of topological spaces (in collaboration with P.S. Aleksandrov) 221
32. The Betti groups of locally bicompact spaces 226
33. Properties of the Betti groups of locally bicompact spaces 229
34. The Betti groups of metric spaces 232
35. Relative cycles. The Alexander duality theorem 234
36. On open mappings 236
37. Skew-symmetric forms and topological invariants 239
38. A study of the diffusion equation with increase in the amount of substance, and its application to a biological problem (in collaboration with I.G. Petrovskii and N.S. Piskunov) 242
39. A simplified proof of the Birkhoff-Khinchin ergodic theorem 271
§1. Introduction (271). §2. Reduction to the discrete case (272). §3. Proof of the main theorem (274).
40. On inequalities for suprema of consecutive derivatives of an arbitrary function on an infinite interval 277
§1.Statement of the problem and results (277)§2.Sufficiency of condition (1) (281). §3. The case $k = 1$ (283). §4. Necessity of condition (1) for any n (289).
41. On rings of continuous functions on topological spaces (in collaboration with I.M. Gel'fand) 291
§1. The case of a bicompact space (292)§2.The ring $C'(S)$ in the general case (294). §3. The ring $C(S)$ in the general case (295).

42. Curves in a Hilbert space invariant with respect to a one-parameter group of motions 298
43. Wiener spirals and some other interesting curves in a Hilbert space 303
44. Points of local topological character of countably-multiple open mappings of compacta 308
45. Local structure of turbulence in an incompressible viscous fluid at very large Reynolds numbers 312
46. On the degeneration of isotropic turbulence in an incompressible viscous fluid 319
47. Dissipation of energy in isotropic turbulence 324
48. Equations of turbulent motion in an incompressible fluid 328
49. A remark on Chebyshev polynomials least deviating from a given function 331
50. On the breakage of drops in a turbulent flow 339
51. On dynamical systems with an integral invariant on a torus 344
52. On the preservation of conditionally periodic motions under small variations of the Hamilton function 349
53. The general theory of dynamical systems and classical mechanics . 355
Introduction (355). §1. Analytic dynamical systems and their stable properties (357). §2. Dynamical systems on a two-dimensional torus and some canonical systems with two degrees of freedom (360). §3. Are dynamical systems on compact manifolds "in general" transitive, and should we regard the continuous spectrum as the "general" case and the discrete spectrum as an "exceptional" case ? (365). §4. Some remarks on the non-compact case (367). §5. Transitive measures, spectra, and eigenfunctions of analytic systems (370).
54. Some fundamental problems in the approximate and exact representation of functions of one or several variables 375
55. On the representation of continuous functions of several variables as superpositions of continuous functions of a smaller number of variables 378
56. On the representation of continuous functions of several variables as superpositions of continuous functions of one variable and addition . 383
§1. Construction of the functions ψ^{pq} (384). §2. Construction of the functions χ^q (385).

57. On the linear dimension of topological vector spaces 388
§1. The linear dimension $\delta(E)$ (389). §2. The approximate dimension $d_a(E)$ (391).
58. A refinement of the concept of the local structure of turbulence in an incompressible viscous fluid at large Reynolds numbers 393
59. P.S. Aleksandrov and the theory of δs-operations 398
60. A qualitative study of mathematical models of population dynamics 403
§1. Introductory notes (403). §2. Volterra's equations and their generalization (404). §3. Investigation of the behaviour of solutions of the system (4) (405).

Commentary 410

On the papers on the theory of functions and set theory (A.N. Kolmogorov) 410
Trigonometric and orthogonal series (P.L. Ul'yanov) 411
A. Divergence of trigonometric Fourier series (411). B. Divergence of orthogonal series (413). C. Convergence of Fourier series (414). D. On Fourier-Lebesgue coefficients (416). E. Conjugate functions and series (418).
Descriptive set theory (I.I. Parovichenko) 424
Measure theory and theory of the integral (V.A. Skvortsov) 428
On the paper "Studies on the concept of integral" (428). On the paper "On the limits of generalization of the integral" (429). On the paper "On the Denjoy integration process" (430). On the paper "On measure theory" (430). On the paper "On the notion of the mean" (431).
Points of discontinuity of functions (E.P. Dolzhenko) 435
Theory of approximation (S.A. Telyakovskii and V.M. Tikhomirov) . . 436
Inequalities for derivatives (V.M. Tikhomirov and G.G. Magaril-Il'yaev) 442
Rings of continuous functions (E.A. Gorin) 447
Curves in a Hilbert space (Yu.A. Rozanov) 451
On the papers on intuitionistic logic (A.N. Kolmogorov) 451
Intuitionistic logic (V.A. Uspenskii and V.E. Plisko) 452
On the papers on homology theory (A.N. Kolmogorov) 466
Homology theory (G.S. Chogoshvili) 467
On the paper "On open mappings" (A.N. Kolmogorov) 476

Topology (A.V. Arkhangel'skii) 476

On the paper "Finite coverings of topological spaces"(476). On the paper "On open mappings" (476).

Axiomatics of projective geometry (A.V. Mikhalev) 479

On the paper on the diffusion equation (A.N. Kolmogorov) 481

The diffusion equation (G.I. Barenblatt) 481

On the papers on turbulence (A.N. Kolmogorov) 487

Turbulence (A.M. Yaglom) 488

On the papers on classical mechanics (A.N. Kolmogorov) 503

Classical mechanics (V.I. Arnol'd) 504

1. The basic problem of dynamics (504). 2. Rigid body and planetoid (504). 3. Limiting degeneration (505). 4. Self-degeneration (506). 5. Diffusion (506). 6. Measure of the exceptional set (507). 7. Smoothness (508). 8. Non-Hamiltonian systems (509). 9. Numerical trials (510). 10. Exponential instability (510). 11. Exchange, capture, and oscillation (512). 12. Spectral properties (513).

On the papers on superpositions (A.N. Kolmogorov) 518

Superpositions (V.I. Arnol'd) 519

1. Representability theorems (519)§2. Non-representability theorems (520).§3. Hilbert's thirteenth problem (521). 4. The resolvent problem (522). 5. Approximations (523).

List of works by A.N. Kolmogorov 528

Editors' Foreword

The Praesidium of the USSR Academy of Sciences has decided to publish three volumes of Selected Works of A.N. Kolmogorov, one of the most prominent mathematicians of the 20th century.

The creative work of A.N. Kolmogorov is exceptionally versatile. In his studies on trigonometric and orthogonal series, theory of measure and integral, mathematical logic, approximation theory, geometry, topology, functional analysis, classical mechanics, ergodic theory, superposition of functions, and information theory, many conceptual and fundamental problems were solved and new questions were posed which gave rise to a great number of investigations. A.N. Kolmogorov is one of the founders of the Soviet school of probability theory, mathematical statistics, and the theory of turbulence. In these areas he obtained a number of basic results, with many applications to mechanics, geophysics, linguistics, biology and other branches of knowledge.

This edition includes the most important papers by A.N. Kolmogorov on mathematics and natural science. It does not include philosophical and pedagogical studies of A.N. Kolmogorov, his articles written for the "Bol'shaya Sov'etskaya Entsiklopediya", papers on prosody and various applications of mathematics and publications on general questions.

The material of this edition was selected and grouped by A.N. Kolmogorov.

The first volume consists of papers on mathematics (except those relating to probability theory and information theory) and also on turbulence and classical mechanics. The second volume is devoted to probability theory and mathematical statistics. The third volume contains papers on information theory and the theory of automata.

Inside each volume the papers are placed in chronological order. The papers are divided by A.N. Kolmogorov into separate sections. Most of the sections are supplied with commentary, in which an attempt is made to reflect upon the influence of A.N. Kolmogorov's creative work on the development of modern mathematics. As a rule, the commentary is preceded by short introductory notes made by A.N. Kolmogorov.

The first volume was prepared by S.M. Nikol'skii and V.M. Tikhomirov, and the other two by Yu.V. Prokhorov and A.N. Shiryaev.

The creative work of A.N. Kolmogorov is a living constituent part of modern science, and we hope that the publication of his Selected Works will arouse great interest among scientists.

Andrei Nikolaevich Kolmogorov

A Brief Biography

Andrei Nikolaevich Kolmogorov was born on 25 April 1903 in Tambov. His father, Nikolai Matveevich Kataev was an algriculturist. His mother, Mariya Yakovlevna Kolmogorova, died in childbirth and his mother's sister, Vera Yakovlevna, took care of him. His early childhood was passed at the estate of his mother's parents in the village of Tunoshna near Yaroslavl'. When he was about six years old, he and his aunt V.Ya. Kolmogorova moved to Moscow. In 1910 A.N. Kolmogorov was accepted into the first preparatory class of E.A. Repman's private gymnasium in Moscow. In the difficult years of 1919–1920 Andrei Nikolaevich worked for the railway. In 1920, upon receiving the certificate of the 90th second-stage school (this was the new name of the gymnasium where he had begun to study), A.N. Kolmogorov entered the Physics-Mathematics Department of Moscow University.

He did not immediately decide to become a mathematician. Among the broad circle of interests of Andrei Nikolaevich, an important part was played by history. Andrei Nikolaevich attended the seminar of the prominent Russian historian, Professor S.V. Bakhrushin. At the seminar, Andrei Nikolaevich carried out research on landowners in Novgorod according to cadasters of the 15th–16th centuries. At the same time Andrei Nikolaevich obtained outstanding results in the theory of trigonometric series and set theory and, as a consequence, his interest in mathematics outweighed the others.

In his first year at the University (1920–1921) A.N. Kolmogorov attended the lectures of N.N. Luzin (theory of functions) and A.K. Vlasov (projective geometry) and participated in the seminar of V.V. Stepanov on trigonometric series. In 1921 N.N. Luzin offered to make A.N. Kolmogorov his student. That same year Andrei Nikolaevich completed his first work on trigonometric series, and at the beginning of 1922 he accomplished his first study in descriptive set theory. In the summer of 1922, A.N. Kolmogorov constructed a Fourier series divergent almost everywhere (the paper was dated 2 June, 1922), and this result immediately gained him worldwide recognition.

In 1925 Andrei Nikolaevich graduated from Moscow University and became N.N. Luzin's post-graduate student. The same year Andrei Nikolaevich began his investigations in probability theory.

In 1929 he took his first long boating cruise on the Volga with Pavel Sergeevich Aleksandrov, marking the beginning of a friendship which lasted to the last days of Pavel Sergeevich (P.S. Aleksandrov died on 16 November, 1982). In 1935 A.N. Kolmogorov and P.S. Aleksandrov acquired a house in Komarovka (on the banks of the river Klyaz'ma, not far from Moscow) where, for the most part, their creative life was spent.

In 1929, on completing his post-graduate course, A.N. Kolmogorov became a senior research worker at the Institute of Mathematics at Moscow University. From June 1930 to March 1931, Andrei Nikolaevich was on his first scientific mission abroad, visiting Göttingen, Munich and Paris. In 1931 Kolmogorov became a professor at Moscow University.

In 1933 A.N. Kolmogorov was appointed Director of the Institute of Mathematics and Mechanics of Moscow University. He held this post until 1939 (and again for a short period from 1951 to 1953). In 1939 A.N. Kolmogorov was made a Full Member of the Academy of Sciences of the USSR and became Secretary-Academician of the Physics-Mathematics Science Section of the Academy. That same year Andrei Nikolaevich was appointed to the Chair of Probability Theory of the Physics-Mathematics Department of Moscow University (until 1966) and chief of the Probability Theory Department of the Steklov Institute of Mathematics of the USSR Academy of Sciences (until 1958). In 1941 A.N. Kolmogorov was awarded the USSR State Prize (together with A.Ya. Khinchin) for work on probability theory.

At the end of the thirties and the beginning of the forties, A.N. Kolmogorov became interested in problems of turbulence and became head of the Laboratory of Atmospheric Turbulence of the Institute of Geophysics of the USSR Academy of Sciences, where he worked until 1949.

During the Second World War A.N. Kolmogorov actively participated in working out problems related to the defence of his motherland.

In the autumn of 1942 Andrei Nikolaevich married Anna Dmitrievna Egorova, a school friend.

In 1949 A.N. Kolmogorov was given the P.L. Chebyshev Prize of the USSR Academy of Sciences (together with B.V. Gnedenko).

From 1954 to 1958 A.N. Kolmogorov was dean of the Mechanics-Mathematics Department of Moscow University. In 1958 in the spring semester, he was a Visiting Professor at the University of Paris. In 1960 A.N. Kolmogorov founded the Laboratory of Probability and Statistical Methods at Moscow University. He was scientific advisor (until 1966) and head of the Laboratory (from 1966 to 1976).

In 1954 Andrei Nikolaevich took part in the International Congress of Mathematics in Amsterdam. There he gave an hour-long review lecture on celestial mechanics. Andrei Nikolaevich was given the honour of closing the scientific program of the Amsterdam Congress with this lecture. In 1954 Andrei Nikolaevich was also, for two months, a Visiting Professor at Humboldt University in Berlin. In 1962, 1966 and 1970 Andrei Nikolaevich took part in the International Congresses of Mathematics in Stockholm, Moscow and Nice.

In 1963 A.N. Kolmogorov was one of the sponsors of the Boarding School at Moscow University and in subsequent years he devoted much effort and energy to this educational institute.

In 1965 A.N. Kolmogorov won the Lenin Prize (together with V.I. Arnol'd) for work on classical mechanics.

In 1970 and in 1971–1972, A.N. Kolmogorov took part in two four-month cruises on the research vessel "Dmitrii Mendeleev".

From 1970 to 1980 A.N. Kolmogorov occupied the newly founded Chair of Mathematical Statistics at the Mechanics-Mathematics Department of Moscow University, and from 1980 on he occupied the Chair of Mathematical Logic at the same Department.

At the meeting of 28 April 1953, A.N. Kolmogorov was elected an Honorary Member of the Moscow Mathematical Society and in the period from 1964 to 1973 he was President of this Society.

Among the students of A.N. Kolmogorov are I.M. Gel'fand, A.I. Mal'tsev, M.D. Millionshchikov, V.S. Mikhalevich, S.M. Nikol'skii, A.M. Obukhov, Yu.V. Prokhorov (Full Members of the USSR Academy of Sciences), V.I. Arnol'd, L.N. Bol'shev, A.A. Borovkov, A.S. Monin, B.A. Sevast'yanov (Corresponding Members of the USSR Academy of Sciences), B.V. Gnedenko, S.Kh. Siradzhinov (Members of the Republican Academies of Sciences),

Yu.A. Rozanov (a Lenin Prize winner) and about sixty doctors and candidates of science.

Many universities and academies throughout the world elected A.N. Kolmogorov as their member. He was made an Honorary Member of the Romanian Academy of Sciences (1965; a Corresponding Member from 1957 onwards), a Foreign Member of the Polish Academy of Sciences (1956), an Honorary Member of the London Royal Statistical Society (1956), a Foreign Member of the Royal Netherlands Academy of Sciences (1963), a Member of the Royal Society of London (1964), an Honorary Member of the Hungarian Academy of Sciences (1965), a Member of the National Academy of Sciences of the United States of America (1967), a Foreign Member of the Paris Academy of Sciences (1968), an Honorary Member of the Calcutta Mathematical Society, etc; he received Honorary Doctorates from the Universities of Paris, Stockholm, Warsaw and Budapest, and he was deemed worthy of the International Prize of the Bolzano Foundation, a Gold Medal from the American Meteorological Society, the Helmholtz Medal, etc.

A.N. Kolmogorov's outstanding achievements were highly appreciated by the Soviet Government. He was awarded many State Orders and Medals, and in 1963 he was given the title of Hero of Socialist Labour.

1985 V.M. Tikhomirov

Preface to the English Edition

On 20 October 1987, one of the most outstanding scientists of the 20th century, Andrei Nikolaevich Kolmogorov, passed away.

A serious illness darkened his last years, did not allow him to move without assistance, hampered his speech, and made it difficult for him to communicate with people. But his intellect did not fail him. The short introductory notes preceding the commentary to some of the sections of papers in his Selected Works are the last words of the great scientist addressed to the mathematical world. At first he tried to write — in huge letters on very large sheets of paper. But later even this became impossible for him and he began to dictate, summoning all his strength to fulfil what he felt was his primary duty. These notes reflect the unique traits of his personality but they are too short to reveal the mystery of his extraordinary creative life.

Andrei Nikolaevich said many times (and not only to me) that at the end of his life he would like to write his mathematical biography. Alas! This intention has not been realized and I believe, could not be realized because his thoughts were always turned to the future and not to the past. So the enigma of his creative work has to be unravelled by his followers. In the preparation of these Selected Works a first attempt has been made to reveal the influence of the most important papers by A.N. Kolmogorov on the development of science. His commentary is meant to give the most essential initial outlines.

Andrei Nikolaevich was eager to see his Selected Works published in the West. He insisted that all his papers be translated into the English language, the main means of communication in modern mathematics. On behalf of his friends and pupils, I should like to express our gratitude to the Publishing House that undertook to fulfil the wishes of Andrei Nikolaevich. Unfortunately, he could not see this book published in English. Thanks are also due to Professor V.M. Volosov who translated the present volume.

1989 V.M. Tikhomirov

1. A FOURIER-LEBESGUE SERIES DIVERGENT ALMOST EVERYWHERE*

The purpose of this paper is to give an *example of a summable*[1] *function with an almost everywhere* (i.e. everywhere except on a set of measure zero) *divergent Fourier series*).

The function constructed in the paper is not square summable, and I know nothing about the order of magnitude of the coefficients of its Fourier series. The methods used here do not allow one to construct an everywhere divergent Fourier series.

1. In the sequel I will prove that there exists a sequence of functions $\phi_1(x), \ldots, \phi_n(x), \ldots$, defined for $0 \le x \le 2\pi$, such that

1° $\phi_n(x) \ge 0, \quad \int_0^{2\pi} \phi_n(x)dx = 2 \quad (n = 1, 2, \ldots);$

2° the partial sums of the Fourier series of $\phi_n(x)$ are bounded;

3° one can associate with each $\phi_n(x)$ a positive number M_n, a set E_n and an integer q_n such that

3a) $\lim M_n = \infty$;

3b) $\lim_{n\to\infty} \operatorname{mes} E_n = 2\pi$;

3c) for every point of E_n there is a partial sum of the Fourier series of ϕ_n whose index does not exceed q_n and whose absolute value is greater than M_n.

Assuming that the functions $\phi_n(x)$ have already been constructed, it is easy to find an increasing sequence of integers $n_1, n_2, \ldots, n_k, \ldots$ such that

A) $\dfrac{1}{\sqrt{M_{n_k}}} \le \dfrac{1}{2^k}$, so that $\displaystyle\sum_{k=1}^{\infty} \frac{1}{\sqrt{M_{n_k}}} \le 1$;

B) $\frac{1}{2}\sqrt{M_{n_k}}$ is greater than the sum of the maxima of the absolute values of the partial sums of the Fourier series of the $k-1$ functions $\phi_{n_1}, \ldots, \phi_{n_{k-1}}$;

C) $q_{n_i} \le \dfrac{1}{2^k}\sqrt{M_{n_k}}$ for all $i < k$.

If the n_i are known for all $i < k$ then n_k satisfying A), B), C) can also be defined.

We now set

$$\Phi(x) = \sum_{k=1}^{\infty} \frac{1}{\sqrt{M_{n_k}}} \phi_{n_k}(x)$$

* 'Une série de Fourier-Lebesgue divergente presque partout', *Fund. Math.* **4** (1923), 324–328.

[1] I.e., Lebesgue integrable.

By 1° and A), this series converges[2] almost everywhere to a summable function $\Phi(x)$ and the Fourier coefficients of $\Phi(x)$ coincide with the sums of the Fourier coefficients of the functions

$$\frac{1}{\sqrt{M_{n_k}}}\phi_{n_k}(x) \quad (k = 1, 2, \ldots).$$

According to 3°, there is a partial sum of the Fourier series of $\phi_{n_k}(x)$ that is greater than M_{n_k} at the points of E_{n_k}. Consider a partial sum of the Fourier series of $\Phi(x)$. We have:

a) for the series of $\phi_{n_k}(x)/\sqrt{M_{n_k}}$ the corresponding sum is greater than $\sqrt{M_{n_k}}$;

b) By B), for the sum of the first $k-1$ terms of the series representing $\Phi(x)$ it is smaller than $\frac{1}{2}\sqrt{M_{n_k}}$;

c) for the terms with indices $s > k$ it is smaller than $6/2^s$.

For by C), the partial sum corresponding to the index $q_{n_k} \leq (1/2^s)\sqrt{M_{n_s}}$ is smaller than $(2q_{n_k}+1)$ times the integral of the absolute value of the function, which in this case is equal to $2/\sqrt{M_{n_s}}$.

It follows from a), b) and c) that the absolute value of the corresponding partial sum of the Fourier series of $\Phi(x)$ is not smaller than

$$\frac{1}{2}\sqrt{M_{n_k}} - \frac{6}{2^k}.$$

From this we conclude that the Fourier series of $\Phi(x)$ is divergent at each point of the set $E = \overline{\lim_{k\to\infty}} E_{n_k}$ while $\operatorname{mes} E = 2\pi$.

2. The construction of $\phi_n(x)$.

Let

$$\lambda_1 = 1, \ \lambda_2, \ldots, \lambda_n$$

be an increasing finite sequence of odd numbers satisfying certain conditions, given below.

Define a sequence $m_1, m_2, \ldots, m_n$ by

$$m_1 = n, \quad 2m_k + 1 = \lambda_k(2n+1). \tag{1}$$

[2] See, for example, *Fund. Math.* **4** (1923), p.211, Fubini's theorem (which reads: an everywhere convergent series of non-decreasing functions can be differentiated termwise).

We set

$$A_k = k\frac{4\pi}{2n+1}, \quad 1 \le k \le n, \quad A_n = 2\pi - \frac{2\pi}{2n+1} \tag{2}$$

Finally, we set $\phi_n(x) = m_k^2/n$ on the interval

$$\Delta_k = \left[A_k - \frac{1}{m_k^2},\ A_k + \frac{1}{m_k^2}\right]. \tag{3}$$

For each x outside the intervals Δ_k we put $\phi_n(x) = 0$.

It is obvious that

$$\phi_n(x) \ge 0, \qquad \int_0^{2\pi} \phi_n(x)dx = 2, \qquad \text{(condition } 1^\circ\text{)}$$

and $\phi_n(x)$ has bounded variation; hence 2° is also satisfied.

Consider the partial sum of the Fourier series of $\phi_n(x)$ corresponding to the index m_k:

$$\frac{1}{\pi}\int_0^{2\pi} \phi_n(\alpha)\frac{\sin((2m_k+1)(\alpha-x)/2)}{2\sin((\alpha-x)/2)}d\alpha. \tag{4}$$

Suppose that x belongs to the interval

$$\sigma_k = \left[A_{k-1} + \frac{2}{n^2},\ A_k - \frac{2}{n^2}\right]. \tag{5}$$

If the λ_i are defined for $i < k$ and, consequently, $\phi_n(x)$ is defined on Δ_i, then we can take λ_k so large that the integral (4) over the union of all intervals Δ_i $(i < k)$ is arbitrarily small for each x belonging to σ_k. I assume that the absolute value of the integral is smaller than 1.

We now consider the integral (4) over the interval Δ_s $(s \ge k)$:

$$\frac{1}{\pi}\int_{\Delta_s}\frac{m_s^2\sin((2m_k+1)(\alpha-x)/2)}{n2\sin((\alpha-x)/2)}d\alpha = \frac{1}{\pi}\int_{\Delta_s}\frac{m_s^2}{n}\frac{\sin((2m_k+1)(A_s-x)/2)}{2\sin((A_s-x)/2)}d\alpha +$$

$$+\frac{1}{\pi}\int_{\Delta_s}\frac{m_s^2}{n}\left[\frac{\sin((2m_k+1)(\alpha-x)/2)}{2\sin((\alpha-x)/2)} - \frac{\sin((2m_k+1)(A_s-x)/2)}{2\sin((A_s-x)/2)}\right]d\alpha. \tag{6}$$

Taking into account the fact that $|\alpha - A_s| \le 1/m_s^2$, we can conclude that the difference inside the square brackets is smaller than

$$\frac{1}{m_s^2}\max\left|\frac{d}{d\alpha}\frac{\sin((2m_k+1)\alpha/2)}{2\sin(\alpha/2)}\right| \le \frac{4m_k^2}{m_s^2} \le 4.$$

Since the length of Δ_s is $2/m_s^2$, the second term in (6) is smaller than $4/n$.

Taking the constant out of the first integral, we find that (6) is equal to

$$\frac{2}{\pi n}\frac{\sin((2m_k+1)(A_s-x)/2)}{2\sin((A_s-x)/2)}+\frac{\tau}{n}, \quad |\tau|\leq 4. \tag{7}$$

The sum of the terms with τ corresponding to $s=k,k+1,\dots,n$ is smaller than 4 in absolute value. We note that

$$\frac{2m_k+1}{2}(A_s-A_k)=(s-k)\lambda_k 2\pi,$$
$$\sin(2m_k+1)(A_s-x)/2)=\sin((2m_k+1)(A_k-x)/2),$$
$$A_s-x<A_s-A_{k-1}=(s-k+1)\frac{4\pi}{2n+1}<(s-k+1)\frac{2\pi}{n},$$
$$\frac{1}{\sin\frac{1}{2}(A_s-x)}>\frac{n}{\pi(s-k+1)},$$

and hence the absolute value of the sum for $s=k,k+1,\dots,n$ of the first terms in (7) is equal to

$$\frac{1}{\pi n}\left|\sin((2m_k+1)(A_k-x)/2)\right|\sum_{s=k}^{n}\frac{1}{\sin\frac{1}{2}(A_s-x)}>$$
$$>\frac{1}{\pi^2}\left|\sin((2m_k+1)(A_k-x)/2)\right|\sum_{r=1}^{n-k}\frac{1}{r}-5. \tag{8}$$

Thus, for any x in σ_k the absolute value of (4) is greater than

$$\frac{1}{\pi^2}\left|\sin((2m_k+1)(A_k-x)/2)\right|\sum_{r=1}^{n-k}\frac{1}{r}-5.$$

Let E_n be the set of all x belonging to the union of the σ_k with $n-k>\sqrt{n}$ and having the property that

$$\frac{1}{\pi^2}\left|\sin((2m_k+1)(A_k-x)/2)\right|>\frac{1}{\sqrt{\sum_{r=1}^{[\sqrt{n}]}\frac{1}{r}}}=\frac{1}{N_n}.$$

One can see that for any point of E_n belonging to σ_k the m_kth partial sum of the Fourier series of $\phi_n(x)$ is greater than $N_n-5=M_n$.

It is easy to show that

$$\lim_{n\to\infty}\operatorname{mes}E_n=2\pi$$

Moscow, 2 June 1922

2. ON THE ORDER OF MAGNITUDE OF THE COEFFICIENTS OF FOURIER-LEBESGUE SERIES*

It is well known that the Fourier coefficients of a summable function tend to zero. In this paper I prove the following proposition on cosine series.

1. *For any sequence* $\{a_n\}_{n=1}^{\infty}$ *converging to zero there is a sequence* $\{a'_n\}_1^{\infty}$ *such that*

1) $|a_n| < a'_n$;

2) $\sum_{n=1}^{\infty} a'_n \cos nx$ *is the Fourier series of a summable function.*

We consider the series

$$a_0/2 + a_1 \cos x + \ldots + a_n \cos nx + \ldots \tag{1}$$

and we set

$$\Delta_n = a_n - a_{n+1}, \quad \Delta'_n = \Delta_n - \Delta_{n+1}.$$

Applying the Abel transformation twice, we obtain the two series

$$\frac{1}{2}\Delta_0 + \sum_{n=1}^{\infty} \Delta_n \frac{\sin((2n+1)x/2)}{2\sin(x/2)}, \tag{2}$$

$$\frac{1}{2}\Delta'_0 + \sum_{n=1}^{\infty} \Delta'_n \frac{1}{2}\Big(\frac{\sin((n+1)x/2)}{\sin(x/2)}\Big)^2. \tag{3}$$

If the condition

$$\lim_{n\to\infty} a_n \frac{\sin((2n+1)x/2)}{2\sin(x/2)} = \lim_{n\to\infty} \Delta_n \Big(\frac{\sin((n+1)x/2)}{\sin(x/2)}\Big)^2 = 0$$

is satisfied, then (1) converges simultaneously with (3). If $a_n \to 0$, the condition is satisfied everywhere except at the points $x \equiv 0 \,(\mathrm{mod}\, 2\pi)$. Hence $\Delta_n \to 0$.

If the series $\sum_{n=1}^{\infty} |\Delta_n|$ converges, then (3) and (1) also converge everywhere to some function $f(x)$, except at $x \equiv 0 \,(\mathrm{mod}\, 2\pi)$. Noting that

$$\frac{1}{2}\int_0^{2\pi} |\Delta'_n| \Big(\frac{\sin((n+1)x/2)}{\sin(x/2)}\Big)^2 dx = \pi(n+1)|\Delta'_n|,$$

we obtain the following assertion.

---

* 'Sur l'ordre de grandeur des coéfficients de la série de Fourier-Lebesgue', *Bull. Acad. Pol. Sci., Ser. A* (1923), 83–86.

2. *If the series* $\sum_{n=1}^{\infty}(n+1)|\Delta'_n|$ *converges and the coefficients* a_n *tend to zero, then* (3) *converges to a summable function. This means that the series* (1), *which converges to a summable function* $f(x)$ *everywhere except at* $x \equiv 0 \pmod{2\pi}$, *is a Fourier-Lebesgue series.*

In particular, if all the Δ'_n are positive, then

$$\sum_{n=0}^{\infty}(n+1)|\Delta'_n| = \sum_{n=0}^{\infty}(n+1)\Delta'_n = a_0.$$

Consequently,

3. *If the coefficients of a cosine series tend to zero and their second differences are positive, then this series is a Fourier-Lebesgue series.*

This proves Proposition 1, since for any sequence $\{a_n\}$ converging to zero there is a sequence $\{a'_n\}$ also converging to zero and such that $a'_n > |a_n|$ and the second differences of a'_n are positive.

Remark 1. Suppose that the series

$$\sum_{n=1}^{\infty} a_n \cos nx = f(x)$$

satisfies the conditions of Proposition 2. We set $x = my + \pi/2$. Then

$$f\left(my + \frac{\pi}{2}\right) = \sum_{n=1}^{\infty}(-a_{4n-3}\sin(4n-3)my + a_{4n-1}\sin(4n-1)my) + \\ + \sum_{n=1}^{\infty}(-a_{4n-2}\cos(4n-2)my + a_{4n}\cos 4nmy).$$

The first sum is the Fourier-Lebesgue series of

$$\frac{f(my+\pi/2) - f(-my+\pi/2)}{2}.$$

In particular, taking $m = 1$ we see that

$$-a_1 \sin y + a_3 \sin 3y - a_5 \sin 5y + a_7 \sin 7y \ldots$$

is a Fourier-Lebesgue series.

Remark 2. The remainder of (1) is equal to

$$R_n = \frac{1}{2}\sum_{k=n+1}^{\infty}\Delta'_k\left(\frac{\sin((k+1)x/2)}{\sin(x/2)}\right)^2 + \frac{1}{2}\Delta_n\left(\frac{\sin((n+1)x/2)}{\sin(x/2)}\right)^2 + \\ + a_n\frac{\sin((2n+1)x/2)}{2\sin(x/2)}.$$

Under the condition of Proposition 2 we have

$$\lim_{n\to\infty}\int_0^{2\pi}\Big|\sum_{k=n+1}^{\infty}\Delta'_k\Big(\frac{\sin((k+1)x/2)^2}{\sin(x/2)}\Big)^2\Big|dx=\lim_{n\to\infty}\sum_{k=n+1}^{\infty}\pi(k+1)|\Delta'_k|=0,$$

$$\lim_{n\to\infty}\int_0^{2\pi}\Big|\Delta_n\Big(\frac{\sin((n+1)x/2)}{\sin(x/2)}\Big)^2\Big|dx=\lim_{n\to\infty}\pi(n+1)|\Delta_n|=$$

$$=\lim_{n\to\infty}\pi(n+1)\Big|\sum_{k=n}^{\infty}\Delta'_k\Big|\leq\lim_{n\to\infty}\pi\sum_{k=n}^{\infty}(k+1)|\Delta'_k|=0.$$

In this case $\int_0^{2\pi}|R_n|dx$ tends to zero together with

$$|a_n|\int_0^{2\pi}\Big|\frac{\sin((2n+1)x/2)}{2\sin(x/2)}\Big|dx$$

and hence together with $|a_n|\log n$.

4. *If the series* $\sum_{n=1}^{\infty} n|\Delta'_n|$ *converges, then the condition*

$$\lim_{n\to\infty} a_n\log n=0$$

is necessary and sufficient for (1) *to converge in the mean* (*in the metric of* L).

Thus, out of the two Fourier-Lebesgue series

$$\sum_{n=2}^{\infty}\frac{\cos nx}{\log n},\quad \sum_{n=2}^{\infty}\frac{\cos nx}{(\log n)^{1+\epsilon}}$$

the second series converges in the mean whereas the first one does not (see [1]).

3 December 1922

References

1. S. Banach and H. Steinhaus, 'Sur la convergence en moyenne', *Bull. Acad. Sci. Cracovie*, 1918.

3. A REMARK ON THE STUDY OF CONVERGENCE OF FOURIER SERIES*

As usual, we set

$$S_n = \frac{a_0}{2} + \sum_{k=1}^{n}(a_k \cos kx + b_k \sin kx),$$
$$\sigma_n = \frac{S_0 + S_1 + \ldots + S_{n-1}}{n}.$$

1. **Theorem**. *Suppose that a sequence of integers n_m $(m = 1, 2, \ldots)$ satisfies the condition*

$$n_{m+1}/n_m > \lambda > 1.$$

Then the sequence S_{n_m} corresponding to the Fourier series of any arbitrary square integrable function converges almost everywhere to the function.

Proof. It is well known that the sequence $\{\sigma_{n_m}\}$ converges almost everywhere to the function, so it suffices to show that the sequence $\{S_{n_m} - \sigma_{n_m}\}$ converges to zero almost everywhere. It is easy to see, however, that the latter property is a consequence of the convergence of the following series:

$$\sum_{m=1}^{\infty} \int_0^{2\pi} (S_{n_m} - \sigma_{n_m})^2 dx. \tag{1}$$

We consider the sum of the first p terms of (1):

$$\frac{1}{\pi}\sum_{m=1}^{p} \int_0^{2\pi} (S_{n_m} - \sigma_{n_m})^2 dx = \sum_{m=1}^{p} \frac{1}{n_m^2} \sum_{k=1}^{k=n_m} k^2(a_k^2 + b_k^2) =$$
$$= \sum_{k=1}^{n_p} k^2(a_k^2 + b_k^2)\Big[\frac{1}{n_{m_k}^2} + \frac{1}{n_{m_k+1}^2} + \ldots + \frac{1}{n_p^2}\Big], \tag{2}$$

where n_{m_k} is determined by the inequality

$$n_{m_k-1} < k \leq n_{m_k}.$$

* 'Une contribution à l'étude de la convergence des séries de Fourier', *Fund. Math.* **5** (1924), 96–97.

It is obvious that

$$\frac{1}{n_{m_k}^2}+\frac{1}{n_{m_k+1}^2}+\ldots+\frac{1}{n_p^2}\le\frac{1}{n_{m_k}^2}\left(1+\frac{1}{\lambda^2}+\ldots+\frac{1}{\lambda^{2p}}\right)<$$
$$<\frac{1}{n_{m_k}^2}\frac{\lambda^2}{\lambda^2-1}\le\frac{1}{k^2}\frac{\lambda^2}{\lambda^2-1};$$

so the sum in (2) does not exceed

$$\frac{\lambda^2}{\lambda^2-1}\sum_{k=1}^{k=n_p}(a_k^2+b_k^2).$$

This implies that (1) converges, which completes the proof.

2. Theorem. *Suppose that only the terms with indices n_m are non-zero in a Fourier-Lebesgue series (with n_m satisfying the inequality in the statement of Theorem 1). Then this series converges almost everywhere.*

Proof. Theorem 2 is an immediate consequence of Theorem 1 for square integrable functions. However, it is true *for all integrable functions.* The sequence σ_{n_m} converges almost everywhere and hence only the difference

$$|S_{n_m-1}-\sigma_{n_m}|\le\sum_{k=1}^{m-1}\frac{n_k}{n_m}(|a_{n_k}|+|b_{n_k}|) \tag{3}$$

needs to be considered.

Since $|a_{n_k}|+|b_{n_k}|$ tends to zero for $k\to\infty$ and, on the other hand,

$$\sum_{k=1}^{m-1}\frac{n_k}{n_m}<\sum_{k=1}^{m-1}\frac{1}{\lambda^{m-k}}<\frac{1}{\lambda-1},$$

it follows that (3) tends to zero as $m\to\infty$.

7 October 1922

4. ON CONVERGENCE OF FOURIER SERIES*

(in collaboration with G.A. Seliverstov)

Hardy [1] proved the following theorem.

If the series

$$\sum_{n=1}^{\infty}(a_n^2+b_n^2)(\log n)^2$$

converges, then the series

$$\sum_{n=1}^{\infty}(a_n \cos nx + b_n \sin nx) \tag{1}$$

converges almost everywhere (except on a set of measure zero).

We know from D.E. Men'shov's work [2] that for general orthogonal series the factor $(\log n)^2$ cannot be replaced by a function $w(n)$ such that $w(n) = o[(\log n)^2]$.

In this paper we prove that the factor $(\log n)^2$ can be replaced by $(\log n)^{1+\epsilon}$ in the case of a trigonometric series.

Lemma. *For a trigonometric sum*

$$S(x) = \sum_{k=1}^{n}(a_k \cos kx + b_k \sin kx) \quad (n > 1)$$

we have

$$\int_0^{2\pi} S_{k(x)}(x)dx \leq \sqrt{C \log n \sum_{p=1}^{n}(a_p^2+b_p^2)}.$$

Here $k(x)$ is an arbitrary integer-valued function with range between 1 and n,

$$S_{k(x)}(x) = \sum_{p=1}^{k(x)}(a_p \cos px + b_p \sin px),$$

and C is an absolute constant.

---

* 'Sur la convergence des séries de Fourier', *C.R. Acad. Sci. Paris* **178** (1924), 303–306.

Proof. Using Schwarz's inequality, we have

$$\int_0^{2\pi} S_{k(x)}(x)dx = \frac{1}{\pi}\int_0^{2\pi}\int_0^{2\pi} S(\alpha)\sum_{p=1}^{k(x)} \cos p(x-\alpha)d\alpha dx =$$

$$= \frac{1}{\pi}\int_0^{2\pi} S(\alpha)\int_0^{2\pi}\sum_{p=1}^{k(x)} \cos p(x-\alpha)dx d\alpha \le$$

$$\le \sqrt{\frac{1}{\pi}\int_0^{2\pi} S^2(\alpha)d\alpha\frac{1}{\pi}\int_0^{2\pi}\Big[\int_0^{2\pi}\sum_{p=1}^{k(x)} \cos p(x-\alpha)dx\Big]^2 d\alpha}.$$

Further,

$$\int_0^{2\pi}\Big[\int_0^{2\pi}\sum_{p=1}^{k(x)} \cos p(x-\alpha)d\alpha\Big]^2 = \int_0^{2\pi}\int_0^{2\pi}\int_0^{2\pi}\sum_{p=1}^{k(x)} \cos p(x-\alpha)\times$$

$$\times\sum_{p=1}^{k(y)} \cos p(y-\alpha)dydxd\alpha =$$

$$= \int_0^{2\pi}\int_0^{2\pi}\sum_{p=1}^{\min[k(x),k(y)]2\pi}\int_0^{2\pi} \cos p(x-\alpha)\cos p(y-\alpha)d\alpha dxdy =$$

$$= \int_0^{2\pi}\int_0^{2\pi}\sum_{p=1}^{\min[k(x),k(y)]} \cos p(x-y)dxdy.$$

It is easy to show that the last expression does not exceed $C\log n$, where C is an absolute constant.

Theorem. *Suppose that* $\tau(n) < \tau(n+1)$ *and that the series*

$$\sum_{n=1}^{\infty}\tau(n)(a_n^2+b_n^2) = A \tag{2}$$

and

$$\sum_{n=1}^{\infty}\frac{1}{n\tau(n)} = N \tag{3}$$

are convergent. Then (1) *converges almost everywhere.*

We first prove that the partial sums of (1) are bounded almost everywhere. We set

$$S_{p,l} = \sum_{q=2^{2^p}}^{l}(a_q\cos qx + b_q\sin qx),\quad p=0,1,2,\ldots, 2^{2^p}\le l < 2^{2^{p+1}},$$

and

$$A_p = \sum_{q=2^{2^p}}^{2^{2^{p+1}}-1} (a_q^2 + b_q^2).$$

By the Lemma we have

$$\int_0^{2\pi} S_{p,l(x)}(x)dx \le \sqrt{C \log 2^{2^{p+1}} A_p} \le C' \sqrt{2^p A_p},$$

where $l(x)$ is an arbitrary integer-valued function satisfying the inequality given for l. Let $\Phi(x)$ be the supremum of the partial sums of (1). Then the above inequality implies that

$$\int_0^{2\pi} \Phi(x)dx \le C' \sum_{p=0}^{\infty} \sqrt{A_p 2^p} + 2\pi(|a_1| + |b_1|). \tag{4}$$

The series on the right-hand side is convergent. Indeed,

$$\sum_{p=0}^{\infty} A_p \tau(2^{2^p}) \le \sum_{n=1}^{\infty} (a_n^2 + b_n^2)\tau(n) = A.$$

By the Cauchy theorem,

$$\sum_{p=1}^{\infty} \frac{2^p}{\tau(2^{2^p})} < 2\sum_{r=1}^{\infty} \frac{1}{\tau(2^r)} < 4\sum_{n=1}^{\infty} \frac{1}{n\tau(n)} = 4N,$$

and consequently

$$\sum_{p=1}^{\infty} \sqrt{A_p \tau(2^{2^p}) \frac{2^p}{\tau(2^{2^p})}} \le 2\sqrt{AN}.$$

The integral (4) can be made arbitrarily small by disregarding certain first terms in (1). This proves that (1) is convergent.

14 January 1924

References

1. G.H. Hardy, 'On the summability of Fourier's series', *Proc. London Math. Soc.* **12** (1913), 365–372.
2. D. Menchoff, 'Sur les séries de fonctions orthogonales', *Fund. Math.* **4** (1923), 82–105.

5. AXIOMATIC DEFINITION OF THE INTEGRAL*

A class K of functions is a class of functions defined almost everywhere on the real line such that the following condition is fulfilled.

Condition K. If f is a function of class K and $\phi(x)$ is a monotone function increasing from $-\infty$ to ∞ with bounded ratios

$$\frac{\phi(x_2) - \phi(x_1)}{x_2 - x_1},$$

then the function $f[\phi(x)]\phi'(x)$ also belongs to K. This expression is assumed to be equal to zero if $\phi'(x) = 0$, even when $f[\phi(x)]$ is not defined.

The class K is said to be X-integrable if a functional (X-integral)

$$\int_0^x f(\alpha)d\alpha$$

can be defined on the class so that the following properties hold.

Properties of the integral.

1) $\int_0^0 f(\alpha)d\alpha = 0$;

2) $\int_0^x f(\alpha)d\alpha$ is a continuous function of x;

3) its X-derivative with respect to x is equal to f almost everywhere;

4) $\int_0^{\phi(x)} f(\alpha)d\alpha = \int_0^x f[\phi(\alpha)]\phi'(\alpha)d\alpha$ if $\phi(\alpha)$ satisfies the above-mentioned condition.

We assume that the X-derivative is a functional with the following properties.

Properties of the derivative.

1) $[F_1(x) - F_2(x)]' = F_1'(x) - F_2'(x)$;

2) the set of values x of F at which $F'(x) = 0$ has measure zero.

Theorem 1. *Given a function $f(x)$ and a differentiation process X, there can be only one function*

$$F(x) = \int_0^x f(\alpha)d\alpha,$$

* 'La définition axiomatique de l'intégrale', *C.R. Acad. Sci. Paris* **180** (1925), 110–111.

even for f in different integrable classes K_1 and K_2.

Scheme of the proof. Assume, on the contrary, that there are two different functions $F_1(x)$ and $F_2(x)$ for integrable classes K_1 and K_2. The X-derivative of

$$\Phi(x) = F_1(x) - F_2(x)$$

is equal to zero almost everywhere. Assume that $\Phi(x_0) > 0, x_0 > 0$ (in the other cases the argument is similar). We set

$$\Psi(x) = x + \max_{0 \leq \alpha \leq x} \Phi(\alpha) \quad \text{if } x > 0,$$
$$\Psi(x) = x \quad \text{if } x \leq 0.$$

The inverse function $\phi(x)$ ($\phi[\Psi(x)] = x$) satisfies the conditions specified in the beginning. The function

$$\Phi[\phi(x)] = F_1[\phi(x)] - F_2[\phi(x)]$$

has a non-zero derivative on a set of positive measure, which is impossible.

Theorem 2. *For a given differentiation process X there is an X-integrable class K^m containing all the other integrable classes.*

Theorem 3. *For asymptotic differentiation, the integrable class K^m coincides with the class of functions totalizable in the sense of Denjoy.*

Remark. We have not given a complete axiomatic definition of differentiation in this paper. Therefore it may happen that different integrals correspond to different differentiation processes.

6. ON THE LIMITS OF GENERALIZATION OF THE INTEGRAL*

I. Introduction

§1. Consider[1] the definition of the classical *Riemann integral*: we partition the interval of integration (a, b) into partial intervals $\Delta_1, \Delta_2, \ldots, \Delta_n$ and find the infimum m_k and supremum M_k of $f(x)$ on each Δ_k. We call

$$s = \sum_{k=1}^{n} \Delta_k m_k$$

the lower Riemann sum and

$$S = \sum_{k=1}^{n} \Delta_k M_k$$

the upper Riemann sum.[2] If the supremum of the quantity s over all possible partitions coincides with the infimum of the quantity S over all possible partitions, then this common value is called the integral of $f(x)$ over the interval (a, b).

The Lebesgue integral of a bounded function can be obtained in a similar way: we split the interval (a, b) into sets $E_1, E_2, \ldots, E_n$ and, as above, define

$$s = \sum_{k=1} E_k m_k \text{ and } S = \sum_{k=1} E_k M_k.$$

If the supremum of s over all possible partitions of (a, b) into measurable sets coincides with the similarly taken infimum of S, then this common value is called the integral

$$\int_a^b f(x)dx.$$

In the case of the Riemann integral each partition into sufficiently small intervals produces sums s and S that differ by a small quantity only. By contrast, in the case of the Lebesgue integral the choice of E_k is determined

---

* 'Sur les bornes de la généralisation de l'intégrale', 1925. Unpublished.

[1] This is a detailed and complete account of my paper [1] (see article No.5 in this volume). Also see the introduction to my memoir [2] (article No.16 of this volume).

[2] Here Δ_k also denotes the length of the interval Δ_k and in what follows E_k also denotes the measure of E_k.

by the behaviour of the function. For example, we can take, using Lebesgue's notation,

$$E_k = E\Big[l_{k-1} < f(x) \leq l_k\Big].$$

The general Lebesgue integral can be defined by means of a method similar to the two above-mentioned methods: we split (a, b) into an infinite sequence of sets $E_1, E_2, \dots, E_k, \dots$ and denote by s and S the sums of the series

$$\sum_{k=1}^{\infty} E_k m_k \quad \text{and} \quad \sum_{k=1}^{\infty} E_k M_k,$$

provided that they converge absolutely. If the series do not converge absolutely, we assume that s and S are not defined, which is natural since the sum of the series can be quite arbitrary for different orders of magnitude of E_k. As above, we define the integral of $f(x)$ over (a, b) as the common value of the supremum s and infimum S over all partitions for which s and S are defined.

If the function is measurable but not summable, the integral can sometimes be obtained by a similar method. Like the above, the presentation immediately below serves as an explanation of the meaning of axiomatic construction considered in this memoir: here we shall give only a *definition of the integral that is equivalent to the one given by Dirichlet.* To this end, we split (a, b) into an infinite sequence of intervals $\Delta_1, \Delta_2, \dots, \Delta_k, \dots$, arbitrarily located. The collection of their end points is a reducible set. For each interval we define m_k and M_k. We call $m_k\Delta_k$ and $M_k\Delta_k$ *elements* of the lower and upper Riemann sums, respectively, *corresponding to* Δ_k. When added together, the elements are taken in accordance with the order of the intervals. Assume that among Δ_k there is a sequence $\Delta_{k_1}, \Delta_{k_2}, \dots, \Delta_{k_m}, \dots$ such that the union of all the Δ_{k_m} is also an interval (to within a countable set of points). We also assume that the series of elements corresponding to Δ_{k_m} converge absolutely to s' and S'. In this case we call s' and S' *elements* corresponding to the union of the Δ_{k_m}. We define Riemann sums by repeating this process transfinitely (if this is possible). And then we define the integral as above, proceeding from the Riemann sums.

The above argument allows one to conclude that the basic idea behind the concept of the ordinary integral

$$\int f(x)dx$$

is retained in the case of the more general definition.

The expression $f(x)dx$ is used to denote an infinitesimal element in the Riemann sum converging to the integral. In the preliminary presentation of the subsequent results we call $f(x)dx$ an *element* of the function.

§**2**. The constructive method that has been described is not sufficient for our purpose, which is to investigate the limits of the possible generalization of the integral. Here an axiomatic approach should be applied.

In the preceding section two parts of the integration theory were outlined: 1) *the theory of absolute integration*, in which the order of summation is unimportant; and 2) *the theory of ordered integration*, in which the result of summation is completely determined by the order of the elements.

The domain of absolute integration can be defined more precisely with the help of the following axiom: *the value of the integral does not change under a measure preserving change of variables.* It is possible to show that for measurable functions Lebesgue integration is the most general process of absolute integration. From a certain point of view, which will be further elaborated in §4 of this introduction, the Lebesgue theory of absolute integration cannot be generalized, not even in a domain containing non-measurable functions.

The theory of ordered integration can be defined with the help of the following axiom: *if a change of variables $y = \phi(x)$ does not violate the order and does not transform sets of measure zero on the x-axis into sets of positive measure on the y-axis, then*

$$\int_0^{\phi(x)} f(\alpha)d\alpha = \int_0^x f[\phi(\alpha)]\phi'(\alpha)d\alpha.$$

In what follows we use a more special class of transformations, corresponding to extension of an interval on the real axis. The Denjoy integral satisfies the indicated axiom. The present memoir is devoted to finding the limits of the domain of ordered integration.

We first show that *ordered integration yields continuous primitive functions.* It may be important in certain areas of mathematics to consider discontinuous primitive functions. In this case, however, one has to reject the above-stated axiom on change of variables, or one has to confine oneself to a certain kind of symmetric deformation in a neighbourhood of each point. In such an integration theory the value of the integral depends not only on the order of the elements of the function but also on the distances between them.

To complete the explanation of the method used subsequently, we need to make a few remarks concerning what we call a *uniqueness principle.*

§**3.** In existing definitions of integral, the integrand function is everywhere, except possibly on a set of measure zero, equal to the derivative of its primitive function. However, in the most general definition of integral the axiom of the existence of a derivative should not be introduced. Indeed, a given definition of integral may give rise to a definition of derivative not introduced beforehand. But every definition of integral must satisfy the following axiom of uniqueness:

Two functions that differ on a set of positive measure cannot have the same primitive function.

Indeed, a theory allowing a non-zero function to have a zero primitive cannot be used either for the definition of Fourier coefficients or for solving other existing problems.

§**4.** Here we point out two applications of the axiom of uniqueness, one in the theory of absolute integration and the other in the theory of ordered integration.

1. It is possible to show that absolute integration yields absolutely continuous primitive functions. But each absolutely continuous function is the primitive of a summable function. It follows that, with the axiom of uniqueness included into the set of axioms, non-summable functions and, in particular, non-measurable functions, cannot be absolutely integrable.

2. In any integration theory it should hold that the inequality

$$|f(x)| \leq K$$

necessarily implies the inequality

$$\left|\int_a^b f(x)dx\right| \leq K|b-a|.$$

It can now be seen that bounded functions have primitive functions with bounded Dini derived numbers. But derivatives of primitive functions are measurable. Hence, a bounded non-measurable function cannot be integrable in any sense.

II. Definition of $\mathcal{K}$ systems

Two classes of functions $K \equiv \{F(x)\}$ and $k \equiv \{f(x)\}$ form a $\mathcal{K}$ system if they satisfy the following conditions.

1. The functions of class K are continuous and equal to zero at $x = 0$.

2. The functions of class k are defined and finite everywhere, except possibly on a set of measure zero. (Two functions that differ only on a set of measure zero are considered equal.)

3. There is a one-to-one correspondence between the classes k and K. By the *primitive* of $f(x)$ is meant the corresponding function $F(x)$, and $f(x)$ is called the *derivative* of $F(x)$. Thus, the $\mathcal{K}$ system under study is a system of pairs of functions

$$[F(x), f(x)].$$

4. If two pairs $[F_1(x),\ f_1(x)]$ and $[F_2(x),\ f_2(x)]$ belong to $\mathcal{K}$, then so does any pair $[l_1F_1(x) + l_2F_2(x),\ l_1f_1(x) + l_2f_2(x)]$, where l_1 and l_2 are arbitrary real numbers.

5. If the pair $[F(x), f(x)]$ belongs to $\mathcal{K}$, then so does the pair

$$[F[\phi(x)],\ f[\phi(x)]\phi'(x)],$$

where $\phi(x)$ is a monotone function that varies from $-\infty$ to ∞ and has bounded ratios $(\phi(x_2) - \phi(x_1))/(x_2 - x_1)$. Here we put $f[\phi(x)]\phi'(x) = 0$ for $\phi'(x) = 0$, even if $f[\phi(x)]$ is not defined.

These five properties of $\mathcal{K}$ systems are not sufficient for a complete axiomatic definition of derivative and integral. Indeed, in different $\mathcal{K}$ systems different primitive functions $F_1(x)$ and $F_2(x)$ may correspond to the same derivative $f(x)$ and, conversely, one and the same primitive function $F(x)$ may have different derivatives. However, these properties allow one to reduce the definition of integral to the definition of derivative.

Indeed, assume that some new definition of derivative has been given, that is, let there be a functional depending on the choice of $F(x)$ and x and defined for certain functions $F(x)$ and certain values of x. We shall call such a functional a *generalized derivative* $F'(x)$. We also assume that the following two conditions are fulfilled.

1) If $F(x)$ is asymptotically differentiable at x, then $F'(x)$ coincides with the asymptotic derivative;

2) $[l_1F_1(x)+l_2F_2(x)]' = l_1F_1'(x)+l_2F_2'(x)$.

We now add the following sixth property to the above-stated five properties of $\mathcal{K}$ systems.

6. If the pair $[F(x), f(x)]$ belongs to $\mathcal{K}$, then $F'(x) = f(x)$ everywhere except possibly on a set of measure zero.

Using these properties we shall prove the following: *if a function $f(x)$ belongs to different $\mathcal{K}$ systems, then the same primitive $F(x)$ corresponds to $f(x)$ in all such systems.*

Proof. Assume the contrary: let two different functions $F_1(x)$ and $F_2(x)$ correspond to $f(x)$ in systems $\mathcal{K}_1$ and $\mathcal{K}_2$. Then putting $\Phi(x) = F_1(x) - F_2(x)$ we have $\Phi'(x) = F_1'(x) - F_2'(x) = 0$. Assume that $\Phi(x_0) > 0$ and $x_0 > 0$ (in the other cases the argument is similar). We set

$$\psi(x) = \begin{cases} x + \max\limits_{0\le\alpha\le x} \Phi(\alpha) & \text{for } x \ge 0, \\ x & \text{for } x < 0. \end{cases}$$

It is clear that $(\psi(x_2)-\psi(x_1))/(x_2-x_1) \ge 1$; hence, for the inverse function $\phi(x)$ $(\phi[\psi(x)] = x)$ we have

$$(\phi(x_2)-\phi(x_1))/(x_2-x_1) \le 1.$$

By Property 5, the functions $F_1[\phi(x)]$ and $F_2[\phi(x)]$ belong to the respective classes $\mathcal{K}_1$ and $\mathcal{K}_2$ of primitive functions, and their derivatives are equal to $f[\phi(x)]\phi'(x)$ except on a set of measure zero. It follows that $\{\Phi[\phi(x)]\}'$ exists and is equal to zero everywhere, except possibly on a set of measure zero.

On the other hand, it is easy to show that the set of points x such that $\Phi(x) = \max\limits_{0\le\alpha\le x}\Phi(\alpha)$ contains a subset P of measure zero such that the set of corresponding values of $\Phi(x)$ (and consequently, of $\psi(x)$) has positive measure. Let Q be the set of values of $\psi(x)$ on P. For $x \in P$ we have

$$\Phi(x) = \psi(x) - x$$

and

$$\Phi[\phi(x)] = \psi[\phi(x)] - \phi(x) = x - \phi(x),$$
$$\{\Phi[\phi(x)]\}' = 1 - \phi'(x).$$

The derivative in the last relation is asymptotic. The derivative $\phi'(x)$ is equal to zero everywhere on Q, except possibly on a set of measure zero, since the set of values of ϕ on Q is the set P, which has measure zero. It follows that $\{\Phi[\phi(x)]\}' = 1$ almost everywhere on Q, which is a contradiction.[3].

III

It is now easy to show that the *union of all $\mathcal{K}$ systems with Property* 6 *is also a $\mathcal{K}$ system having the properties* 1–6.

Properties 1, 2, 5 and 6 are obviously fulfilled. Property 3 is also fulfilled, since a certain derivative corresponds to each primitive and vice versa, as has been shown above.

Proof of Property 4. We shall first prove the following: if two systems

$$\mathcal{K}_1 \equiv \{[F_1(x), f_1(x)] \quad \text{and} \quad \mathcal{K}_2 = \{[F_2(x), f_2(x)]\}$$

have the properties 1–6, then the system

$$\mathcal{K} = \{[l_1 F_1(x) + l_2 F_2(x),\ l_1 f_1(x) + l_2 f_2(x)]\}$$

also has these properties. The verification of properties 1, 2, 4, 5, 6 does not require a difficult argument. It remains to verify property 3, that is, the fact that a derivative corresponds to each primitive. To this end we have to show the following: if a derivative can be represented in two forms $l_{11}f_{11}(x) + l_{21}f_{21}(x)$ and $l_{12}f_{12}(x) + l_{22}f_{22}(x)$, where f_{11} and f_{12} belong to the class k_1 and f_{21} and f_{22} belong to the class k_2, then for the corresponding primitive functions we have $l_{11}F_{11}(x) + l_{21}F_{21}(x) = l_{12}F_{12}(x) + l_{22}F_{22}(x)$.

[3] The following assertion can be proved in a similar way: *if two functions f_1 and f_2 coinciding on an interval Δ belong to classes k_1 and k_2 of systems $\mathcal{K}_1$ and $\mathcal{K}_2$ possessing Property 6, then the primitive functions F_1 and F_2 can differ on the interval only by a constant.*
The proof is essentially the same. Only the interval Δ is considered and the following changes are introduced into the proof. Denote by a and b the end points of Δ. We assume that $\Phi(x_0) > \Phi(a)$ (the other cases are similar), where $x_0 \in \Delta$, and put

$$\psi(x) = x + \max_{a \leq \alpha \leq x} [\Phi(\alpha)] \text{ for } x > a,$$

$$\psi(x) = x + \Phi(x) \text{ for } x \leq a.$$

We have $l_{11}f_{11}(x)-l_{12}f_{12}(x)=l_{22}f_{22}(x)-l_{21}f_{21}(x)$. The first term in the relation is an element of k_1 while the second is an element of k_2. By what has been proved above, their primitive functions satisfy the relation $l_{11}F_{11}(x)-l_{12}F_{12}(x)=l_{22}F_{22}(x)-l_{21}F_{21}(x)$. The desired relation is a consequence of the above. This obviously implies that Property 4 holds for the union of the systems.

Thus, the union of all $\mathcal{K}$ systems with Property 6 forms a *maximal system* $\mathcal{K}_m$. For all functions of the corresponding class k_m the integral is defined uniquely by axioms 1–6. Such an integral does not, however, exist for any function that does not belong to k_m. This means that the system of axioms is complete.

IV

If we want to obtain a definition of integral using the above-described method, we must specify a certain concept of derivative. It seems natural to consider the asymptotic derivative. In this case Property 6 reads as follows.

6^a. If the pair $[F(x),f(x)]$ belongs to $\mathcal{K}$, then $f(x)$ coincides with the asymptotic derivative $F'(x)$ everywhere, with the possible exception of a set of measure zero. (In what follows $F'(x)$ denotes the asymptotic derivative.)

By what was said in the preceding section, there is a maximal system $\mathcal{K}_m^a$. *This system consists of Denjoy totalizable functions and their primitives.*

Proof. It is well known that the system of totalizable functions and their primitives (solvable functions) has Properties 1–6^a. It remains to show that the system is maximal. Assume the contrary: let there be an unsolvable function $F(x)$ belonging to the class K of a system $\mathcal{K}$ with Property 6^a. There are two possible cases here.

1) On a certain perfect set $\mathcal{P}$ of measure zero the function $F(x)$ has non-zero simple variation.[4] We denote by $V(F,\mathcal{P},a,b)$ the simple variation of $F(x)$ on the portion of $\mathcal{P}$ contained between a and b. Suppose that a portion of $\mathcal{P}$ lies between A and B and that $V(F,\mathcal{P},A,B)>0$ for it (the case $V(F,\mathcal{P},A,B)<0$

[4] Simple variation was introduced by Denjoy [4]. For a perfect set $\mathcal{P}$ it is equal to $F(\sup\mathcal{P})-F(\inf\mathcal{P})-\sum_k(F(b_k)-F(a_k))$, where (a_k,b_k) are complementary intervals of $\mathcal{P}$ (on the interval $[\inf\mathcal{P},\ \sup\mathcal{P}]$) and the series is absolutely convergent.

is similar). We set

$$\psi(x) = \begin{cases} x + \max_{A \le \alpha \le x} V(F, \mathcal{P}, A, a) & \text{if } A \le x \le B, \\ x + F(A) & \text{if } x < A, \\ x + \max_{A \le \alpha \le B} V(F, \mathcal{P}, A, a) & \text{if } x > B. \end{cases}$$

The end of this proof is similar to the previous one. Property 5 holds for the inverse function $\phi(x)$. Denote by Q the set of values of $\psi(x)$ on the portion of $\mathcal{P}$ under consideration. It can easily be seen that $F[\phi(x)]$ has positive asymptotic derivative on a subset of Q of positive measure. On the other hand, $\{F[\phi(x)]\}' = f[\phi(x)]\phi'(x) = 0$ everywhere on Q except possibly on a set of measure zero. We have arrived at a contradiction.

2) The variation of $F(x)$ is irreducible on a certain perfect set $\mathcal{P}$ of measure zero.[5] In this case we can find a function $\phi(x)$ satisfying Property 5 and such that $F[\phi(x)]$ has no asymptotic derivative on a certain set of positive measure.

Construction of $\phi(x)$. The set Q is obtained by a change of variable on a set $\mathcal{P}_1$ which is a part of $\mathcal{P}$. We define this set using a *regular system of intervals.* This is a system of closed intervals

$$S = \{\Delta_{i_1 i_2 \ldots i_{n-1} i_n}\}$$

with the following properties. Assume that $i_1, i_2, \ldots, i_{n-1}$ are fixed and i_n runs from 1 to $N_{i_1 i_2 \ldots i_{n-1}}$. Intervals of the same rank (with a fixed n) do not intersect and are located in order of increasing indices i_n. Each interval $\Delta_{i_1 i_2 \ldots i_n}$ is contained in the corresponding interval $\Delta_{i_1 i_2 \ldots i_{n-1}}$ of the preceding rank. Intervals of the first rank lie in the interval Δ whose end points are a and b. $\mathcal{P}_1$ is the set of all points belonging to infinitely many intervals Δ.

The interval $\Delta_{i_1 i_2 \ldots i_n}$ is transformed to an interval $\delta_{i_1 i_2 \ldots i_n}$. The portion of $\mathcal{P}_1$ contained in the former interval is transformed into a portion of Q contained in the latter interval. Let $e_{i_1 i_2 \ldots i_n}$ be the measure of the latter portion. These numbers e must satisfy the following condition:

$$\sum_{k=1}^{N_{i_1 i_2 \ldots i_n}} e_{i_1 i_2 \ldots i_n k} = e_{i_1 i_2 \ldots i_n}.$$

5 The concept of reducibility of variation is also due to Denjoy [4]. The variation of $F(x)$ is reducible on a perfect set $\mathcal{P}$ if for any portion of $\mathcal{P}$ there is a subportion for which the simple variation of $F(x)$ exists.

We also assume that

$$\sum_{k=1}^{N} e_k = e = 1.$$

If all the numbers e are defined, then the measure of the portion of Q corresponding to a portion of $\mathcal{P}_1$ is defined uniquely. Denote by $e(x)$ the measure of the portion of Q corresponding to the portion of $\mathcal{P}_1$ that is contained between a and x. More precisely, for any x lying between two intervals of rank n we define $e(x)$ as the sum of all the numbers e corresponding to the intervals of rank n lying to the left of x. Therefore $e(x)$, which is a continuous and increasing function, is defined on all complementary intervals of $\mathcal{P}_1$. We can define $e(x)$ on $\mathcal{P}_1$ by continuity.

The function inverse to $\psi(x) = x + e(x)$ is the desired function $\phi(x)$; it is obvious that it satisfies Property 5.

The intervals $\Delta_{i_1 i_2 \dots i_n}$ and the numbers $e_{i_1 i_2 \dots i_n}$ are defined by induction in the following way. Assume that $\Delta_{i_1 i_2 \dots i_n}$ and $e_{i_1 i_2 \dots i_n}$ have already been defined. We now define an integer $N_{i_1 i_2 \dots i_n}$, intervals $\Delta_{i_1 i_2 \dots i_n k}$, and numbers $e_{i_1 i_2 \dots i_n k}$. Taking a sufficiently large $N_{i_1 i_2 \dots i_n}$ and choosing the intervals accordingly, we can guarantee that the following conditions are satisfied.

1. Each $\Delta_{i_1 i_2 \dots i_n k}$ is contained in $\Delta_{i_1 i_2 \dots i_n}$ and contains a perfect portion of $\mathcal{P}$.

2. The length of $\Delta_{i_1 i_2 \dots i_n k}$ is smaller than $1/n$.

3. The intervals $\Delta_{i_1 i_2 \dots i_n k}$ do not overlap and are located in order of increasing index k.

4. Denote by $w'_{i_1 i_2 \dots i_n k}$ and $w''_{i_1 i_2 \dots i_n k}$ the minimum value of $|F(x) - F(y)|$, where x belongs to $\Delta_{i_1 i_2 \dots i_n k}$ and y belongs to $\Delta_{i_1 i_2 \dots i_n k-1}$ in the case of w' and belongs to $\Delta_{i_1 i_2 \dots i_n k+1}$ in the case of w''. Further, denote by $w_{i_1 i_2 \dots i_n}$ the maximum of w', w''. We now require that

$$\sum_{k=1}^{N_{i_1 \dots i_n}} w_{i_1 i_2 \dots i_n k} = W_{i_1 i_2 \dots i_n} \geq n e_{i_1 i_2 \dots i_n}.$$

We also set

$$e_{i_1 i_2 \dots i_n k} = \frac{w_{i_1 i_2 \dots i_n k}}{W_{i_1 i_2 \dots i_n}} e_{i_1 i_2 \dots i_n}.$$

For $n = 1$ we take a single interval $\Delta_1 = \Delta$ and $e_1 = 1$. The numbers e thus defined satisfy the above-stated conditions.

Having defined $\phi(x)$ in this way we now show that $F[\phi(x)]$ has no asymptotic derivative at each density point of Q.

Indeed, such a point x is contained in a sequence of intervals $\delta_{i_1 i_2 \ldots i_n}$ with increasing n. Since x is a density point of Q, the measure $e_{i_1 i_2 \ldots i_n}$ of the portion of Q contained in the interval $\delta_{i_1 i_2 \ldots i_n}$ differs from the length of the interval by a higher-order infinitesimal relative to the length. Assume that the corresponding number $w''_{i_1 i_2 \ldots i_n}$ is not smaller than $w'_{i_1 i_2 \ldots i_n}$ (the opposite case is similar). Then $w_{i_1 i_2 \ldots i_n} = w''_{i_1 i_2 \ldots i_n} \leq w_{i_1 i_2 \ldots i_n + 1}$, which, according to the construction of e implies that $e_{i_1 i_2 \ldots i_n} \leq e_{i_1 i_2 \ldots i_n} \leq e_{i_1 i_2 \ldots i_n + 1}$. It follows that the length of $\delta_{i_1 i_2 \ldots i_n}$ may exceed that of $\delta_{i_1 i_2 \ldots i_n + 1}$ only by an infinitesimal of a higher order relative to the lengths of both intervals. The distance between the intervals is also infinitely small relative to their lengths. On the other hand, Property 4 of intervals Δ and the definitions of e and w imply that

$$|F[\phi(x)] - F[\phi(y)]| \geq w_{i_1 i_2 \ldots i_n} \geq e_{i_1 i_2 \ldots i_n}(n-1)$$

for any y belonging to $\delta_{i_1 i_2 \ldots i_n + 1}$. It readily follows that $F[\phi(x)]$ has no asymptotic derivative at x (see Fig. 1).

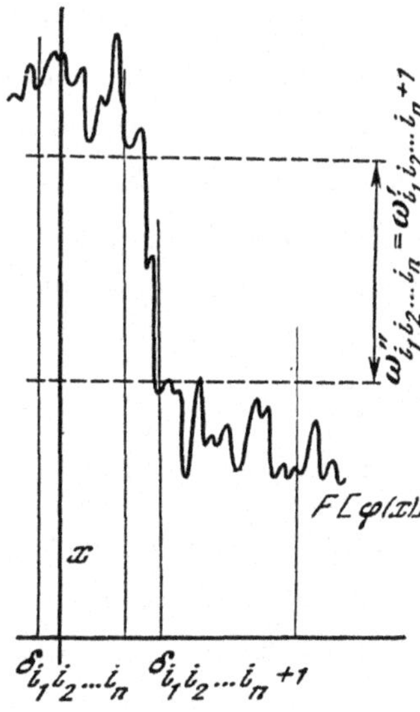

Fig. 1

Replacing 6^a by Property $6°$, which requires that $f(x)$ coincide with the ordinary derivative $F'(x)$ everywhere except possibly on a set of measure zero, we obtain a definition of integral which is equivalent to the first definition of Denjoy.

V

Thus, to obtain further generalizations of the ordered integral we must define a derivative more general than the asymptotic derivative. I shall give here a suitable definition only to demonstrate that a generalization of the Denjoy integral that preserves all the properties of the ordered integral is possible. A thorough study of this kind of definition is yet to be carried out.

First, the concept of density of a set should be generalized. Without loss of generality we confine ourselves to studying density on the right of the origin. Let $y = \phi(x)$ be a function having two first derivatives such that

$$\phi(0) = 0, \ \phi'(0) = 0, \text{ and } \phi''(x) > 0 \text{ for } x > 0.$$

In this case a set E_x on the x-axis corresponds to each set E_y on the y-axis and vice versa. The following assertions can be proved:

1) if E_y has a certain right density at 0, then E_x also has a density, equal to the density of E_y;

2) it is possible that E_x has a density whereas E_y has not.

In view of this, the following definition of generalized density seems natural: to take the ordinary density of E_x (if it exists) as the generalized density of E_y at 0. But this definition depends on the choice of $\phi(x)$. The situation is similar to that arising in the summation of divergent series: it may happen that the densities corresponding to two different functions do not coincide. Therefore we have to choose a definite function $\phi(x)$. To avoid ambiguity, we confine ourselves to the function

$$y = \phi(x) = e^{-1/x},$$

which satisfies the above conditions for $x < 1$. The choice is motivated by the following property of the function: if E_y is transformed by means of the function $y = e^{-1/x}$ into a set E_x having a density, then using some other arbitrary function ϕ it can be transformed either into a set whose density is equal to the previous one or into a set of indeterminate density.

This shows that this method of defining the generalized density gives the same result as all other methods. To make a step forward it seems to be necessary to introduce new axioms for density.

A new definition of derivative more general than that of the asymptotic derivative can easily be obtained using the above definition of generalized density. It is possible to show that the application of the method of §3 with such a generalization of derivative leads to a definition of integral which is more general than the one stated by Denjoy. This means that the maximal system $\mathcal{K}_m$ corresponding to the new derivative is more extensive than $\mathcal{K}_m^a$. Here is an example of a primitive function $F(x)$ in our system that is not primitive in $\mathcal{K}_m^a$:

$$f_\Delta(x) = \begin{cases} x/\Delta & \text{if } 0 \le x \le \Delta; \\ 1 & \text{if } \Delta \le x \le \frac{1}{2}; \\ 1-(x-\frac{1}{2})/\Delta & \text{if } \frac{1}{2} \le x \le \frac{1}{2}+\Delta; \\ 0 & \text{if } \frac{1}{2}+\Delta \le x \le 1; \end{cases}$$

$$f_\Delta(x+1) = f_\Delta(x);$$

$$\phi_\Delta(x) = \Delta f_\Delta(x/\Delta),$$

$$\psi_n(x) = \begin{cases} \psi_{\Delta_n}(x) & \text{if } \psi_k'(x) = 0 \text{ for all } k < n; \\ 0 & \text{if } \psi_k'(x) \ne 0 \text{ for some } k < n; \end{cases}$$

$$F(x) = \sum_{n=1}^{\infty} \psi_n(x).$$

The constants Δ_n must decrease sufficiently fast and the ratio Δ_n/Δ_{n+1} must be an integer (see Fig. 2).

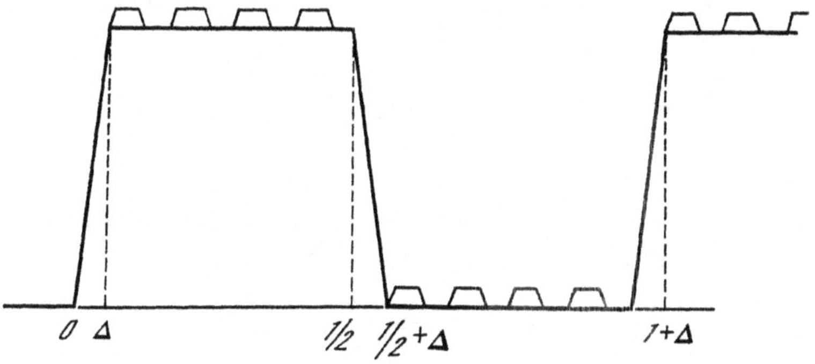

Fig. 2

VI

Thus, any generalization of the Denjoy integral is related to a generalization of the asymptotic derivative. To establish the limits of such a generalization we replace Property 6 of $\mathcal{K}$ systems by the following property.

$\overline{6}$. If $F(x)$ has an asymptotic derivative $F'(x)$ on E, then it is equal to $f(x)$. A $\mathcal{K}$ system with Property $\overline{6}$ is denoted in what follows by $\overline{\mathcal{K}}$.[6]

The system of axioms 1–$\overline{6}$ is not complete (see the end of §3), that is, in two different systems $\overline{\mathcal{K}}_1$ and $\overline{\mathcal{K}}_2$ two different primitive functions may correspond to the same derivative. But this system of axioms is sufficient for determining the limits of the domain of ordered integration. More precisely, it is possible to show that a *function $f(x)$ that is not summable on any subinterval of Δ cannot be a derivative in a $\overline{\mathcal{K}}$ system.*

Proof. We say that $F(x)$ grows on $\Delta = (a,b)$ if $F(a) < F(x) < F(b)$ for any x between a and b. (This definition does not agree with the ordinary one but for us it is suitable.) If $F(x)$ is continuous and $F(a) < F(b)$ for the given interval (a,b), then there is a subinterval $\Delta' = (a',b')$ of Δ such that $F(a') = F(a)$, $F(b') = F(b)$ and F grows on Δ'.

Assume that F_1 grows on $\Delta_1 = (a_1,b_1)$ and F_2 grows on $\Delta_2 = (a_2,b_2)$. In this case we denote by

$$R[F_1,F_2,\Delta_1,\Delta_2]$$

the interval $\delta = [F_1(a_1) + F_2(a_2),\ F_1(b_1) + F_2(b_2)]$.

Assume that there is a system $S_1 = [\Delta_1^1, \Delta_1^2, \dots, \Delta_1^n]$ of finitely many non-intersecting intervals whose location on the line corresponds to the order of their indices and that F_1 grows on each of the intervals Δ_1^k. Also assume that a similar system $S_2 = [\Delta_2^1, \Delta_2^2, \dots, \Delta_2^n]$ with the same number of intervals exists for F_2. Then an interval

$$\delta^k = (\alpha^k, \beta^k) = R[F_1, F_2, \Delta_1^k, \Delta_2^k]$$

corresponds to each pair Δ_1^k, Δ_2^k.

Finally, assume that for the interval δ^k we have $\beta^{k+1} > \beta^k$ and $\alpha^{k+1} \le \beta^k$. Then the union of the δ^k is again an interval. We denote it by

$$R[F_1, F_2, S_1, S_2] = \delta.$$

Lemma 1. *Assume that the interval $R[F_1, F_2, S_1', S_2'] = \delta'$ contains (a,b) and the interval $R[F_1, F_2, S_1'', S_2''] = \delta''$ contains (b,c), where $a < b < c$. Also*

[6] If a function f belongs to the class k of a $\overline{\mathcal{K}}$ system and coincides on Δ with a function f^a belonging to k_m^a (which means that it is totalizable), then the primitive F can differ on Δ from the primitive F^a only by a constant. Thus, every integral satisfying the axioms 1–$\overline{6}$ coincides with the Denjoy integral if both integrals exist. This follows immediately from the footnote in Section III.

assume that the systems S'' lie entirely to the right of the corresponding systems S'. Then there are a system S_1 contained in the union of S_1' and S_1'' and a system S_2 contained in the union of S_2' and S_2'' such that the interval

$$R[F_1, F_2, S_1, S_2] = \delta$$

contains (a, c).

Lemma 2. *If $R[F_1, F_2, \Delta_1, \Delta_2] = \delta$, where the functions F_1 and F_2 are continuous and each of them has unbounded variation on any interval contained in the corresponding interval, then there are systems S_1 and S_2 of arbitrarily small measure contained in Δ_1 and Δ_2 respectively, such that*

$$R[F_1, F_2, S_1, S_2] = \delta.$$

Proof. By the hypothesis, F_1 grows on $\Delta_1 = (a_1, b_1)$ from $F(a_1)$ to $F(b_1)$. We divide the interval into two parts (a_1, c_1) and (c_1, b_1) such that F grows on (a_1, c_1). This can be done in such a way that (a_1, c_1) is arbitrarily small. We also divide $\Delta_2 = (a_2, b_2)$ into two parts (a_2, c_2) and (c_2, b_2) such that (c_2, b_2) is sufficiently small and F_2 grows on (c_2, b_2).

For (a_1, c_1) it is possible to find a system $S_{11} = [\Delta_{11}^1, \Delta_{11}^2, \ldots, \Delta_{11}^n]$ of intervals $\Delta_{11}^k = (a_{11}^k, b_{11}^k)$ located in accordance with the order of their indices and satisfying the following:

$$a_{11}^1 = a_1, \quad b_{11}^n = c_1,$$
$$F(b_{11}^k) > F_1(a_{11}^k), \quad F(b_{11}^k) > F_1(a_{11}^{k+1}),$$
$$\sum_{k=1}^{k=n} F_1(b_{11}^k) - F_1(a_{11}^k) > [F_1(c_1) + F_2(c_2)] - [F_1(a_1) - F_2(a_2)].$$

This is possible because F_1 has unbounded variation on (a_1, c_1). It can be assumed in addition that F_1 grows on each Δ_{11}^k. We can now find a corresponding system S_{12} of arbitrarily small measure on (a_2, c_2) such that

$$R[F_1, F_2, S_{11}, S_{12}] = \delta_1 = \{[F_1(a_1) + F_2(a_2)],\ [F_1(c_1) + F_2(c_2)]\}$$

(see Fig. 3).

In the same way we can construct systems S_{21} and S_{22} on (c_1, b_1) and (c_2, b_2) such that

$$R[F_1, F_2, S_{21}, S_{22}] = \delta_2 = \{[F_1(c_1) + F_2(c_2)], [F_1(b_1) + F_2(b_2)]\}.$$

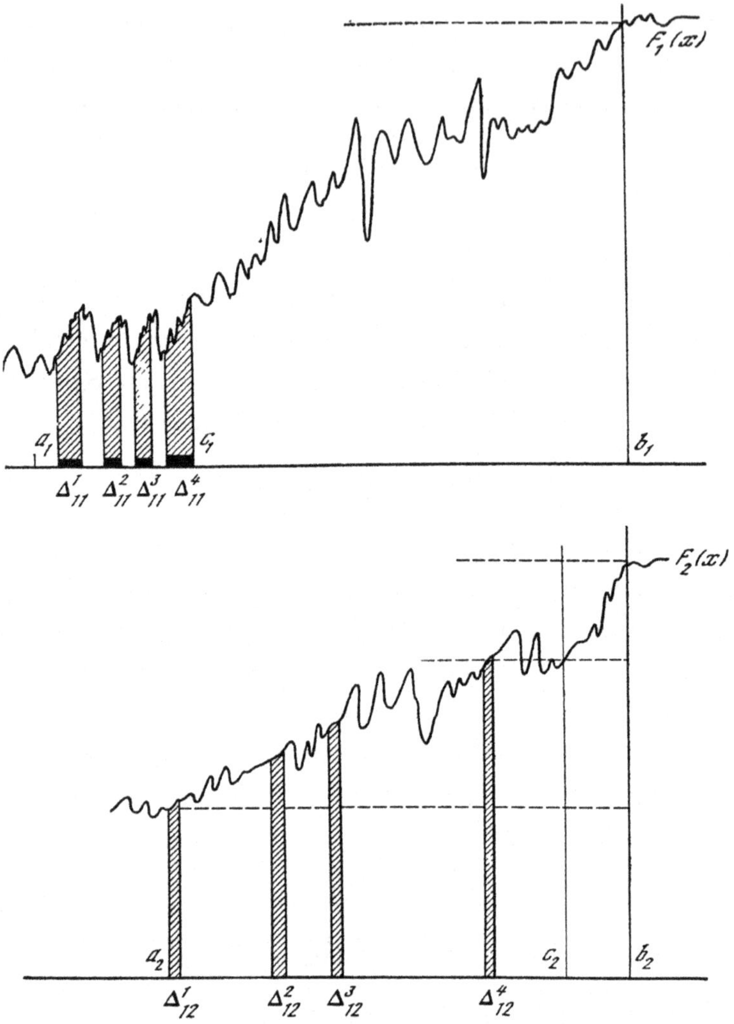

Fig. 3

The construction of S_{22} is similar to that of S_{11} and the construction of S_{21} is similar to that of S_{12}. Lemma 1 allows one to find the required systems S_1 and S_2 from $S_{11}, S_{12}, S_{21}, S_{22}$.

Coming back to the main subject of this section we assume, contrary to what we want to prove, that there is a derivative $f(x)$ in the system $\overline{\mathcal{K}}$ which is not summable on each interval belonging to Δ. The corresponding primitive function $F(x)$ cannot have bounded variation on any interval contained in Δ. Choose an interval $\Delta' = (a', b')$ in Δ on which F grows from $F(a')$ to $F(b')$ and apply Lemma 2 to the functions $F_1 = F, F_2 = F$ and the intervals $\Delta_1 = \Delta'$ and $\Delta_2 = \Delta'$. We then obtain two systems S_1 and S_2 of arbitrarily small measure for which

$$R[F, F, S_1, S_2] = R[F, F, \Delta', \Delta'] = \delta.$$

If we apply the same lemma to each pair of corresponding intervals of the systems S_1 and S_2, then we obtain two new systems S_1^1 and S_2^1 contained in S_1 and S_2 such that

$$R[F, F, S_1^1, S_2^1] = \delta.$$

Repeating this procedure, we obtain a sequence of pairs of systems S_1^n and S_2^n with measures tending to zero. Let $\mathcal{P}_1$ denote the perfect set that is the limit of S_1 and let $\mathcal{P}_2$ be the analogous set for S_2. Both sets have measure zero. By definition, there is a one-to-one correspondence between the sets, preserving the order and continuity. We extend this correspondence to the complementary intervals. Let $\xi_1(x)$ be a function that maps $\mathcal{P}_1$ onto $\mathcal{P}_2$ and let $\xi_2(x)$ be a function that maps $\mathcal{P}_2$ onto $\mathcal{P}_1$. It can easily be seen that $F(x) + F[\xi_1(x)]$ grows on $\mathcal{P}_1$ and that $F(x) + F[\xi_2(x)]$ grows on $\mathcal{P}_2$.

We set

$$\zeta_1(x) = \max_{a \le \alpha \le x, \alpha \in \mathcal{P}_1} \{F(\alpha) + F[\xi_1(\alpha)]\},$$
$$\zeta_2(x) = \max_{a \le \alpha \le x, \alpha \in \mathcal{P}_2} \{F(\alpha) + F[\xi_2(\alpha)]\}.$$

The functions

$$\psi_1(x) = x + \xi_1(x) + \zeta_1(x)$$

and

$$\psi_2(x) = x + \xi_2(x) + \zeta_2(x)$$

transform $\mathcal{P}_1$ and $\mathcal{P}_2$ into the same set Q of positive measure δ. The inverse functions $\phi_1(x)$ and $\phi_2(x)$ satisfy Property 5. It is possible to show that the function

$$\Phi(x) = F[\phi_1(x)] + F[\phi_2(x)]$$

has asymptotic derivative equal to 1 almost everywhere on Q.

Thus, we have arrived at a contradiction, because a primitive in $\mathcal{K}$ must have derivative equal to zero almost everywhere on Q. Indeed,

$$\Phi'(x) = f[\phi_1(x)]\phi_1'(x) + f[\phi_2(x)]\phi_2'(x) = 0.$$

The following stronger theorem is likely to be true: *a derivative that is not summable on every portion of a perfect set cannot exist in* $\overline{\mathcal{K}}$. Our methods, however, are not sufficient for proving this theorem.

Moscow, 14 February 1925

References

1. A. Kolmogoroff, 'La définition axiomatique de l'intégrale', *C.R. Acad. Sci. Paris* **180** (1925), 119–112. (No. 5)
2. A. Kolmogoroff, 'Untersuchungen über den Integralbegriff', *Math. Ann.* **103** (1930), 659–696. (No. 16)
3. H. Lebesgue, *Leçons sur l'intégration et la recherche des fonctions primitives*, Paris 1904.
4. A. Denjoy, 'Sur la dérivation et son calcul inverse', *C.R. Acad. Sci. Paris* **162** (1916), 377–380.

7. ON THE POSSIBILITY OF A GENERAL DEFINITION OF DERIVATIVE, INTEGRAL AND SUMMATION OF DIVERGENT SERIES*

I. Any method of generalized differentiation that can be used in applications must satisfy the following conditions.

1°. The generalized derivative must coincide with the ordinary derivative if the latter exists.

2°. If $\phi(x) = af(x)$, then $\phi(x)$ is differentiable at the same points as $f(x)$ and $\phi'(x) = af'(x)$.

3°. If $\phi(x) = f(ax)$ and $f(x)$ is differentiable at ax_0, then $\phi'(x_0) = af'(ax_0)$.

4°. If $\phi(x) = f(x+h)$ and f is differentiable at $x_0 + h$, then $\phi'(x_0) = f'(x_0 + h)$.

5°. If both f_1 and f_2 are differentiable at x_0, then $\phi(x) = f_1(x) + f_2(x)$ is also differentiable at x_0 and

$$\phi'(x_0) = f_1'(x_0) + f_2'(x_0).$$

The following conditions must hold if we want to introduce infinite derivatives:

$$\text{if } a > 0, \text{ then } a \cdot (+\infty) = +\infty, \quad a \cdot (-\infty) = -\infty,$$
$$\text{if } a < 0, \text{ then } a \cdot (+\infty) = -\infty, \quad a \cdot (-\infty) = +\infty,$$
$$+\infty + a = +\infty, \quad -\infty + a = -\infty.$$

If all the above conditions are satisfied and the function

$$F(x) = \sum_{n=1}^{\infty} \frac{\cos 3^n x}{3^n}$$

has a derivative, either finite or infinite, on a set of positive measure, then this derivative is not a measurable function.

II. Assume that there is a method for the summation of divergent series that satisfies the conditions

$$(1) \quad \sum_{n=1}^{\infty} v_n = v_1 + \sum_{n=1}^{\infty} v_{n+1},$$

$$(2) \quad \sum_{n=1}^{\infty} av_n = a \sum_{n=1}^{\infty} v_n,$$

* 'Sur la possibilité de la définition générale de la dérivée, de l'intégrale et de la sommation des séries divergentes', *C.R. Acad. Sci.* Paris **180** (1925), 362–364.

and assume that using this method a finite or infinite sum can be defined for each series of the form $\sum \sin 3^n x$. Then it is possible to construct an effective example of a function that is not Lebesgue measurable.

III. Assume that the following conditions hold for a given definition of integral:

$$(1)\quad \int_a^b f(x)dx + \int_b^c f(x)dx = \int_a^c f(x)dx;$$

$$(2)\quad \int_a^b kf(x)dx = k\int_a^b f(x)dx;$$

$$(3)\quad \int_{ka}^{kb} f\left(\frac{x}{k}\right)dx = k\int_a^b f(x)dx.$$

If under these conditions the integrals

$$\int_0^1 f_t(x)dx$$

(with functions f_t defined below), can be defined either finite or infinite, for each point t of a set of positive measure, then it is possible to construct an example of a function that is not Lebesgue measurable. Here

$$f_t = 2^n \sin 3^n t \quad \text{for } \frac{1}{2^{n-1}} \geq x > \frac{1}{2^n}.$$

The proof is rather lengthy but almost obvious.

26 January 1925

8. ON CONJUGATE HARMONIC FUNCTIONS AND FOURIER SERIES*

Privalov [1] proved the following theorem:

Assume that $f(\theta)$ is a summable function and

$$f(\rho,\theta) = \frac{1}{2\pi}\int_{-\pi}^{\pi} f(\alpha)\frac{1-\rho^2}{1+\rho^2-2\rho\cos(\alpha-\theta)}d\alpha.$$

If z tends to $e^{i\theta}$ along any path that is not tangent to the unit circle, then for almost all θ the harmonic function $g(z)$ conjugate to $f(z)$ tends to a definite limit

$$g(\theta) = -\frac{1}{2\pi}\int_{-\pi}^{\pi}\frac{f(\theta+\alpha)}{\tan(\alpha/2)}d\alpha,$$

where the integral is understood as

$$\lim_{\epsilon+0}\int_{-\pi}^{-\epsilon}+\int_{\epsilon}^{\pi}.$$

In general, $g(\theta)$ is not a summable function [2]. In this paper I will prove that $[g(\theta)]^{1-\epsilon}$ is summable for $\epsilon > 0$ (Theorem 2). As a direct consequence, I will prove Theorem 3 on the mean convergence of Fourier series, which implies that every Fourier-Lebesgue series converges in measure.

Theorem 1. *Let E be the set of all θ such that $|g(\theta)| < R$, where R is an arbitrary number. Then*

$$R\,\mathrm{mes}\,E \le C\int_{-\pi}^{\pi}|f(\theta)|d\theta,$$

where C is an absolute constant.

Proof. Assume that there exists no such constant C. Then for any n there are a function $f_n(\theta)$, a set E_n and a constant R_n such that

1) $|g_n(\theta)| > R_n$ if θ belongs to E_n;

2) $R_n\,\mathrm{mes}\,E_n > n$;

3) $\int_{-\pi}^{\pi}|f_n(\theta)|d\theta < 1$.

* 'Sur les fonctions harmoniques conjuguées et les séries de Fourier', Fund. Math. 7 (1925), 24–29.

Assume, in addition, that $f_n(\theta)$ and $g_n(\theta)$ are continuous functions; this is possible if we put, for example, $\bar{f}_n(z) = f_n(z\rho)$, $\bar{g}_n(z) = g_n(z\rho)$, where $\rho < 1$ and $1 - \rho$ is so small that the conditions 1)– 3) are retained.

It is easy to prove the existence of a sequence $n_1, n_2, \dots, n_k, \dots$, possibly with some coincident consecutive n_k's such that

$$\lim_{k\to\infty} n_k = \infty,$$

(A) the series $\sum_{k=1}^{\infty} \operatorname{mes} E_{n_k}$ is divergent; the series $\sum_{k=1}^{\infty} \operatorname{mes} E_{n_k}/n_k$ is convergent.

Therefore, setting $a_k = \sqrt{n_k}/R_{n_k} < \operatorname{mes} E_{n_k}/\sqrt{n_k}$ we have

(B) the series $\sum_{k=1}^{\infty} a_k$ is convergent;

(C) $\lim_{k\to\infty} a_k R_{n_k} = \infty.$

We consider the sum (with integers p_n)

$$\sum_{k=1}^{\infty} a_k f_{n_k}(z^{p_k}) = \phi(z).$$

This series converges almost everywhere on the circle $\rho = 1$ to a summable function since, by 3),

$$\sum_{k=1}^{\infty} \int_{-\pi}^{\pi} |a_k f_{n_k}(e^{ip_k\theta})| d\theta < \sum_{k=1}^{\infty} a_k,$$

and to the Poisson integral of the function inside the circle. The series

$$\sum_{k=1}^{\infty} a_k g_{n_k}(z^{p_k}) = \psi(z)$$

converges inside the circle to the conjugate function.

We now define two sequences

$$p_1 < p_2 < \dots < p_k < \dots \to \infty,$$

$$\rho_1 < \rho_2 < \dots < \rho_k < \dots \to 1$$

with the following properties.

I. Let $\widetilde{E}_k$ be the set of points $e^{i\theta}$ such that $e^{ip_k\theta}$ belongs to E_{n_k} and let E be the limes superior of $\widetilde{E}_k$. Then

$$\operatorname{mes} E = 2\pi.$$

II. For all θ and all $1 > \rho \geq \rho_k$ we have

$$\left|\sum_{q=1}^{k-1} a_q g_{n_q}[(\rho e^{i\theta})^{p_q} - \sum_{q=1}^{k-1} a_q g_{n_q}[(e^{i\theta})^{p_q}]\right| < 1,$$

$$\left|a_k g_{n_k}[(\rho e^{i\theta})^{p_k}] - a_k g_{n_k}[(e^{i\theta})^{p_k}]\right| < 1.$$

III. For all θ and all $\rho \leq \rho_k$ we have

$$\left|\sum_{q=k+1}^{\infty} a_q g_{n_q}[(\rho e^{i\theta})^{p_q}]\right| < 1.$$

Indeed: 1) it is possible to prove that condition I is fulfilled if the numbers p_k increase sufficiently rapidly;[1] 2) given $p_1, p_2, \ldots, p_k$, it is possible to choose ρ_k such that II is satisfied, by virtue of the continuity of $g_n(\theta)$; 3) to satisfy III, we take $p_{k+1}, p_{k+2}, p_{k+3}, \ldots$ so large that $\rho_k^{p_{k+1}}, \rho_k^{p_{k+2}}, \rho_k^{p_{k+3}}, \ldots$ are sufficiently small, and in this case III holds since $g_n(0) = 0$.

We consider a point θ belonging to $\widetilde{E}_k$. By III and II,

$$\psi(\rho_k, \theta) = \sum_{q=1}^{k-1} a_q g_{n_q}[(\rho_k e^{i\theta})^{p_q}] + a_k g_{n_k}[(\rho_k e^{i\theta})^{p_k}] +$$

$$+ \sum_{q=k+1}^{\infty} a_q g_{n_q}[(\rho_k e^{i\theta})^{p_k}] = \sum_{q=1}^{k-1} a_q g_{n_q}[e^{ip_q\theta}] + a_k g_{n_k}[e^{ip_k\theta}] + \tau, \quad |\tau| < 3,$$

$$\psi(\rho_{k-1}, \theta) = \sum_{q=1}^{k-1} a_q g_{n_q}[(\rho_{k-1} e^{i\theta})^{p_q}] + \sum_{q=k}^{\infty} a_q g_{n_q}[(\rho_{k-1} e^{i\theta})^{p_k}] =$$

$$= \sum_{q=1}^{k-1} a_q g_{n_q}[e^{ip_q\theta}] + \tau', \quad |\tau'| < 2.$$

Consequently,

$$|\psi(\rho_k, \theta) - \psi(\rho_{k-1}, \theta)| > |a_k g_{n_k}[e^{ip_k\theta}]| - 5 > |a_k R_{n_k}| - 5.$$

It follows from this result (see I and (C)) that for any $\theta \in E$ the function $\psi(\rho, \theta)$ does not tend to a limit for ρ tending to one. The contradiction to Privalov's result proves the theorem.

---

[1] The proof is lengthy; it uses the fact that (A) is divergent and the following remark: for given measurable sets $\widetilde{E}_1, \widetilde{E}_2, \ldots, \widetilde{E}_{k-1}$ one can take p_k so large that $\widetilde{E}_k$ turns out to be rather uniformly distributed on the circle and the measure of the intersection of $\widetilde{E}_1 \widetilde{E}_2 \ldots \widetilde{E}_{k-1}$ with $\widetilde{E}_k$ is close to the product of the measures divided by 2π.

Theorem 2. *For* $1 > \epsilon > 0$ *we have*

$$\int_{-\pi}^{\pi} |g(\theta)|^{1-\epsilon} d\theta \leq \frac{C}{\epsilon}\Big(\int_{-\pi}^{\pi} |f(\theta)| d\theta\Big)^{1-\epsilon},$$

where C *is an absolute constant.*

Proof. The proof is based on the following remarks: a) the measure of the set on which $|g(\theta)| > R$ is smaller than the measure of the set on which $|1/\theta| > R$ times $C\int_{-\pi}^{\pi}|f(\theta)|d\theta$; b) the function $|1/x|^{1-\epsilon}$ is summable.

Theorem 3. *If a function* $f(x)$ *is summable and* S_n *is the sum of the first* n *terms of its Fourier series, then*

$$\lim_{n\to\infty}\int_0^{2\pi} |f(x) - S_n(x)|^{1-\epsilon} dx = 0, \tag{1}$$

provided that $1 > \epsilon > 0$.[2]

Proof. It is well known that

$$S_n = \frac{1}{\pi}\int_{-\pi}^{\pi} \frac{f(x+\alpha)}{\alpha} \sin n\alpha d\alpha + \omega_n(x),$$

where $\omega_n(x)$ tends uniformly to zero. Similarly, we can write

$$S_n = \frac{1}{2\pi}\int_{-\pi}^{\pi} \frac{f(x+\alpha)}{\tan(\alpha/2)} \sin n\alpha d\alpha + \omega_n'(x),$$

where $\omega_n' \to 0$ uniformly.

Consequently

$$\begin{aligned} S_n - \omega_n' &= \frac{1}{2\pi}\int_{-\pi}^{\pi} \frac{f(x+\alpha)}{\tan(\alpha/2)} \sin n\alpha d\alpha = \\ &= \frac{\cos nx}{2\pi}\int_{-\pi}^{\pi} \frac{f(x+\alpha)\sin n(x+\alpha)}{\tan(\alpha/2)} d\alpha - \\ &\quad - \frac{\sin nx}{2\pi}\int_{-\pi}^{\pi} \frac{f(x+\alpha)\cos n(x+\alpha)}{\tan(\alpha/2)} d\alpha, \end{aligned}$$

where the integrals are understood as $\lim_{\epsilon\to+0}(\int_{-\pi}^{-\epsilon} + \int_{\epsilon}^{\pi})$.

[2] This is not true for $\epsilon = 0$ (see [3]).

If $\phi(x)$ and $\psi(x)$ are the functions conjugate to $f(x)\sin nx$ and $f(x)\cos nx$, then

$$\int_{-\pi}^{\pi} |S_n(x)|^{1-\epsilon}dx \le \int_{-\pi}^{\pi} (|\cos nx\phi(x)| + |\sin nx\psi(x)| + |\omega_n'(x)|)^{1-\epsilon}dx \le$$

$$\le \int_{-\pi}^{\pi} (|\phi(x)|^{1-\epsilon} + |\psi(x)|^{1-\epsilon} + |\omega_n'(x)|^{1-\epsilon})dx \le$$

$$\le \frac{C}{\epsilon}\Big(\int_{-\pi}^{\pi} |f(x)\sin nx|dx\Big)^{1-\epsilon} + \frac{C}{\epsilon}\Big(\int_{-\pi}^{\pi} |f(x)\cos nx|dx\Big)^{1-\epsilon} +$$

$$+ \int_{-\pi}^{\pi} |\omega_n'(x)|^{1-\epsilon}dx \le \frac{2C}{\epsilon}\Big(\int_{-\pi}^{\pi} |f(x)|dx\Big)^{1-\epsilon} + \int_{-\pi}^{\pi} |\omega_n'(x)|^{1-\epsilon}dx. \qquad (2)$$

For any $h > 0$ we construct two functions $f'(x)$ and $f''(x)$ such that

1. $f'(x) + f''(x) = f(x)$;
2. $\Big(\int_{-\pi}^{\pi} |f''(x)|dx\Big)^{1-\epsilon} < \frac{h\epsilon}{4C}$, $\int_{-\pi}^{\pi} |f''(x)|^{1-\epsilon}dx < \frac{h}{2}$;
3. $f'(x)$ is continuous, and in this case it satisfies (1).

If we apply (2) to $f''(x)$ and denote by S_n' and S_n'' the partial sums of the Fourier series of f' and f'', then we obtain

$$\limsup_{n\to\infty} \int_{-\pi}^{\pi} |f(x) - S_n(x)|^{1-\epsilon}dx \le \limsup_{n\to\infty} \int_{-\pi}^{\pi} |f'(x) - S_n'(x)|^{1-\epsilon}dx +$$

$$+ \int_{-\pi}^{\pi} |f''(x)|^{1-\epsilon}dx + \limsup_{n\to\infty} \int_{-\pi}^{\pi} |S_n''|^{1-\epsilon}dx < h.$$

This proves the theorem.

Moscow, February 1923

References

1. I.I. Privalov, *The Cauchy integral*, Saratov 1919 (in Russian).
2. N.N. Luzin, *Integral and trigonometric series*, Moscow 1915 (in Russian).
3. S. Banach and H. Steinhaus, *Sur la convergence en moyenne*, Bull. Acad. Cracovie, 1918.

9. ON THE TERTIUM NON DATUR PRINCIPLE*

Introduction

In the works of Brouwer it has been discovered that the application of the *tertium non datur* principle[1] in the domain of transfinite deduction can be illegitimate. Our aim is to find why this illegitimate use has not up to now led to contradictions and why the illegitimacy itself has often remained unnoticed.

Only finitistic mathematical inferences can be important in applications. However, to justify finitistic inferences use is often made of transfinite deduction. Therefore Brouwer believes that even those who are interested only in finitistic mathematical results should not ignore the intuitionistic criticism of the *tertium non datur* principle.

We prove that all finitistic inferences obtained by means of a transfinite application of the *tertium non datur* principle are true and can also be proved without using the principle.

The following question naturally arises: is there any meaning in those transfinite premises which serve for the derivation of true finitistic inferences?

We prove that every inference obtained by means of the *tertium non datur* principle is true if each proposition involved in its statement is replaced by the proposition asserting its double negation. The double negation of a proposition will be called its "pseudo-truth". Hence, in pseudo-truth mathematics the application of the *tertium non datur* principle is legitimate.

The necessity for the introduction of such notions as "pseudo-existence" and "pseudo-truth" has long been felt in mathematics, for example, in relation to the Zermelo axiom. But only now does one of the types of "pseudo-truth" receive a strict definition and justification in the form of axioms applicable in the domain of "pseudo-truth" and inapplicable to real truth.

I. Formal and Intuitionistic Points of View

§1. From the formal standpoint, mathematics is a collection of formulas (see [1], p. 152). Formulas are combinations of a definite set of elementary symbols. Mathematics is based on a certain group of formulas called axioms, and

* *Mat. Sb.* **32**:4 (1925), 646–667.

[1] *Tertium non datur* principle = principle of the excluded middle (Lat.)

on certain definite rules for constructing new formulas proceeding from given formulas; at the present time such rules are the derivation according to the scheme $\mathfrak{S}, \mathfrak{S} \to \mathfrak{T} | \mathfrak{T}$ and the rules for replacing symbols of various kinds of variables by their particular values.

In contrast to the axioms, which are assumed to be unconditionally "true", a certain group of formulas is regarded as being knowingly "false". A system of axioms is said to be "consistent" if none of the formulas considered "false" can be derived from the axioms according to the given rules.

§2. From the formal viewpoint, the choice of axioms underlying mathematics is arbitrary and must satisfy only demands of practical convenience which lie outside mathematics, and are of course more or less conventional. [2] In this approach the only absolute requirement on any mathematical doctrine is that the axioms it is based on be consistent.

By true and false formulas are meant, respectively, those proved proceeding from the axioms and those leading to a contradiction. The question of truth or falsity of a consistent but unprovable formula is meaningless from the formal viewpoint. The existence of such formulas indicates the incompleteness of the system of axioms. An incomplete system of axioms can be completed, if desirable, by taking one of the unprovable and consistent formulas or, rightfully, the negated formula as a new axiom. Thus, the choice of a formula taken as a new axiom among the various formulas contrary to one another must satisfy only the requirement of convenience.

§3. The intuitionistic point of view is based on the assumption of a real significance of mathematical propositions. Axioms forming the basis of mathematics are regarded as expressions of facts that are given to us. This approach allows the formal method for studying mathematical constructions as one of the possible methods, but contradicts the formal interpretation of mathematics as a whole.

---

[2] See [2], the introduction. Hilbert's approach is also close to this point of view, namely, by absolute truths (absolute Wahrheiten) he means only propositions of "metamathematics", that is, assertions about consistency. However, on the other hand, he believes that formulas of ordinary mathematics (eigentliche Mathematik) are nevertheless expressions of certain thoughts (Gedanken) (see [1], pp. 152–153).

The question of the nature of unprovable but consistent propositions is treated from the intuitionistic viewpoint in a completely different way as compared to the purely formal viewpoint. Suppose that we are given a system of axioms for a certain area of mathematics, for example, geometry. The axioms express properties of the object under study, which in this specific case is space. Suppose further, that a proposition belonging to the chosen area is unprovable using the given axioms, but does not lead to a contradiction. From the intuitionistic viewpoint there can be two cases here. First, it may happen that the truth or falsity of the proposition under consideration follows from a direct consideration; in this case one can take as a new axiom this proposition if it is true or the negated proposition if the former is false. Second, it may happen that the proposition is indeterminate, that is, its truth or falsity cannot be inferred from direct consideration; in this case one can only try to deduce this proposition from some other true propositions that are apparently true; if this attempt fails one must regard the proposition as being indeterminate since it is possible that later in time one will have to accept some axioms to be apparently true which will allow one to derive the truth or falsity of the proposition and it is unknown in advance which of these alternatives holds.

§4. The formal viewpoint is also put forward in mathematical logic. It is in the framework of mathematical logic where we encounter this viewpoint in the present paper. Nevertheless, in mathematical logic the formal viewpoint is based on disregarding the real significance of mathematical propositions. Indeed, nobody would suggest applying logical formulas of no real significance to reality. Therefore, since mathematical logic is considered a mere formal system whose formulas have no real significance, it is separated from general logic because the formal viewpoint can exist only in mathematics and mathematical logic but not in ordinary logic which has a claim on significance in application to reality.

But we do not separate out a special "mathematical logic" from general logic, and we only believe that the features of mathematics as a science pose specific problems in logic which are investigated using some special "logic of mathematics". And it is only in this kind of logic that doubts arise concerning the unconditional applicability of the *tertium non datur* principle.

§5. The distinction between the two above-mentioned viewpoints manifests itself even in propositional logic. In what follows, by general propositional logic is meant a science investigating properties of arbitrary propositions in relation to their truth, falsity, and the derivation process irrespective of their content (each proposition is regarded as an irreducible object of investigation). A formal expression in general propositional logic is realized using symbols for arbitrary propositions $A, B, C, \ldots$, the implication symbol $A \to B$, and the negation symbol $\bar{A}$.

Hilbert suggested the following system of axioms of propositional logic (see [1], p.153).

Axioms of implication

$$
(1) \qquad \begin{aligned}
&1.\quad A \to (B \to A).\\
&2.\quad \{A \to (A \to B)\} \to (A \to B).\\
&3.\quad \{A \to (B \to C)\} \to \{B \to (A \to C)\}.\\
&4.\quad (B \to C) \to \{(A \to B) \to (A \to C)\}.
\end{aligned}
$$

Axioms of negation

$$
(2) \qquad \begin{aligned}
&5.\quad A \to (\overline{A} \to B).\\
&6.\quad (A \to B) \to \{(\overline{A} \to B) \to B\}.
\end{aligned}
$$

The proof of the consistency of these axioms is quite elementary (see [3]). From the formal viewpoint, this allows one to take them as a basis for general propositional logic.

Moreover, Hilbert's system is complete in the sense that it can not be completed by a new independent axiom without contradiction. More precisely, each formula written using the symbols of propositional logic, for example,

$$\overline{\overline{(A \to B)}} \to (\overline{\overline{A}} \to \overline{\overline{B}}),$$

either can be proved on the basis of Hilbert's axioms or can be used to derive, on the basis of the same axioms, the consequence,

$$A$$

that is, the truth of an arbitrary proposition.

§6. From the intuitionistic viewpoint, the consistency of Hilbert's axioms is far from being a sufficient reason for adopting it. In the next chapter we will analyze reasons for their significance for general propositions and particular types of proposition.

Of the two Hilbert axioms of negation, Axiom 6 expresses the *tertium non datur* principle in a somewhat unusual form. The illegitimacy of the application of the principle to arbitrary propositions was shown by Brouwer. [3] Axiom 5 is used only in the symbolic representation of propositional logic, and therefore Brouwer's criticism is irrelevant to it; nevertheless, it has no intuitionistic foundation either.

Thus, along with the criticism of Hilbert's axioms, we will have to put forward new axioms of negation whose applicability to arbitrary propositions must be justified.

II. Axioms of Propositional Logic

§1. Axioms of general propositional logic have a claim on significance for all propositions; consequently, they must follow from general properties of propositions. Of course, all the further investigation is by no means a question of defining the basic notions and proving the axioms of propositional logic, but is a research on their intuitive sources using all the notions and methods of logic.

[3] See [4], p. 252. Hilbert also believes that the *tertium non datur* principle in application to infinite sets of objects is not intuitively obvious. In this case he expresses it symbolically by the two formulas

$$\overline{(a)}\ A(a) \text{ äq. } (Ea)\ \overline{A}(a),$$

$$\overline{(Ea)}A(a) \text{ äq. } (a)\ \overline{A}(a)$$

(see [1], p.153). As to the *tertium non datur* principle in general propositional logic (Axiom 6), Hilbert says nothing about its intuitive obviousness; probably he considers it indubitable. These ideas of Hilbert are not inseparably related to his purely formal aim of investigating consistency. We believe the axioms are wrong.

First, Axiom 6 is not intuitively obvious. Its relation to finitistic logic (finite Logik) is merely apparent, since the justification of the truth of Axiom 6, as will be shown in the next chapter, requires the consideration of the content of propositions, which can be transfinite, whereas the truth of the Axioms of implication 1–4 can be established irrespective of the content of the propositions.

Second, if Axioms 1–6 are accepted, then the two formulas given above can be proved using certain axioms whose intuitive obviousness cannot be considered doubtful. The proofs will be given in Chapter 5; to justify the investigation below it suffices to use the first argument.

A proposition in propositional logic is regarded as a finite object of investigation. When a proposition is considered irrespective of the synthesis of the subject and predicate it contains, the only remaining characteristic property of the proposition distinguishing it from other types of judgement is, as Aristotle says, [4] that it admits a consideration from the standpoint of truth or falsity. It seems natural to try to derive the axioms of general propositional logic without going outside its own limits, that is, only from the above-mentioned characteristic property of a proposition. In the sections below we will investigate to what extent this is possible.

§2. The meaning of the symbol $A \rightarrow B$ consists in that after it has been proved that A is true, one must also recognize that B is true. Formally, if formula A is written, then formula B can also be written. [5] Hence, the relation of implication between two propositions does not establish any relation between their content.

The first Hilbert axiom of implication, meaning "what is true follows from everything", is implied by the following formal interpretation of implication: if B is itself true, then, on finding that A is true, we must also regard B as being true. The truth of the other three axioms of implication is also readily seen based on this interpretation of the notion of implication. Here the character of the propositions under consideration is completely irrelevant; consequently, no doubts in the applicability of the axioms to arbitrary propositions can arise.

Of interest is the question of completeness of the whole Hilbert system of four axioms of propositional logic; the question must be stated in the following way: by a true formula is meant a formula proved using the axioms of implication and the axioms of negation; can any true formula written only using symbols for arbitrary propositions and the implication symbol, without the negation symbol, be proved based on only the four axioms of implication?

§3. In relation to a completed proposition regarded as a whole, negation is merely the interdiction of recognition of the proposition as being true. A more complete interpretation of negation can be gained by regarding a proposition

[4] De interp. 4; De anima III, 6 (in Latin) (see [8, 9]).

[5] It is this fact that is expressed by the scheme $\mathfrak{S}, \mathfrak{S} \rightarrow \mathfrak{T} \mid \mathfrak{T}$ in Hilbert's metamathematics. Sigwart also believes that this is the most general scheme for any deduction (see [5]).

as a statement about the subject made by the predicate; then negation is interpreted as an assertion that the predicate is inadequate to the subject.

The symbol $\overline{A}$ in propositional logic expresses the first interpretation of negation as interdiction of the assumption that the proposition A is true. However, the ordinary logical tradition is to reduce the first interpretation to the second one for the reason that the latter is more primary. [6] This proves impossible in application to mathematical propositions.

Indeed, since the negation of a proposition is the result of direct consideration, the second interpretation, which proceeds from the idea of unrealizability of the synthesis created by the proposition, is closer to the essence than the first one which is based on the purely formal idea of interdiction. However, when negation results from deduction, the reduction of the first interpretation to the second one is no longer necessary, and, in the case of mathematical propositions, it sometimes is even impossible. Indeed, many negated mathematical propositions are proved by contradiction according to the scheme [7] $\mathfrak{S} \to \mathfrak{T}, \overline{\mathfrak{T}} \mid \overline{\mathfrak{S}}$ and cannot be proved in a different way.

Thus, the first interpretation of negation is of independent value. It was first indicated by Brouwer, who defines negation as absurdity (see [4]). It is based on the second interpretation since to derive a negated proposition by contradiction one must have negated propositions at one's disposal, but at the same time it is broader than the latter.

§4. Hilbert's first axiom of negation, which reads "what is false implies everything", appeared only with the creation of symbolic logic; the same is true, by the way, for the first axiom of implication. However, the axiom in question does not have and cannot have any intuitive grounds, since it asserts something about consequences of what is impossible in the sense that we must assume B to be true if a true proposition A is assumed to be false whereas the first axiom of implication follows with intuitive obviousness from the correct understanding of the idea of logical implication.

Hence, Hilbert's first axiom of negation cannot be an axiom of intuitionistic propositional logic whatever the interpretation of negation we proceed from. Of course, this does not exclude the possibility that the axiom can be a formula

---

[6] For example, see [5].

[7] See §6.

that can be proved based on other axioms.

§5. Hilbert's second axiom of negation expresses the *tertium non datur* principle. It presents the principle in the form in which it is applied for deduction: if B is implied by A and by $\overline{A}$, then B is true. Its ordinary form is "any proposition is either true or false" [8] and it is equivalent to the above. [9]

It is clear that the first interpretation of negation, as the interdiction of the assumption that the proposition is true, does not make one confident that the *tertium non datur* principle is true; by the way, no such attempts were ever made. Consequently, to justify the principle it is necessary to consider the content of the proposition, that is, the relation of the predicate to the subject. Even in the simplest case of a proposition of the type "all A are B", the consideration inevitably involves the relation of all possible A to the predicate B and the set of A's can be infinite. Brouwer showed [10] that in the case of such transfinite propositions the *tertium non datur* principle cannot be regarded as being obvious from this viewpoint either.

§6. Thus, from the intuitionistic viewpoint, neither of the two Hilbert axioms of negation can be taken as an axiom in general propositional logic. Here we put forward an axiom which we call the principle of contradiction:

$$5. \qquad (A \to B) \to \{(A \to \overline{B}) \to \overline{A}\}.$$

Its meaning is the following: if A implies both the truth and falsity of a proposition B, then the proposition A itself is false.

The usual principle of contradiction reading "a proposition cannot be true and false simultaneously" cannot be stated in terms of arbitrary propositions, implication and negation. Our principle contains something more; namely,

[8] This is Leibniz' simplest formulation (*Nouveaux Essais*, IV, 2; (in French) (see [10]).

[9] The second form is expressed symbolically thus: $A \vee \overline{A}$, where $\vee$ means "or". The equivalence of the two forms is readily proved based on the axioms of implication and the axioms below taken from the work by Ackermann [3], which define the meaning of the symbol $\vee$:

1. $A \to A \vee B$.
2. $B \to A \vee B$.
3. $(A \to C) \to (B \to C) \to (A \vee B) \to C$.

[10] See [4], or the detailed presentation in [6], for an example of a proposition that cannot be proved unless the *tertium non datur* principle is illegitimately applied.

in combination with the first axiom of implication it implies the *reductio ad absurdum* principle as well: if B is true and A implies that B is false, then A is false.

The truth of the suggested axiom follows from the simplest interpretation of negation as interdiction of the assumption that the proposition is true and is irrespective of the content of the propositions.

The system of five axioms, including the four axioms of implication (1) and the above axiom of negation, will be called the system $\mathfrak{B}$. We do not know any formulas of general propositional logic possessing intuitive obviousness in application to arbitrary propositions that cannot be proved on the basis of this system of axioms. However, the question whether this is a complete system of axioms of general intuitionistic propositional logic remains open.

§7. As has been seen, although the *tertium non datur* principle cannot be taken as an axiom of general propositional logic, in a limited domain of propositions (Brouwer calls them finitistic propositions) it is valid. Here we will investigate the limits of the domain of finitistic propositions; this problem is not as simple as one might think; therefore, we confine ourselves to the assumption that a domain of this kind exists.

Along with the *tertium non datur* principle, in the domain of finitistic propositions the law of double negation is valid, which is expressed symbolically as

(4) $$6. \quad \overline{\overline{A}} \to A.$$ [11]

It is, of course, obvious that all five axioms of general propositional logic (the system $\mathfrak{B}$) are valid in the domain of finitistic propositions as well. The system consisting of the axioms of the system $\mathfrak{B}$ (that is, (1) and (3)) and the axiom of double negation (4) will be called the system $\mathfrak{H}$.

We claim that the system $\mathfrak{H}$ is equivalent to the system of Hilbert's axioms (1) and (2). The two systems have the same axioms of implication. Consequently, to prove our claim it suffices to prove formulas (3) and (4) based on formulas (2) and conversely; we use the axioms of implication in both cases. We shall not present the proofs of (3) and (4) based on Hilbert's axioms (1)

[11] The formula $A \to \overline{\overline{A}}$ can be proved based on the system $\mathfrak{B}$. See formula (34) below.

and (2); the proof of the converse based on the axioms (3) and (4) which are introduced for the first time here, will be given in the next section.

For our purposes the system $\mathfrak{H}$ has the advantage that it is obtained from the system $\mathfrak{B}$ of general propositional logic by adding the single axiom of double negation; this substantially facilitates further investigation.

It is obvious that the system $\mathfrak{H}$ as well as Hilbert's system is complete. All formulas of traditional propositional logic can be derived from it. They all are true provided that the symbols $A, B, C, \ldots$ for arbitrary propositions are replaced in them by symbols $A^f, B^f, C^f, \ldots$ for arbitrary finitistic propositions. The proof of this fact meets with some difficulties, which will be elucidated and overcome in the next chapter [12].

§8. We shall denote the axioms (1), (3) and (4) of the system $\mathfrak{H}$ by their numbers 1–6. The numbers of the formulas based on Axiom 6 are doubly underlined since they are valid only in the domain of finitistic propositions, whereas the others are applicable to arbitrary propositions.

$$(5)\qquad \frac{A \to (B \to A)}{\overline{B} \to (A \to \overline{B})} \qquad \text{Axiom 1}$$

$$(6)\qquad \frac{\{A \to (B \to C)\} \to \{B \to (A \to C)\}}{[(B \to C) \to \{(A \to B) \to (A \to C)\}] \to [(A \to B) \to \{(B \to C) \to (A \to C)\}]} \qquad \text{Axiom 3}$$

$$[(B \to C) \to \{(A \to B) \to (A \to C)\}] \to [(A \to B) \to \{(B \to C) \to (A \to C)\}] \qquad (6)$$

$$(7)\qquad \frac{(B \to C) \to \{(A \to B) \to (A \to C)\}}{(A \to B) \to \{(B \to C) \to (A \to C)\}} \qquad \text{Axiom 4}$$

$$(8)\qquad \frac{(A \to B) \to \{(B \to C) \to (A \to C)\}}{\{\overline{B} \to (A \to \overline{B})\} \to [\{(A \to \overline{B}) \to \overline{A}\} \to (\overline{B} \to \overline{A})]} \qquad (7)$$

$$\{(\overline{B} \to (A \to \overline{B})\} \to [\{(A \to \overline{B}) \to \overline{A}\} \to (\overline{B} \to \overline{A})] \qquad (8)$$

$$(9)\qquad \frac{\overline{B} \to (A \to \overline{B})}{\{(A \to \overline{B}) \to \overline{A}\} \to (\overline{B} \to \overline{A})} \qquad (5)$$

$$(A \to B) \to \{(A \to \overline{B}) \to \overline{A}\} \qquad \text{Axiom 5}$$

[12] See §4 in Chapter 3.

(10) $$\frac{\{(A \to \overline{B}) \to \overline{A}\} \to (\overline{B} \to \overline{A})}{(A \to B) \to (\overline{B} \to \overline{A})}$$ (9)

(11) $$\frac{(A \to B) \to (\overline{B} \to \overline{A})}{(B \to A) \to (\overline{A} \to \overline{B})}$$ (10)

$$A \to (B \to A)$$ Axiom 1

(12) $$\frac{(B \to A) \to (\overline{A} \to \overline{B})}{A \to (\overline{A} \to \overline{B})}$$ (11)

(13) $$\frac{A \to (\overline{A} \to \overline{B})}{A \to (\overline{A} \to \overline{\overline{\overline{B}}})}$$ (12)

(14) $$\frac{\overline{\overline{A}} \to A}{\overline{\overline{\overline{B}}} \to B}$$ Axiom 6

(15) $$\frac{(B \to C) \to \{(A \to B) \to (A \to C)\}}{(\overline{\overline{\overline{B}}} \to B) \to \{(A \to \overline{\overline{\overline{B}}}) \to (A \to B)\}}$$ Axiom 4

$$(\overline{\overline{\overline{B}}} \to B) \to \{(A \to \overline{\overline{\overline{B}}}) \to (A \to B)\}$$

(16) $$\frac{\overline{\overline{\overline{B}}} \to B}{(A \to \overline{\overline{\overline{B}}}) \to (A \to B)}$$ (14)

(17) $$\frac{(A \to \overline{\overline{\overline{B}}}) \to (A \to B)}{(\overline{A} \to \overline{\overline{\overline{B}}}) \to (\overline{A} \to B)}$$ (16)

$$A \to (\overline{A} \to \overline{\overline{\overline{B}}})$$ (13)

(18) $$\frac{(\overline{A} \to \overline{\overline{\overline{B}}}) \to (\overline{A} \to B)}{A \to (\overline{A} \to B)}$$ (16)

Thus Hilbert's first axiom of negation is proved.

(19) $$\frac{(A \to B) \to (\overline{B} \to \overline{A})}{(\overline{A} \to B) \to (\overline{B} \to \overline{\overline{A}})}$$ (10)

(20) $$\frac{(A \to B) \to \{(A \to \overline{B}) \to \overline{A}\}}{(\overline{B} \to \overline{A}) \to \{(\overline{B} \to \overline{\overline{A}}) \to \overline{\overline{B}}\}}$$ Axiom 5

$$(A \to B) \to (\overline{B} \to \overline{A}) \qquad (10)$$

$$(21) \qquad \frac{(\overline{B} \to \overline{A}) \to \{(\overline{B} \to \overline{\overline{A}}) \to \overline{\overline{B}}\}}{(A \to B) \to \{(\overline{B} \to \overline{\overline{A}}) \to \overline{\overline{B}}\}} \qquad (20)$$

$$(22) \qquad \frac{(A \to B) \to \{(B \to C) \to (A \to C)\}}{\{(\overline{A} \to B) \to (\overline{B} \to \overline{\overline{A}})\} \to [\{(\overline{B} \to \overline{\overline{A}}) \to \overline{\overline{B}}\} \to \{(\overline{A} \to B) \to \overline{\overline{B}}\}]} \qquad (7)$$

$$\{(\overline{A} \to B) \to (\overline{B} \to \overline{\overline{A}})\} \to [\{(\overline{B} \to \overline{\overline{A}}) \to \overline{\overline{B}}\} \to \{(\overline{A} \to B) \to \overline{\overline{B}}\}] \qquad (22)$$

$$(23) \qquad \frac{(\overline{A} \to B) \to (\overline{B} \to \overline{\overline{A}})}{\{(\overline{B} \to \overline{\overline{A}}) \to \overline{\overline{B}}\} \to \{(\overline{A} \to B) \to \overline{\overline{B}}\}} \qquad (19)$$

$$(A \to B) \to \{(\overline{B} \to \overline{\overline{A}}) \to \overline{\overline{B}}\} \qquad (21)$$

$$(24) \qquad \frac{\{(\overline{B} \to \overline{\overline{A}}) \to \overline{\overline{B}}\} \to \{(\overline{A} \to B) \to \overline{\overline{B}}\}}{(A \to B) \to \{(\overline{A} \to B) \to \overline{\overline{B}}\}} \qquad (23)$$

$$\underline{\underline{(25)}} \qquad \frac{(A \to \overline{\overline{B}}) \to (A \to B)}{\{(\overline{A} \to B) \to \overline{\overline{B}}\} \to \{(\overline{A} \to B) \to B\}} \qquad \underline{\underline{(16)}}$$

$$(A \to B) \to \{(\overline{A} \to B) \to \overline{\overline{B}}\} \qquad (24)$$

$$\underline{\underline{(26)}} \qquad \frac{\{(\overline{A} \to B) \to \overline{\overline{B}}\} \to \{(\overline{A} \to B) \to B\}}{(A \to B) \to \{(\overline{A} \to B) \to B\}} \qquad \underline{\underline{(25)}}$$

Thus, Hilbert's second axiom is also proved.

Among the formulas provable by means of Axioms of $\mathfrak{H}$ without the double negation axiom, formulas (12) and (24) are most similar to Hilbert's axioms of negation. The second of them, which is close to the *tertium non datur* principle, means the following: if B is implied by both the truth and falsity of A, then B cannot be false. Indeed, assume that B is false; then A cannot be true since the truth of A would imply that of B, and the falsity of A would imply the truth of B

III. Particular Propositional Logic and its Domain of Applicability

§1. The formulas proved on the basis of axioms of $\mathfrak{B}$ constitute general propositional logic. The set of formulas proved on the basis of the six axioms of the system $\mathfrak{H}$ will be called particular propositional logic. [13] The content of particular propositional logic is broader than that of general propositional logic, while its domain of applicability is narrower. All that follows is devoted to the investigation of the domain of applicability of particular propositional logic. This domain may turn out to be somewhat narrower than the domain of applicability of the *tertium non datur* principle in Hilbert's form.

§2. We introduce the symbols $A^{\cdot}, B^{\cdot}, C^{\cdot}, \ldots$ denoting an arbitrary proposition for which its double negation implies the proposition itself. Finitistic propositions are of this type. All true propositions also possess this property; however, this fact is not used in what follows. Brouwer showed that all negated propositions also belong to this type (see [6]). The proof below is based only on axioms of the system $\mathfrak{B}$.

Proceeding from the axioms of implication one can easily prove the following formulas:

$$A \to A \tag{27}$$

$$\frac{(A \to B) \to \{(A \to \overline{B}) \to \overline{A}\}}{(\overline{A} \to A) \to \{(\overline{A} \to \overline{A}) \to \overline{\overline{A}}\}} \qquad \text{Axiom 5} \tag{28}$$

$$\frac{A \to (B \to A)}{A \to (\overline{A} \to A)} \qquad \text{Axiom 1} \tag{29}$$

$$A \to (\overline{A} \to A) \qquad (29)$$

$$\frac{(\overline{A} \to A) \to \{(\overline{A} \to \overline{A}) \to \overline{\overline{A}}\}}{A \to \{(\overline{A} \to \overline{A}) \to \overline{\overline{A}}\}} \qquad (28) \tag{30}$$

$$\frac{\{A \to (B \to C)\} \to \{B \to (A \to C)\}}{[A \to \{(\overline{A} \to \overline{A}) \to \overline{\overline{A}}]\} \to \{(\overline{A} \to \overline{A}) \to (A \to \overline{\overline{A}})\}} \qquad \text{Axiom 3} \tag{31}$$

$$[A \to \{(\overline{A} \to \overline{A}) \to \overline{\overline{A}}]\} \to \{(\overline{A} \to \overline{A}) \to (A \to \overline{\overline{A}})\} \qquad (31)$$

$$\frac{A \to \{(\overline{A} \to \overline{A}) \to \overline{\overline{A}}\}}{(\overline{A} \to \overline{A}) \to (A \to \overline{\overline{A}})} \qquad (30) \tag{32}$$

[13] General propositional logic can also be given another (significant) definition (see §5 of Chapter 1). At present, particular propositional logic can be defined only formally, since the real content of its formulas will only be established later.

(33) $$\frac{A \to A}{\overline{A} \to \overline{A}} \qquad (27)$$

$$(\overline{A} \to \overline{A}) \to (A \to \overline{\overline{\overline{A}}}) \qquad (32)$$

(34) $$\frac{\overline{A} \to \overline{A}}{A \to \overline{\overline{\overline{A}}}} \qquad (33)$$

(35) $$\frac{(A \to B) \to (\overline{B} \to \overline{A})}{(A \to \overline{\overline{A}}) \to (\overline{\overline{\overline{A}}} \to \overline{A})} \qquad (10)$$

$$(A \to \overline{\overline{A}}) \to (\overline{\overline{\overline{A}}} \to \overline{A}) \qquad (35)$$

(36) $$\frac{A \to \overline{\overline{A}}}{\overline{\overline{\overline{A}}} \to \overline{A}} \qquad (34)$$

Formula (36) proves that all negated propositions are propositions of type $A^{\cdot}$.

The system of axioms $\mathfrak{H}$ differs from the system $\mathfrak{B}$, which is universally applicable, only in the axiom of double negation. For propositions of the type $A^{\cdot}$ it is expressed by the formula

(37) $$\overline{\overline{A}}^{\cdot} \to A^{\cdot}.$$

Only this formula is regarded as being true, while formula (4) is considered unjustified.

However, what has been said does not yet imply that all formulas of particular propositional logic are true for propositions of type $A^{\cdot}$; indeed, in their derivation the axiom of double negation (4) is applied not only to elementary propositions, which in the case of propositions of the type $A^{\cdot}$ is legitimate according to (37), but also to complex formulas; however, it is not, for instance, known whether a formula of the type $A^{\cdot} \to B^{\cdot}$ is of type $A^{\cdot}$.

§3. We shall now prove that any formula expressed in terms of the symbols $A^{\cdot}, B^{\cdot}, C^{\cdot}, \ldots$ and the symbols of implication and negation is a formula of type $A^{\cdot}$. To this end it suffices to consider the two simple cases below.

First, each negated proposition is a proposition of type $A^{\cdot}$ by virtue of the Brouwer-like formula (36).

Second, we shall prove that propositions of type $A \to B^{\cdot}$ are also propositions of type $A^{\cdot}$.

$$(38) \qquad \frac{A \to A}{(A \to B) \to (A \to B)} \qquad (27)$$

$$(39) \qquad \frac{\{A \to (B \to C)\} \to \{B \to (A \to C)\}}{\{(A \to B) \to (A \to B)\} \to [A \to \{(A \to B) \to B\}]} \qquad \text{Axiom 3}$$

$$\{(A \to B) \to (A \to B)\} \to [A \to \{(A \to B) \to B\}] \qquad (39)$$

$$(40) \qquad \frac{(A \to B) \to (A \to B)}{A \to \{(A \to B) \to B\}} \qquad (38)$$

$$(41) \qquad \frac{(A \to B) \to (\overline{B} \to \overline{A})}{(\overline{B} \to \overline{A}) \to (\overline{\overline{A}} \to \overline{\overline{B}})} \qquad (10)$$

$$(A \to B) \to (\overline{B} \to \overline{A}) \qquad (10)$$

$$(42) \qquad \frac{(\overline{B} \to \overline{A}) \to (\overline{\overline{A}} \to \overline{\overline{B}})}{(A \to B) \to (\overline{\overline{A}} \to \overline{\overline{B}})} \qquad (41)$$

$$(43) \qquad \frac{(A \to B) \to (\overline{\overline{A}} \to \overline{\overline{B}})}{\{(A \to B) \to B\} \to \{(\overline{\overline{A \to B}}) \to \overline{\overline{B}}\}} \qquad (42)$$

$$A \to \{(A \to B) \to B\} \qquad (40)$$

$$(44) \qquad \frac{\{(A \to B) \to B\} \to \{(\overline{\overline{A \to B}}) \to \overline{\overline{B}}\}}{A \to \{(\overline{\overline{A \to B}}) \to \overline{\overline{B}}\}} \qquad (43)$$

$$(45) \qquad \frac{\{A \to (B \to C)\} \to \{B \to (A \to C)\}}{[A \to \{(\overline{\overline{A \to B}}) \to \overline{\overline{B}}\}] \to \{(\overline{\overline{A \to B}}) \to (A \to \overline{\overline{B}})\}} \qquad \text{Axiom 3}$$

$$[A \to \{(\overline{\overline{A \to B}}) \to \overline{\overline{B}}\}] \to \{(\overline{\overline{A \to B}}) \to (A \to \overline{\overline{B}})\} \qquad (45)$$

$$(46) \qquad \frac{A \to \{(\overline{\overline{A \to B}}) \to \overline{\overline{B}}\}}{(\overline{\overline{A \to B}}) \to (A \to \overline{\overline{B}})} \qquad (44)$$

This formula is true for arbitrary propositions A and B. Replacing A and B by $A^{\cdot}$ and $B^{\cdot}$ and using (37), one readily derives the formula

$$(47) \qquad (\overline{\overline{A^{\cdot} \to B^{\cdot}}}) \to (A^{\cdot} \to B^{\cdot}),$$

which shows that propositions of type $A^{\cdot} \to B^{\cdot}$ are of type $A^{\cdot}$.

Passing gradually to more complicated formulas one can prove the assertion at the beginning of the section.

§4. We can now assert that all formulas of particular propositional logic are true for propositions of type $A^{\cdot}$, which include all finitistic and negated propositions. Indeed, the symbols $A^{\cdot}, B^{\cdot}, C^{\cdot}, \ldots, A^{\cdot} \to B^{\cdot}$, and $\overline{A}^{\cdot}$ admit of all operations applicable to the symbols of general propositional logic, namely, replacement of symbols $A^{\cdot}, B^{\cdot}, C^{\cdot}, \ldots$ by any formula written using these symbols and the derivation according to the scheme $\mathfrak{S} \to \mathfrak{T}$, $\mathfrak{S}|\mathfrak{T}$; moreover, all six axioms of $\mathfrak{H}$ are true for them.

We have thus found the precise limits of the domain of applicability of particular propositional logic; namely, this domain coincides with the domain of applicability of the double negation formula (4).

IV. Pseudo-Truth Mathematics

§1. In the preceding chapter we established that all formulas of traditional propositional logic can in fact be proved as formulas of particular propositional logic. To this end, it is only necessary to recognize that they are related only to propositions of type $A^{\cdot}$. In this case the formulas themselves turn out to be formulas of type $A^{\cdot}$.

We now pose the following question: is it possible, by imposing certain restrictions on the real interpretation, to reconstruct the significance of all those mathematical formulas that are proved by an illegitimate application of formulas of particular propositional logic and, specifically, the *tertium non datur* principle outside the domain of their applicability? This problem turns out to be solvable.

§2. Along with ordinary mathematics, we construct a new "pseudo-mathematics", in such a way that to each formula of the former there corresponds a formula of the latter so that each pseudo-mathematical formula is a formula of type $A^{\cdot}$. At present we do not consider the question of truth of pseudo-mathematical formulas; it will be treated in §5 of this chapter.

By a formula is meant a simple or complex symbol expressing a proposition. By elementary or first-order formulas we shall mean formulas none of whose parts is a formula, such as the formula $a = a$. By an nth order formula we shall mean a formula whose parts are formulas of order not exceeding $n - 1$. For example, the formula

$$a = b \to \{A(a) \to B(a)\}$$

is a formula of the third order since its constituent part $A(a) \to B(a)$ is a formula of the second order.

The pseudo-mathematical formula corresponding to an elementary formula $\mathfrak{S}$ is the formula

$$\mathfrak{S}^* \equiv \overline{\overline{\mathfrak{S}}}, \tag{48}$$

expressing the double negation of $\mathfrak{S}$. For the sake of convenience, in what follows the double negation of $\mathfrak{S}$ will be denoted $n\mathfrak{S}$.

The pseudo-mathematical formula corresponding to an nth-order formula $F(\mathfrak{S}_1, \mathfrak{S}_2, \dots, \mathfrak{S}_k)$, where $\mathfrak{S}_1, \mathfrak{S}_2, \dots, \mathfrak{S}_k$ are formulas of orders not exceeding $(n - 1)$, is written

$$F(\mathfrak{S}_1, \mathfrak{S}_2, \dots \mathfrak{S}_k)^* \equiv nF(\mathfrak{S}_1^*, \mathfrak{S}_2^*, \dots, \mathfrak{S}_k^*), \tag{49}$$

where $\mathfrak{S}_1^*, \mathfrak{S}_2^*, \dots, \mathfrak{S}_k^*$ are regarded as being defined. For instance, the pseudo-mathematical formula corresponding to the formula

$$a = b \to \{A(a) \to B(a)\}$$

is the formula

$$n[n(a = b) \to n\{nA(a) \to nB(a)\}].$$

To each symbol that is not a formula there also corresponds a definite pseudo-mathematical symbol. To a simple or complex symbol none of whose parts is a formula there corresponds in pseudo-mathematics a symbol identical to the former. To a complex symbol that includes formulas there corresponds a symbol in which all formulas $\mathfrak{S}$ are replaced by formulas $\mathfrak{S}^*$.

§3. All formulas in mathematics are deduced from axioms,[14] which we denote by $\mathfrak{U}_1, \mathfrak{U}_2, \dots, \mathfrak{U}_k$, by using the operations of replacement of variables by their

[14] Here the axioms of mathematics include all the axioms of logic.

particular values and the derivation according to the scheme $\mathfrak{S}, \mathfrak{S} \to \mathfrak{T} | \mathfrak{T}$. To the axioms in pseudo-mathematics there correspond formulas $\mathfrak{U}_1^*, \mathfrak{U}_2^*, \ldots, \mathfrak{U}_k^*$. We shall prove that each pseudo-mathematical formula corresponding to a formula proved on the basis of the axioms $\mathfrak{U}$ is a consequence of the formulas $\mathfrak{U}^*$. To this end it suffices to establish the following two facts.

First, if replacement of variables by particular values in a formula $\mathfrak{S}$ results in a formula $\mathfrak{T}$, then the insertion of the corresponding formulas and symbols into the corresponding places in the formula $\mathfrak{S}^*$ yields the formula $\mathfrak{T}^*$.

Second, by analogy with the scheme $\mathfrak{S}, \mathfrak{S} \to \mathfrak{T} | \mathfrak{T}$, the scheme

$$\mathfrak{S}^*, (\mathfrak{S} \to \mathfrak{T})^* | \mathfrak{T}^* \tag{50}$$

is true.

Indeed, we have

$$(\mathfrak{S} \to \mathfrak{T})^* \equiv (\overline{\overline{\mathfrak{S}^* \to \mathfrak{T}^*}}). \tag{51}$$

Since $\mathfrak{S}^*$ and $\mathfrak{T}^*$ are formulas of type $A^{\cdot}$, formula (47) implies

$$(\overline{\overline{\mathfrak{S}^* \to \mathfrak{T}^*}}) \to (\mathfrak{S}^* \to \mathfrak{T}^*), \tag{52}$$

$$\frac{(\mathfrak{S} \to \mathfrak{T})^* \xrightarrow{\mathfrak{S}^*} (\mathfrak{S}^* \to \mathfrak{T}^*)}{\mathfrak{T}^*}. \tag{53}$$

We thus see that to each correct proof in ordinary mathematics there corresponds a correct proof in pseudo-mathematics. This implies the truth of the proposition stated at the beginning of the section.

§4. The formulas of pseudo-mathematics corresponding to the five axioms of general propositional logic are the following:

1. $n\{nA \to n(nB \to nA)\}$;

2. $n[n\{nA \to n(nA \to nB)\} \to n(nA \to nB)]$;

(54) 3. $n[n\{nA \to n(nB \to nC)\} \to n\{nB \to n(nA \to nC)\}]$;

4. $n[n(nB \to nC) \to n\{n(nA \to nB) \to n(nA \to nC)\}]$;

5. $n[n(nA \to nB) \to n\{n(nA \to n(\overline{nB})) \to n(\overline{nA})\}]$.

These formulas can be obtained by substituting nA, nB, nC for $A^{\cdot}, B^{\cdot}, C^{\cdot}$ into the formulas

1. $n\{A^{\cdot} \to n(B^{\cdot} \to A^{\cdot})\}$

(55) ....

5. $n[n(A^{\cdot} \to B^{\cdot}) \to n\{n(A^{\cdot} \to n\overline{B}^{\cdot}) \to n\overline{A}^{\cdot}\}]$.

Since (55) are formulas of particular propositional logic, it is legitimate to prove them using all the axioms of $\mathfrak{H}$ or all Hilbert's axioms. Their proof encounters no difficulties. Hence, all formulas (54) turn out to be true. It follows that all formulas of pseudo-mathematics corresponding to true formulas of general propositional logic are true.

§5. All the axioms of mathematics known to us have the same property as the axioms of general propositional logic; namely, the formulas of pseudo-mathematics corresponding to them are true. For instance, the formula corresponding to the axiom

$$(a)A(a) \to A(a)$$

is the true formula

$$n\{n(a)nA(a) \to nA(a)\}.$$

Axioms with the above-stated property will be called axioms of type $\mathfrak{N}$. Further, by formulas of type $\mathfrak{N}$ we shall mean formulas proved using axioms of type $\mathfrak{N}$. All axioms and formulas of mathematics known to us [15] are of type $\mathfrak{N}$.

By what has been said, the part of pseudo-mathematics whose formulas correspond to formulas of type $\mathfrak{N}$ acquires real significance in the sense that all its formulas are true, since they are consequences of true formulas corresponding in pseudo-mathematics to axioms of type $\mathfrak{N}$. The name "pseudo-mathematics" in relation to this part, which is the only one existent at present, is no longer appropriate, since it is a collection of true formulas and thus belongs to real mathematics.

[15] Sometimes, in mathematics as axioms one takes formulas whose truth is not obvious; such is, for instance, the so-called Zermelo axiom. However, they also have the property $\mathfrak{U} \to \mathfrak{U}^*$.

We shall say that a proposition is pseudo-true if its double negation is true. Hence, a proposition of type $n\mathfrak{S}$ asserts that the proposition $\mathfrak{S}$ is pseudo-true. Formulas of pseudo-mathematics always express only propositions on pseudo-truth. The part of pseudo-mathematics having a real significance can therefore rightfully be called pseudo-truth mathematics.

§6. In an ordinary presentation of mathematics, some conclusions are drawn by illegitimately using formulas of particular propositional logic, for example the *tertium non datur* principle. As has been shown, all these cases can be reduced to the application of the law of double negation

$$(4) \qquad 6. \quad \overline{\overline{A}} \to A.$$

Among these conclusions, let us consider those which, apart from the illegitimate formula (4), are based only on axioms of type $\mathfrak{R}$. The formulas expressing them will be called formulas of type $\mathfrak{R}'$.

We now construct the pseudo-mathematical formulas corresponding to formulas of type $\mathfrak{R}'$. They will all follow from formulas $\mathfrak{U}^*$ corresponding to axioms of type $\mathfrak{R}$ and from the formula

$$(56) \qquad n\{n(\overline{\overline{nA}}) \to nA\},$$

corresponding to formula (4).

Formula (56) is true. Indeed, by virtue of (34), it is implied by the formula

$$(57) \qquad n\ (\overline{\overline{nA}}) \to nA.$$

Formula (57) can be obtained by substitution from the formula

$$(58) \qquad n\ (\overline{\overline{A^{\cdot}}}) \to A^{\cdot}$$

of particular propositional logic and, as is well known, in particular propositional logic repetition of negation an even number of times leads to a positive (that is, non-negated) assertion.

Thus, formulas of pseudo-mathematics are based on the true formula (56), whereas the corresponding formulas of ordinary mathematics are based on an illegitimate use of formula (4).

Hence, we finally state the following.

All conclusions of ordinary mathematics based on the application of the formula of double negation and other formulas depending on it (such as the *tertium non datur* principle) outside the domain of finitistic propositions cannot be regarded as being justified rigorously.

However, since, besides the formula of double negation, they are based only on axioms of type $\mathfrak{R}$, and since no other axioms are yet known, the corresponding formulas of pseudo-mathematics are true and consequently, belong to pseudo-truth mathematics.

In other words, all conclusions based on axioms of type $\mathfrak{R}$ and the formula of double negation are true if each proposition contained in them is understood in the sense of the assertion that it is pseudo-true, that is, as its double negation.

V. Applications

§1. On discovering the illegitimacy of the application of the *tertium non datur* principle to transfinite propositions, Brouwer posed the problem of justifying mathematics without this principle and, to a considerable extent, realized this.[16] However, it turned out that there are a number of mathematical propositions that cannot be proved without the *tertium non datur* principle; these propositions are thus rejected by Brouwer. Below we consider some examples of this kind of proposition.

In the previous chapters we have shown that, along with the presentation of mathematics without the *tertium non datur* principle, the ordinary presentation can also be retained. However, to this end all propositions should be given a limited interpretation in the sense that each proposition of ordinary mathematics should be replaced by the assertion that it is pseudo-true. However, this kind of presentation nevertheless retains the two remarkable properties below.

1. If an argument based on the application of the *tertium non datur* principle, even in the domain of transfinite propositions, yields a finitistic conclusion, then it is true in the ordinary sense. Indeed, by what has been established, it can be proved as a conclusion concerning pseudo-truth; in the domain of finitistic propositions, pseudo-truth coincides with ordinary truth.

2. Application of the *tertium non datur* principle never leads to a contradiction. Indeed, if a false formula were obtained using it, then the correspond-

[16] For example, see [7].

ing pseudo-mathematical formula would be provable without the principle and would nevertheless lead to a contradiction. [17]

The first of these statements contradicts a remark by Brouwer, [18] who believes that finitistic conclusions based on transfinite application of the *tertium non datur* principle must also be considered unproved.

§2. Propositions that we cannot prove without using illegitimately the *tertium non datur* principle are usually based not directly on the *tertium non datur* principle of propositional logic but on another principle with the same name. Indeed, the *tertium non datur* principle in the form characteristic of propositional logic: "each proposition is either true or false", implies the subsequent conclusions according to Hilbert's formula: if B is implied by both A and $\overline{A}$, then B is true. However, in the case of transfinite propositions we are interested in, it is difficult to obtain a positive conclusion B from a purely negated proposition $\overline{A}$; to this end it is first necessary to transform $\overline{A}$ to some different proposition.

The most ordinary type of transfinite proposition reads: for all a the proposition $A(a)$ is true; symbolically this is written as $(a)\,A(a)$. When the negation of this proposition $\overline{(a)}\,A(a)$ is intended to receive a positive conclusion, it is reduced to the form $(Ea)\overline{A}(a)$, that is, there is (a) such that $A(a)$ is false for it. The equivalence of the latter proposition to the simple negation of the proposition $(a)\,A(a)$ is expressed by the following two formulas:

$$\overline{(a)}\,A(a) \rightarrow (Ea)\overline{A}(a), \tag{59}$$

$$(Ea)\overline{A}(a) \rightarrow \overline{(a)}\,A(a). \tag{60}$$

In §4 of the present chapter we will show that, besides certain formulas whose intuitive obviousness is indubitable, including axioms of general propositional logic, the proof of formula (59) requires only the assumption that the law of double negation (4) is valid. Formula (60) can also be proved without this principle.

If formulas (59) and (60) are assumed to be valid, then we have a right to state the *tertium non datur* principle for propositions of the type $(a)\,A(a)$

---

[17] It is assumed that all axioms under consideration are axioms of type $\mathfrak{R}$ and moreover, that all formulas $\mathfrak{S}$ that are considered automatically false are such that the corresponding formulas $\mathfrak{S}^*$ are also false.

[18] See [4], footnote on p. 252.

thus:

$$(a)\,A(a) \lor (Ea)\,\overline{A}(a),$$

that is, either $A(a)$ is true for all a or there exists a for which $A(a)$ is false.

Hilbert (see [1], p.157) supplements (59) and (60) with the following formulas:

$$\overline{(Ea)}\;A(a) \to (a)\,\overline{A}(a), \tag{61}$$

$$(a)\,\overline{A}(a) \to \overline{(Ea)}\;A(a). \tag{62}$$

In accordance with what has been said above, he believes that formulas (59)–(62) justify the application of the *tertium non datur* principle to infinite sets of objects (a). In contrast to (59) and (60), the two formulas (61) and (62) can be proved only on the basis of axioms that are intuitively true. The proof is given in §4.

Thus, formulas (60)–(62) are simply true formulas; formula (59) is proved on the basis of the law of double negation (4); consequently, all that has been presented in the previous chapter applies to conclusions derived on the basis of (59).

§3. First we note that when defining the meaning of the symbol $(a)\,A(a)$ as "for all (a) the proposition $A(a)$ is true" we understand "for all (a)" in the same way as "for each (a)", that is, in the sense that, whatever the given (a), it can be asserted that $A(a)$ is true.

Each formula of general propositional logic, when written independently, means that it is true for all possible propositions $A, B, C, \ldots$. For example, the formula $A \to \overline{\overline{A}}$ means that for any proposition its truth implies its double negation. Thus, one cannot assert that the introduction of the symbol $(a)\,A(a)$ for the first time brings us outside the domain of finitistic propositions, since the notion "for all (a)" is implicitly contained in all formulas involving symbols of variables.

Generally, the formula $A(a)$ written independently means that, whatever the particular value (a), the proposition $A(a)$ is true. This implies the following principle, which cannot be expressed symbolically: if a formula $\mathfrak{S}$ is written independently, then the formula [19] $(a)\mathfrak{S}$ can be written. When referring to this principle we denote it by P.

[19] The symbol (a) can also stand in front of a formula not involving the variable

Further, we introduce the following axioms:

$$(63)\quad \begin{array}{ll} \text{I.} & (a)\,\{A(a) \to B(a)\} \to \{(a)\,A(a) \to (a)\,B(a)\}. \\ \text{II.} & (a)\,\{A \to B(a)\} \to \{A \to (a)\,B(a)\}. \\ \text{III.} & (a)\,\{A(a) \to C\} \to \{(Ea)\,A(a) \to C\}. \\ \text{IV.} & A(a) \to (Ea)\,A(a). \end{array}$$

We assume all these axioms to be intuitively obvious. Their choice and their number is determined exclusively by our aim to prove formulas (59)–(62).

§4.

$$(A \to B) \to \to (\overline{B} \to \overline{A}) \qquad (11)$$

$$(64)\quad \frac{A(a) \to (Ea)A(a)}{\overline{(Ea)}A(a) \to \overline{A}(a)} \qquad \text{Axiom IV}$$

$$(65)\quad (a)\{\overline{(Ea)}\,A\,(a) \to \overline{A}\,(a)\} \qquad (64)\mathbf{P}$$

$$(a)\,\{A \to B(a)\} \to \to \{A \to (a)\,B(a)\} \qquad \text{Axiom II}$$

$$(66)\quad \frac{(a)\,\{\overline{(Ea)}\,A(a) \to \overline{A}(a)\}}{\overline{(Ea)}\,A(a) \to (a)\overline{A}(a)} \qquad (65)$$

Thus, formula (61) is proved.

$$\{A \to (B \to C)\} \to \to \{B \to (a \to C)\} \qquad \text{Axiom 3}$$

$$(67)\quad \frac{A \to (\overline{A} \to \overline{B})}{\overline{A} \to (A \to \overline{B})} \qquad (12)$$

(a) at all. Instead of the principle P one can introduce the axiom

$$(a)\,V,$$

where V denotes a true proposition, and the following substitution rule, which cannot be expressed symbolically either: any independently written formula can be substituted for V.

(68) $$\frac{\overline{A} \to (A \to \overline{B})}{\overline{A}(a) \to \{A(a) \to \overline{(Ea)}A(a)\}}$$ (67)

(69) $$(a)[\overline{A}(a) \to \{A(a) \to \overline{(Ea)}A(a)\}]$$ (68)**P**

$$(a)\,\{A(a) \to B(a)\} \to \\ \to \{(a)\,A(a) \to (a)\,B(a)\}$$ Axiom I

(70) $$\frac{(a)[\overline{A}(a) \to \{A(a) \to \overline{(Ea)}\,A(a)\}]}{(a)\,\overline{A}(a) \to (a)\,\{A(a) \to \overline{(Ea)}A(a)\}}$$ (69)

(71) $$\frac{(a)\{A(a) \to C\} \to \{(Ea)A(a) \to C)\}}{(a)\{A(a) \to \overline{(Ea)}A(a)\} \to \{(Ea)A(a) \to \overline{(Ea)}A(a)\}}$$ Axiom III

$$(A \to B) \to \\ \to \{(A \to \overline{B}) \to \overline{A}\})$$ Axiom 5

(72) $$\frac{A \to A}{(A \to \overline{A}) \to \overline{A}}$$ (27)

(73) $$\frac{(A \to \overline{A}) \to \overline{A}}{\{(Ea)A(a) \to \overline{(Ea)}A(a)\} \to \overline{(Ea)}A(a)}$$ (72)

$$(a)\overline{A}(a) \to (a)\{A(a) \to \overline{(Ea)}A(a)\}$$ (70)

$$(a)\{A(a) \to \overline{(Ea)}A(a)\} \to \{(Ea)A(a) \to \overline{(Ea)}A(a)\}$$ (71)

(74) $$\frac{\{(Ea)A(a) \to \overline{(Ea)}A(a)\} \to \overline{(Ea)}A(a)}{(a)\overline{A}(a) \to \overline{(Ea)}A(a)}$$ (73)

We have thus proved formula (62).

(75) $$\frac{(a)\overline{A}(a) \to \overline{(Ea)}A(a)}{(a)\overline{\overline{A}}(a) \to \overline{(Ea)}\,\overline{A}(a)}$$ (74)

(76) $$\frac{A \to \overline{\overline{A}}}{A(a) \to \overline{\overline{A}}(a)}$$ (34)

(77) $$(a)\{A(a) \to \overline{\overline{A}}(a)\}$$ (76)**P**

$$(a)\{A(a) \to B(a)\} \to \\ \to \{(a)A(a) \to (a)B(a)\}$$ Axiom II

(78) $$\frac{(a)\{A(a) \to \overline{\overline{A}}(a)\}}{(a)A(a) \to (a)\overline{\overline{A}}(a)}$$ (77)

$$(a)A(a) \to (a)\overline{\overline{A}}(a)$$ (78)

(79) $$\frac{(a)\overline{\overline{A}}(a) \to (\overline{Ea})\overline{A}(a)}{(a)A(a) \to (\overline{Ea})\overline{A}(a)}$$ (75)

$$(A \to B) \to \qquad \to (\overline{B} \to \overline{A})$$ (11)

(80) $$\frac{(a)A(a) \to (\overline{Ea})\overline{A}(a)}{(\overline{\overline{\overline{Ea}}})\bar{A}(a) \to \overline{(a)}A(a)}$$ (79)

(81) $$\frac{A \to \overline{\overline{\overline{A}}}}{(Ea)\overline{A}(a) \to (\overline{\overline{\overline{Ea}}})\overline{A}(a)}$$ (34)

$$(Ea)\overline{A}(a) \to (\overline{\overline{\overline{Ea}}})\overline{A}(a)$$ (81)

(82) $$\frac{(\overline{\overline{\overline{Ea}}})\overline{A}(a) \to \overline{(a)}\, A(a)}{(Ea)\overline{A}(a) \to \overline{(a)}\, A(a)}$$ (80)

Formula (60) is thus proved.

The proof of formula (59) can be carried out without application of the axiom of double negation. The numbers of the formulas that are based on this axiom are underlined.

(83) $$\frac{(\overline{Ea})A(a) \to (a)\overline{A}(a)}{(\overline{Ea})\overline{A}(a) \to (a)\overline{\overline{A}}(a)}$$ (66)

$\underline{\underline{(84)}}$ $$\frac{\overline{\overline{A}} \to A}{\overline{\overline{A}}(a) \to A(a)}$$ $\underline{\underline{\text{Axiom 6}}}$

$\underline{\underline{(85)}}$ $$(a)\{\overline{\overline{A}}(a) \to A(a)\}$$ $\underline{\underline{(84)\mathbf{P}}}$

$$(a)\{A(a) \to B(a)\} \to \qquad \to \{(a)A(a) \to (a)B(a)\}$$ Axiom II

$\underline{\underline{(86)}}$ $$\frac{(a)\{\overline{\overline{A}}(a) \to A(a)\}}{(a)\overline{\overline{A}}(a) \to (a)A(a)}$$ $\underline{\underline{(85)}}$

$$(\overline{Ea})\overline{A}(a) \to (a)\overline{\overline{\overline{A}}}(a) \qquad (83)$$

(87) $$\frac{(a)\overline{\overline{\overline{A}}}(a) \to (a)A(a)}{(\overline{Ea})\overline{A}(a) \to (a)A(a)}$$ (86)

$$(A \to \overline{B}) \to \to (B \to \overline{A}) \qquad (11)$$

(88) $$\frac{(\overline{Ea})\overline{A}(a) \to (a)A(a)}{\overline{(a)}A(a) \to (\overline{\overline{\overline{Ea}}})\overline{A}(a)}$$ (87)

(89) $$\frac{\overline{\overline{\overline{A}}} \to A}{(\overline{\overline{\overline{Ea}}})\overline{A}(a) \to (Ea)\overline{A}(a)}$$ Axiom 6

$$\overline{(a)}A(a) \to (\overline{\overline{\overline{Ea}}})\overline{A}(a) \qquad (88)$$

(90) $$\frac{(\overline{\overline{\overline{Ea}}})\overline{A}(a) \to (Ea)\overline{A}(a)}{\overline{(a)}\,A(a) \to (Ea)\overline{A}(a)}$$ (89)

We see that formula (59) has been proved using the formula of double negation.

§5. A beautiful example of a proposition that cannot be proved without illegitimate use of the *tertium non datur* principle was given by Brouwer (see [6]); namely, the fact that each real number can be represented as an infinite decimal fraction cannot be regarded as being proved. Brouwer even indicated a definite number for which it is unknown what its first digit in the decimal representation is.

Another example of this kind is the proposition asserting that the complement of a closed set is an open set, that is, each point not belonging to the closed set is contained in an interval none of whose points belongs to the closed set. [20] As is known, the proof of the proposition is carried out in the following way: by the *tertium non datur* principle in the form given in §2 of the present chapter, either all intervals containing a chosen point contain points belonging to the closed set or at least one of them does not contain them; the

[20] This example was indicated by P.S. Novikov.

first assumption leads to a contradiction since it implies that the point belongs to the closed set and consequently, the other assumption is true. In contrast to Brouwer's example, we cannot indicate a definite closed set and a definite point belonging to its exterior for which the existence of the required interval is doubtful.

§6. The following example is more interesting: it is impossible to prove, without using the *tertium non datur* principle, all propositions whose proof can be reduced to application of the principle of transfinite induction. For example, such is the proposition asserting that each closed set is a union of a perfect set and a countable set.

The proof of such propositions is often carried out without using the principle of transfinite induction. But all such proofs are based on the *tertium non datur* principle as applied to infinite sets or on the law of double negation.

It is important to note that the principle of transfinite induction itself can be derived without any new assumptions as compared to those in the theory of point sets, but necessarily involves the *tertium non datur* principle. To this end it is only required to state the principle of transfinite induction without using the term "transfinite number", whose introduction would require new axioms. Instead, we will consider well-ordered (from left to right) sets of rational numbers. By a segment of such a set we mean the part of it lying to the left of a point belonging or not belonging to it. A segment will always be a well-ordered set. The set of segments itself is also a well-ordered set. A segment of a set is said to be regular if there are points of the set not belonging to the segment. The principle of transfinite induction is now stated in the following way.

Let a property J which well-ordered sets of rational numbers may or may not possess satisfy the conditions below.

1. The sets consisting of a single point possess the property J.

2. If all regular segments of a set possess the property J, then the set itself possesses it.

Under these conditions, all well-ordered sets of rational numbers possess the property J.

The principle of transfinite induction thus stated can be used in the same cases where the ordinary principle is applicable. It is proved in the following way: either all sets possess property J or there is a set E not possessing it; the latter assumption leads to a contradiction, since among the segments of E there must exist a first segment not possessing the property J, and the existence of such a segment contradicts the hypotheses.

The above examples convincingly show that, along with the Brouwerian presentation of mathematics without the *tertium non datur* principle, it is necessary to retain the ordinary presentation using this principle, albeit only as a presentation of pseudo-truth mathematics.

Moscow, 30 September 1925

References

1. D. Hilbert, *Math Ann.* **88** (1923), 151–165.
2. A.N. Whitehead and B. Russel, *Principia mathematica* 1–3, Cambridge, 1910–1913.
3. W. Ackermann, *Math. Ann.* **93** (1924), 1–36.
4. L.E.J. Brouwer, *Jahresber. Deutschen Math, Vereinigung* **33** (1925), 251–256.
5. Ch. Sigwart, *Logik* 1, 2, Freiburg, 1908.
6. L.E.J. Brouwer, *Math. Ann.* **83** (1921), 201–210.
7. L.E.J. Brouwer, *Verh. Kon. Ned. Akad. Wetensch.* Amsterdam (I) **12**:5 (1918); (II) **12**:7 (1919).
8. Aristotle, *Categoriae et liber de interpretatiore*, Oxford University Press, 1949; Russian transl., *Ob istolkovanii: Sochineniya*, Vol. 2, 91–116, Mysl', Moscow, 1978.
9. Aristotle, *De Anima*, Oxford University Press, 1956; Russian transl., *O dushe: Sochineniya*, Vol. 1, 369–448, Mysl', Moscow, 1975.
10. G.W. Leibniz, *New essays on human understanding*, Cambridge University Press, 1981; Russian transl., *Novye opyty o chelovecheskom razume*, Sotsekgiz, Moscow-Leningrad, 1936.

10. ON CONVERGENCE OF FOURIER SERIES *

(In collaboration with G.A. Seliverstov)

In the previous paper [4] we have stated a convergence condition for a trigonometric series which is more general than the similar conditions given earlier by Fatou [1], Weyl [6], Hobson [3], Plancherel [5], and Hardy [2].

We proved that the series

$$\sum_{n=2}^{\infty}(a_n \cos nx + b_n \sin nx) \tag{1}$$

converges almost everywhere if the series

$$\sum_{n=2}^{\infty}(\log n)^{1+\epsilon}(a_n^2 + b_n^2)$$

converges for some positive ϵ.

It was pointed out that all the conditions given earlier (the most general of them, due to Hardy, differs from our condition in the factor $\log^2 n$ instead of $(\log n)^{1+\epsilon}$) are valid for any orthogonal system of functions, whereas our condition does not hold for some of them. This follows from Men'shov's results [7].

In this paper we give a simpler proof of the following stronger assertion: *the series* (1) *converges almost everywhere provided that the series*

$$\sum_{n=2}^{\infty}\log n(a_n^2 + b_n^2) \tag{2}$$

converges.

1. Lemma. *Let a trigonometric sum be given:*

$$S(x) = \sum_{k=2}^{n}(a_k \cos kx + b_k \sin kx).$$

Then the following inequality holds:

$$\int_0^{2\pi} \frac{S_{k(x)}(x)}{\sqrt{\log k(x)}}dx \leq \sqrt{C\sum_{p=2}^{n}(a_p^2 + b_p^2)}, \tag{3}$$

* 'Sur la convergence des séries de Fourier', *Atti Accad. Naz. Lincei* **3** (1926), 307–310.

where $k(x)$ is an arbitrary integer-valued function ranging from 2 to n; $S_{k(x)}(x)$ is the sum of the first $k(x)$ terms of $S(x)$; and C is an absolute constant.

Proof. By the Schwarz inequality,

$$\int_0^{2\pi} \frac{S_{k(x)}(x)}{\sqrt{\log k(x)}} dx =$$

$$= \int_0^{2\pi} \frac{1}{\sqrt{\log k(x)}} \frac{1}{\pi} \int_0^{2\pi} S(\alpha) \sum_{p=1}^{k(x)} \cos p(x-\alpha) d\alpha dx =$$

$$= \frac{1}{\pi} \int_0^{2\pi} S(\alpha) \int_0^{2\pi} \frac{\sum_{p=1}^{k(x)} \cos p(x-\alpha)}{\sqrt{\log k(x)}} dx\, d\alpha \leq$$

$$\leq \sqrt{\frac{1}{\pi^2} \int_0^{2\pi} S^2(\alpha) d\alpha \int_0^{2\pi} \Big[\int_0^{2\pi} \frac{\sum_{p=1}^{k(x)} \cos p(x-\alpha)}{\sqrt{\log k(x)}} dx\Big]^2 d\alpha} \leq$$

$$\leq \sqrt{\frac{1}{\pi} \sum_{p=2}^{n} (a_p^2 + b_p^2) I}.$$

To prove (3) we have only to establish that I is bounded. Integrating by parts with respect to α and using the fact that the trigonometric system is orthogonal, we have

$$I = \int_0^{2\pi} \Big[\int_0^{2\pi} \frac{\sum_{p=1}^{k(x)} \cos p(x-\alpha)}{\sqrt{\log k(x)}} dx\Big]^2 d\alpha =$$

$$= \int_0^{2\pi}\int_0^{2\pi}\int_0^{2\pi} \frac{\sum_{p=1}^{k(x)} \cos p(x-\alpha) \sum_{q=1}^{k(y)} \cos q(y-\alpha)}{\sqrt{\log k(x) \log k(y)}} dx\, dy\, d\alpha =$$

$$= \pi \int_0^{2\pi}\int_0^{2\pi} \frac{\sum_{p=1}^{k(x,y)} \cos p(x-y)}{\sqrt{\log k(x) \log k(y)}} dx\, dy,$$

where $k(x,y)$ is the minimum of $k(x)$ and $k(y)$.

Let P be the set of those (x,y) for which $k(x) \leq k(y)$ and consequently, $k(x,y) = k(x)$, $\log k(y) \geq \log k(x)$, and let Q be the set of those (x,y) for which $k(y) < k(x)$, $k(x,y) = k(y)$, and $\log k(x) > \log k(y)$. Then we have

$$\frac{I}{\pi} \le \iint_P \frac{\left|\sum_{p=1}^{k(x)} \cos p(x-y)\right|}{\log k(x)} dx\, dy + \iint_Q \frac{\left|\sum_{p=1}^{k(y)} \cos p(x-y)\right|}{\log k(y)} dx\, dy \le$$

$$\le \int_0^{2\pi} \frac{1}{\log k(x)} \int_0^{2\pi} \left|\sum_{p=1}^{k(x)} \cos p(x-y)\right| dy\, dx +$$

$$+ \int_0^{2\pi} \frac{1}{\log k(y)} \int_0^{2\pi} \left|\sum_{p=1}^{k(y)} \cos p(x-y)\right| dx\, dy.$$

It can easily be seen that the last expression does not exceed an absolute constant C.

2. It follows from the condition (2) (satisfied by (1)) that there is a sequence $\tau(n)$ increasing to infinity such that the series

$$\sum_{n=2}^{\infty} \tau(n) \log n (a_n^2 + b_n^2) \tag{4}$$

converges.

We set

$$A_n = \sqrt{\tau(n) \log n}\, a_n, \quad B_n = \sqrt{\tau(n) \log n}\, b_n.$$

Then (4) reduces to

$$\sum_{n=2}^{\infty} (A_n^2 + B_n^2).$$

We introduce the following notation:

$$S_n(x) = \sum_{k=2}^{n} (A_k \cos kx + B_k \sin kx), \quad \sigma_n(x) = \frac{1}{n} \sum_{k=2}^{n} S_k(x),$$

$$\Delta_n = \frac{1}{\sqrt{\tau(n) \log n}} - \frac{1}{\sqrt{\tau(n+1) \log(n+1)}}, \quad \Delta_n' = \Delta_n - \Delta_{n+1}.$$

The sequence $\tau(n)$ can be defined in such a way that all the Δ_n' are positive.

Then the series [1]

$$\sum_{n=2}^{\infty} n\Delta_n' = \frac{1}{\sqrt{\tau(2)}\log 2} + \Delta_2$$

converges absolutely and consequently

$$n\Delta_n = n\sum_{k=n}^{\infty} \Delta_k' < \sum_{k=n}^{\infty} k\Delta_k' \to 0.$$

Applying Abel's transformation twice to (1) we obtain

$$\sum_{n=2}^{\infty}(a_n \cos nx + b_n \sin nx) = \sum_{n=2}^{\infty} \frac{1}{\sqrt{\tau(n)}\log n}(A_n \cos nx + B_n \sin nx) =$$

$$= \sum_{n=2}^{\infty} \Delta_n S_n(x) = \sum_{n=2}^{\infty} n\Delta_n' \sigma_n(x). \tag{5}$$

To justify the application of Abel's transformation it suffices to prove the relations

$$\frac{S_n(x)}{\sqrt{\tau(n)}\log n} \to 0, \quad n\Delta_n \sigma_n \to 0.$$

The second of them holds if σ_n are bounded, that is, it holds almost everywhere. The first relation holds almost everywhere, as follows from our lemma. Finally, if the sequence σ_n converges (which is true almost everywhere), then so does the series (5). Thus, we have proved that the series (1) converges almost everywhere.

7 February 1926

References

1. P. Fatou, 'Séries trigonométriques et séries de Taylor', *Acta Math.* **30** (1906), 335–400.
2. G.H. Hardy, 'On the summability of Fourier series', *Proc. London Math. Soc.* **12** (1913), 365–372.

[1] We have

$$\sum_{n=2}^{\infty} n\Delta_n' = \sum_{n=2}^{\infty}\sum_{k=n}^{\infty} \Delta_k' + \sum_{k=2}^{\infty} \Delta_k' = \sum_{n=2}^{\infty} \Delta_n + \Delta_2 = \frac{1}{\sqrt{\tau(2)}\log 2} + \Delta_2.$$

3. E.W. Hobson, 'On the convergence of series of orthogonal functions', *Proc. London Math. Soc.* **12** (1913), 297–308.
4. A. Kolmogoroff and G. Seliverstoff, 'Sur la convergence des séries de Fourier', *C.R. Acad. Sci. Paris* **178** (1924), 303–306. (No. 4)
5. M. Plancherel, 'Sur la convergence des séries de fonctions orthogonales', *C.R. Acad. Sci. Paris* **156** (1913), 539–541.
6. H. Weyl, 'Über die Konvergenz von Reihen, die nach Orthogonalfunktionen fortschreiten', *Math. Ann.* **67** (1909), 225–245.
7. D. Menshoff, 'Sur les séries de fonctions orthogonales', *Fund. Math.* **4** (1923), 82–105.

11. A FOURIER-LEBESGUE SERIES DIVERGENT EVERYWHERE*

The purpose of this paper is to give an example of a summable function with an everywhere divergent Fourier series.

1°. For each integer n we define the function $\phi_n(x)$ by

$$\sigma_m(x) = \tfrac{1}{2} + \sum_{k=1}^{m} \frac{m-k}{m} \cos kx,$$

$$m_i = n^{4(i+1)}, \quad A_i = \frac{4i\pi}{2n+1} \quad (0 \le i \le n),$$

$$\phi_n(x) = \frac{1}{n+1} \sum_{i=0}^{n} \sigma_{m_i}(A_i - x).$$

It is obvious that

$$\phi_n(x) \ge 0, \quad \int_0^{2\pi} \phi_n(x)dx = \pi.$$

2°. It is possible to represent $\phi_n(x)$ in the form

$$\phi_n(x) = \tfrac{1}{2} + \sum_{k=1}^{m_n} a_k \cos(kx + \lambda_k).$$

We consider a partial sum of $\phi_n(x)$:

$$S_k(x) = \tfrac{1}{2} + \sum_{l=1}^{k} a_l \cos(lx + \lambda_l) = \frac{1}{n+1} \sum_{i=0}^{j} \sigma_{m_i}(A_i - x) + \\ + \frac{1}{n+1} \sum_{i=j+1}^{n} \frac{k}{m_i} \sigma_k(A_i - x) + \\ + \frac{1}{n+1} \sum_{i=j+1}^{n} \frac{m_i - k}{m_i} \frac{\sin((2k+1)(A_i - x)/2)}{2\sin\frac{1}{2}(A_i - x)}, \tag{1}$$

where j is assumed to satisfy $m_j \le k < m_{j+1}$. The first two terms in (1) are non-negative. Therefore

$$S_k(x) \ge \frac{1}{n+1} \sum_{i=j+1}^{n} \frac{m_i - k}{m_i} \frac{\sin((2k+1)(A_i - x)/2)}{2\sin\frac{1}{2}(A_i - x)}. \tag{2}$$

---

* 'Une série de Fourier-Lebesgue divergente partout', *C.R. Acad. Sci. Paris* **183** (1926), 1327–1328.

3°. It is possible to prove that the following inequality holds for all sufficiently large n and all x in the interval $[A_j - 1/n^3,\ A_j + 1/n^3]$:

$$S_{n^2}(x) > C_1 n, \tag{3}$$

where C_1 is an absolute constant.

4°. Assuming that n is sufficiently large, we can find for any x in the interval $[A_j - 1/n^3,\ A_j + 1/n^3]$ an index k such that $2k+1$ is divisible by $2n+1$ and the following inequalities are satisfied:

$$m_j \leq k \leq \tfrac{1}{2} m_{j+1}, \quad -\sin((k+\tfrac{1}{2})x) \geq \tfrac{1}{2}.$$

In this case (2) implies that

$$\begin{aligned} S_k(x) &\geq -\frac{1}{n+1} \sin((k+\tfrac{1}{2})x) \sum_{i=j+1}^{n} \frac{m_i - k}{m_i} \frac{1}{2\sin\frac{1}{2}(A_i - x)} \geq \\ &\geq C_2 \log(n-j). \end{aligned} \tag{4}$$

5°. In view of (3) and (4), if x satisfies the inequality

$$0 \leq x \leq 2\pi - 1/\sqrt{n},$$

then an index k can be defined such that

$$S_k(x) \geq C_3 \log n.$$

We finally set

$$\Phi(x) = \sum_{m=1}^{\infty} M_m \phi_{n_m}(x).$$

If $\sum M_n$ is an absolutely convergent series, then $\Phi(x)$ is summable. If n_m increases sufficiently fast, then the Fourier series of Φ diverges everywhere. These two facts can be proved using results in my paper [1].

27 December 1926

References

1. A.Kolmogoroff, 'Une série de Fourier-Lebesgue divergente presque partout', *Fund. Math.* **4** (1923), 324–328. (No. 1)

12. ON CONVERGENCE OF SERIES OF ORTHOGONAL FUNCTIONS *

(In collaboration with D.E. Men'shov).

The following theorem contains the main result of the paper.

Theorem 1. *There exist a system of functions $\phi_n(x)$ that are orthogonal on $(0,1)$ and assume only the values ±1, and a sequence of numbers a_n, such that the series*

$$\sum_{n=1}^{\infty} a_n^2 \tag{1}$$

converges, whereas the series

$$\sum_{n=1}^{\infty} a_n \phi_n(x) \tag{2}$$

diverges everywhere.

It is well known (see [1, 2]) that for any system of orthogonal functions the convergence of the series

$$\sum_{n=1}^{\infty} a_n^2 \log_n^2 \tag{3}$$

is a sufficient condition for (2) to converge almost everywhere. It is also well known (see [2]) that $(\log n)^2$ cannot be replaced by any other more slowly increasing multiplier.

In the case when the functions $\phi_n(x)$ are equibounded, we cannot give an equally exhaustive answer. We can only prove the following theorem.

Theorem 2. *For any positive function $W(n)$ satisfying $W(n) = o(\log n)$ there exist a system of functions $\phi_n(x)$, $n = 1, 2, 3, \dots$, that are orthogonal on $(0,1)$ and assume only the values ±1 , and a sequence of real numbers a_n, such that the series (2) diverges everywhere, whereas the series*

$$\sum_{n=1}^{\infty} a_n^2 W(n) \tag{4}$$

converges.

---

* 'Sur la convergence des séries de fonctions orthogonales', *Math. Z.* **26** (1927), 432–441.

Thus, it may happen that for equibounded functions $\phi_n(x)$ the convergence of the series

$$\sum_{n=1}^{\infty} a_n^2 \log n \tag{5}$$

is a sufficient condition for (2) to converge. This is precisely the case for trigonometric series (see [3], [4]).

It is also possible to construct an almost everywhere divergent series (2) with trigonometric functions as the $\phi_n(x)$. But in this case it is impossible to approach the limit $W(n) = \log n$. For this reason we merely formulate the following theorem without proof.

Theorem 3. *There exists an almost everywhere divergent series of the form*

$$\sum_{n=1}^{\infty} a_n \cos(m_n x + \lambda_n)$$

such that the integers m_n are all different and the series (1) *converges.*[1]

Theorem 1 is a direct consequence of Theorem 2. The following lemma is needed in the proof of the latter theorem.

Main lemma. *For any positive integer n there is a sum*

$$\sum_{k=1}^{n} a_k \phi_k(x) \tag{6}$$

with the following properties.

1. *The functions $\phi_k(x)$ are pairwise orthogonal on* $(0,1)$, *assume only the values* ± 1, *and are constant on each of the intervals δ forming a suitable finite partition of* $(0,1)$.

2. $\sum_{k=1}^{n} a_k^2 = 1$.

3. *For any point of a set E with* $\operatorname{mes} E = 1/2$, *there is a number $p \leq n$ such that*

$$\sum_{k=1}^{p} a_k \phi_k(x) \geq C\sqrt{\log n},$$

where C is an absolute constant.

[1] The theorem was proved by A.N. Kolmogorov.

Proof of the main lemma

1°. We consider only the case $n = 12(4^m - 1)$, where m is a positive integer. The general case is a simple consequence of this case. We replace $C\sqrt{\log n}$ by $(1/3)\sqrt{m}$ in the inequality in 3, which is legitimate.

Below we will construct the terms of (6) for the numbers $12(4^m - 1)$ and then in 11° we will define their indices. We introduce first some auxiliary functions.

2°. Let $S_l(i,r)$ be a function of integral arguments with the following properties:

a) it is defined for $1 \leq i \leq 3.2^l$, $1 \leq r \leq 3.2^l - 1$;

b) for a fixed i it assumes each value $1, 2, 3, \dots, 3.2^l$, except i, once and only once;

c) for a fixed r it assumes all integral values from 1 to 3.2^l;

d) $S_l[S_l(i,r),r] = i$;

e) to each pair of integers i_1, i_2 ($i_1 \neq i_2$, $1 \leq i_1 \leq 3.2^l$, $1 \leq i_2 \leq 3.2^l$) there corresponds a unique r such that

$$S_l(i_1,r) = i_2, \quad S_l(i_2,r) = i_1.$$

Property e) is a consequence of b) and d). We define $S_l(i,r)$ by means of recurrence relations.

A. For $l = 1$ the values of $S_l(i,r)$ are defined by the table

i					
6	5	4	2	3	1
5	6	2	3	1	4
4	3	6	1	2	6
3	4	1	5	6	2
2	1	5	6	4	3
1	2	3	4	5	6
	1	2	3	4	5 r

B. The values of $S_{l+1}(i,r)$ are obtained from the values of $S_l(i,r)$ in the following way:

a) if $r < 3.2^l$, $i \leq 3.2^l$, then $S_{l+1}(i,r) = S_l(i,r)$;

b) if $r < 3.2^l, i > 3.2^l$, then $S_{l+1}(i,r) = S_l(i - 3.2^l, r) + 3.2^l$;

c) if $r \geq 3.2^l$, $i \leq 3.2^l$, then

$$S_{l+1}(i,r) = \begin{cases} i+r & \text{for } i+r \leq 3.2^{l+1}, \\ i+r-3.2^l & \text{for } i+r > 3.2^{l+1}; \end{cases}$$

d) if $r \geq 3.2^l$, $i > 3.2^l$, then the values of $S_{l+1}(i,r)$ are uniquely determined by means of property d) in 2° from the values defined above.

Remark. For $l = 2$ the values of $S_l(i,r)$ are defined by the following table:

i												
12	11	10	8	9	7	6	5	4	3	2	1	
11	12	8	9	7	10	5	4	3	2	1	6	
10	9	12	7	8	11	4	3	2	1	6	5	
9	10	7	11	12	8	3	2	1	6	5	4	
8	7	11	12	10	9	2	1	6	5	4	3	
7	8	9	10	11	12	1	6	5	4	3	2	
6	5	4	2	3	1	12	7	8	9	10	11	
5	6	2	3	1	4	11	12	7	8	9	10	
4	3	6	1	2	5	10	11	12	7	8	9	
3	4	1	5	6	2	9	10	11	12	7	8	
2	1	5	6	4	3	8	9	10	11	12	7	
1	2	3	4	5	6	7	8	9	10	11	12	
	1	2	3	4	5	6	7	8	9	10	11	r

The structure of this table can also be explained in terms of the correspondence between i_1 and i_2 (property e) in 2°): the correspondence between i_1 and i_2 ranging between 1 and 6 is guaranteed by the left lower quarter of the table; the correspondence between i_1 and i_2 ranging from 7 to 12 is established by the left upper quarter in a similar way; and finally, the correspondence between i_1 ranging from 1 to 6 and i_2 ranging from 7 to 12 is established by the right half of the table.

3° Let $S_{l,q}(i,r)$ be a function having the properties b), c), d) and e) for the function $S_l(i,r)$ and also the following property:

a′) it is defined for $1 \leq i \leq 3.2^l$, $1 \leq r \leq 3.2^l$ $(r \neq q)$.

We define $S_{l,q}(i,r)$ as follows: $S_{l,q}(i,r) = S_l(i, r-q)$ if $r-q > 0$; $S_{l,q}(i,r) = S_l(i, r-q+3.2^l)$ if $r-q < 0$. It follows that

$$S_l(i,r) = S_{l,3.2^l}(i,r).$$

4°. We set

$$U_{l,q}(i,r) = \min[i;\ S_{l,q}(i,r)].$$

It follows from property d) in 2° that

$$U_{l,q}(i,r) = U_{l,q}[S_{l,q}(i,r),r]; \tag{7}$$

and property e) implies that the following relation holds for r corresponding to the pair i_1, i_2:

$$U_{l,q}(i_1,r) = U_{l,q}(i_2,r) = \min(i_1,i_2). \tag{8}$$

5°. We set $\psi_t(x) = +1$ for $x \in ((k-1)/(3.2^t),\ (2k-1)/(6.2^t))$ and $\psi_t(x) = -1$ for $x \in ((2k-1)/(6.2^t), k/(3.2^t))$, where k is an arbitrary positive integer.

6°. Definition of the functions ϕ_{lq_i}. Let

$$1 \le l \le m, \quad 1 \le q \le 3.2^l, \quad 1 \le i \le 3.2^l.$$

In this case the total number of functions ϕ_{lq_i} is equal to

$$n = \sum_{l=1}^{m}(3.2^l)^2 = 12(4^m - 1).$$

We set $\phi_{lq_i} = \phi_l$ on the interval $((q-1)/(3.2^l),\ q/(3.2^l))$ and

$$\phi_{lq_i} = \pm\psi_\alpha, \quad \alpha = m + 9.4^m + 3.2^l.q + U_{l,q}(i,r),$$

on the interval $((r-1)/(3.2^l), r/(3.2^l)),\ r \neq q$, where we write (+) if $U_{l,q}(i,r) = i$ and (−) if $U_{l,q}(i,r) = S_{l,q}(i,r)$.

7°. Definition of the coefficients a_{lq_i}:

$$a_{lq_i} = \pm\frac{1}{\sqrt{m}\cdot 3.2^l},$$

where we write (+) if $1 \le i \le 2.2^l$ and (−) if $2.2^l < i \le 3.2^l$.

8°. In order that the sum (6) have the properties described in the main lemma, we take $a_{lq_i}\phi_{lq_i}(x)$ as its terms, with their total number determined in 6°. We have:

$$\sum a_{lq_i}^2 = \sum_{l=1}^{m}\sum_{q=1}^{3.2^l}\sum_{i=1}^{3.2^l}\frac{1}{m\cdot 9.4^l} = 1,$$

which means that condition 2 of the main lemma is fulfilled.

9°. The following three lemmas, giving some properties of the functions $\psi_t(x)$ (see 5°), are required in order to prove that the functions ϕ_{lq_i} are pairwise orthogonal.

Lemma 1.

$$\int_{(k-1)/(3.2^n)}^{k/(3.2^n)} \psi_t(x)dx = 0 \quad \text{for } t \geq n.$$

Lemma 2.

$$\int_{(k-1)/(3.2^n)}^{k/(3.2^n)} \psi_{t_1}(x)\psi_{t_2}(x)dx = 0 \quad \text{if } t_1 \neq t_2,\ t_1 \geq n.$$

Lemma 3.

$$\int_{(k-1)/(3.2^n)}^{k/(3.2^n)} [\psi_t(x)]^2 dx = \frac{1}{3.2^n}.$$

In all three lemmas, k and n are positive integers.

We shall also need

Lemma 4. *The set of points x belonging to the interval $((g-1)/(3.2^l),\ g/(3.2^l))$ and such that the sum*

$$\sigma(x) = \sum_{k=1}^{n} a_k \psi_{t_k}(x)$$

is positive, and the set of those points of this interval at which the sum is negative have equal measures. It is supposed here that g is an integer and that all the t_k are not smaller than t (both t and t_k are positive integers).

The proof of the lemma follows immediately from the obvious relation

$$\sigma\Big(\frac{g}{3.2^t} - y\Big) = -\sigma\Big(\frac{g-1}{3.2^t} + y\Big),$$

which holds automatically for all irrational y.

10°. The following cases must be considered separately in order to prove that two given functions $\phi_{l_1q_{1i_1}}$ and $\phi_{l_2q_{2i_2}}$ are orthogonal.

First case: $l_1 \neq l_2$. Assume for the sake of definiteness that $l_1 > l_2$ and consider the interval $((k-1)/(3.2^{l_1}),\ k/(3.2^{l_1}))$. On this interval, $\phi_{l_1q_{1i_1}} = \psi_{t_1}$

and $\phi_{l_2q_2i_2} = \psi_{t_2}$, where, in general, t_1 and t_2 depend on k. As follows from the definition of ϕ_{lq_i} (see 6°), $t_1 \neq t_2$ and $t_1 \geq l_1$ for any k; hence, Lemma 2 implies that on each of the intervals the integral of the product of $\phi_{l_1q_1i_1}$ by $\phi_{l_2q_2i_2}$ is equal to zero, whence we conclude that the functions are orthogonal on (0,1).

Second case: $l_1 = l_2 = l$; $q_1 \neq q_2$. In this case $\phi_{lq_1i_1} = \psi_{t_1}$ and $\phi_{lq_2i_2} = \psi_{t_2}$ on $((k-1)/(3.2^l), k/(3.2^l))$, where $t_1 \geq l$, $t_2 \geq l$, and $t_1 \neq t_2$. Therefore the integral of the product $\phi_{lq_1i_1} \cdot \phi_{lq_2i_2}$ is also equal to zero on the interval and the two functions $\phi_{l_1q_1i_1}$ and $\phi_{l_2q_2i_2}$ are orthogonal on (0,1).

Third case: $l_1 = l_2 = l$, $q_1 = q_2 = q$, $i_1 \neq i_2$. Repeating the argument in the two preceding cases, we see that on each interval $((k-1)/(3.2^l), k/(3.2^l))$, except for two specific intervals, the integral of the product $\phi_{lqi_1} \cdot \phi_{lqi_2}$ is equal to zero. One of the exceptional intervals corresponds to the case $k = q$, and we have $\phi_{lqi_1} = \phi_{lqi_2} = \psi_l$ on this interval (see 6°). Consequently, by Lemma 3, the integral of the product is equal to $1/(3.2^l)$.

The other exceptional interval is the one for which, according to (8), $U_{lq}(i_1, k)) = U_{lq}(i_2, k)$. On this interval we have $\phi_{lqi_1} = -\phi_{lqi_2} = \pm\psi_t$ (with t determined as described in 6°). Therefore, by Lemma 3, the integral of the product is equal to $-1/(3.2^l)$. Thus, in the third case the two functions $\phi_{l_1q_1i_1}$ and $\phi_{l_2q_2i_2}$ are also orthogonal on (0,1).

11°. For each ϕ_{lqi} we define a fundamental point x_{lqi} by means of the conditions:

$$x_{lqi} = \frac{2q-1}{6.2^l} \quad \text{if } i \leq 2.2^l$$

and

$$x_{lqi} = \frac{q}{3.2^l} \quad \text{if } i > 2.2^l.$$

We number the functions ϕ_{lqi} in order of decreasing x_{lqi}. Functions with coinciding fundamental points can be numbered arbitrarily. Further, we number the coefficients a_{lq} and the points x_{lqi} in the same order as the corresponding functions ϕ_{lqi}. Consider the sum

$$\sum_{k=1}^{n} a_k \phi_k(x), \quad n = 12(4^m - 1), \tag{6'}$$

where a_k and ϕ_k are equal to a_{lqi} and ϕ_{lqi} numbered as explained above.

We will prove that the sum (6′) possesses all three properties mentioned in the statement of the main lemma. The first two properties have already been obtained in 8° and 10°. It remains to prove that the third property holds for (6′).

12°. Let g be an integer. Denote by x_k $(1 \leq k \leq n)$ the points x_{lqi} numbered in the same order as $a_k\phi_k(x)$, that is, from right to left, and consider the following partial sum on the interval $\Delta = ((g-1)/(6.2^m), g/(6.2^m))$:

$$\sum_{k=1}^{p_g} a_k\phi_k(x), \tag{9}$$

consisting of all terms $a_k\phi_k(x)$ such that the corresponding fundamental points x_k lie to the right of Δ. (Since the x_k are numbered from right to left, there is an integer p_g not depending on the points of the interval Δ such that (9) has the above-mentioned property.)

For any l $(1 \leq l \leq m)$ there is exactly one integer q such that Δ is contained in $((q-1)/(3.2^l), q/(3.2^l))$. For some l the interval Δ entirely belongs to the left half of $((q-1)/(3.2^l), q/(3.2^l))$, whereas for other values of l it lies in the right half. Let l be such that Δ is contained in the left half of the interval. In this case the points x_{lqi} lie to the right of Δ for all i $(1 \leq i \leq 3.2^l)$ (see 11°), and consequently, (9) contains all terms of the form $a_{lqi}\phi_{lqi}$. Among them the first 2.2^l terms are equal to $+1/(\sqrt{m} \cdot 3.2^l)$, whereas the other 2^l terms are equal to $-1/(\sqrt{m} \cdot 3.2^l)$ (see 6° and 7°).

Thus, the sum of all terms under consideration is equal to $\frac{1}{3}\sqrt{m}$.

We now consider a value of l such that Δ is contained in the right half of $((q-1)/(3.2^l), q/(3.2^l))$. In this case only the points x_{lq_i} with $i > 2.2^l$ lie to the right of Δ. Therefore only 2^l terms of the form $a_{lqi}\phi_{lqi}$ are contained in (9) and all of them are positive since both a_{lqi} and ϕ_{lqi} are negative. The sum of the terms under consideration is equal to $\frac{1}{3}\sqrt{m}$. Thus, we obtain the same value as in the previous case.

We denote by $\sum'$ the sum of all those terms in (9) that have been considered so far. It follows from the above argument that $\sum' = \frac{1}{3}\sqrt{m}$ for all the points of Δ.

We next consider the sum $\sum''$ of all those terms of (9) that do not enter into $\sum'$. For $x \in \Delta$ each term $a_k\phi_k(x)$ in $\sum''$ is equal to some $a_k\phi_{t_k}(x)$, where $t_k > m$ (see the definition of ϕ_{lqi} in 6°). It follows from Lemma 4 that $\sum''$ is non-negative on a set of measure equal to half the length of Δ, and hence the entire sum (9) is greater than or equal to $\frac{1}{3}\sqrt{m}$ on at least a set of measure $\frac{1}{2}|\Delta|$. Property 3 in the main lemma has thus been proved.

Proof of Theorem 2

1°. Given some n, we denote by a_{kn} and $\phi_{kn}(x)$ respectively, the numbers a_k and the functions $\phi_k(x)$ satisfying the condition of the main lemma. For an arbitrary interval $\delta = (a, b)$, we set

$$\phi_{kn}(x, \delta) = \phi_{kn}\left(\frac{x-a}{b-a}\right).$$

The sum

$$\sum_{k=1}^{n} a_{kn}\phi_{kn}(x, \delta)$$

has the same properties as the sum (6) in the main lemma. We merely have to replace (0,1) by δ and E by a set $E(\delta)$ with $\operatorname{mes} E(\delta) = \frac{1}{2}|\delta|$.

2°. In order to prove Theorem 2 we consider a positive function $W(n)$ satisfying the sole requirement that

$$\lim_{n\to\infty} \frac{W(n)}{\log n} = 0. \tag{10}$$

It is obvious that there is an infinite sequence of positive integers n_ν ($\nu = 1, 2, 3, \ldots$) such that

$$\sum_{i=1}^{\nu-1} n_i < n_\nu, \tag{11}$$

$$\frac{W(n)}{\log n} < \frac{1}{\nu^2} \quad \text{for } n \geq n_\nu. \tag{12}$$

We set

$$N_\nu = \sum_{i=1}^{\nu} n_i. \tag{13}$$

According to (11),

$$N_\nu < 2n_\nu. \tag{14}$$

3°. We define numbers a_n and functions $\phi_n(x)$ as follows.

a) If $n \le n_1$, we put

$$a_n = \frac{a_{nn_1}}{\sqrt{\log n_1}}, \quad \phi_n(x) = \phi_{nn_1}(x).$$

b) Assuming that a_n and $\phi_n(x)$ have already been defined for $n \le N_{\nu-1}$ and that the interval (0,1) has been partitioned into a finite number of intervals δ such that all the functions $\phi_n(x)$ $(n < N_{\nu-1})$ are constant on each δ, we put, for $N_{\nu-1} < n \le N_\nu$,

$$a_n = \frac{a_{kn_\nu}}{\sqrt{\log n_\nu}}, \quad \phi_n(x) = \phi_{kn_\nu}(x, \delta)$$

on each δ, where $k = n - N_{n-1}$.

The property that the functions $\phi_n(x)$ be constant on certain intervals δ obtained in a finite partition of (0,1) is retained at each step; therefore it is possible to determine $\phi_n(x)$ and a_n consecutively for all n.

4°. It can easily be seen that $\phi_n(x)$ are pairwise orthogonal and the series

$$\sum_{n=1}^{\infty} a_n \phi_n(x)$$

diverges almost everywhere. By changing the values of some of the functions $\phi_n(x)$ on a set of measure zero it is possible to make (5) diverge everywhere.

The proof of Theorem 2 will be complete once we have shown that the series

$$\sum_{n=1}^{\infty} a_n^2 W(n)$$

converges. This results from the following calculation:

$$\sum_{n=N_1+1}^{\infty} a_n^2 W(n) = \sum_{\nu=2}^{\infty} \sum_{n=N_{\nu-1}+1}^{N_\nu} a_n^2 \frac{W(n)}{\log n} \log n \le$$

$$\le \sum_{\nu=2}^{\infty} \log N_\nu \sum_{n=N_{\nu-1}+1}^{N_\nu} a_n^2 \frac{W(n)}{\log n} \le$$

$$\le C' \sum_{\nu=2}^{\infty} \log n_\nu \cdot \frac{1}{(\nu-1)^2} \sum_{k=N_{\nu-1}+1}^{N_\nu} a_n^2 = C' \sum_{\nu=2}^{\infty} \frac{1}{(\nu-1)^2} = C,$$

where C' and C are constants.[3]

[3] The proof of Theorem 2 is similar to an argument already used by D.E. Men'shov in [2] (see [2], p.99).

5 March 1926

References

1. H. Rademacher, 'Einige Sätze über Reihen von allgemeinen Orthogonalfunktionen', *Math. Ann.* **87** (1922), 112–138.
2. D. Menchoff, 'Sur la série de fonctions orthogonales', *Fund. Math.* **4** (1923), 82–105.
3. A. Plessner, 'Über die Konvergenz von trigonometrischen Reihen', *J. Reine Angew. Math.* **155** (1925), 15–25.
4. A. Kolmogoroff and G. Seliverstoff, 'Sur la convergence des séries de Fourier', *Atti Acad. Naz. Lincei Rend.* **3**:10 (1926), 307–310. (No. 10)

13. ON OPERATIONS ON SETS*

In 1921 N.N. Luzin delivered a series of lectures on the theory of functions at Moscow University, in which he defined the *class of C-sets* and posed the problem of comprehensive study of such sets. Recently E.A. Selivanovskii [1] has published several results relating to this class and has given it a rather complete characterization. In this connection, the investigation presented below would seem to be of interest, since I obtain some of these results proceeding from very general considerations. The class of C-sets, as well as the class of B-measurable sets splits into non-empty $\aleph_1$ classes, and it turns out that the same is true for an arbitrary class of sets generated by iteration of operations of a very general type, defined below, and the operation of taking a complement.

The operations themselves were defined by Hausdorff in the second edition of his "Set Theory". But neither the definition of the complementation operation nor the main theorem on complements was stated there.

The following material is only a slight modification of a manuscript completed on 3 January 1922.

I.

All sets to be considered lie in the open interval $(0,1)$ of the real line. The *closed sets* relative to this interval, including the "empty set", form the basic class of elementary sets.

The operations we are going to study are defined by specifying a collection of parts of the natural scale

$$1, 2, 3, 4, \ldots,$$

called *number chains*. An operation X is considered completely defined if the collection $\{U^X\}$ of its number chains is specified, where each chain U^X is simply a set of natural numbers. We call $\{U^X\}$ the *defining system* of the operation X.

Assume now that a sequence of sets $E_1, E_2, E_3, \ldots$ is given. Then to each chain

$$U^X \equiv \{n_1, n_2, n_3, \ldots\}$$

* *Mat. Sb.* **35** (1928), 414–422.

[1] 'Sur une classe d'ensembles définis par une infinité dénombrable de conditions', *C.R. Acad. Sci. Paris* **184** (1927), 1311–1314.

there corresponds a certain chain of sets

$$\mathfrak{U}^X \equiv \{E_{n_1}, E_{n_2}, E_{n_3}, \ldots\}.$$

By the *nucleus* $R[\mathfrak{U}^X]$ of the chain $\mathfrak{U}^X$ is meant the product of all its constituent sets. The result of an operation X on the given sequence of sets is defined as the sum of the nuclei of all the corresponding chains:

$$X(E_1, E_2, E_3, \ldots) = \sum R[\mathfrak{U}^X].$$

We consider some specific operations of the above-defined kind. First of all, there is the operation of *summation*:

$$\sum\{E_n\} = E_1 + E_2 + \ldots$$

This operation is defined by means of a countable collection of chains each of which consists of a single element:

$$U_1^{\sum} \equiv \{1\},\ U_2^{\sum} \equiv \{2\}, \ldots.$$

The operation of *multiplication* $\prod$ is defined by means of a single chain consisting of the whole natural scale:

$$U^{\Pi} \equiv \{1, 2, 3, \ldots\}.$$

It is also easy to define the operation of taking the *upper limit* and *lower limit* of a sequence of sets. Naturally the operation of taking the limit of a sequence of sets cannot be defined in this way because this does not yield a definite result for every sequence of sets .

The *A-operation* is usually defined for a countable collection of sets indexed not simply by natural numbers but by all possible ordered tuples of natural numbers

$$n_1, n_2, \ldots, n_k.$$

By a chain of ordered tuples U^A we mean any sequence of ordered tuples

$$U^A_{n_1 n_2 \ldots} \equiv \{n_1, n_1 n_2, n_1 n_2 n_3, \ldots\}.$$

The set

$$A\{E_{n_1 n_2 \ldots n_k}\}$$

which is obtained by means of the A-operation from the sets $E_{n_1 n_2 \dots n_k}$ corresponding to all possible ordered tuples $n_1, n_2, \dots, n_k$ is defined as the sum of the nuclei of all the possible chains

$$\mathfrak{U}^A_{n_1 n_2 \dots} \equiv \{E_{n_1}, E_{n_1 n_2}, E_{n_1 n_2 n_3} \dots\}.$$

Since the ordered tuples $n_1 n_2, \dots, n_k$ can be indexed by fixed natural numbers $\mu(n_1, n_2, \dots, n_k)$, this definition can easily be reduced to the above form, since chains of ordered tuples can be replaced by chains of their numbers μ.

An operation X is said to be *normal* if the set resulting from its twice repeated application

$$E = X\{E_n\}_n, \ E_n = X\{E_{nm}\}_m$$

can be obtained by means of the X-operation on the sets E_{nm} arranged in a suitable way:

$$E = X\{E_{n_k m_k}\}, \ m_k = \phi(k), \ n_k = \psi(k).$$

Generally speaking, $\phi(k)$ and $\psi(k)$ can depend on the family of sets under consideration, but for all normal operations known to us they can be chosen depending only on the operation itself.

The summation and multiplication operations are obviously normal; the operations of taking the upper and lower limits are not normal.

The proof of the fact that the A-operation is normal, using the same method as in the proof of the fact that the A-operation applied to A-sets again yields an A-set, is much more complicated.

II.

Consider the system of number chains $\{U^X\}$ of the operation X. A collection of natural numbers is said to be a *complementary* chain to the given system if it has at least one common element with any chain U^X. The system of all complementary chains $\{U^{\overline{X}}\}$ determines an operation that will be called the *complementary* operation to X and denoted $\overline{X}$.

The following basic property of the operation $\overline{X}$ explains why it has been introduced as the operation complementary to X: if $\overline{E}$ denotes the geometric complement to E, then

$$E = X(E_1, E_2, \dots)$$

implies that

$$\overline{E} = \overline{X}(\overline{E}_1, \overline{E}_2, \ldots).$$

The operation complementary to $\overline{X}$, naturally denoted by $\overline{\overline{X}}$, when applied to any system of sets will obviously give the same result as X; this can be written as follows:

$$\overline{\overline{X}} \sim X.$$

In particular, the operation complementary to summation is precisely multiplication, whereas the operation complementary to multiplication, although it is equivalent to summation, is defined by a different system of chains. Obviously, the operation complementary to a normal operation is itself normal.

The operation complementary to the A-operation is called the Γ-*operation*. Like the A-operation, it can also be naturally defined by means of chains of ordered tuples of natural numbers. P.S. Aleksandrov offered the following elegant geometric description for Γ-chains: consider the ordered tuples $n_1, n_2, \ldots, n_k$ as groups in the zero-dimensional Baire space; then a system of ordered tuples is a Γ-chain if and only if it completely covers the whole space.

III.

Sets obtained by means of the application of the X-operation to a sequence of closed sets will be called X-*sets*. The sets complementary to X-sets can be obtained from open sets (domains) by means of the $\overline{X}$-operation.

Main Theorem. *There exists an X-set whose complement is not an X-set.*

Consider the infinite-dimensional cube whose points are determined by the coordinates

$$0 \leq a_{nm} \leq 1, \quad 0 \leq b_{nm} \leq 1,$$

where the indices n and m assume all possible values. We construct a Peano curve inside the cube, that is, functions $\phi_{nm}(t)$ and $\psi_{nm}(t)$ that are continuous and such that the curve

$$a_{nm} = \phi_{nm}(t), \quad b_{nm} = \psi_{nm}(t)$$

covers each point of the cube at least once as t runs over the interval from zero to one.

We now define the set $P_n(t)$ as the closed set containing all points of (0,1) except those belonging to the intervals $\Delta_{nm}(t)$ with end points $\phi_{nm}(t)$ and $\psi_{nm}(t)$ corresponding to all possible values of m. It is clear that for any sequence $P_1, P_2, P_3, \ldots$ of closed sets there exists t such that

$$P_1 = P_1(t),\ P_2 = P_2(t), \ldots .$$

Further, we put

$$E_t = X\{P_n(t)\}.$$

It follows that to any X-set E there corresponds t such that

$$E = E_t.$$

Finally, let Q_n be the set of all t that belong to the corresponding $P_n(t)$. It is not too difficult to prove that Q_n is a closed set. The set

$$E = X\{Q_n\}$$

has the following property: if t belongs to E_t, then it also belongs to E, while if t does not belong to E_t, then it does not belong to E either.

It is clear that E is an X-set. However, if $\overline{E}$ were also an X-set, then for some t we would have

$$\overline{E} = E_t,$$

whence a contradiction can easily be obtained. This proves the main theorem.

IV.

We define the $W(X)$-*domain* as the minimal collection of sets containing all closed sets and complements of all these sets, and invariant with respect to the X-operation. Obviously, $W(\Sigma)$ and $W(\Pi)$ coincide with the collection of B-measurable sets, and $W(A)$ and $W(\Gamma)$ coincide with Luzin's C-*domain*. We will develop a classification of sets of the $W(X)$-domain similar to the Baire classification and prove, under certain very general assumptions, that all classes in this classification are non-empty.

We now define the classes $W_\alpha(X)$ and $\overline{W}_\alpha(X)$, where α is an ordinal number of the second class, using the following induction procedure.

1. $W_0(X)$ is the collection of all closed sets; $\overline{W}_0(X)$ is the collection of all open sets.

2. If $W_\alpha(X)$ and $\overline{W}_\alpha(X)$ have already been defined for $\alpha < \beta$, then $W_\beta(X)$ is defined as the collection of sets obtained by means of the $\overline{X}$-operation on the sets each of which belongs to one of the classes $\overline{W}_\alpha(X)$, $\alpha < \beta$. Accordingly, $\overline{W}_\beta(X)$ is the class of sets obtained from the sets of the classes $W_\alpha(X)$, $\alpha < \beta$, by means of the X-operation.

Each operation on a sequence of identical sets gives the original set:

$$X(E, E, \ldots) = E.$$

Therefore $W_\alpha(X)$ obviously contains all $\overline{W}_\beta(X)$ for $\beta < \alpha$. On the other hand, it is clear that $W_\alpha(X)$ also contains all $W_\beta(X)$ for $\beta < \alpha$. Accordingly, $\overline{W}_\alpha(X)$ contains all $\overline{W}_\beta(X)$ and $W_\beta(X)$.

Obviously, the complement of a set of class $W_\alpha(X)$ always belongs to $\overline{W}_\alpha(X)$. Together with the above-mentioned facts this implies that

$$W(X) = \sum W_\alpha(X) = \sum \overline{W}_\alpha(X),$$

where the summation is extended over all ordinal numbers of the second class.

Indeed, closed sets, their complements, and open sets belong to $W_\alpha(X)$; using transfinite induction it is possible to prove that all sets in $W_\alpha(X)$ and $\overline{W}_\alpha(X)$ also belong to $W(X)$. On the other hand, it is easy to see that the sums

$$\sum W_\alpha(X) = \sum \overline{W}_\alpha(X)$$

satisfy the requirements we imposed at the beginning of this section on the classes among which $W(X)$ is the minimal class.

We now set

$$V_\alpha(X) = W_\alpha(X) - \sum_{\beta<\alpha} W_\beta(X), \quad \overline{V}_\alpha(X) = \overline{W}_\alpha(X) - \sum_{\beta<\alpha} \overline{W}_\beta(X).$$

Then

$$W(X) = \sum V_\alpha(X) = \sum \overline{V}_\alpha(X),$$

where the terms of the last two sums do not intersect.

Theorem (non-emptiness theorem for the classes). *Assume that the summation operation can be replaced by a combination of the X- and $\overline{X}$-operations. Then none of the classes $V_\alpha(X)$ and $\overline{V}_\alpha(X)$ can be empty.*

1. Assume that

$$V_\alpha(X) = 0;$$

then

$$W_\alpha(X) = \sum_{\beta<\alpha} W_\beta(X), \quad \overline{W}_\alpha(X) = \sum_{\beta<\alpha} \overline{W}_\beta(X) \subset W_\alpha(X).$$

It can easily be seen that in this case

$$W(X) = W_\alpha(X).$$

We will show that this relation is impossible under the condition of the theorem.

2. Assume that an operation Y and a sequence of operations $Y_1, Y_2, Y_3, \ldots$ are given. We define an operation Z, denoted

$$Z = (Y | Y_1, Y_2, \ldots),$$

at first, as is explained below, as an operation on a doubly-indexed system of sets:

$$\{E_{nm}\}.$$

By a *chain* U^Z we mean any collection of pairs of indices (nm) such that the set of values of the first index in these pairs forms a chain U^Y and the set of values of the second index in the pairs with a given first index n forms a chain U^{Y_n}.

It is easy to see that

$$E = Y\{E_n\}, \quad E_n = Y_n\{E_{nm}\}_m$$

implies that

$$E = Z\{E_{nm}\}_{nm},$$

and, conversely, the last relation implies the existence of sets E_n satisfying the first two relations. Thus, the Z-operation replaces the application of the Y-operation to the results of Y_n-operations and, conversely, the result of the Z-operation can be obtained by consecutive application of Y_n- and Y-operations.

On renumbering the pairs (nm) by means of a single index k, we readily obtain an operation with one index:

$$Z\{E_k\}_k$$

having the same properties.

In particular, the operation

$$Z \equiv (\Sigma|Y_1, Y_2, \ldots)$$

applied to the case where

$$E_{nm} = 0 \quad \text{for all } n \neq n_0$$

and

$$E_{nm} = E_m \quad \text{for } n = n_0,$$

results in

$$Z\{E_{nm}\}_{nm} = Y_{n_0}\{E_m\}.$$

In other words, all the sets produced by any of the operations Y_n can be obtained using Z.

3. We define operations X_α and $\overline{X}_\alpha$ by means of the following induction procedure. Assuming that all transfinite numbers ν,

$$\nu < \beta < \alpha,$$

are arranged in some way as a sequence $\nu_1, \nu_2, \nu_3, \ldots$, we put

$$\overline{X}_1 = X, \quad X_1 = \overline{X},$$

and, assuming that X_ν and $\overline{X}_\nu$ are already defined for $\nu < \beta$, we put

$$X_\beta \equiv (\overline{X}|\overline{X}_{\beta-1}, \overline{X}_{\beta-1}, \overline{X}_{\beta-1}, \ldots),$$
$$\overline{X}_\beta \equiv (X|X_{\beta-1}, X_{\beta-1}, X_{\beta-1}, \ldots)$$

if β is not a limit ordinal.

If β is a limit ordinal, we put

$$X'_\beta \equiv (\Sigma|X_{\nu_1}, X_{\nu_2}, X_{\nu_3}, \ldots),$$
$$X_\beta \equiv (\overline{X}|X'_\beta, X'_\beta, X'_\beta, \ldots),$$
$$\overline{X}_\beta \equiv (X|X'_\beta, X'_\beta, X'_\beta, \ldots).$$

4. It is not difficult to prove by induction that any set of class $W_\alpha(X)$ can be obtained from closed and open sets by means of the operation X_α. By the main theorem, a set not belonging to $W_\alpha(X)$ can be constructed using $\overline{X}_\alpha$. However, it can easily be proved that such a set must belong to $W_\alpha(X)$, which proves the theorem.

V.

The application to the theory of B- and C-sets needs no explanation.

14 December 1928

14. ON THE DENJOY INTEGRATION PROCESS *

Assume that $f(x)$ is a function with period $b-a$. Let

$$a \leq \xi_1 \leq x_1 \leq \ldots \leq \xi_i \leq x_i \leq \ldots \leq \xi_n \leq b \ (x_0 = a, \ x_n = b)$$

be a partition of the interval $[a, b]$. According to the definition given by Denjoy, $f(x)$ is (B) integrable and the integral

$$B \int_a^b f(x) dx$$

is equal to I if for any positive R not depending on the norm ω of the partition x_i, the measure of the set of those t that satisfy

$$|I - \phi(t)| = |I - \sum (x_i - x_{i-1}) f(x_i + t)| > R, \quad 0 < t < b - a,$$

tends to zero as $\omega \to 0$.

Denjoy proved that all summable functions are (B) integrable. Our purpose is to show that *all expressions*

$$g(x) = \frac{1}{2\pi} \int_0^{2\pi} \frac{f(x+\alpha)}{\tan(\alpha/2)} d\alpha$$

which are conjugate functions of summable functions $f(x)$ of period 2π are also (B) integrable. It is well known that among these functions $g(x)$ there are such that are not summable on any interval.

We earlier established the inequality (see [1])

$$\text{mes}\{|g(x)| > R\} < \frac{C}{R} \int_0^{2\pi} |f(x)| dx, \tag{1}$$

where C is an absolute constant. Consider the sum $\phi(t)$ corresponding to $g(x)$. Then $\phi(t)$ is the conjugate function of

$$\sum (x_i - x_{i-1}) f(\xi_i + t).$$

By (1),

$$\text{mes}\{|\phi(t)| > R\} < \frac{2\pi C}{R} \int_0^{2\pi} |f(x)| dx. \tag{2}$$

* 'Sur un procédé d'intégration de M. Denjoy', *Fund. Math.* **11** (1928), 27–28.

For arbitrarily small ϵ we can represent f in the form

$$f(x) = f_1(x) + f_2(x),$$

where $f_1(x)$ is bounded and

$$\int_0^{2\pi} |f_2(x)|dx < \frac{\epsilon R}{8\pi C}.$$

The function $g_1(x)$ conjugate to $f_1(x)$ is summable on $(-\pi, \pi)$ and the integral is equal to zero. Therefore, for sufficiently small ω we have

$$E_1 = \text{mes}\{|\phi_1(t)| > \tfrac{1}{2}R\} < \epsilon/2.$$

On the other hand, by virtue of (2),

$$E_2 = \text{mes}\{|\phi_2(t)| > \tfrac{1}{2}R\} < \epsilon/2.$$

If ω is sufficiently small, the two inequalities imply

$$\text{mes}\{|\phi(t)| > R\} \leq E_1 + E_2 < \epsilon,$$

which proves that $g(x)$ is (B) integrable.

It is possible to show in a similar way that $g(x) \cdot \cos nx$ and $g(x) \cdot \sin nx$ are also (B) integrable and that the Fourier-(B) series of $g(x)$ is conjugate to the Fourier-Lebesgue series of $f(x)$.

5 April 1927

References

1. A. Kolmogoroff, 'Sur les fonctions harmoniques conjuguées et les séries de Fourier', *Fund. Math.* 7:8 (1925), 23–28. (No. 8)

15. ON THE TOPOLOGICAL GROUP-THEORETIC FOUNDATION OF GEOMETRY *

It is well known that n-dimensional spaces of constant curvature (that is, hyperbolic, Euclidean, elliptic and spherical spaces) have $(n+1)n/2$-dimensional continuous groups of motions. All the other n-dimensional geometric formations that up to now are known admit lesser freedom of motion. [1] It seems natural to pose the problem of characterizing spaces of constant curvature as the only type of topological spaces with a sufficient amount of freedom of motion. To this end, we consider a topological space R and a group Γ of single-valued continuous mappings of R onto itself. To make Γ similar to a group of motions, it is natural to require that elements of Γ be uniformly equicontinuous. [2]

Condition 1. For any pair of points (x, y) and any neighbourhood $U(y)$ there are neighbourhoods $V(x)$ and $W(y)$ such that for any mapping of Γ that transforms at least one point of $V(x)$ into a point in $W(y)$ the entire image of $V(x)$ lies in $U(y)$.

If R is metrizable and locally compact, then Condition 1 is necessary and sufficient for the existence of a metric (distance function) in R such that all mappings belonging to Γ preserve distances between two points, that is, are congruent mappings. [3]

Condition 2. R is metrizable and locally compact.

Condition 3. R is connected.

Condition 4. Γ is transitive. [4]

---

* 'Zur topologisch-gruppentheoretischen Begründung der Geometrie', *Nachr. Ges. Wiss. Göttingen* **8** (1930), 208–210.

[1] With the remarkable exception of van Danzig solenoids in the one-dimensional case; see *Fund. Math.* **15** (1929), 102–105.

[2] I believe that this definition of uniform continuity of a system of mappings in a neighbourhood of a pair of points (x, y) also remains interesting to a certain extent in the general case of mappings from one space R into another space R'.

[3] If R is a compact space, such a distance function $\rho^*(x, y)$ can be defined as the upper limit of $\rho(x', y')$ for all images (x', y') of pairs of points (x, y) under mappings belonging to Γ, where $\rho(x, y)$ is an arbitrary distance function in R.

[4] Van Danzig solenoids satisfy Conditions 1–4; Condition 3 can be replaced by a stronger condition (not fulfilled for solenoids):

Condition 3$'$. R is connected and locally connected.

Conditions 1, 2, 3$'$ and 4 hold, for example, for the group of motions of a circular

The mappings belonging to Γ and preserving a given point x form a subgroup Γ_x of rotations about x. We denote by $S(x|y)$ the collection of images of y under the mappings belonging to Γ_x, and call it a sphere around x. If z belongs to $S(x|y)$, then $S(x|y)$ and $S(x|z)$ coincide. In what follows we denote spheres with centre at x simply by $S(x)$.

Condition 5. Given two arbitrary distinct spheres $S'(x)$ and $S''(x)$ with the same centre x, one of them separates the other from the centre.

It is possible to show using Conditions 1–5 that there exist on R a convex [5] distance function $\rho(x,y)$ which is invariant with respect to Γ. This function is determined by Γ to within a positive multiplier. If R is one-dimensional, it follows from 1–5 that R is homeomorphic to either an ordinary circle or a line. Moreover, it is possible to map R onto a circle (or a line) in such a way that Γ becomes the group of rotations and reflections (the group of translations and reflections). There is also a possibility that in the general case as well, a space R satisfying 1–5 is necessarily homeomorphic to a finite-dimensional space of constant curvature. But here Conditions 1–5 are no longer sufficient for obtaining the complete group of motions or motions and reflections: even in the four-dimensional Euclidean space there exists a group of motions satisfying 1–5 whose dimension is equal only to seven. [6]

Any mapping belonging to Γ_x transforms the sphere $S(x)$ together with its centre x onto itself. These mappings of $S(x)$ onto itself form a group $\Gamma\{S(x)\}$. It is now possible to define for any $y \in S(x)$ the group $\Gamma_y\{S(x)\}$ of rotations of $S(x)$ about y. By a second-order sphere we mean the collection $S(x,y|z)$ (also simply denoted by $S(x,y)$) of images of a point $z \in S(x)$ under the mappings belonging to $\Gamma_y\{S(x)\}$. Generally, if an nth-order sphere $S(x_1,x_2,\dots,x_n)$ and the group $\Gamma\{S(x_1,x_2,\dots,x_n)\}$ are given, then for any $y \in S(x_1,x_2,\dots,x_n)$ the group of rotations $\Gamma_y\{S(x_1,x_2,\dots,x_n)\}$ and the $(n+1)$th-order spheres $S(x_1,x_2,\dots,x_n,y|z)$, or simply $S(x_1,x_2,\dots,x_n,y)$, are defined in a similar way. Finally, the groups $\Gamma\{S(x_1,x_2,\dots,x_n,y)\}$ are also defined.

Condition 5′. Given two distinct spheres $S'(x_1,x_2,\dots,x_n)$ and $S''(x_1,x_2,\dots$

---

cylinder. The question arises whether R is necessarily a manifold under these conditions.

[5] $\rho(x,y)$ is called convex if for any x,y there is at least one z such that $\rho(x,y) = 2\rho(x,z) = 2\rho(y,z)$.

[6] This group corresponds to the unimodular group of linear quaternary substitutions $x' = ax+1$, $|a|=1$.

$\ldots, x_n)$, one of them separates the other from the centre x_n in $S(x_1, x_2, \ldots$ $\ldots, x_{n-1}|x_n)$.

Condition 5′ becomes meaningless for higher-order spheres belonging to the sphere $S(x_1, x_2, \ldots, x_n)$ when the latter is zero-dimensional. It is possible to show that all spheres of a sufficiently high order are in fact zero-dimensional if the conditions 1–4 and 5′ are fulfilled. It follows that one-dimensional spheres also exist and that they are necesarily homeomorphic to ordinary circles. Using induction we finally obtain

Main theorem. *Under Conditions* 1, 2, 3, 4 *and* 5′, *R is homeomorphic to a finite-dimensional space of finite curvature and R can be mapped onto this space in such a way that Γ goes into the complete group of motions and reflections.*

18 July 1930

16. STUDIES ON THE CONCEPT OF INTEGRAL *

This paper is primarily written with the purpose of elucidating the logical nature of integration processes. And the reason why here, in addition to the unification of various approaches, a new generalization of the concept of integral arises, seems to lie in the fact that generalizations often help to grasp the very essence of the concept in question (cf. the preface to the second edition of [1]). It is quite possible that all the generalizations can also be of interest for applications, but I see the advantages of a more general approach mainly in the simplicity and clarity the new concepts bring along.

My interest in the general problems of integration theory was stimulated by Professor N.N. Luzin. I am also grateful to B.I. Glivenko for useful discussions.

Chapter 1. Introduction

§1. What is at present understood by "integration" is not, rigorously speaking, a logical concept. Rather it is a collective name for several operations each of which has its own definition. But in most cases the operations have the common property that integration of a continuous function results in the classical integral. I do not think that the whole variety of definitions can be reduced in a natural manner to a single definition. In other words, the problem of finding a natural definition of integral that contains all the earlier definitions as special cases seems to be absolutely hopeless.

In particular, I believe that an attempt to unify within a general concept, on the one hand, the Stieltjes-type integral for functions on abstract sets (in the form recently suggested by Fréchet [2]) and, on the other hand, the Denjoy integral must inevitably encounter substantial difficulties. Indeed, such a unified concept can be applied to functions defined on arbitrary abstract sets if it involves constructions based only on general properties of sets and functions. At the same time, it must incorporate peculiarities of the Denjoy integration method closely related to the order relation on the number line.

It is not for the first time that a process of generalizing a mathematical concept simultaneously in several directions has been considered; a similar situation appears, for instance, when different properties of natural numbers are used as a starting point for essentially different generalizations. Here we

---

* 'Untersuchungen über Integralbegriff', *Math. Ann.* **103** (1930), 654–696.

obtain, on the one hand, ordinal and cardinal numbers and on the other hand, real numbers.

Integration in the sense of Denjoy (he calls it "totalization") is a generalization of the classical integration method, interpreted as finding the primitive of a given function. This is a quite natural and logical method of generalization that can be regarded, say, as the last link in the chain that begins with the Newtonian concept of integral.

Another and even more traditional way originates from the general and rather ambiguous idea that integration is a summation method related to an infinite number of infinitesimals corresponding to infinitely small portions of the integration domain. This idea, to which the very word "integral" is due, was introduced by Leibniz in a vague but extremely general form. It was first interpreted (in a more or less satisfactory way to a modern mathematician) by Cauchy. Of the many subsequent works on the problem of integration those devoted to the development of Leibniz's idea form the majority. The Fréchet definition already mentioned (leading to a different direction as compared to the Denjoy integral) is related to the same idea.

The aim of this work can be formulated as the elaboration of the broadest possible definition of integral that realizes Leibniz's concept in a form satisfying modern requirements of rigorousness and to the largest possible extent, preserves its universal nature.

The two directions are not the only ones in which a reasonable development preserving sufficiently many classical properties of integration is possible. In particular, I want to note that a new and very general operation, which, by the way, is applicable in various problems of probability theory, can be obtained by proceeding from the problem of finding the mean value of a function (whose solution in classical situations is given by the integral). I hope to consider this operation (which of course contains classical integration as a particular case) in another paper. Here I want only to mention the remarkable fact that even in the case of a real function $f(x)$ of one real variable, a natural and uniquely determined mean value on the unit interval may differ from the Denjoy integral (over the same interval).

In my opinion, the fact that we have two definitions of integral each of which is an important and natural generalization of the classical integral but yielding different values for the same function, justifies the above-mentioned idea that it is hopeless to try to find a universal concept of integral that would embrace all reasonable definitions.

Even in this work we will deal with two definitions, in Chapters 2 and 3, which yield different values for certain functions (see no. 14 in Chapter 3).

It should be noted that these two definitions yield different results only for set functions constructed ad hoc, for which it is not clear whether integration of such functions is of interest at all.

§**2.** Let x be a variable ranging over the interval (a, b). Leibniz assumed that to every x there corresponds an infinitesimal dy. The sum of all these dy gives the so-called *Summa Omnium* of Leibniz, or the integral of dy over (a, b). The meaning of this definition was not clear enough even for mathematicians of the 18th century and before the time of Cauchy, they had to reject Leibniz's approach and define integration as the inverse of differentiation.

Cauchy was the first to translate the definition of Leibniz from the language of metaphysics into the language of mathematics and thus to open the possibility, at least for the particular case $dy = f(x)dx$, actually to calculate the integral

$$\int_a^b f(x)dx \tag{1}$$

as the limit of the known integral sums

$$\sum f(\xi_i)(x_i - x_{i-1}). \tag{2}$$

We denote by Δ_i the intervals (x_{i-1}, x_i), by $l(\Delta)$ the length of the interval Δ, and by $f(\Delta)$ the many-valued interval function whose value on Δ is the collection of all values that $f(x)$ assumes on Δ. Then (2) takes the form

$$\sum f(\Delta_i)l(\Delta_i) = \sum \phi(\Delta_i). \tag{3}$$

The idea of Leibniz is implemented as follows. To find an infinitesimal dy for each value of x in (a, b) means to define an interval function $\phi(\Delta)$ such that $\phi(\Delta)$ becomes *infinitesimally small* as Δ contracts towards the point x. To find the sum of all dy means to determine the limit of integral sums (3) for an infinite refinement (extension) of the partition of the interval into subintervals.

The following four generalizations of the Cauchy integration method are possible.

1. It is possible to consider arbitrary interval functions $\phi(\Delta)$ instead of $\phi(\Delta) = f(\Delta)l(\Delta)$.

2. The domain of integration can be partitioned into a finite or infinite number of arbitrary sets (not necessarily into intervals).

3. The domain of integration need not necessarily be an interval on the number line. Sets of elements of arbitrary nature can be taken as integration domains.

4. Finally, the function $\phi(\Delta)$ can be allowed to take values in an arbitrary vector space, not necessarily in the real line.

Substantial progress in the first of the four directions was made in connection with the Stieltjes integral and its direct generalizations. Instead of the length $l(\Delta)$, an arbitrary additive interval function was introduced. But the general formulation of the problem was first suggested in 1924 by Burkill [3], who defined the integral

$$\int_a^b d\phi(\Delta)$$

for an arbitrary interval function $\phi(\Delta)$ as the limit of the sums

$$\sum \phi(\Delta_i)$$

for indefinitely refined partitions of (a, b) into subintervals. Burkill considered only single-valued functions $\phi(\Delta)$, which did not allow him to obtain the classical Cauchy definition of the integral as a special case of his own definition.

In this work we systematically use many-valued functions. Moreover, we consecutively replace point functions $f(x)$ by interval functions and, even more generally, by set functions $f(E)$ whose values on E are all the values that $f(x)$ assumes on the set E. This is the only way that leads to a general theory of integral with only *one* set function under the sign of the integral. The splitting of the function subject to integration that has up to now dominated in all integration theories is therefore an artificial and unjustified restriction, which both reduces the generality of the problem and affects the simplicity of the method.

Essential progress in the second direction was achieved, as is well known, by Lebesgue, who defined the integral (1) as the limit of the sums

$$\sum f(E_i)m(E_i), \tag{4}$$

where the E_i are measurable sets into which the interval (a, b) is partitioned and $m(E)$ is the measure of E. Lebesgue required in his definition that the sums (4) tend to a limit for partitions of a special kind. A general definition of the limit for expressions depending on partitions of the interval when they are "indefinitely refined" ("extended") (cf. no.8 in Chapter 2) is due to Moore [4]. Moore himself points out that the concept of limit he introduces makes it possible to give a particularly simple form of the definition of the Lebesgue integral and leads to a generalization of the latter. In this way certain generalizations of the classical Stieltjes integral were found by Smith [5].

As to the generality of admissible integration domains, it was finally established by Fréchet, in the above-mentioned paper, after a number of papers by other authors, that a general theory of integration can be developed without any restrictions on the domains. Until now, all that has been done in these first three directions in integration theory is taken into account when the definition in this paper is constructed. The remaining fourth direction (integration of functions with values in an arbitrary vector space) is deliberately not considered here. Most of the results can be extended to this case as well, but the inclusion of this generalization would overburden the presentation.

§3. In the previous section we discussed the development of the concept of integral beginning with Leibniz's *Summa Omnium*. We see that this development is a natural and inevitable consequence of Leibniz's idea, although some recent definitions seem to be very distant from the original idea.

The situation is different in relation to the concept of indefinite integral. The classical interpretation of the indefinite integral as a point function is essentially connected with integration over an interval on the real line and cannot be generalized even to the case of multiple integrals. Only recently has a more general and, at the same time, simpler idea been developed, which considers an indefinite integral as a *set function* on the set over which integration extends. In what follows we use the latter approach only.

In the case of the Stieltjes-Radon integral we deal with two additive set functions $\phi(E)$ and $\psi(E)$ and with a point function $f(x)$, related as follows:

$$\psi(E) = \int_E f(x)d\phi(x), \tag{1}$$

$$d\psi(x)/d\phi(x) = f(x). \tag{2}$$

Here the second relation holds only "almost everywhere". It should also be noted that the concept of derivative of a set function with respect to another set function depends essentially on the geometric properties of the spaces in which the corresponding sets lie. However, we will demonstrate in the second appendix to this paper that even in the case of a Stieltjes-type integral (with no geometrical concepts involved), the derivative (2) can be introduced in such a way that (2) implies (1). The general concept of integration connected with the formulas (1), (2) was many times advocated by Lebesgue, who emphasized its fundamental importance for ***physics***. However, we shall see that this concept of integration is a ***purely logical*** consequence of a general method of integration:

$$\psi(E) = \int_E \phi(dE), \tag{3}$$

where $\phi(E)$ is not necessarily an additive set function. In one of my future papers I hope to show that this even more general concept is also justified mathematically and proves particularly useful in measure theory and in the general theory of quadrature of surfaces.

If in (3) only one of the two set functions is additive, then integration cannot of course be related to any differentiation method. But even in this case, ϕ and ψ are connected by a relation of a differential character, which will be introduced in what follows under the name of differential equivalence.

Chapter 2. The First Integration Theory (Countable Partitions)

§1. Definition and introductory remarks

1. A system of sets having the property that the intersection of any two elements of the system also belongs to the system is called an $\mathfrak{M}$-*system* (multiplicative system) and is denoted by $\mathfrak{M}$. [1]

2. We write

$$DE = \sum_n E_n,$$

[1] If $\mathfrak{M}$ contains two disjoint sets, then, by the assumption, the *empty* set also belongs to $\mathfrak{M}$.

to denote a *partition* of a set $E \in \mathfrak{M}$ into a finite or countable number of pairwise disjoint elements of the system:

$$E = \sum_n E_n.$$

The sets E_n are called elements of the partition D.

3. Each partition DE with elements E_n induces a unique partition

$$DE' \equiv \sum_n E'_n \equiv \sum_n E'E$$

of each subset E' of E. This partition will be denoted by the same symbol D.

4. A partition $D'E$ is called an *extension* of DE if each element E'_n of D' is contained in an element E_m of D. In this case we write $D' > D$. It is obvious that $D'' > D'$ and $D' > D$ imply $D'' > D$.

5. By the *product* of two partitions $D'E$ and $D''E$ we mean the partition

$$[D'D'']E = \sum_{nm} E'_n E''_m.$$

Here E'_n and E''_m are elements of D' and D'', respectively. By the basic property of $\mathfrak{M}$-systems, the elements of the *product of partitions* belong to the $\mathfrak{M}$-system. The product is obviously an *extension of each of the partitions.*

6. We also define the *sum* of partitions. Given

$$DE \equiv \sum_n E_n, \quad D^{(n)}E_n \equiv \sum_m E_{nm},$$

and

$$D'E \equiv \sum_{nm} E_{nm} \equiv \sum_n \sum_m E_{nm},$$

we put, by definition,

$$D'E = \sum_n D^{(n)}E_n.$$

Each extension $D'E$ of DE can be uniquely represented in this form; in this case the partitions $D^{(n)}E_n$ are called *components* of $D'E$. The converse is also true: each partition $\sum D^{(n)}E_n$ is an extension of DE.

7. We denote by $\mathfrak{M}E$ the collection of elements of all partitions DE of E. The system of sets $\mathfrak{M}E$ is itself an $\mathfrak{M}$-system. Indeed, any two elements E' and E'' of $\mathfrak{M}E$ are elements of certain partitions $D'E$ and $D''E$ and the intersection $E'E''$ is an element of $[D'D'']E$.

If G belongs to $\mathfrak{M}E$, then $\mathfrak{M}G$ is a part of $\mathfrak{M}E$.

8. Assume that a single-valued or many-valued function $f(D)$ is defined on a certain set of partitions DE of E. Following Moore, we say that a number I is the *limit of* $f(DE)$ *under indefinite extension of the partition* DE, and write

$$I = l[f(DE)],$$

if for any $\epsilon > 0$ there is a partition DE such that for all $D' > D$ the function $f(D')$ is defined and the following inequality holds (see [4]):

$$\sup |f(D') - I| < \epsilon.$$

It is easy to see that there can be at most one limit. Assuming that I' and I'' are two limits, we find two partitions $D'E$ and $D''E$ such that for their product (which is also an extension of each of them) we have

$$\sup |f([D'D'']) - I'| < \epsilon/2, \quad \sup |f([D'D'']) - I''| < \epsilon/2,$$

whence $|I' - I''| < \epsilon$ and, since $\epsilon > 0$ is arbitrary, $I' = I''$.

9. The limit of $f(DE)$ exists if and only if the *Cauchy convergence criterion* is fulfilled: for any $\epsilon > 0$ there is an "ϵ-regular partition DE", that is, a partition such that $D' > D$ always implies [3]

$$\sup |f(DE) - f(D'E)| < \epsilon.$$

10. The above definition can easily be extended to the case $I = \pm\infty$. We put

$$l[f(DE)] = +\infty$$

if for any $H < +\infty$ there is DE such that

$$\inf[f(D'E)] > H$$

[3] The difference of two many-valued expressions consists of the differences of all values of the expressions.

for $D' > D$. The definition for minus infinity is similar.

§2. Definition and elementary properties of the integral

11. Assume that a function $f(E)$ is defined for all elements of DE. We introduce the notation

$$(Rf)(DE) = \sum_n f(E_n).$$

If f is many-valued, then so is (Rf) (*different values for* (Rf) *are obtained if for any* E_n *we use all the possible values of* $f(E_n)$ *in the process of summation*). For the above notation to make sense we assume [4] that the series on the right-hand side is absolutely convergent for any choice of values of $f(E_n)$.

The expression $(Rf)(DE)$ may be called the *Riemann sum* for f corresponding to the partition DE.

12. *We define the integral of* f *on* E *with respect to the given system* $\mathfrak{M}$ *by putting*

$$(\mathfrak{M}) \int_E f(dE) = l[(Rf)(DE)],$$

where l *is the limit introduced in* no. 8.

The letter $(\mathfrak{M})$ before the integral can usually be omitted, since each time several integrals are considered simultaneously, they all relate to the same $\mathfrak{M}$-system.

13. It is clear that the integral cannot exist if (Rf) is not defined for all extensions of at least one partition DE. As to the function f, only its values on elements of $\mathfrak{M}E$ are used in the definition of the integral. Moreover, it is sufficient to take only the values the function assumes on the elements of all extensions of a certain partition DE. We denote by $\mathfrak{M}DE$ the collection of these sets. It is evident that $\mathfrak{M}DE$ is a subsystem of $\mathfrak{M}E$, and it is also an $\mathfrak{M}$-system. We will say that a set function is defined on $\mathfrak{M}E$ in a differential way or, simply, that it is a *differentially defined function* if it is defined on all elements of a certain system $\mathfrak{M}DE$. Then a necessary condition for the integral of f on E to exist is that f is defined on $\mathfrak{M}E$ in a differential way.

---

[4] It should be stipulated here that all functions must assume the zero value on the empty set. This convention is necessary in any additive theory of set functions, since otherwise no integrable function can exist.

14. Properties of the integral.

I. *Assume that the integral of f on E exists. Then it exists on each set belonging to $\mathfrak{M}E$.*

This property will be proved along with the following.

II. *The integral is an additive set function, that is, for any partition DE we have*

$$\int_E f(dE) = \sum_n \int_{E_n} f(dE_n), \tag{1}$$

and the existence of the integral on the left-hand side of the relation implies the existence of all integrals on the right-hand side (which coincides with the assertion in I) *and the absolute convergence of the series of the integrals.*

Proof. Indeed, assume that there exists an integral of f on E, which means that $(Rf)(DE)$ converges to a definite limit under indefinite extension of DE. Then, according to no. 9, for any $\epsilon > 0$ there is an ϵ-regular partition $D'E$ such that for any extension $D''E$ of $D'E$ we have

$$\sup |(Rf)(D''E) - (Rf)(D'E)| < \epsilon.$$

I claim that $D'E_n$ is also an ϵ-regular partition. For let $D'''E$ be an extension of $D'E_n$. Then, as can be easily seen,

$$\begin{aligned} &\sup |(Rf)(D'''E_n) - (Rf)(D'E_n)| \leq \\ &\leq \sup |(Rf)(D'''E_n + \sum_{m\neq n} D'E_m) - (Rf)\sum_m (D'E_m)| = \\ &= \sup |(Rf)(D^{\text{iv}}E) - (Rf)(DE)| < \epsilon, \end{aligned}$$

where $D^{\text{iv}}E$ is obviously an extension of $D'E$.

Since an ϵ-regular partition $D'E_n$ exists for any $\epsilon > 0$, the limit of $(Rf)(DE_n)$, that is, the integral I_n of f on E_n, exists for each n. We choose a partition D'_nE_n for each n such that for any extension D''_nE_n we have

$$\sup |(Rf)(D''_nE_n) - I_n| < \epsilon/2^n.$$

We now put

$$D'E \equiv \sum_n D'_nE_n.$$

If $D''E = \sum D''_n E_n$ is an arbitrary extension of $D'E$, then the inequality

$$\sup|(Rf)(D''E) - \sum_n I_n| \le \sum_n \sup|(Rf)(D''_n E_n) - I_n| < \epsilon$$

is obviously fulfilled, and it follows that the integral of f on E coincides with $\sum_n I_n$, which is the desired result.

III. $$\int_E [f_1(dE) + f_2(dE)] = \int_E f_1(dE) + \int_E f_2(dE),$$

and the existence of the expression on the right-hand side implies the existence of that on the left-hand side.

IV. $$\int_E kf(dE) = k\int_E f(dE).$$

V. *If $f_1(E) > f_2(E)$ for all sets on which both functions f_1 and f_2 are defined, then* [5]

$$\int_E f_1(dE) \le \int_E f_2(dE).$$

VI. *The integral of an additive function on $\mathfrak{M}E$ coincides with the function.*

VII. *Assume that a function f and a sequence of functions $f_1, \dots, f_n, \dots$ are given such that* [6]

$$\sup[(R|f - f_n|)(DE)] \to 0 \quad (n \to \infty).$$

Then

$$\int_E f_n(dE) \to \int_E f(dE) \quad (n \to \infty),$$

and the existence of the integrals on the left-hand side of the last relation implies the existence of the integral on the right-hand side.

All these properties are obvious without proof.

§3. Differential equivalence and a second definition of the integral

15. We see that if a function f is integrable on E (and hence is defined on some $\mathfrak{M}DE$), then the integral

$$F(G) = \int_G f(dG)$$

[5] The inequality $f_1(E) > f_2(E)$ means that each value of $f_1(E)$ is greater than every value of $f_2(E)$.

[6] Here sup denotes the supremum of the values of $(R|f - f_n|)$ over DE for a fixed value of n.

is an additive function defined on all elements G of the system $\mathfrak{M}E$. It is natural to inquire about the relationship between f and F. The following definition helps to clarify this relationship.

Definition. Functions $f(G)$ and $g(G)$ differentially defined on $\mathfrak{M}E$ are said to be differentially equivalent if for any $\epsilon > 0$ there is a partition DE such that for any $D' > D$ we have

$$\sup \sum |f(E'_n) - g(E'_n) < \epsilon.$$

The property of differential equivalence can be expressed in a more compact form in two versions:

$$l[(R|f - g|)(DE)] = 0$$

or

$$\int_E |f(dE) - g(dE)| = 0.$$

Finally, we introduce the following notation for differential equivalence of f and g on $\mathfrak{M}E$:

$$f(dE) = g(dE)(\mathfrak{M}E),$$

where $(\mathfrak{M}E)$ can be omitted when no confusion is possible.

16. *Differential equivalence is a transitive property, that is, the relations*

$$f(dE) = g(dE), \quad g(dE) = h(dE)$$

imply that

$$f(dE) = h(dE).$$

Indeed, there are partitions D_1E and D_2E such that the inequalities

$$\sup(R|f - g|)(D'_1E) < \epsilon/2, \quad \sup(R|g - h|)(D'_2E) < \epsilon/2$$

hold for their extensions D'_1E and D'_2E. It can now easily be verified that the inequality

$$\sup(R|f - h|)(D'_3E) < \epsilon$$

holds for an arbitrary extension D'_3E of the partition $D_3E = [D_1D_2]E$.

17. We also prove the following proposition, in order to obtain a more complete characterization of the concept of differential equivalence.

Proposition. *Assume that a function $\phi(x_1, \ldots, x_n)$ of n real variables satisfies the Lipschitz condition*

$$|\phi(x_1, x_2, \ldots, x_n) - \phi(y_1, y_2, \ldots, y_n)| < K \sum_n |x_k - y_k|.$$

Also assume that functions $f_1, f_2, \ldots, f_n$ and $f'_1, f'_2, \ldots, f'_n$ are differentially defined on $\mathfrak{M}E$ and $f_k(dE) = f'_k(dE)$, $k = 1, \ldots, n$. Then

$$\Phi(dE) = \phi[f_1(dE), f_2(dE), \ldots, f_n(dE)] =$$
$$= \Phi'(dE) = \phi[f'_1(dE), f'_2(dE), \ldots, f'_n(dE)].$$

Proof. For any $\epsilon > 0$ there are partitions $D_k E$ having the property that for all $D'_k > D_k$ we have

$$\sup \sum_m |f'_k(E'_{km}) - f_k(E'_{km})| < \epsilon/nK.$$

Any extension $D'E$ of the partition

$$DE = [D_1, D_2, \ldots, D_n]E$$

is also an extension of each $D_k E$, which implies that

$$\sup \sum_m |\Phi'(E'_m) - \Phi(E'_m)| < \epsilon,$$

whence the desired assertion follows.

18. It follows from (1) that for differentially equivalent functions f and g and any $G \in \mathfrak{M}E$ we have

$$\int_G |f(dG) - g(dG)| = 0,$$

and, by Property V of the integral (no. 14),

$$\int_G [f(dG) - g(dG)] \leq 0, \quad \int_G [g(dG) - f(dG)] \leq 0,$$
$$\int_G [f(dG) - g(dG)] = 0. \tag{2}$$

If one of the functions f, g is integrable, then evidently

$$\int_G g(dG) = \int_G g(dG).$$

We now wish to show that the converse is also true: if (2) holds for any G, then f and g are differentially equivalent. Thus, *a necessary and sufficient condition for two functions to be differentially equivalent is that the integral of their difference vanishes on each $G \in \mathfrak{M}E$.*

Putting $h = f - g$ we reduce the assertion to the following.

19. *For any function $h(E)$, the fulfilment of the condition*

$$\int_G h(dG) = 0$$

for all $G \in \mathfrak{M}E$ implies

$$\int_E |h(dE)| = 0$$

or, equivalently,

$$h(dE) = 0.$$

Proof. Since h is integrable on E, there is an ϵ-regular partition DE such that for any extension of DE we have

$$\sup |(Rh)(D'E)| < \epsilon.$$

I claim that in this case

$$\sup(R|h|)(D'E) < 4\epsilon$$

for any $D' > D$, which, obviously, proves the proposition.

Assume, on the contrary, that there exists an extension $D'E$ such that

$$\sup(R|h|)(D'E) = \sup \sum_n |h(E'_n)| \geq 4\epsilon.$$

Then the collection N of all indices n contains a subsystem M such that

$$\sup \Big| \sum_M h(E'_n) \Big| \geq 2\epsilon.$$

For any n belonging to $N - M = L$ we choose $D^{(n)}E'_n$ such that

$$\sup |(Rh)(D^{(n)}E'_n)| < \epsilon/2^n$$

and we set

$$D''E \equiv \sum_M E'_n + \sum_L D^{(n)} E'_n.$$

It can now be seen that

$$\sup |(Rh)(D''E)| \geq \sup |\sum_m h(E'_n)| - \sum_L \sup |(Rh)(D^{(n)} E'_n)| > \epsilon,$$

which is a contradiction since $D''E$ is an extension of DE.

20. *A function $f(G)$ that is integrable on E is differentially equivalent to its indefinite integral on $\mathfrak{M}E$:*

$$F(G) = \int_G f(dG).$$

Indeed,

$$\int_G F(dG) = F(G) = \int_G f(dG),$$

and, by no 18, we have

$$f(dE) = F(dE).$$

The converse is also true: if $f(G)$ is differentially equivalent on $\mathfrak{M}E$ to a certain additive function $F(G)$, then, as follows from no. 18,

$$\int_E f(dE) = \int_E F(dE) = F(E).$$

21. Thus, we have

Second definition of the integral. *The integral of f on E is an additive function on $\mathfrak{M}E$ differentially equivalent to f.*

It follows that such a function, if it exists, is *determined uniquely* and that the second definition of the integral is equivalent to the first one.

§4. A remark on infinite values of the integral

22. In no. 10 we explained the meaning of the expression "a function of partitions DE converges to $\pm\infty$ under indefinite extension of the partitions". But if we want to obtain a definition for infinite values of the integral, it would hardly be reasonable to confine ourselves to replacing, in the definition of no. 12, the limit defined in no. 8 by that in no. 10. A more natural way is to find a generalization of the concept of absolute convergence that would allow us to take into account infinite values of Riemann sums $(Rf)(DE)$ as well. It seems also to be appropriate to drop the requirement that the integrand function be finite and thus to admit $\pm\infty$ as possible values for it.

Thus, we introduce the following definition: a series

$$\sum_n u_n$$

is said to be absolutely convergent in the generalized sense in the following three cases:

a) if it absolutely converges in the ordinary sense;

b) if the sum of its positive terms is equal to $+\infty$ while the series of negative terms is absolutely convergent in the ordinary sense. In this case, by definition the series is convergent to $+\infty$. Changing the roles of the plus and minus signs we obtain series absolutely convergent to $-\infty$.

c) if some terms are equal to $+\infty$ while the others converge to a finite value or to $+\infty$ in the sense of a) or b). The case of $-\infty$ is treated in a similar way.

23. Except for the modification of the definition of absolute convergence, everything remains the same in the definition of $(Rf)(DE)$ while the definition of the integral using the infinite limit defined in no. 10 is similar to that in no. 12.

24. As to the properties I–VII of the integral, only three of them, namely, V, VI and VII, remain unchanged; and in VI the notion of an additive function should be understood in the following natural generalized sense: a single-valued function $f(G)$ on $G \in \mathfrak{M}E$ which can assume infinite values for some G is said to be additive in the generalized sense if for any partition DG we have

$$f(G) = \sum_n f(G_n),$$

where the sum is understood in the sense that the series is absolutely convergent according to the generalized definition in no. 22.

25. Property I no longer holds in general; the following example demonstrates this. Assume that $\mathfrak{M}E$ consists of two disjoint sets E_1 and E_2, $E_1 + E_2 = E$, and let

$$f(E_1) = \pm 1, \quad f(E_2) = +\infty.$$

It is obvious that the integral of f on E is equal to $+\infty$ whereas the integral on E_1 does not exist at all.

Property II is retained in a restricted sense since in this case the existence of the integral on the left-hand side does not imply that the right-hand side makes sense.

Property III is retained, except for the case when the right-hand side takes the form $(+\infty)+(-\infty)$. Property IV holds if k is not equal to 0 or $\pm\infty$.

26. The second definition of integral based on the concept of differential equivalence cannot be extended to the case when infinite values of the integral are allowed. But the following proposition remains true.

If $f_1(dE) = f_2(dE)$, then the two functions f_1 and f_2 are simultaneously either integrable or not integrable, and in the former case we have

$$\int_E f_1(dE) = \int_E f_2(dE).$$

§5. Some special cases of integrability. Semi-additive functions and functions of bounded variation

27. A single-valued function (which can assume infinite values as well) is said to be *upper semi-additive* if for any partition DG of a set $G \in \mathfrak{M}E$ we have

$$f(G) \geq \sum_n f(G_n),$$

where the series on the right-hand side must be absolutely convergent in the generalized sense of no. 22. If under the same conditions we have

$$f(G) \leq \sum_n f(G_n),$$

then f is said to be *lower semi-additive.*

If $f(G)$ is upper semi-additive, then $-f(G)$ is lower semi-additive, and vice versa; hence in most cases we can confine ourselves to studying one of these two classes of functions.

28. *For an upper semi-additive function f we have for all $D' > D$ the inequality*

$$(Rf)(DG) \geq (Rf)(D'G).$$

Indeed, if

$$DG \equiv \sum_n G_n, \quad D'G_n \equiv \sum_m G_{nm},$$

then

$$(Rf)(DG) = \sum_n f(G_n) \geq \sum_n \sum_m f(G_{nm}) = (Rf)(D'G).$$

The opposite inequality obviously holds for a lower semi-additive function.

29. *A semi-additive function on $\mathfrak{M}E$ is integrable on E.*

Indeed, assume that f is upper semi-additive. Then, as can can easily be seen, the infimum

$$I = \inf[(Rf)(DE)]$$

satisfies the definition of the integral $\int_E f(dE)$. For a lower semi-additive function we have, accordingly,

$$\int_E f(dE) = \sup[(Rf)(DE)].$$

These facts are special cases of the following assertion.

If a single-valued function $f(DE)$ does not increase, that is, $D'E > DE$ implies $f(D'E) \leq f(DE)$, then

$$l[f(DE)] = \inf[f(DE)],$$

whereas for a non-decreasing function the following relation holds:

$$l[f(DE)] = \sup[f(DE)].$$

30. If $f, f_1, \ldots, f_n$ are additive functions, then the functions

$$|f(G)|, \quad +\sqrt{f_1^2(G) + f_2^2(G) + \ldots + f_n^2(G)}$$

are lower semi-additive. The proof requires a very simple calculation (with special regard for the cases when infinite values appear).

It follows that for additive functions $f, f_1, \ldots, f_n$ on $\mathfrak{M}E$ the following integrals exist:

$$\int_E |f(dE)|, \quad \int_E \sqrt{f_1^2(dE) + \ldots + f_n^2(dE)}.$$

The first of them is the *total variation* of $f(G)$ on E. The idea of integral representation of the total variation of an additive function goes back to Radon, who, however, did not indicate the precise meaning of the integral. As to the other integral, we call it the *joint total variation* of $f_1, \ldots, f_n$.

31. If $\mathfrak{M}E$ is an *additive system* of sets (that is, a system containing the differences and finite or countable sums of its elements) and if a finite additive function $f(G)$ is defined on $\mathfrak{M}E$, then, as is well known, the total variation of f on E is also finite. On the other hand, the following simple example shows that for a general system $\mathfrak{M}E$, a finite and additive function $f(G)$ on $\mathfrak{M}E$ can have infinite total variation.

Let E be the semi-open interval $0 \leq x < 1$ and let $\mathfrak{M}E$ consist of all intervals $a \leq x < b$ with $0 \leq a < b \leq 1$. We consider a continuous function $f(x)$ having infinite variation at 0 (say, $f(x) = x\sin(1/x)$) and define the interval function $f[a,b)$ as the increment of $f(x)$ on the interval $[a,b)$:

$$f[a,b) = f(b) - f(a).$$

It is easy to see that $f[a,b)$ is additive on $\mathfrak{M}E$ (the only doubt that can arise here is related to partitions of the interval $[0,b)$; for this, notice that any partition of this interval must contain an element of the form $[0,a]$, and since all the other elements are separated from the dangerous point 0, the function remains additive in this case as well). Since it is obvious that

$$(\mathfrak{M}E)\int_E |f(dE)| = +\infty,$$

we obtain the desired result.

32. *A necessary and sufficient condition for each of the additive functions* $f_1(G), \ldots, f_n(G)$ *to have bounded variation is that the integral*

$$I = \int_E \sqrt{f_1^2(dE) + f_2^2(dE) + \ldots + f_n^2(dE)}$$

be finite.

For the proof it suffices to observe that, on the one hand

$$|f_k| \leq \sqrt{f_1^2 + \ldots + f_n^2}$$

implies that

$$\int_E |f_k(dE)| \leq I,$$

and on the other hand,

$$I = \sup[(R\sqrt{f_1^2 + \ldots + f_n^2})(DE)] \leq \sum_k \sup[(R|f_k|)(DE)] =$$
$$= \sum_k \int_E |f_k|(dE).$$

33. Let $f(G)$ be an additive function of bounded variation. Setting

$$\phi(G) = \tfrac{1}{2}\Big[\int_G |f(dG)| + f(G)\Big], \quad \psi(G) = \tfrac{1}{2}\Big[\int_G |f(dG)| - f(G)\Big]$$

we obtain the ordinary representation of $f(G)$ as a difference of two non-negative functions:

$$f(G) = \phi(G) - \psi(G).$$

34. If two functions $f'(G)$ and $f''(G)$ are differentially equivalent, then it follows immediately that their absolute values $|f'(G)|$ and $|f''(G)|$ are also differentially equivalent. Further, this implies that

$$\int_E |f(dE)| = \int_E |F(dE)|$$

provided that $f(G)$ has a finite integral $F(G)$ on the sets $G \in \mathfrak{M}E$, which is differentially equivalent to $f(G)$ (by no. 20). Thus, the total variation can be defined for each function having a finite integral.

§6. Some results on homogeneous functions

35. I. *Assume that $f_1(G), \ldots, f_n(G)$ are finite additive functions of bounded variation on $\mathfrak{M}E$ and $\phi(x_1, \ldots, x_n)$ is a positively homogeneous function of degree one:*

$$\phi(tx_1, tx_2, \ldots, tx_n) = t\phi(x_1, x_2, \ldots, x_n), \quad t \geq 0, \tag{1}$$

with bounded second derivatives in any direction:

$$\left|\frac{\partial^2 \phi}{\partial s^2}\right| \leq \frac{K}{\sqrt{x_1^2 + x_2^2 + \ldots + x_n^2}}. \tag{2}$$

Then the function

$$\phi(G) = \phi[f_1(G), f_2(G), \ldots, f_n(G)]$$

has a finite integral on E.

Proof. We set

$$\rho = \sqrt{x_1^2 + x_2^2 + \ldots + x_n^2}, \quad \psi = \phi + 2K\rho.$$

We claim that the cone V in the $(n+1)$-dimensional space defined by the relation

$$z = \psi(x_1, x_2, \ldots, x_n)$$

is convex downward. Indeed, a sufficient condition for the cone to be convex is that

$$\partial^2\psi/\partial s^2 \geq 0$$

for any direction orthogonal to the radius-vector joining 0 to the given point of the cone. This follows from the fact that for these directions we have

$$\partial^2\rho/\partial s^2 = 1/\rho,$$

which, by (2), implies that

$$\partial^2\psi/\partial s^2 = \partial^2\phi/\partial s^2 + 2K\partial^2\rho/\partial s^2 > 0.$$

It follows from the convexity of V that the mid-point of the segment with end points $z', x_1', \ldots, x_n'$ and $z'', x_1'', \ldots, x_n''$ lying on V is located above the cone, that is,

$$\frac{\psi(x') + \psi(x'')}{2} > \psi\left(\frac{x' + x''}{2}\right),$$

whence, by (1),

$$\psi(x') + \psi(x'') > \psi(x' + x''). \tag{3}$$

Simple induction with subsequent passage to the limit shows that

$$\sum_m \psi(x^{(m)}) > \psi(x) \tag{4}$$

when the series

$$\sum_m x_k^{(m)} = x_k$$

is absolutely convergent for all k.

It follows directly from (4) that $\psi[f_1(G), \ldots, f_n(G)]$ is lower semi-additive and hence integrable on E. Since $\rho[f_1(G), \ldots, f_n(G)]$ is also integrable, this proves the existence of the integral

$$\int_E \phi(dE) = \int_E \psi(dE) - 2K \int_E \rho(dE).$$

36. II. *Let $f_1(G), \ldots, f_n(G)$ be additive functions of bounded variation on $\mathfrak{M}E$ and let $\phi(x_1, \ldots, x_n)$ be a continuous, positively homogeneous function of degree one. Then the function*

$$\phi(G) = \phi[f_1(G), f_2(G), \ldots, f_n(G)]$$

has a finite integral on E.

Proof. We represent ϕ as the limit of a sequence of functions:

$$\phi_m(x_1, x_2, \ldots, x_n) \to \phi, \; m \to \infty,$$

satisfying the conditions in no. 35 and such that for any $\eta > 0$ and sufficiently large m we have

$$|\phi_m - \phi| \leq \eta\delta.$$

We further put

$$K = \int_E \sqrt{f_1^2(dE) + \ldots + f_n^2(dE)} = \sup\left[\left(R\sqrt{f_1^2 + \ldots + f_n^2}\right)(DE)\right].$$

Then for large m we have

$$(R|\phi_m - \phi|)(DE) \leq \eta K.$$

By Property VII (no. 14), the existence of the integrals for ϕ_m implies the existence of the integral

$$\int_E \phi(dE)$$

on the set E.

37. III. *The result of no. 35 also remains valid under the assumption that the functions $f_1, \ldots, f_n$ have finite integrals on E but are not necessarily additive.*

Indeed, by no. 20, the functions $f_k(G)$ are differentially equivalent to their indefinite integrals $F_k(G)$. It now follows from no. 17 that the functions $\phi(G) = \phi[f_1(G), \ldots, f_n(G)]$ and $\Phi(G) = \Phi[F_1(G), \ldots, F_n(G)]$ are also differentially equivalent. Since the second of these functions is integrable by no. 35, so is the first one.

§7. Integrals of the type $\int_E f(x)\phi(dE)$

38. **Definition.** Given a function $f(x)$ (possibly many-valued), we denote by $f(E)$ the set function whose values on each set E coincide with the collection of all values that $f(x)$ assumes when x runs over E.

If $f(x)$ is defined for all $x \in E$ and $\phi(G)$ is differentially defined on $\mathfrak{M}E$, then we put

$$(\mathfrak{M}) \int_E f(x)\phi(dE) = (\mathfrak{M}) \int_E f(dE)\phi(dE). \tag{1}$$

Our primary interest lies in the case when $f(x)$ is single-valued and $\phi(G)$ is finite and additive on $\mathfrak{M}E$. In this case we say that (1) is an *integral of Stieltjes type*. However, certain results will be proved for the general case as well.

39. The relation

$$\int_E f(x)[\phi_1(dE) + \phi_2(dE)] = \int_E f(x)\phi_1(dE) + \int_E f(x)\phi_2(dE) \tag{2}$$

follows directly from Property III (no. 14), and the existence of the right-hand side of (2) implies the existence of the left-hand side. The following formula requires a slightly more complicated proof:

$$\int_E [f_1(x) + f_2(x)]\phi(dE) = \int_E f_1(x)\phi(dE) + \int_E f_2(x)\phi(dE). \tag{3}$$

It follows from Property III that

$$\int_E [f_1(dE) + f_2(dE)]\phi(dE) = \int_E f_1(dE)\phi(dE) + \int_E f_2(dE)\phi(dE), \tag{4}$$

and, according to our definition, the right-hand side of (4) is equal to the right-hand side of (3) while the left-hand side of (3) coincides with

$$\int_E (f_1 + f_2)(dE)\phi(dE),$$

where $(f_1 + f_2)(G)$ is the collection of values that $f_1(x) + f_2(x)$ assumes when x runs over the entire set G.

If $f(G)$, regarded as a set according to the definition in no. 38, is contained in $f'(G)$, then we write (using standard set-theoretic notation) $f(G) \subset f'(G)$. We use a similar notation for functions of partitions. It is evident that $f(DE) \subset f'(DE)$ implies the relation

$$l[f(DE)] = l[f'(DE)], \tag{5}$$

where, again, the existence of the right-hand side implies the existence of the left-hand side.

The inclusion $f(G) \subset f'(G)$ on $\mathfrak{M}E$ implies the analogous inclusion

$$(Rf)(DE) \subset (Rf')(DE)$$

for Riemann sums, whence, in view of (5), we conclude that

$$(\mathfrak{M}) \int_E f(dE) = (\mathfrak{M}) \int_E f'(dE).$$

Further, since

$$(f_1 + f_2)(G) \subset f_1(G) + f_2(G),$$

$$(f_1 + f_2)(G)\phi(G) \subset [f_1(G) + f_2(G)]\phi(G),$$

it follows from (4) that

$$\int_E [f_1(x) + f_2(x)]\phi(dE) = \int_E (f_1 + f_2)(dE)\phi(dE) =$$

$$= \int_E [f_1(dE) + f_2(dE)]\phi(dE) = \int_E f_1(x)\phi(dE) + \int_E f_2(x)\phi(dE),$$

that is, we obtain (3), where again the existence of the right-hand side implies the existence of the left-hand side.

40. Assume now that $\phi(G)$ is non-negative and additive on $\mathfrak{M}E$. Consider a partition DG of a set $G \in \mathfrak{M}E$. It is easy to show that for each function $f(G)$ the following relations hold:

$$\sup\{[R(f\phi)(DG)]\} = \sup\{\sum_n f(G_n)\phi(G_n)\} \leq$$

$$\leq \sup\{f(G) \sum_n \phi(G_n)\} \leq \sup[f(G)\phi(G)],$$

and a similar relation holds for infima:

$$\inf\{[R((f\phi)(DG)]\} \geq \inf[f(G)\phi(G)].$$

Accordingly, for any extension $D'E$ of DE we have

$$\sup\{[R(f\phi)(D'E)\} \leq \sup\{[R(f\phi)(DE)]\} \tag{6}$$

and

$$\inf\{[R(f\phi)(D'E)\} \geq \inf\{[R(f\phi)(DE)]\}. \tag{7}$$

Thus, the Riemann sums of $f\phi$ for an extension of DE are contained within narrower bounds than the sums for the partition DE itself. Further, from the definition of the integral as the limit of Riemann sums we conclude that for any partition we have

$$\inf\{[R(f\phi)(DE)]\} \leq \int_E f(x)\phi(dE) \leq \sup\{[R(f\phi)(DE)]\} \tag{8}$$

Based on this result, in the case of a non-negative additive function $\phi(G)$ on $\mathfrak{M}E$ we can define the integral

$$\int_E f(x)\phi(dE)$$

as the number I having the property that for any $\epsilon > 0$ there is a partition DE such that

$$|[R(f\phi)(DE)] - I| < \epsilon. \tag{9}$$

Indeed, if there is a partition with the required properties, then any extension of the partition possesses these properties (by virtue of (6), (7)). This leads to the original definition of the integral. The converse passage is trivial since the present condition is a *weakened* form of the original one.

41. If $\phi(G)$ is an additive function of bounded variation, then the difference representation (no. 33)

$$\phi(G) = \psi(G) - \chi(G)$$

leads, by (2), to an analogous representation for integrals:

$$\int_E f(x)\phi(dE) = \int_E f(x)\psi(dE) - \int_E f(x)\chi(dE). \tag{10}$$

This formula can also serve as a definition of the integral of $f\phi$ if the integrals on the right-hand side are defined as in no. 40. As a special case of the definition, we obtain the definition of the Stieltjes integral given by Fréchet and some other authors.

It should also be noted that, as follows from (6)–(8), the application of the notion of limit in the sense of Moore can be avoided in the case of a non-negative function $\phi(G)$, since the very existence of a partition with small oscillation of the corresponding Riemann sums guarantees a good approximation to the integral. However, the definition of integral by means of (10) is, generally, a roundabout way, and in this case the Moore concept of limit agrees better with the essence of the problem.

42. A function $f(x)$ is said to be *measurable* on $\mathfrak{M}$ if for any two real numbers a and b the set $E(a \leq f(x) < b)$ of all x for which $f(x)$ satisfies the inequality inside the parentheses belongs to $\mathfrak{M}$. It can easily be proved that the integral (1) exists if $f(x)$ is bounded and measurable on $\mathfrak{M}$ and $\phi(G)$ has bounded variation on the same system of sets.

43. If $f(x)$ is smaller in absolute value than a constant K and the functions $\phi(E)$ and $\psi(E)$ are differentially equivalent, then so are the functions $f(E)\phi(E)$ and $f(E)\psi(E)$. Further, since the integral is differentially equivalent to the integrand function, we conclude that

$$\int_E f(x)\phi(dE) = \int_E f(x)\Big[\int_E \phi(dE)\Big]$$

if $f(x)$ is bounded and $\phi(G)$ has a finite integral on E.

Thus, the existence of a finite integral of bounded variation for $\phi(E)$ implies that $f(x)\phi(E)$ is integrable if f is bounded and measurable. In particular, the integral

$$\int_E f(x)\phi[f_1(dE), f_2(dE), \ldots, f_n(dE)]$$

exists, provided that the conditions of at least one of the propositions in §6 are fulfilled for $f_1, f_2, \ldots, f_n$. As another particular case of our definition we obtain the integral in the sense of Hellinger [6]:

$$\int_a^b u(x)\frac{df(x)df_1(x)}{dg(x)}.$$

§8. Examples of $\mathfrak{M}$-systems

44. Let the system $\mathfrak{M}_1$ consist of all open intervals (infinite intervals can also be included) and all one-point sets on the real line. Here DE are partitions of E into at most *countably many* intervals and individual points. In many cases (in particular, when the function subject to integration is equal to zero on one-point sets) only intervals are essential.

45. Integrals of the form

$$(\mathfrak{M}_1)\int_\Delta f(x)l(d\Delta),$$

where $l(\Delta)$ is the length of the interval Δ, are slight generalizations of the Cauchy-Riemann integral: functions that are integrable in this sense are those and only those that are integrable by the method of Dirichlet.

46. If $F(x)$ has only jump discontinuities, then $(VF)(E)$ is defined as the difference $F(b-0)-F(a+0)$ when E is the interval (a,b); when E consists of a single point a, we put $(VF)(E)=F(a+0)-F(a-0)$. By definition, we write

$$(\mathfrak{M}_1)\int_\Delta f(x)dF(x)=(\mathfrak{M}_1)\int_\Delta f(x)(VF)(d\Delta). \tag{1}$$

This concept is close to the notion of the Stieltjes integral, and differs from the latter in that not only finite but also countable partitions are admissible here. Besides, the concept of limit in the sense of Moore has been used here. The second distinction is important for the following result, which can be easily proved:

If $f(x)$ and $F(x)$ are of bounded variation on Δ, then the integral (1) *exists.*

If both functions $f(x)$ and $F(x)$ are discontinuous at the same point, then the ordinary definition by Stieltjes is insufficient. Such integrals are very important, in particular, in some problems of probability theory.

47. Let $\mathfrak{M}_2$ consist of all Lebesgue measurable sets on the real line. The integral

$$(\mathfrak{M}_2)\int_E f(x)m_1(dE),$$

where m_1 is the linear Lebesgue measure, coincides with the Lebesgue integral. As was mentioned in no. 40, here the concept of limit in the sense of Moore does not bring about any new properties.

48. Let $\mathfrak{M}_3$ consist of all Borel measurable sets on the real line. If $F(x)$ is of bounded variation, then, as is known, $(VF)(E)$ can be defined for each $E \subset \mathfrak{M}_3$. By analogy with no. 44, we define

$$(\mathfrak{M}_3) \int_E f(x)dF(x) = (\mathfrak{M}_3) \int_E f(x)(VF)(dF).$$

For Borel measurable functions this definition of the Stieltjes integral coincides with the one given by Lebesgue (see [1]).

49. We say that an $\mathfrak{M}$-system is a Z-system (a "decomposable system") if for any two sets $E_1 \subset E$ it is possible to find a partition DE such that E_1 is the first element of the partition and all the other elements belong to $\mathfrak{M}$.

Examples of Z-systems are the systems $\mathfrak{M}E$ (no. 7) and $\mathfrak{M}DE$ (no. 13). Thus, the definition of integral depends only on the values of the integrand function in some Z-system. Accordingly, we are first of all interested in those cases where integration is performed with respect to systems $\mathfrak{M}$ that are themselves Z-systems. The above systems $\mathfrak{M}_1, \mathfrak{M}_2, \mathfrak{M}_3$ satisfy this condition. If instead of $\mathfrak{M}_1$ we take the system $\mathfrak{M}'$ consisting only of intervals (which is not a Z-system) as the basis for integration theory, then such a theory, which admits of a formally consistent development, is in fact meaningless since an interval cannot be partitioned into smaller disjoint intervals in such a way that the integral of any interval function coincides with that function.

By definition, for a Z-system the system ZE coincides with the collection of all subsets of E that belong to the given Z-system.

50. If $\mathfrak{M}'$ is contained in another system $\mathfrak{M}$, then the existence of the integrals

$$(\mathfrak{M}') \int_E f(dE), \quad (\mathfrak{M}) \int_E f(dE)$$

does not imply that they coincide. This can be shown by means of simple examples. Let $\mathfrak{M}'$ consist of a single set $E \neq \varnothing$ and let $\mathfrak{M}$ consist of sets E, E_1 and E_2 such that $E = E_1 + E_2$, $E_1E_2 = \varnothing$. Let $f(E) = 0$, $f(E_1) = f(E_2) = 1$. Then

$$(\mathfrak{M}') \int_E f(dE) = 0, \quad (\mathfrak{M}) \int_E f(dE) = 2.$$

However, the formula

$$(\mathfrak{M}') \int_E f(dE) = (\mathfrak{M}) \int_E f(dE)$$

can be proved in the following special case.

If $\phi(E)$ is non-negative and additive on $\mathfrak{M}$ then

$$(\mathfrak{M}') \int_E f(x)\phi(dE) = (\mathfrak{M}) \int_E f(x)\phi(dE) \tag{1}$$

for any subsystem $\mathfrak{M}'$ of $\mathfrak{M}$ provided that both integrals exist.

By no. 40, for any $\epsilon > 0$ there exists a partition DE in $\mathfrak{M}'$ satisfying the inequality (9) in no. 40, where I denotes the integral on the left-hand side of (1). Since DE can also be regarded as a partition in $\mathfrak{M}$, it follows from formula (8) of the same no. 40 that the integral on the right-hand side of (1) cannot differ from I by more than ϵ. Since ϵ is arbitrary, the desired result follows.

51. According to formula (10) in no. 41, (1) holds for Stieltjes-type integrals if these can be defined by means of (10), that is, if $\phi(G)$ has bounded variation on $\mathfrak{M}E$.

But, generally, the above discussed question remains open for Stieltjes integrals.

Chapter 3. The Second Integration Theory (Finite Partitions)

§1. Introductory remarks

1. The integration theory developed in Chapter 2 is based on a systematic study of partitions DE of a set E into a finite or countable number of subsets. In this chapter we study the changes the theory undergoes if only finite partitions are allowed (that is, partitions of E into finitely many sets).

One might think that the definition of the integral that arises in this way is narrower than our previous definition. But this is not so since there are important cases in which the integral defined by means of finite partitions exists, whereas no integrability follows from definitions based on infinite partitions. Moreover, the point of view to which we now adhere directly follows the idea of the classical Cauchy-Riemann theory and makes it possible to elaborate a complete and visual construction.

In the case of finite partitions we retain the same notation as that used for similar purposes for infinite partitions. When objects of both theories appear simultaneously, then those related to infinite partitions (that is, to the theory of Chapter 2) are marked by an asterisk.

2. All definitions in §1 of Chapter 2 remain valid under the present assumptions as well, and the notion of a sum is naturally introduced only for finite partitions. It is clear that, in general, $\mathfrak{M}E$ does not here coincide with the corresponding system $\mathfrak{M}E^*$ for the same $\mathfrak{M}$ and E may turn out to be a narrower system. For finite partitions the limit $l[f(DE)]$ may differ from the analogous expression $l[f(D^*E)]$ even when the initial system $\mathfrak{M}$ is the same under the passage to the limit in both cases.

3. Riemann sums (Chapter 2, no. 11) now exist for any partition provided that the function subject to integration is finite, whereas the absence of absolute convergence of the corresponding series in Chapter 2 inevitably leads to difficulties.

As to the definition of the integral itself, its formulation naturally remains unchanged.

Properties I–VI, with the exception of Property II, also remain valid in the case of finite partitions. Property II is replaced by *finite additivity*, which means that (1) (see Chapter 2, no. 14) holds for finite partitions; and from the existence of the integrals on the right-hand side of the formula it follows that the integral on the left-hand side also exists.

4. The definition of differential equivalence and the related results also do not change under the natural stipulation that all partitions are finite, and additivity of the corresponding functions is to be understood as finite additivity.

5. Infinite values for the integral are introduced as in §4, and here there is no need to extend the concept of absolute convergence. Instead, we agree that the sum of any finite number of infinities of the same sign is equal to the infinity of the same sign, and that this result does not change when several finite terms are added.

§2. Decomposable systems

6. By analogy with no. 49 in Chapter 2, we say that an $\mathfrak{M}$-system is a Z-system if for any set $E \in \mathfrak{M}$ and for any subset E_1 of it belonging to $\mathfrak{M}$ there is a partition DE whose first element is E_1 and the other elements (here their number is only finite) belong to $\mathfrak{M}$. It is clear that a Z-system thus defined must be a Z^*-system. The latter will be considered below under the name of extended Z-system.

7. *Let $E_1, E_2, \ldots, E_n$ be subsets of Z such that*

$$\sum_{m=1}^{n} E_m \subset E, \quad E_i E_j = \varnothing \text{if } i \neq j.$$

Then there is a partition $DE = \sum_{m=1}^{n+p} E_m$ whose first n elements coincide with the given E_m.

By definition, there are partitions in Z,

$$D_m E = E_m + \sum_{s=1}^{t_m} E_s^{(m)} \quad (m = 1, 2, \ldots, n),$$

that contain E_m as the first element. Then

$$DE = \sum_{m=1}^{n} E_m + \sum_{s_1, s_2, \ldots, s_n} E_{s_1}^{(1)} E_{s_2}^{(2)} \ldots E_{s_n}^{(n)}$$

is the desired partition.

8. As in the case of a Z^*-system, ZE is the collection of all subsets of E belonging to Z. It follows that $ZE = Z^*E$, which does not necessarily hold for a general $\mathfrak{M}$-system. As an example, we consider an $\mathfrak{M}^*$-system consisting of E and countably many elements of the infinite partition DE. Then $\mathfrak{M}E$ consists of the single set E and $\mathfrak{M}^*E$ consists of E and all E_n.

9. Further, let KZ be the minimal field of sets containing the system of sets Z. We are about to prove that KZ *is the collection of sets that admit finite partitions into elements of* Z. To this end, we denote by σZ the system of all such sets. This system is obviously contained in KZ, and it remains to show that σZ is itself a field. Let

$$E = \sum_{m=1}^{n} E_m, \quad E' = \sum_{q=1}^{p} E'_q$$

be two sets belonging to σZ and represented as sums of pairwise disjoint elements of Z. Since each of the sets $E_m E'_q$ belongs to Z, it follows from no. 7 that there exist partitions of the form

$$D_m E_m \equiv \sum_q E_m E'_q + \sum_{s=1}^{t_m} E_s^{(m)}.$$

It is clear that the sets

$$EE' = \sum_m \sum_q E_m E'_q,$$

$$E + E' = \sum_q E'_q + \sum_m \sum_s E_s^{(m)},$$

$$E - E' = \sum_m \sum_s E_s^{(m)}$$

belong to σZ, since the right-hand sides of these relations are partitions into pairwise disjoint elements of Z. Thus, σZ is a field and, consequently, coincides with KZ.

10. *An additive function $f(E)$ on Z has a unique additive extension to KZ.*

It is clear that (under the assumption that such an extension is possible)

$$f(E) = (Rf)(DE) \tag{1}$$

for any partition of a set E belonging to KZ into elements of Z. Since f is additive on Z, the following relation holds for any partitions DE and $D'E$:

$$(Rf)(DE) = (Rf)\{[DD']E\} = (Rf)(D'E),$$

which shows that the definition of $f(E)$ given by (1) for an arbitrary element E of KZ does not depend on the choice of the partition DE. The function $f(E)$ thus extended to the field of sets KZ is additive.

11. *For $E \in KZ$ we set*

$$(Z)\int_E f(dE) = \sum_{m=1}^{n} (Z)\int_{E_m} f(dE_m),$$

where $DE \equiv \sum E_m$ is a partition of E in Z.

According to what has been proved, this definition does not depend on the choice of the partition. The integral thus defined possesses Properties I–VI.

12. We give the following examples of Z-systems: the systems $\mathfrak{M}_1$ consisting of all open intervals and one-point sets on the real line; it has already been considered in no. 44 in Chapter 2; the system $\mathfrak{M}_4$ of all "half-open" intervals $[a, b)$ (we used this system in no. 31 in Chapter 2); the system $\mathfrak{M}_5$ of all intervals of the form (a, b), $[a, b)$, $(a, b]$, and $[a, b]$, with one-point sets considered as closed intervals $[a, a]$.

It is also obvious that each field (in particular, the systems $\mathfrak{M}_2$ and $\mathfrak{M}_3$ in nos. 47 and 48, Chapter 2) is a Z-system. Z-systems are a natural basis for constructing an integration theory.

§3. On the interrelation between the two methods of integration

13. We will consider only the integrals

$$(Z)\int_E f(dE), \quad (Z)\int_E^* f(dE) \tag{1}$$

with respect to Z-systems. As has been mentioned in no. 8, ZE and ZE^* coincide in this case. Generally, $\mathfrak{M}E$ and $\mathfrak{M}E^*$ can be different, and this fact alone already causes difficulties in constructing a more or less simple theory.

14. We first give an example showing that the integrals (1) can in fact be different. To this end, we take the system $\mathfrak{M}_1$ of open intervals and one-point sets, which is a Z-system (by no. 12), and define $f(E)$ by putting $f(E) = 1$ for any interval of the form $E = (0, a)$ and $f(E) = 0$ for the other elements of $\mathfrak{M}_1$. It is clear that $f(E)$ is additive (that is, finitely additive) and that it is only upper semi-additive in the sense of infinite partitions. For the unit interval $\Delta = (0, 1)$ we obviously have

$$(\mathfrak{M}_1)\int_\Delta f(d\Delta) = 1, \quad (\mathfrak{M}_1)\int_\Delta^* f(d\Delta) = 0$$

so that the two definitions of the integral yield different values.

15. Following Fréchet, we say that a set function defined on Z is *continuous on Z* if for any decreasing sequence of sets E_n with empty intersection the sequence $f(E_n)$ converges to zero for $n \to \infty$. It is easy to see that the function constructed in no. 14 is not continuous.

16. *If $f(E)$ is continuous on a field K, then*

$$(K)\int_E f(dE) = (K)\int_E^* f(dE),$$

and the existence of the integral on the left-hand side implies the existence of the integral on the right-hand side.

Assume that the left-hand integral I exists and is finite. Then for any $\epsilon > 0$ there exists a finite partition $DE \equiv \sum_{m=1}^n E_m$ such that for any extension $D'E > DE$ of it we have

$$|(Rf)(D'E) - I| < \epsilon. \tag{2}$$

We now consider an arbitrary infinite extension

$$D^*E = \sum_{k=1}^{\infty} E_k^*$$

of DE and show that the series

$$\sum_{k=1}^{\infty} f(E_k^*) = (Rf)(D^*E) \tag{3}$$

is absolutely convergent. The sequence of sets

$$G_{mk} = E_m \sum_{i=k}^{\infty} E_i^* = E_m - \sum_{i=1}^{k-1} E_i^* \quad (k = 1, 2, \ldots)$$

(whose elements obviously belong to K) satisfies the continuity condition in no. 15, so that $\lim f(G_{mk}) = 0$ We can therefore take k_0 so large that for all $m \leq n$ we have

$$f(G_{mk}) < \epsilon/n, \quad k > k_0. \tag{4}$$

The finite partition

$$D_k' E \equiv \sum_{i=1}^{k-1} E_i^* + \sum_{m=1}^{n} G_{mk}$$

is an extension of DE and, consequently, satisfies (2). It follows from (2) and (4) that for sufficiently large k we have

$$\left|\sum_{i=1}^{k} f(E_i^*) - I\right| < 2\epsilon. \tag{5}$$

This implies that the series (3) must be absolutely convergent, since otherwise it would be possible to renumber E_k^* in such a way that (5) would be violated for an arbitrarily large k.

This proves that the Riemann sum $(Rf)(D^*E)$ exists and, by (5), satisfies the inequality

$$|(Rf)(D^*E) - I| \leq 2\epsilon. \tag{6}$$

Since a stronger inequality (2) holds for finite partitions, any extension of DE satisfies (6) (since ϵ is arbitrary), which yields the desired formula

$$(K) \int_E^{**} f(dE) = I.$$

The case of an integral with infinite values can be treated in a similar manner.

17. The above proposition is no longer valid when we deal with integrals in a Z-system that is not a field of sets. To demonstrate this we consider the following example. We introduce a system $\mathfrak{M}_6$ whose elements are sets of all irrational points lying inside squares with rational vertices (for simplicity, we call each such set also a square). By a rational (irrational) point we mean any point in the plane both coordinates of which are rational (irrational). We put $f(E)$ equal to the length of the side of the square for the squares E adjoining the y-axis on the right. For all other squares we put $f(E) = 0$. It can be easily seen that for the unit square E_1 we have

$$(\mathfrak{M}_6) \int_{E_1} f(dE_1) = 1, \quad (\mathfrak{M}_6) \int_{E_1}^{*} f(dE_1) = 0,$$

although $f(E)$ is continuous on $\mathfrak{M}_6$ and $\mathfrak{M}_6$ is a Z-system.

18. Here is an example of a function that is continuous on a Z-system, is integrable in the sense of finite partitions, and is not integrable in the sense of infinite partitions. We consider the variation $(VF)(E)$ of a continuous (in the sense of no. 46, Chapter 2) point function $F(x)$ on the system $\mathfrak{M}_1$. As is known, (VF) is continuous on $\mathfrak{M}_1$ in the sense of no. 15 and $\mathfrak{M}_1$ is a Z-system. Nevertheless,

$$(\mathfrak{M}_1) \int_{E} (VF)(dE) = (VF)(E),$$

since (VF) is additive, whereas

$$(\mathfrak{M}_1) \int_{E}^{*} (VE)(dE)$$

exists only if $F(x)$ has a bounded variation on E.

19. A function $f(E)$ on Z is said to be *completely continuous* if for any sequence

$$E_1 \supset E_2 \supset \ldots \supset E_n \supset \ldots$$

of elements of KZ with empty intersection and for any sequence of partitions

$$D_1E_1, \ D_2E_2, \ \ldots, \ D_nE_n, \ldots$$

of the sets E_n into elements of Z the sums $(Rf)(D_n E_n)$ become infinitely small as n increases (the partitions are of course assumed to be finite).

It is obvious that a completely continuous function is continuous in the sense of Fréchet (see no. 15).[7]

20. *If $f(E)$ is completely continuous on Z, then*

$$(Z)\int_E f(dE) = (Z)\int_E^* f(dE),$$

and the existence of one of the integrals implies the existence of the other.

Assume now that the first of the integrals exists and is equal to a finite number I. Then for any $\epsilon > 0$ we can find a partition $DE = \sum_{m=1}^n E_m$ such that for any extension $D'E > DE$ of it we have

$$|(Rf)(D'E) - I| < \epsilon. \tag{2}$$

Let $D^*E = \sum_{k=1}^n E_k^*$ be an infinite extension of DE. The elements of the sequence of sets

$$G_k = \sum_{i=k}^{\infty} E_i^*$$

belong to KZ. Since f is completely continuous, for any partitions $D_n E_n$ we have

$$(Rf)(D_n G_n) \to 0.$$

Let k be so large that for any DG_k we have

$$|(Rf)(DG_k)| < \epsilon. \tag{3}$$

By no. 7, there is a partition of E (into sets belonging to Z) having the form

$$D_k E = \sum_{i=1}^{k-1} E_i^* + \sum_{m=1}^{n} \sum_{q=1}^{p_k} E_m E_{kq},$$

where, clearly, $\sum_q E_{kq} = G_k$. The partition

$$D'_k E = \sum_{i=1}^{k-1} E_i^* + \sum_{m=1}^{n} \sum_{q=1}^{p_k} E_m E_{kq}$$

[7] The two concepts coincide for additive functions; that is why complete continuity was not needed in the previous theories. But in the general case it is complete continuity that serves as a natural generalization to non-additive functions of the concept of continuity for additive functions.

is an extension of DE; hence it satisfies (2). On the other hand, by (3),

$$\left|\sum_{m=1}^{n}\sum_{q=1}^{p_k} f(E_m E_{kq})\right| = |(Rf)(D'_k G_k)| < \epsilon, \tag{4}$$

which together with (2) implies

$$\left|\sum_{i=1}^{k-1} f(E_i^*) - I\right| < 2\epsilon. \tag{5}$$

As in no. 16, we conclude that the series

$$\sum_{k=1}^{\infty} f(E_k^*) = (Rf)(D^*E)$$

is absolutely convergent and

$$|(Rf)(D^*E) - I| \leq 2\epsilon.$$

This inequality is valid for an arbitrary infinite extension of DE, and for a finite extension even (2) holds; hence, as in no. 16, we obtain

$$I^* = (Z)\int_E^* f(dE) = I. \tag{6}$$

When I is infinite, the proof is completely similar.

Assume now that I^* exists and is finite. Then for any $\epsilon > 0$ there is a partition D^*E such that for any extension D_1^*E of it we have

$$|(Rf)(D_1^*E) - I^*| < \epsilon. \tag{7}$$

Our aim is to construct, proceeding from D^*E, a finite partition DE such that for any $D_1E > DE$,

$$|(Rf)(D_1E) - I^*| < 3\epsilon. \tag{8}$$

If D^*E is indeed an infinite partition, that is,

$$D^*E = \sum_{k=1}^{\infty} E_k^*,$$

then there exists k such that (we use our previous notation) (3) holds and, in addition,

$$\Big|\sum_{i=k}^{\infty} f(E_i^*)\Big| < \epsilon. \tag{9}$$

The latter property is also a consequence of the absolute convergence of the series $(Rf)(D^*E)$, which we assumed in (7).

Next we set

$$DE \equiv D_k E \equiv \sum_{i=1}^{k-1} E_i^* + \sum_{q=1}^{p_k} E_{kq}.$$

Let $D_1 E$ be a finite extension of DE. The partition

$$D_1^* E \equiv \sum_{i=1}^{k-1} D_1 E_i^* + \sum_{i=k}^{\infty} E_i^*$$

is an extension of D^*E; hence (7) holds. On the other hand,

$$|(Rf)(D_1E) - (Rf)(D_1^*E)| \le \Big|\sum_{q=1}^{p_k} f(E_{kq}\Big| + \Big|\sum_{i=k}^{\infty} f(E_i^*)\Big| < 2\epsilon$$

by (3) and (9), which together with (7) shows that (8) holds for an infinite D^*E. If D^*E is finite, we put $DE = D^*E$, and then (8) follows immediately from (7). Finally, since ϵ is arbitrarily small, we obtain

$$I = (Z)\int_E f(dE) = I^*.$$

In the case of an inifnite I^* a similar argument applies. This completes the proof.

§4. Examples

21. By analogy with §8 of Chapter 2, we now consider our new concept of integral for specific $\mathfrak{M}$-systems. First we consider the system $\mathfrak{M}_1$ (Chapter 2, no. 45) consisting of intervals and one-point sets on the real line.

As before, we denote the length of the interval Δ by $l(\Delta)$. Then the definition we obtain for

$$(\mathfrak{M}_1)\int_\Delta f(x)l(d\Delta)$$

coincides with that of the Cauchy integral

$$\int_\Delta f(x)dx.$$

22. Assume that we are given two continuous functions $x = \phi(t)$ and $y = \psi(t)$. We can consider them as a parametric representation of a continuous curve. Let $s(\Delta)$ denote the distance between the points of the curve corresponding to the end points of the interval Δ (the length of the chord). Then the length of the curve itself is defined by the integral

$$(\mathfrak{M}_1) \int_{\Delta} s(d\Delta).$$

23. Using the same notation as in no. 46, Chapter 2, we define in the same way

$$(\mathfrak{M}_1) \int_{\Delta} f(x)dF(x) = (\mathfrak{M}_1) \int_{\Delta} f(x)(VF)(d\Delta). \tag{1}$$

This definition coincides with the definition of the Stieltjes integral given by Smith. The integral (1) exists when $F(x)$ and $f(x)$ are of bounded variation (as in Chapter 2). But in contrast to Chapter 2, (1) exists even when $F(x)$ is only continuous while $f(x)$, as before, is of bounded variation. This is one of the important cases when our present definition is broader than that in Chapter 2.

24. The integral

$$(\mathfrak{M}_2) \int_{E} f(x)m_1(dE)$$

is similar to that studied in no. 47, Chapter 2. It exists only for bounded functions and coincides with the Lebesgue integral in this case.

Appendix 1. The integral as a many-valued function

1. Up to now we have systematically used many-valued functions under the limit (or integral) sign under the sole assumption that the value of the limit (or the integral) is uniquely determined. However, the argument can be made both simpler and more general if we drop the condition of single-valuedness in this latter case as well. A substantial advantage obtained here is that now every differentially defined function becomes integrable, and we can say that functions integrable in the previous sense are uniquely integrable. The supremum and infimum of a many-valued integral are now interpreted as upper and lower integrals. In this appendix we give a brief survey of such an integration theory. We use the same constructions as in Chapter 2, that is, infinite partitions. We leave it to the reader to consider the changes that are necessary in the case of finite partitions.

2. Assume that we are given a single-valued or many-valued function $f(DE)$ defined on partitions DE of E into elements of a system $\mathfrak{M}$ and that f can assume finite or infinite values. By a *limit of $f(DE)$ under indefinite extension of DE* we now mean any number

$$I = L[f(DE)]$$

having the property that for an arbitrary DE and any $\epsilon > 0$ there is a partition $D'E > DE$ such that

$$\inf |f(D'E) - I| < \epsilon.$$

Here all values that f assumes (for the given $D'E$) are taken into account when the infimum is calculated. We set $L[f(DE)] = +\infty$ if for each DE and any $H < +\infty$ there exists $D'E > DE$ such that $\sup f(D'E) > H$. The definition of $L[f(DE)] = -\infty$ is similar.

3. It is easy to verify that an arbitrary function $f(DE)$ defined on all partitions DE has at least one limit and that the set of all limits of $f(DE)$ is closed. Therefore the values of $\sup L[f(DE)]$ and $\inf L[f(DE)]$ are uniquely determined and themselves are values of $L[f(DE)]$.

If

$$\sup L[f(DE)] = \inf L[f(DE)]$$

(that is, if $L[f(DE)]$ is defined uniquely), then we have

$$L[f(DE)] = l[f(DE)],$$

which leads to the concept of limit given in no. 8, Chapter 2.

4. The definition of the sum of a series stated in no. 22, Chapter 2 must now be modified. If the series is not absolutely convergent, any value between $-\infty$ and $+\infty$ is allowed to be a value of the sum. After that, the symbol $(Rf)(DE)$ is defined word for word as in no. 11, Chapter 2, and the concept of integral is introduced by means of the formula

$$(\mathfrak{M}) \overline{\underline{\int_E}} f(DE) = L[(Rf)(DE)].$$

5. Further, we set

$$\overline{\int_E} f(dE) = \sup \overline{\underline{\int_E}} f(dE), \quad \underline{\int_E} f(dE) = \inf \overline{\underline{\int_E}} f(DE).$$

If $\overline{\int} = \underline{\int}$, then we simply write $\overline{\int} = \underline{\int} = \int$, which is the integral in the sense of Chapter 2.

6. As to the properties of the integral stated in no. 14 in Chapter 2, the following changes should be made.

Property I loses its meaning, because any differentially defined function is now integrable. If the integral is finite and uniquely determined, then it is defined uniquely on each element of $\mathfrak{M}E$.

Property II remains valid with the summation rule for many-valued functions given in Chapter 2 (no. 11).

Property III no longer holds, whereas Properties IV, V, VI, and VII remain valid.

7. We will not present a detailed study of the properties of the lower and upper integrals here, but merely note that the following relations hold for any partition DE:

$$\overline{\int_E} F(dE) = \sup \sum_n \overline{\int_{E_n}} f(dE_n), \quad \underline{\int_E} f(dE) = \inf \sum_n \underline{\int_{E_n}} f(dE_n).$$

If the corresponding series are absolutely convergent, then the supremum and infimum signs can be omitted. Otherwise the integrals on the left-hand sides are equal to plus and minus infinity, respectively. Thus, a *finite upper or lower integral is necessarily an additive set function.*

Appendix 2. On the justification of differentiation on abstract sets

1. Here we are going to show that the definition of derivative of an additive set function with respect to another set function can be introduced irrespective of the geometrical properties of their domains. Since we will not develop a general theory, we introduce, for clarity, some restrictive assumptions.

Assume that F is an additive system of sets, $\phi(E)$ is a finite non-negative and additive function defined on the system and $f(x)$ is a point function, measurable with respect to F (cf. Chapter 2, no. 42) *and bounded in absolute value by a constant, say* 1. *Then for any E belonging to F the integral*

$$\psi(E) = (F) \int_E f(x) d\phi(dE) \tag{1}$$

exists and, as can easily be seen,

$$|\psi(E)| \leq \phi(E). \tag{2}$$

A function $f(x)$ that is measurable with respect to F is called the derivative of ψ with respect to ϕ:

$$f(x) = \psi(dE)/\phi(dE),$$

if it satisfies (1) for all E belonging to F.

2. We first prove that the derivative is defined uniquely (to within sets E with $\phi(E) = 0$). In other words, if both $f_1(x)$ and $f_2(x)$ satisfy (1) and if G is the set where $f_1(x) \neq f_2(x)$, then $\phi(G) = 0$. Indeed, we have

$$\int_E (f_1 - f_2)(x)\phi(dE) = 0$$

for all E belonging to F, and according to no. 19 in Chapter 2, we also have

$$\int_E |f_1 - f_2|(x)\phi(dE) = 0.$$

It readily follows that for any $\epsilon > 0$ the set G_ϵ on which $|f_1 - f_2| > \epsilon$ satisfies the condition $\phi(G_\epsilon) = 0$. Therefore $\phi(G) = 0$, as required.

3. We now prove that

$$f(x) = \psi(dE)/\phi(dE)$$

exists, provided that $\psi(E)$ is additive and (2) holds. To this end we will describe a method for constructing f which can be used in the definition of derivative in more general situations.

The function $f(x)$ will be constructed for a specific set E, but the construction can easily be extended to all other elements of F. First we construct a function $f(DE, x)$ depending on the point x and the partition $DE = \sum_n E_n$. To do this, we put

$$f(DE, x) = \psi(E_n)/\phi(E_n), \quad x \in E_n,$$

and $f(DE, x) = 0$ if $\phi(E_n) = 0$. For any $\epsilon > 0$ there is a partition DE such that any extension $D'E$ of it satisfies the inequality

$$I = \int_E [f(DE, x) - f(D'E, x)]^2 \phi(dE) < \epsilon. \tag{3}$$

Indeed, let

$$D'E \equiv \sum_n D'E_n \equiv \sum_n \sum_m E_{nm}.$$

Then

$$\begin{aligned} I &= \sum_n \sum_m \Big(\frac{\psi(E_{nm})}{\phi(E_{nm})} - \frac{\psi(E_n)}{\phi(E_n)}\Big)^2 \phi(E_{nm}) = \\ &= \sum_n \Big(\sum_m \frac{\psi^2(E_{nm})}{\phi(E_{nm})} - 2\frac{\psi(E_n)}{\phi(E_n)} \sum_m \psi(E_{nm}) + \frac{\psi^2(E_n)}{\phi^2(E_n)} \sum_m \phi(E_{nm})\Big) = \quad (4) \\ &= \sum_n \Big(\sum_m \frac{\psi^2(E_{nm})}{\phi(E_{nm})} - \frac{\psi^2(E_n)}{\phi(E_n)}\Big) = \Big(R\frac{\psi^2}{\phi}\Big)(D'E) - \Big(R\frac{\psi^2}{\phi}\Big)(DE). \end{aligned}$$

The function $\psi^2(G)/\phi(G)$ is integrable on E since, by virtue of (2), it coincides with the function $\psi^2(G)/\max[\phi(G), \psi(G)]$, which satisfies the conditions of no. 36, Chapter 2. Therefore the last difference in (4) can be made smaller than ϵ by means of an appropriate choice of DE. This proves our assertion.

As in the ordinary theory of mean convergence, it is easy to show that (3) implies the existence of $f(x)$ such that

$$l \int_E [f(x) - f(DE, x)]^2 \phi(dE) = 0$$

(as above, the limit is understood in the sense of Moore's definition for functions of partitions DE). The function $f(x)$ satisfies all the requirements of the formal definition of derivative stated in no. 1.

6 February, 1930

References

1. H. Lebesgue, *Leçons sur l'intégration et la recherche des fonctions primitives*, 2nd ed. Paris, 1928.
2. M. Fréchet, 'Sur l'intégrale d'une fonctionelle étendue à un ensemble abstrait', *Bull. Soc. Math. France* **43** (1915), 248.
3. J.S. Burkill, 'The derivatives of functions of intervals', *Fund. Math.* **5** (1924), 321.
4. E.H. Moore and H.L. Smith, 'A general theory of limits', *Amer. J. Math.* **44** (1922), 102; cf. also S.O. Shatunovskii, *Introduction to Analysis*, Odessa 1923 (in Russian).

5. H.L. Smith, 'On the extension of the Stieltjes integral', *Trans. Amer. Math. Soc.* **27** (1925), 491–515.
6. E. Hellinger, 'Neue Begründung der Theorie quadratischer Formen von unendlichvielen Veränderlichen', *J. Reine Angew. Math.* **136** (1909), 234.

17. ON THE NOTION OF MEAN*

All known types of mean, such as the arithmetic, geometric and harmonic means and the root-mean-square are of the form

$$M(x_1, x_2, \dots, x_n) = \psi\Big(\frac{\phi(x_1) + \phi(x_2) + \dots + \phi(x_n)}{n}\Big), \tag{1}$$

where ϕ is a continuous strictly monotone function and ψ is its inverse function.

It will be shown in this paper that each type of mean is necessarily of the form (1) whenever it satisfies some natural conditions (axioms of the mean).

We assume that the function $M_n(x_1, x_2, \dots, x_n)$ or simply $M(x_1, \dots, x_n)$, is defined for any $n \geq 1$ and all values of $x_1, \dots, x_n$ in the interval $a \leq x \leq b$.

The result established below can also be extended to means defined on the infinite intervals $-\infty < x < \infty$, $a \leq x < \infty$, $-\infty < x \leq b$ (for example, the arithmetic and geometric means). In such a case the proof needs only a minor modification.

A sequence of functions M_n determines a *regular* type of mean if the following conditions hold:

i) $M(x_1, \dots, x_n)$ is continuous and monotone in each variable. For the sake of being specific, we will assume that M increases in each variable.

ii) $M(x_1, \dots, x_n)$ is a symmetric function.

iii) The mean of identical numbers is equal to their common value: $M(x, \dots, x) = x$.

iv) A subset of values can be replaced by their mean with no effect on the total mean:

$$M(x_1, \dots, x_m, y_1, \dots, y_n) = M_{n+m}(x, \dots, x, y_1, \dots, y_n),$$

where $x = M(x_1, \dots, x_n)$.

Theorem. *If conditions* i) *to* iv) *hold, then the mean* $M(x_1, \dots, x_n)$ *has the form* (1) *where* ϕ *is a continuous increasing function and* ψ *is its inverse function.*

Proof. Let $M(mx, ny)$ be the mean of m identical values of x and n identical values of y, that is,

$$M(mx, ny) = M(x_1, \dots, x_m, y_1, \dots, y_n), \tag{2}$$

---

* 'Sur la notion de la moyenne', *Atti Accad. Naz. Lincei. Rend.* **12**:9 (1930), 388–391.

$$x_1 = x_2 = \ldots = x_m = x, \quad y_1 = y_2 = \ldots = y_n = y.$$

Then by virtue of iv) and iii),

$$M(pmx, pny) = M(pM(mx, ny)) = M(mx, ny).$$

Hence, if $mn' = nm'$, then

$$M(mx, ny) = M(mn'x, nn'y) = M(nm'x, nn'y) = M(m'x, n'y). \tag{3}$$

We can now define the function

$$\psi(z) = M(pb, (q-p)a) \tag{4}$$

for each rational z such that $0 \leq z = p/q \leq 1$.

This function is well defined, since for two different representations of z: $z = p/q = p'/q'$ there is the obvious relation $p(q'-p') = (q-p)p'$ and, in view of (3), there is a unique value of $\psi(z)$. The function $\psi(z)$, as a function of rational z, is increasing. Indeed, let $z' > z$. We represent z' and z as fractions with the same denominator: $z' = p'/q$, $z = p/q$, $p' > p$ and in view of the monotonicity property i), we obtain the inequality

$$\psi(z') = M(p'b, (q-p')a) > M(pb, (q-p)a) = \psi(z).$$

It is thus seen that the inverse function $\phi(x)$ can be uniquely defined for all values of x such that z is rational, ϕ being an increasing continuous function.

It is now easy to calculate the mean $M(x_1, \ldots, x_n)$ for the values $x_i = \psi(z_i)$ where the z_i are rational. For this purpose we represent the z_i as fractions with the same denominator: $z_i = p_i/q$. Then $x_i = M(p_i b,\ (q-p_i)a)$ and

$$\begin{aligned} M(x_1, \ldots, x_n) &= M(qx_1, \ldots, qx_n) = M((p_1 + \ldots + p_n)b, (np - p_1 - p_2 - \ldots - p_n)a) \\ &= \psi\Big(\frac{p_1 + p_2 + \ldots + p_n}{nq}\Big) = \psi\Big(\frac{z_1 + \ldots + z_n}{n}\Big) = \\ &= \psi\Big(\frac{\phi(x_1) + \ldots + \phi(x_n)}{n}\Big). \end{aligned} \tag{5}$$

Formula (1) is thus proved for the special values of the variables which were introduced above.

We now prove that the function ψ is continuous for each value in the interval $0 \leq y \leq 1$, either rational or irrational. Assuming the contrary, we find a point y where ψ is discontinuous. If we assume that $y \neq 0$ and $y \neq 1$, then the limits $\psi(y-0) = u$ and $\psi(y+0) = v$ are different, and hence $u < v$. According to (5), we have

$$M(\psi(z_1), \psi(z_2)) = \psi\Big(\frac{z_1 + z_2}{2}\Big)$$

for two rational numbers z_1 and z_2.

If we now make z_1 and z_2 approach y from the left and from the right respectively, then clearly

$$\lim \psi\Big(\frac{z_1 + z_2}{2}\Big) = M(u, v) > u.$$

However, it is possible to arrange the points so that $(z_1 + z_2)/2$ always remains to the left of y. Then $\lim \psi(\frac{1}{2}(z_1 + z_2)) = u$. The contradiction shows that the assumption of discontinuity at the point y is false. The same conclusion can also be derived for the points $y = 1$ and $y = 0$.

It is thus seen that the values of $x = \psi(z)$ associated with rational z form a dense set in the interval between $a = \psi(0)$ and $b = \psi(1)$. Hence, we can define the functions $\psi(z)$ and $\phi(x)$ by continuity for all points of the intervals $0 \leq z \leq 1$ and $a \leq x \leq b$ respectively. Therefore, formula (5) holds by continuity for all values of $x, a \leq x \leq b$. This proves the theorem.

3 October 1930

18. ON THE COMPACTNESS OF SETS OF FUNCTIONS IN THE CASE OF CONVERGENCE IN MEAN*

Let G be a set of functions $f(x)$ defined at each point x of a bounded set F in the n-dimensional Euclidean space R^n and such that

$$\int_F |f(x)|^p \, d\sigma < \infty,$$

where $p > 1$.

A sequence $\{f_m(x)\}$ is said to converge in mean to $f(x)$ if

$$\lim_{m\to\infty} \int_F |f_m(x) - f(x)|^p \, d\sigma = 0. \tag{1}$$

The set G is said to be compact if each sequence $\{f_m(x)\}$ of functions belonging to G contains a subsequence $\{f_{m_k}(x)\}$ that converges in mean. In this paper we present a necessary and sufficient condition for G to be compact.

If we put

$$\rho(f,g) = \left\{\int_F |f(x) - g(x)|^p \, d\sigma\right\}^{1/p}, \tag{2}$$

then (1) can be written as

$$\lim \rho(f_m, f) = 0 \quad \text{as } m \to \infty.$$

The set of all functions whose pth power is absolutely integrable and endowed with the "distance" (2) is known to be a complete metric space.

In what follows, it will always be assumed that $f(x) = 0$ at each point of R^n lying outside F. We set

$$f_\epsilon(x) = \frac{1}{V(\epsilon)} \int_{S(x,\epsilon)} f(y) d\sigma,$$

where $S(x,\epsilon)$ denotes the ball of radius ϵ with centre at x and of volume $V(\epsilon)$. It is easily seen that we always have

$$\rho(f_\epsilon, g_\epsilon) \leq \rho(f,g). \tag{3}$$

We prove that

$$\lim \rho(f_\epsilon, f) = 0 \quad \text{as } \epsilon \to 0$$

* 'Über Kompaktheit der Funktionenmengen bei der Konvergenz im Mittel', *Nachr. Ges. Wiss. Göttingen* **9** (1931), 60–63.

for any function f. Indeed for any $\delta > 0$ there exists a function $u(x)$ that is uniformly continuous throughout the space R^n and possesses the property

$$\rho(u, f) \leq \delta.$$

For this uniformly continuous function we clearly have

$$\lim \rho(u_\epsilon, u) = 0 \quad \text{as } \epsilon \to 0,$$

and since (3) implies

$$\rho(u_\epsilon, f_\epsilon) \leq \rho(u, f) \leq \delta,$$

we thus obtain

$$\overline{\lim}\, \rho(f_\epsilon, f) \leq 2\delta \quad \text{as } \epsilon \to 0.$$

Since δ is an arbitrary positive number, the assertion is proved.

Theorem. *For G to be compact it is necessary and sufficient that the following two conditions hold.*

i) *There exists a constant K such that*

$$\int_F |f(x)|^p \, d\sigma \leq K$$

for all functions belonging to G.

ii) *For any $\delta > 0$ there exists $\epsilon > 0$ such that*[1]

$$\int_F |f_\epsilon(x) - f(x)|^p \, d\sigma \leq \delta$$

for all functions belonging to G.

Proof of the necessity. Assume that condition i) is violated. Then there exists a sequence $\{f_m\}$ of functions belonging to G such that the distance

$$\rho(f_m, 0) = \left\{ \int_F |f_m(x)|^p \, d\sigma \right\}^{1/p}$$

tends to $+\infty$, therefore the distance

$$\rho(f_m, f) \geq \rho(f_m, 0) - \rho(0, f)$$

[1] This form of condition ii) was suggested by Courant.

measured from any other point f increases without bound as m increases. But then the sequence $\{f_m\}$ and hence, the set G are not compact.

We now assume that condition ii) does not hold. Then there exist $\delta > 0$, a sequence

$$f_1, f_2, \ldots, f_m \ldots$$

of functions belonging to G, and a sequence $\epsilon_m > 0$,

$$\lim \epsilon_m = 0 \text{ as } m \to \infty,$$

such that

$$\rho(f_{m\epsilon_m}, f_m) \geq \delta \tag{4}$$

for any m. Clearly

$$\begin{aligned}\delta \leq \rho(f_{m\epsilon_m}, f_m) \leq \rho(f_{m\epsilon_m}, f_{\epsilon_m}) + \rho(f_{\epsilon_m}, f) + \rho(f, f_m) \leq \\ \leq 2\rho(f, f_m) + (f_{\epsilon_m}, f).\end{aligned}$$

Since

$$\lim \rho(f_{\epsilon_m}, f) = 0,$$

it follows from (4) that

$$\underline{\lim}\, \rho(f_m, f) \geq \tfrac{1}{2}\delta.$$

Consequently, the sequence $\{f_m\}$ and the set G are not compact.

Proof of the sufficiency. Denote by G_ϵ the set of functions $f_\epsilon(x)$ corresponding to the functions $f(x)$ belonging to G. The functions belonging to G_ϵ are uniformly bounded and equicontinuous. Indeed, for any set J of volume $V(J)$ we have

$$\begin{aligned}\int_J f(x)d\sigma = \int_J 1 \cdot f(x)d\sigma \leq \Big\{\int_J 1 \cdot d\sigma\Big\}^{1/q}\Big\{\int_J |f(x)|^p d\sigma\Big\}^{1/p} \leq \\ \leq \{V(J)\}^{1/q} K^{1/p}, \quad 1/p + 1/q = 1.\end{aligned} \tag{5}$$

Applying (5) to the set $J = S(x, \epsilon)$ we obtain

$$f_\epsilon(x) = \frac{1}{V(\epsilon)} \int_{S(x,\epsilon)} f(y)d\sigma \leq \{V(\epsilon)\}^{1/q-1} K^{1/p},$$

which proves the uniform boundedness of $f_\epsilon(x)$. Similarly, putting

$$J = \{S(x', \epsilon) \backslash S(x'', \epsilon)\} \cup \{S(x'', \epsilon) \backslash S(x', \epsilon)\},$$

we obtain the inequality

$$|f_\epsilon(x'') - f_\epsilon(x')| = \left|\frac{1}{V(\epsilon)}\int_J f(y)d\sigma\right| \le \{V(J)\}^{1/q}\{V(\epsilon)\}^{-1}K^{1/p}.$$

Since $V(J)$ tends uniformly to zero together with $|x-x'|$, the uniform continuity of $f(x)$ is also proved.

Hence, G is compact in the sense of uniform convergence and, moreover, in the sense of convergence in mean, since if a sequence of functions converges, then it converges in mean as well.

By virtue of condition ii), the sets G_ϵ approximate the set G to within any prescribed degree of accuracy, that is, if ϵ is sufficiently small, then G lies in the union of the balls $S(G_\epsilon, \rho)$ of arbitrarily small radius ρ. This immediately implies the compactness of G.

Göttingen, 6 February 1931

19. ON THE INTERPRETATION OF INTUITIONISTIC LOGIC*

This paper can be considered from two completely different viewpoints.

1. If the intuitionistic cognitive presuppositions are not accepted, then one should take into account only the first section. The conclusions of this section can roughly be summarized as follows.

Along with the development of theoretical logic, which systematizes the schemes of proofs of theoretical truths, it is also possible to systematize the schemes of solution of problems, for example, geometrical construction problems. In this case the syllogism principle can be formulated, for example, as follows. *If we can reduce the solution of problem* b *to the solution of problem* a *and the solution of problem* c *to the solution of problem* b, *then the solution of* c *can also be reduced to the solution of* a.

By introducing an appropriate system of symbols, we can develop a formal calculus enabling us to construct symbolically systems of such solution schemes. Thus, a new *calculus of problems* arises along with theoretical logic. Note that in this case there is no need for special (for example, intuitionistic) cognitive presuppositions.

The following remarkable fact holds: *the calculus of problems coincides in form with the Brouwerian logic recently formalized by Heyting* [1], [2].

2. The second section in which the general intuitionistic presuppositions are accepted, presents a critical analysis of intuitionistic logic. It is shown that this logic should be replaced by the calculus of problems, since the objects under consideration are in fact problems, rather than theoretical propositions.

§1

We do not define the notion of a *problem* but explain it by means of some examples:

1. Find four integers x, y, z and n such that

$$x^n + y^n = z^n, \quad n > 2. \tag{1}$$

2. Prove that Fermat's theorem is false.

3. Construct a circle [1] passing through three given points (x, y, z).

---

* 'Zur Deutung der intuitionistischen Logik', *Math. Z.* **35** (1932), 58–65.

[1] To be precise, the permissible means of construction must be indicated when stating this problem.

4. Given one root of the equation $ax^2 + bx + c = 0$, find the other root.

5. Assuming that the number π has a rational expression

$$\pi = m/n$$

find a similar expression for the number e.

There is an obvious distinction between the first and the second problems which however, is not the subject of a specifically intuitionistic proposition. [2] The fourth and fifth problems are examples of *conditional* problems; the premise of the latter is false, and hence the fifth problem is meaningless or empty. Here and elsewhere the proof of the fact that a problem is *meaningless* is considered as a solution of this problem.

We believe that these examples and explanations allow us to use unambiguously the notions of "problem" and "solution of a problem" in all the cases encountered in specific fields of mathematics. [3] In what follows problems will be denoted by lower case italic letters $a, b, c, \ldots$.

If a and b are problems, then $a \wedge b$ denotes the problem "solve both problems a and b", while $a \vee b$ stands for "solve at least one of the problems a and b". Further, $a \supset$b is the problem "given a solution to problem a, solve problem b" or, which is the same, "reduce the solution of problem b to the solution of problem a".

We never assume a problem to be solvable. Suppose, for example, that Fermat's theorem is true. Then the solution of the first problem is contradictory. Accordingly, $\neg a$ denotes the problem "assuming that there is a solution to problem a, derive a contradiction". [4]

According to these definitions, if a, b, c, d are problems, then each formula $p(a, b, c, \ldots)$ formed from the symbols $\wedge$, $\vee$, $\supset$, and $\neg$ also denotes a problem.

---

[2] By contrast, from the point of view of classical logic the propositions "Fermat's theorem is not true" and "there exist four numbers satisfying (1)" are equivalent.

[3] The basic notions of the logic of propositions, "proposition" and "proof of a proposition" present the same difficulties.

[4] We note that $\neg a$ should not be read as the problem "prove the unsolvability of problem a". In the general case, if the "unsolvability of problem a" is considered as a completely defined notion, we only obtain that $\neg a$ implies the unsolvability of a, and not the converse assertion. If, for example, it were proved that a realization of the well-ordering of the continuum is beyond our possibilities, it would not be possible to assert that the existence of such a well-ordering implies a contradiction.

If $a, b, c, \ldots$ are merely symbols of undefined problems, then it can be said that $p(a, b, c, \ldots)$ is a function of the given variables $a, b, c, \ldots$. In the general case, if x is a variable (of any kind) and $a(x)$ denotes a problem whose meaning depends on the values of x, then $(x)a(x)$ denotes the problem "find a general method for solving the problem $a(x)$ for each specific value of x". This should be understood as follows: the problem $(x)a(x)$ is solved if the problem $a(x_0)$ can be solved for each given specific value of x_0 of the variable x by means of a finite number of steps which are fixed in advance (before x_0 is set). [5]

For a function $p(a, b, c, \ldots)$ of undefined problems $a,\ b,\ c,\ \ldots$ we simply write [6]

$$\vdash p(a, b, c, \ldots)$$

instead of

$$(a)(b)(c) \ldots p(a, b, c, \ldots).$$

Hence, $\vdash p(a, b, c, \ldots)$ denotes the problem "find a general method for solving the problem $p(a, b, c, \ldots)$ for each individual choice of the problems $a,\ b,\ c,\ \ldots$".

Problems of the form $\vdash p(a, b, c, \ldots)$ where p is expressed by means of the symbols $\vee, \wedge, \supset$, and $\neg$ constitute the subject of the *elementary calculus of problems.* [7]

The corresponding functions $p(a, b, c, \ldots)$ are *elementary problem functions.*

The fact that I have solved a problem is a purely subjective one of no general interest in itself. Logical and mathematical problems, however, possess a special property of *universal validity of their solutions*, that is, if I have solved a logical or a mathematical problem, then I can present this solution in a commonly accepted way, and this solution must *necessarily* be recognized as being correct, although this necessity is of a somewhat ideal nature since the reader is assumed to have adequate qualification. [8]

---

[5] As above, we hope that this definition will not lead to misunderstanding in specific areas of mathematics.

[6] This interpretation of the symbol $\vdash$ completely differs from that suggested by Heyting, although it leads to the same rules of the calculus.

[7] This definition is similar to that of the elementary propositional calculus. In propositional calculus, however, logical functions expressed by the symbols $\wedge$, $\vee$, $\supset$ and $\neg$ can be expressed in terms of two of the symbols. In the calculus of problems these four symbols are independent.

[8] The same is literally true for proofs of theoretical propositions. It is, however,

The aim of the calculus of problems is to develop a method allowing one to apply automatically a number of simple computational rules for solving a problem $\vdash p(a, b, c, \ldots)$ where $p(a, b, c, \ldots)$ is an elementary problem function. In order to reduce the whole problem to these computational rules we have, however, to assume that the solutions of certain elementary problems are already known. We *postulate* that the following two groups (A) and (B) have already been solved. The presentation below is meant only for a reader who has already solved all these problems. [9]

2.1 $\vdash a \supset a \wedge a$;

2.11 $\vdash a \wedge b \supset b \wedge a$;

2.12 $\vdash (a \supset b) \supset (a \wedge c \supset b \wedge c)$;

2.13 $\vdash (a \supset b) \wedge (b \supset c) \supset (a \supset c)$;

2.14 $\vdash b \supset (a \supset b)$;

(A) 2.15 $\vdash a \wedge (a \supset b) \supset b$;

3.1 $\vdash a \supset a \vee b$;

3.11 $\vdash a \vee b \supset b \vee a$;

3.12 $\vdash (a \supset c) \wedge (b \supset c) \supset (a \vee b \supset c)$;

4.1 $\vdash \neg a \supset (a \supset b)$;

4.11 $\vdash (a \supset b) \wedge (a \supset \neg b) \supset \neg a$.

Thus, we assume that, given any problems a, b, c, the reader can solve all problems above behind the symbol $\vdash$. This presents no difficulties. For example, in problem 2.12, assuming that the solution of b has already been reduced to the solution of a, one should reduce the solution of $b \wedge c$ to that of $a \wedge c$. Let a solution of $a \wedge c$ be given. This means that we are given both a solution of a and a solution of c. By the hypothesis, we can derive a solution of b from that of a, and, since a solution of c is known, we obtain solutions of both problems b and c and hence a solution of problem $b \wedge c$.

---

essential that every proved proposition is *true*; for problems there is no such notion of truth.

[9] In the case of propositional calculus, if one wants to establish the truth of certain consequences of the axioms, one has to show first that the axioms are true. As to the numbering of the formulas, see [1].

This argument contains a general method for solving the problem

$$(a \supset b) \supset (a \wedge c \supset b \wedge c),$$

which is valid for any a, b, c. Thus, the problem

2.12 $\vdash (a \supset b) \supset (a \wedge c \supset b \wedge c)$

(under the generality symbol $\vdash$) can now be regarded as being solved.

The second group, (B), for which we have postulated the existence of solutions consists of only three problems. [10] Namely, our considerations are based on the fact that knowing a general method we can always solve the following problems for given elementary problem functions p, q, r, s, ...

I. If $\vdash p \wedge q$ is solved, solve $\vdash p$.

II. If $\vdash p$ and $\vdash p \supset q$ are solved, solve $\vdash q$.

III. If $\vdash p(a, b, c, \ldots)$ is solved, solve $\vdash p(q, r, s, \ldots)$.

We can now formulate the *rules* of our calculus of problems.

1. First, we include the problems of group (A) in the list of solved problems.

2. If the list includes $\vdash p \wedge q$, then we are allowed to replace it by $\vdash p$.

3. If both formulas $\vdash p$ and $\vdash p \supset q$ are in the list, then we can replace them by $\vdash q$.

4. If $\vdash p(a, b, c, \ldots)$ is in the list and $q, r, s, \ldots$ are arbitrary problem functions, then we are allowed to replace it by $\vdash p(q, r, s, \ldots)$ in the list.

Based on the above postulates, it is easily seen that the formal calculus does in fact guarantee the solution of the corresponding problems.

We are not going to develop this calculus further here, since all formal rules and *a priori* formulas above coincide with the computational rules and axioms suggested by Heyting [1]. Hence, we can interpret all formulas of this paper as problems and assume that all problems are solved.

Here we only note some particularly interesting problems (which are also regarded as being solved):

4.3 $\vdash a \supset \neg\neg a$;

4.2 $\vdash (a \supset b) \supset (\neg b \supset \neg a)$;

4.3.2 $\vdash \neg\neg\neg a \supset \neg a$.

[10] These problems cannot, however, be expressed using symbols of the elementary calculus of problems.

The solutions of problems 4.3 and 4.2 are clear without calculation. The solution of problem 4.3.2 is obtained from 4.3 and 4.2 if b is replaced by $\neg\neg a$ in 4.2.

If the *a priori* accepted formulas in group (A) are supplemented by the formula

$$\vdash a \vee \neg a \tag{2}$$

(which is the law of the excluded middle in propositional calculus), then we obtain the complete set of axioms of classical propositional calculus. In our interpretation of problems, formula (2) reads as follows: find a general method which for any problem a allows one either to find its solution or to derive a contradiction from the existence of such a solution! In particular, if the problem a consists of proving a proposition, one must have a general method which allows one either to prove each proposition or to reduce it to a contradiction. Unless the reader considers himself omniscient, he will perhaps agree that (2) cannot be in the list of problems that he has solved.

Astonishingly, however, the problem: [11]

4.8 $\vdash \neg\neg(a \vee \neg a)$

can be solved, as is seen from Heyting's calculus.

The formula

$$\vdash \neg\neg a \supset a$$

(known in propositional calculus as the law of double negation) cannot appear in our calculus of problems either, since in view of 4.8, it implies (2).

It is thus seen that in contrast to the formulas obtained in Heyting's intuitionistic logic, even some of the simplest formulas of classical propositional logic cannot appear in our calculus of problems.

It should also be mentioned that if $\vdash p$ is false in classical propositional logic, then the corresponding problem $\vdash p$ cannot be solved. Indeed, in view of the earlier accepted formulas and rules of the calculus of problems, this formula $\vdash p$ readily implies the contradictory formula $\vdash a \wedge \neg a$ (cf. [3]).

[11] In propositional logic, 4.8 represents Brouwer's theorem on the consistency of the law of excluded middle.

§2.

The intuitionistic criticism of logical and mathematical theories is based on the following principle: *each meaningful proposition must indicate one or several completely definite situations accessible by our experience.* [12]

If a is a general proposition of the type "each element of a set K possesses the property A" and if, moreover, K is an infinite set, then the negation of a, that is, the proposition "a is false", does not satisfy the stated principle. To overcome this difficulty, Brouwer suggests a new definition of negation, namely "a is false" should be understood as "a leads to a contradiction". Thus, the negation of a proposition a is transformed into the *existential sentence* "there exists a chain of logical inferences leading to a contradiction if a is assumed to be true".

Existential propositions were, however, profoundly criticised by Brouwer. Namely, from the intuitionistic point of view there is no sense in saying simply that "there is at least one element of a set K possessing a property A" without specifying that element.

Brouwer does not, however, intend to exclude existential propositions from mathematics completely. He only explains that an existential proposition should not be stated without presenting the corresponding construction. At the same time, according to Brouwer, an existential proposition is not a mere indication of the fact that we have already found the desired element of K. In this case the existential proposition would be false *prior to* the invention of the construction and true *after* that. Thus, propositions of a completely new type arise, which, although their content does not change in time, can nevertheless be stated only under certain conditions.

The natural question which can arise is whether this specific type of proposition is a mere fiction. Indeed, the problem "find an element of a set K possessing a property A" is posed. This problem actually has a certain sense independent of the state of our knowledge. If this problem has been *solved*, that is, if the corresponding element x is found, we obtain the empirical proposition "our problem is now solved". Thus, Brouwer's existential proposition is partitioned into two elements: an objective component (problem) and a subjective component (its solution). So there remains nothing that can be interpreted as

[12] Cf. [4]. The investigation below of negated and existential propositions is in essence close to this paper by Weyl.

an existential proposition in the proper sense of this term.

Therefore the major result of the intuitionistic criticism of negated propositions should be formulated in the following simple way: *in general it is meaningless to consider the negation of a general proposition as a definite proposition.* But then the subject of intuitionistic logic disappears, since the law of the excluded middle becomes true for all propositions whose negations make sense. [13]

As to mathematics, it follows that the solution of a problem must be considered as an independent task (along with the proof of theoretical propositions). As was shown in §1, the formulas of intuitionistic logic also acquire a new meaning in the area of problems and solutions. [14]

Göttingen, 15 January 1931

References

1. A. Heyting, 'Die formalen Regeln der intuitionistischen Logik', *Sitzungsber. Preuss. Akad. Wiss., Phys-math. Kl.* (1930), 42–56.
2. A. Heyting, 'Die formalen Regeln der intuitionistischen Mathematik', *Sitzungsber. Preuss. Akad. Wiss., Phys-math. Kl.* (1930), 57–71; 158–169.
3. V. Glivenko, 'Sur quelques points de la logique de M. Brouwer', *Bull. Cl. Sci. Acad. Roy. Belg., Sér.* 5, **15** (1929), 183–188.
4. H. Weyl, 'Über die neue Grundlagenkrise der Mathematik', *Math. Z.* **10** (1921), 39–79.
5. A. Heyting, 'Die intuitionistische Grundlegung der Mathematik', *Erkenntnis* **2** (1931), 106–115.

[13] There arises, however, a new question: which logical laws are true for propositions whose negations are meaningless.

[14] This interpretation of intuitionistic logic is closely related to Heyting's idea [5]. Heyting does not, however, distinguish explicitly between a proposition and a problem.

20. ON THE FOUNDATION OF PROJECTIVE GEOMETRY *

In this paper we are going to consider an interesting application of Pontryagin's theorem on continuous algebraic fields. [1] If the foundation of projective geometry is based only on the (three-dimensional) axioms of incidence, then, as is well known, there appear various types of space. These different types of projective space correspond to different types of abstract algebraic field. In fact, each projective space can be endowed with the usual measurement of distances (using the Desargues theorem), and the resulting field of distances (generally non-commutative) is defined uniquely up to an isomorphism, using the initial space. Conversely, given an abstract algebraic field, one can construct projective spaces of a unique type corresponding to this field.

In order to describe the ordinary real projective geometry one would have to introduce two additional groups of axioms in the spirit of the classical method, namely, ordering axioms and axioms of continuity. Only after that would it be possible to provide a foundation of complex projective goemetry by adding complex elements. However, abstract topology can now help us to study the properties of continuity irrespective of ordering properties and, hence, to justify axiomatically complex geometry in a direct manner without using ordering axioms which are valid only in real geometry. In what follows we suggest one such direct approach.

Let three systems of elements be given, called points, lines and planes respectively. Let the elements satisfy the following conditions.

I. Each of the three systems is a connected bicompact topological space.

II. The points, lines and planes satisfy the ordinary projective axioms of incidence.

III. The incidence relation is continuous, that is, a line passing through two points is a continuous function of these points (if they do not coincide), etc.

If three arbitrary points on a line in a projective space that satisfies I–III are chosen to be zero, unity and infinity and if the corresponding measurement of distances is introduced, then it can easily be shown that the resulting field of distances satisfies all the conditions of Pontryagin's theorem. Hence, this

---

* 'Zur Begründung der projektiven Geometrie', *Ann. of Math.* **33** (1932), 175–176.

[1] 'Über stetige algebraische Körper', *Ann. of Math.* **33** (1932), 163–174.

field of distances is isomorphic either to the field of real numbers, the field of complex numbers, or the skew field of quaternions. Therefore there exist spaces of only three different types satisfying conditions I–III, namely ordinary real projective space, complex projective space, and the space constructed on the skew field of quaternions by using homogeneous cooordinates.

If Pascal's theorem (and hence commutativity of multiplication) is taken as a new axiom, this third possibility is excluded. Finally, complex geometry can be obtained by using the completeness axiom.

26 May 1931

21. ON MEASURE THEORY *

Introduction

Let $\mu(E)$ be a non-negative (not necessarily finite-valued) set function defined on all A-sets (analytic sets) of an n-dimensional Euclidean space R^n. The function $\mu(E)$ is said to be a measure function (or simply a measure) if it satisfies the following conditions.

I. If the union of a finite or a countable number of sets E_m covers E, that is, if

$$E \subset \sum E_m$$

(this is denoted as $\{E_m\} = \mathfrak{U}(E)$), then

$$\mu(E) \leq \sum \mu(E_m).$$

II. If a finite or countable number of sets E_m are disjoint and all of them lie in E, that is, if

$$E_m \subset E, \quad E_i E_j = \varnothing, \ i \neq j$$

(this is denoted as $\{E_m\} = \mathfrak{T}(E)$), then

$$\sum \mu(E_m) \leq \mu(E).$$

To state condition III we need the following definition.

Definition. [1] A set E is an unexpanded image of a set E' if there exists a one-to-one mapping from E' onto E under which the distance between any two points of E' is no less than the distance between their images in E.

An unexpanded image of E will always be denoted by $\pi(E)$, where $\pi(x)$ is the corresponding continuous mapping.

III. If E' is an unexpanded image of E, then

$$\mu(E') \leq \mu(E).$$

IV. For a set J (the standard of the measure) we have

$$\mu(J) = 1.$$

---

* 'Beiträge zur Maßtheorie', *Math. Ann.* **107** (1933), 351–366.

[1] The notion of a non-expanding mapping was first used by E. Schmidt for determining the length of a curve (see [2]).

Remark 1. Throughout this paper we assume that $\mu(E)$ is defined for all A-sets in R^n and only for them.

Remark 2. It follows from I and II that for

$$E = \sum E_m, \quad E_i E_j = \varnothing, \ i \neq j,$$

the relation

$$\mu(E) = \sum \mu(E_m)$$

holds, that is, $\mu(E)$ is additive.

Remark 3. It follows from III that

III′ If E' and E'' are congruent, then

$$\mu(E') = \mu(E'').$$

Remark 4. It follows from I that if $E \subset E'$, then

$$\mu(E) \leq \mu(E'),$$

that is, $\mu(E)$ is a monotone set function.

The most important case arises when a k-dimensional cube J^k $(k \leq n)$ is chosen as the standard of the measure J. In this case $\mu(E)$ is called k-dimensional measure and is denoted by $\mu^k(E)$.

It is known, however, [2] that if $k = n$, then the Lebesgue measure $m^n(E)$ is the only set function that is defined on all A-sets in R^n and satisfies conditions I, II, III′ and IV. In the sequel we will see that the Lebesgue measure $m^n(E)$ also satisfies condition III (in addition to III′); hence, Lebesgue measure is the unique solution of our problem of measures, where the n-dimensional unit cube is chosen as the measure standard. If $k < n$, then for any set E in R^k we have $\mu^k(E) = m^k(E)$.

The case $k < n$ was first considered by Carathéodory in his paper [3], which has become fundamental for all further studies in this area and has led

[2] According to Lebesgue, this assertion is true for any additive function $\mu(E)$ that satisfies conditions III and IV and is defined on all Borel sets. However, for any A-set E (as well as for any Lebesgue measurable set) there exist Borel sets E' and E'' such that $E' \subset E \subset E''$ and $m^n(E') = m^n(E) = m^n(E'')$, and hence our assertion is also true for functions defined on all A-sets.

to a number of different definitions of measure. The present paper suggests a principle for ordering this variety of definitions.

Our basic result on k-dimensional measures can be formulated in the following way.

Main theorem. *For any natural number $k < n$ there exist two special measure functions, namely the maximum measure $\overline{\mu}^k(E)$ and the minimum measure $\underline{\mu}^k(E)$. For any A-set E the values of any other k-dimensional measure function $\mu^k(E)$ lie between $\overline{\mu}^k(E)$ and $\underline{\mu}^k(E)$:*

$$\overline{\mu}^k(E) \geq \mu^k(E) \geq \underline{\mu}^k(E).$$

§1. Non-expanding mappings

First we prove that for sets in R^n the Lebesgue measure $m^n(E)$ satisfies condition III. This fact follows from the general theorem below.

Theorem 1. *For any unexpanded image $\pi(E)$ of an arbitrary set E (both sets lie in R^n) the inequality*

$$m^n_S(E) \geq m^n_S\{\pi(E)\}$$

holds, where $m^n_S(E)$ is the outer Lebesgue measure of E.

The proof is based on the following assertion.

Auxiliary theorem. *Let E be a set in R^n belonging to the interior of a ball S^n of volume V. If $\pi(E)$ is a non-expanding mapping from E onto a set $E' = \pi(E)$ that also lies in R^n, then*

$$m^n_S(E') \leq V.$$

It is clear that the diameter d' of E' is no greater than the diameter of S^m. The convex hull F of E' has the same diameter d', and its volume U is no greater than the volume V' of a ball of diameter d'. Therefore

$$m^n_S(E') \leq U \leq V' \leq V,$$

which is the desired result.

Proof of Theorem 1. Let us consider a covering of E by balls S_k^n, $k = 1, 2, 3, \ldots$. These balls can be chosen so that

$$\sum V_k \leq m_S^n(E) + \epsilon,$$

where V_k is the volume of S_k^n. Obviously, such a covering exists for any $\epsilon > 0$. Denoting ES_k^n by E_k we obtain

$$E = \sum E_k, \quad E_k \subset S_k^n, \quad m_S^n\{\pi(E_k)\} \leq V_k,$$
$$m_S^n\{\pi(E)\} \leq \sum m_S^n\{\pi(E_k)\} \leq \sum V_k \leq m_S^n(E) + \epsilon,$$

which proves the theorem.

We would also like to mention the following well-known property of continuous (in particular, non-expanding) mappings.

If $E' = \phi(E)$ is a continuous mapping of an A-set E onto E', then the images of A-subsets contained in E (in particular, the set E' itself) and the pre-images (in E) of the A-subsets of E' are themselves A-sets (see [1]).

If $E' = \phi(E)$, then we will denote the pre-image in E of a set $E'' \subset E'$ by $\phi^{-1}(E'')$.

§2. Maximum measure $\overline{\mu}^k(E)$

We now define the set function $\overline{\mu}^k(E)$ on all A-sets as the infimum of the Lebesgue measure $m^k\pi^{-1}(E)$ for all A-sets $\pi^{-1}(E)$ contained in R^k which can be mapped in a non-expanding manner onto E:

$$\overline{\mu}^k(E) = \inf m^k\pi^{-1}(E).$$

If there are no sets $\pi^{-1}(E)$ at all, then we put

$$\overline{\mu}^k(E) = +\infty.$$

We will prove that $\overline{\mu}^k(E)$ is a k-dimensional measure function and that it is the greatest measure among all k-dimensional measure functions. Thus, the first part of the main theorem will be proved.

I. $\overline{\mu}^k(E)$ satisfies condition I in the Introduction.

If at least one of the summands $\overline{\mu}^k(E_m)$ is equal to $+\infty$, then our assertion is sure to be true. Otherwise, there exist sets E'_m in R_k such that

$$\pi^m(E'_m) = E_m, \quad m^k(E'_m) \leq \overline{\mu}^k(E_m) + \epsilon/2^m.$$

Let us consider a partition of R^k into cubes each of which is smaller than the unit cube. Denoting by E'_{mr} $(r = 1, 2, 3, \ldots)$ the intersection of E'_m with the cubes of this partition we obtain

$$m^k(E'_m) = \sum m^k(E'_{mr}).$$

Further,

$$E_m = \sum \pi^{(m)}(E'_{mr}) = \sum E_{mr}.$$

Here we note that the diameters of E_{mr} are also less than 1. We arrange all the E_{mr} as an ordered sequence

$$E_{m_1 r_1},\ E_{m_2 r_2},\ E_{m_3 r_3},\ \ldots,\ E_{m_i r_i} \ldots$$

and set $\rho_i = \sup \rho(x, y)$ where x runs over all points of the set $E_{m_i r_i}$ and y takes all the values belonging to the union of all $E_{m_j r_j}$ for $j < i$. We now define subsets E^*_{mr} in R^k such that E^*_{mr} is congruent to E'_{mr} and, moreover, the inequality

$$\rho(E^*_{m_i r_i}, \sum_{j<i} E^*_{m_j r_j}) \geq \rho_i$$

holds.

Finally, we define

$$E^* = \sum E^*_{mr}.$$

Each E^*_{mr} can be mapped onto the corresponding E'_{mr} by a congruent mapping:

$$E'_{mr} = K_{mr}(E^*_{mr}).$$

Further,

$$E_{mr} = \pi^{(m)}(E'_{mr}) = \pi^{(m)} K_{mr}(E^*_{mr}).$$

Thus, E^* is mapped onto

$$E = \sum E_m = \sum E_{mr}.$$

This is a non-expanding mapping. Indeed, if two points x_1^* and x_2^* belong to the same set E^*_{mr}, then

$$x = \pi^*(x^*) = \pi^{(m)} K_{mr}(x^*),$$

and, since K_{mr} is a congruent mapping and $\pi^{(m)}$ is a non-expanding mapping, we obtain

$$\rho(x_1^*, x_2^*) \geq \rho(x_1, x_2).$$

On the other hand, if x_1^* and x_2^* belong to different sets, for example, to $E^*_{m_i r_i}$ and $E^*_{m_j rj}$ respectively (with $i > j$), then

$$\rho(x_1^*, x_2^*) \geq \rho_i \geq \rho(x_1, x_2).$$

Finally, we have

$$\overline{\mu}^k(E) \leq m^k(E^*) = \sum m^k(E^*_{mr} = \sum m^k(E'_{mr}) = \sum m^k(E'_m) \leq$$

$$\sum(\overline{\mu}^k(E_m) + \epsilon/2^m) = \sum \overline{\mu}^k(E_m) + \epsilon.$$

Since ϵ is an arbitrary positive number, we obtain

$$\overline{\mu}^k(E) \leq \sum \overline{\mu}^k(E_m),$$

which is the desired result.

II. $\overline{\mu}^k(E)$ satisfies condition II in the Introduction.

Indeed, if $\overline{\mu}^k(E) = +\infty$, then this is trivially true, while if $\overline{\mu}^k(E) < +\infty$, then there exists a set $\pi^{-1}(E)$ in R^k such that

$$\overline{\mu}^k(E) > m^k \pi^{-1}(E) - \epsilon.$$

Therefore

$$\sum \overline{\mu}^k(E_m) \leq \sum m^k \pi^{-1}(E_m) \leq m^k \pi^{-1}(E) \leq \overline{\mu}^k(E) + \epsilon.$$

Since ϵ is an arbitrary positive number, this proves the assertion.

III. The inequality

$$\overline{\mu}^k \pi(E) \leq \overline{\mu}^k(E)$$

(condition III) always holds.

This follows immediately from the fact that the composite of two non-expanding mappings $\pi_1 \pi_2(E)$ is again a non-expanding mapping.

IV. For sets lying in R^k we have

$$\overline{\mu}^k(E) = m^k(E).$$

This follows from Theorem 1 (§1). In particular, $\overline{\mu}^k(J^k) = 1$ (condition IV).

V. For any other k-dimensional measure $\mu^k(E)$ the inequality

$$\mu^k(E) \leq \overline{\mu}^k(E)$$

holds.

Indeed, for any $\pi^{-1}(E) \subset R^k$ we have

$$\mu^k(E) \leq \mu^k \pi^{-1}(E) = m^k \pi^{-1}(E),$$

and consequently

$$\mu^k(E) \leq \inf m^k \pi^{-1}(E) = \overline{\mu}^k(E).$$

§3. Minimum measure $\underline{\mu}^k(E)$

We put

$$\underline{\mu}^k(E) = \sup \sum m^k \pi_i(G_i) \quad (G_i \subset E;\ G_iG_j = \varnothing,\ i \neq j),$$

where the supremum extends over all possible countable sequences of A-sets that satisfy the conditions inside the parentheses and whose images lie in R^k. As in the case of $\overline{\mu}^k(E)$, we prove the following assertions.

I. $\underline{\mu}^k(E)$ satisfies condition I in the Introduction.

Indeed, for any $\epsilon > 0$ there exists a sequence of A-sets G_i ($G_i \subset E$, $G_iG_j = \varnothing$, $i \neq j$) and mappings π_i such that

$$\underline{\mu}^k(E) \leq \sum m^k \pi_i(G_i) + \epsilon.$$

Further, we have

$$\underline{\mu}^k(E_m) \geq \sum m^k \pi_i(E_m G_i),$$

$$\sum \underline{\mu}^k(E_m) \geq \sum \sum m^k \pi_i(E_m G_i) \geq \sum m^k \pi_i(G_i) \geq \underline{\mu}^k(E) - \epsilon.$$

Since ϵ is arbitrary, our assertion is proved.

II. $\underline{\mu}^k(E)$ satisfies condition II.

Indeed, for any m there exists a sequence of A-sets G_{mi} ($G_{mi} \subset E_m$, $G_{mi}G_{mj} = \varnothing$, $i \neq j$) and mappings π_{mi} such that

$$\underline{\mu}^k(E_m) \leq \sum m^k \pi_{mi}(G_{mi}) - \epsilon/2^m.$$

Since the sets G_{mi} lie in E and are disjoint, we have

$$\underline{\mu}^k(E) \geq \sum\sum m^k \pi_{mi}(G_{mi}) \geq \sum \underline{\mu}^k(E_m) - \epsilon,$$

which proves our theorem.

III. The inequality

$$\underline{\mu}^k \pi(E) \leq \underline{\mu}^k(E)$$

always holds.

IV. For sets in R^k we have

$$\underline{\mu}^k(E) = m^k(E).$$

V. For any k-dimensional measure $\mu^k(E)$, we have the inequality

$$\underline{\mu}^k(E) \leq \mu^k(E).$$

Proof.

$$\underline{\mu}^k(E) = \sup \sum m^k \pi_i(G_i) \leq \sup \sum \mu^k(G_i) \leq \mu^k(E).$$

This completes the proof of the main theorem.

§4. Some special properties of the measure functions $\overline{\mu}^k(E)$ and $\underline{\mu}^k(E)$

The following theorem elucidates the interrelation between measure functions of different dimensions.

Theorem 1. *Let $i < k$. Then $\underline{\mu}^k(E) > 0$ implies that $\overline{\mu}^i(E) = +\infty$, and from $\overline{\mu}^k(E) > 0$ it follows that $\overline{\mu}^i(E) = +\infty$.*

Indeed, if $\underline{\mu}^k(E) > 0$, then there exists a subset E' of E that can be mapped in a non-expanding manner onto a subset E'' of R^k with $m^k(E'') > 0$. However, in this case one can find a sequence of subsets E''_m of E'' that lie in different (for example, parallel) subspaces R^i so that

$$\sum m^i(E''_m) = +\infty.$$

Therefore

$$\underline{\mu}^i(E) \geq \underline{\mu}^i(E') \geq \underline{\mu}^i(E'') \geq \sum \underline{\mu}^i(E''_m) = \sum m^i(E''_m) = +\infty,$$

which proves the first part of our theorem.

If we assume that $\overline{\mu}^i(E) < +\infty$ for $i < k$, then there exists $E' = \pi^{-1}(E)$ in R^i. Moreover, E' lies in some R^k, and we have

$$\overline{\mu}^k(E) \leq m^k(E') = 0,$$

as required.

We now denote by $S_a(x) = y$ the similarity mapping of R^n onto itself with ratio a. The measure function $\mu(E)$ is said to have a well-defined order c if for any set E in R^n and for any $S_a(x)$ the relation

$$\mu\{S_a(E)\} = a^c \mu(E)$$

holds. Here c can be an arbitrary positive number.

It is easily seen that the measure functions $\overline{\mu}^k(E)$ and $\underline{\mu}^k(E)$ have order $c = k$. However, we do not know whether each $\mu^k(E)$ has a well-defined order and whether it is necessarily equal to k.

Theorem 2. *Let $\phi(E) = E'$ be a continuous mapping of a set E onto a set E' such that*

$$\rho(x', y') \leq a\rho(x, y),$$

where $x' = \phi(x)$, $y' = \phi(y)$ and $\rho(x, y)$ is the distance between x and y. Further, let $\mu(E)$ be a measure function of order c. Then

$$\mu(E') \leq a^c \mu(E). \tag{1}$$

Indeed, let

$$E'' = S_{a^{-1}}(E').$$

It is always true that

$$\rho(x'', y'') = a^{-1}\rho(x', y') \leq \rho(x, y).$$

Therefore $E'' = S_{a^{-1}}\phi(E)$ is a non-expanding mapping. Furthermore, since the function $\mu(E)$ satisfies condition III, we have

$$\mu(E'') \leq \mu(E)$$

and since in addition, $\mu(E'') = a^{-c}\mu(E')$, we obtain the inequality (1).

§5. Cases when a linear measure is uniquely defined

It is obvious that a linear, that is, one-dimensional, measure is uniquely defined for E when $\overline{\mu}^1(E) = \underline{\mu}^1(E)$. Here we present a sufficient condition for uniqueness.

Theorem 1. *For any simple Jordan curve S the relation*

$$\underline{\mu}^1(S) = \overline{\mu}^1(S) = L(S)$$

holds, where $L(S)$ is the length of S.

Auxiliary theorem. *For any continuum K we have*

$$\underline{\mu}^1(E) \geq d(K),$$

where $d(K)$ is the diameter of K.

Indeed, let a and b be points in K such that $\rho(a,b) = d(K)$. The function $f(x) = \rho(a,x)$ defines a non-expanding mapping of K onto the closed interval Δ on the number axis lying between 0 and $d(K)$ for which $m^1(\Delta) = d(K)$; this proves our assertion.

Proof of Theorem 1. By the definition of $L(S)$, there exists a sequence of points $x_1, x_2, \dots, x_n$ on S such that

$$\sum \rho(x_{i-1}, x_i) \geq L(S) - \epsilon.$$

Let S_i be the part of S between x_{i-1} and x_i. By the auxiliary theorem, $\underline{\mu}^1(S_i) \geq \rho(x_{i-1}, x_i)$ and $\underline{\mu}^1(S) = \sum \underline{\mu}^1(S_i) \geq L(S) - \epsilon$, therefore

$$\underline{\mu}^1(S) \geq L(S). \tag{1}$$

Now let y_0 be the initial point of S. The function $\pi^{-1}(y) = L(y_0, y)$ (where (y_0, y) is the part of S with end points y_0 and y) defines a mapping from S onto the interval Δ of length $L(S)$. The inverse function $\pi(x)$ maps Δ onto S in a non-expanding manner, since for the points x' and x'' corresponding to points y' and y'' on S we clearly have

$$\rho(x', x'') = L(y', y'') \geq \rho(y', y'').$$

It follows that

$$\overline{\mu}^1(S) \leq L(S). \tag{2}$$

Since $\overline{\mu}^1(S) \geq \mu^1(S)$, the validity of the theorem follows from inequalities (1) and (2).

Remark. It can be seen from the above discussion that $L(S)$ is the infimum of the lengths of the intervals Δ that can be mapped in a non-expanding manner onto S. This definition of length goes back to E. Schmidt (cf. [2]).

We can prove a more general assertion.

Theorem 2. *For any continuum K we have $\overline{\mu}^1(K) = \underline{\mu}^1(E)$.*

We first assume that K is not locally connected. Then there exist infinitely many disjoint continua $K_1, K_2, \ldots, K_m, \ldots$ in K such that their diameters exceed a fixed positive constant D. Since $\underline{\mu}^1(K_m) > D$, we obtain

$$\underline{\mu}^1(K) \geq \sum \underline{\mu}^1(K_m) = +\infty$$

and hence $\underline{\mu}^1(K) = \overline{\mu}^1(K) = +\infty$.

Now let K be locally connected. Then it follows from well-known theorems that for any point $a \in K$ and any closed set $F \subset K$ which does not contain a one can find a Jordan curve $(a, F) \in K$ such that a is its end point while the other end point is the only point of this curve that belongs to F. We now give a proof by induction, in which F_0 is a point $a_0 \in K$ and assume that the closed sets $F_0, F_1, \ldots, F_m$ have already been constructed. We select a point a_m in K that is one of the most distant points from F_m, construct the curve $(a_m, F_m) \in K$, and put $F_{m+1} = F_m + (a_m, F_m)$.

If the series

$$\sum L(a_m, F_m) \tag{3}$$

is divergent, then by Theorem 1 we have

$$\underline{\mu}^k(K) \geq \sum \underline{\mu}^1(a_m, F_m) = \sum L(a_m, F_m) = +\infty,$$

and therefore $\underline{\mu}^1(K) = \overline{\mu}^1(K) = +\infty$.

But if the series (3) converges to a finite sum L, then $L(a_m, F_m)$ tends to zero as m increases, so that, in view of the definition of (a_m, F_m), the continuum K can be approximated by the sets F_m as accurately as desired.

We now define continuous curves $S_0, S_1, \ldots, S_m, \ldots$ such that the union of the points of S_m coincides with F_m. The curves S_m are constructed consecutively in the following way: S_0 coincides with $a_0 = F_0$. Assume that S_m has already been constructed and choose the first point x_m on this curve that coincides with the terminal point of (a_m, F_m) belonging to F_m. We now define S_{m+1} as a continuous curve that coincides with S_m from the initial point of S_m up to the point x_m, then runs along the curve (a_m, F_m) forward and backward, and on returning to x_m, runs over the rest of S_m from x_m to the terminal point.

Obviously, all initial and terminal points of the curves S_m coincide with the point a_0. The total length of these curves is

$$L(S_m) = 2 \sum_{k<m} L(a_k, F_k).$$

Assuming that the series (3) is convergent, one can easily see that the curves S_m approximate a continuous curve S whose length is exactly equal to $2L$. The curve S (as a set of points) coincides with the continuum K.

The continuum K can be represented as

$$K = \sum (a_m, F_m) + E, \tag{4}$$

where E is the set of points not belonging to the curves (a_m, F_m). We map S onto the interval Δ of length $2L$ as in the proof of Theorem 1 in such a way that the inverse mapping is non-expanding. In this case the twice swept curves (a_m, F_m) go into the segments of Δ with total length $2L$, while E is mapped onto a set E' of measure $2L - 2L = 0$. Furthermore, since E' is mapped onto E in a non-expanding manner, we have $\overline{\mu}^1(E) = 0$ and hence $\underline{\mu}^1(E) = 0$ which, in view of (4), implies

$$\overline{\mu}^1(K) = \sum \overline{\mu}^1(a_m, F_m) = L = \sum \underline{\mu}^1(a_m, F_m) = \underline{\mu}^1(K).$$

§6. Measure theory in general metric spaces

Throughout the foregoing presentation we could assume that the set functions $\mu(E)$ were defined on all metric A-sets. By a metric set we mean a set of arbitrary elements in which for any pair of elements (x, y) a distance is defined satisfying the usual axioms (see [1]). A metric set is called an (absolute) A-set if it is a continuous image of the set of all irrational numbers.

The set functions $\overline{\mu}^k(E)$ and $\underline{\mu}^k(E)$ are therefore defined for all metric A-sets and possess all the above-mentioned properties in this new domain of definition. In particular, the results of §5 concerning the length of a curve and a linear measure of a continuum are also true for general metric spaces. The possibility of constructing measure theory in general metric spaces was first indicated by P. Urysohn in 1921 in an unpublished paper. Urysohn considered sets E in a compact metric space R and suggested the following construction.

Let $\mathfrak{U}$ be a covering of E consisting of a finite or countable number of closed sets $F_k \subset R$ with diameters $d_k \leq \epsilon$. The measure $m_\sigma(E)$ of order σ ($\sigma > 0$) is defined as

$$m_\sigma(E) = \liminf m(\sigma) \sum d_k^\sigma, \quad m(\sigma) = \frac{\pi^{\sigma/2}}{2^\sigma \Gamma(\sigma/2+1)}.$$

For point sets in R^n this definition was suggested by Hausdorff [4], among many other definitions which do not make sense for general metric spaces. [3]

The measure function $m_\sigma(E)$ satisfies conditions I, II, III and if σ is an integer it also satisfies condition IV with the unit cube J^σ as standard of measure. Consequently, for integers σ we have

$$\underline{\mu}^\sigma(E) \leq m_\sigma(E) \leq \overline{\mu}^\sigma(E).$$

If $\sigma = 1$, then $m_\sigma(E)$ is identified with the linear Carathéodory measure. It seems to me that the relation $m_1(E) = \underline{\mu}^1(E)$ is likely to hold always.

§7. Uniqueness theorem for point sets in Euclidean spaces and its relation with surface measure theory

We now return to sets lying in R^n. The following theorem is true.

Uniqueness theorem. *The relation*

$$\overline{\mu}^k(E) = \underline{\mu}^k(E)$$

holds for any set $E \subset R^n$ that can be represented as an unexpanded image of an A-set $E' \subset R^k$.

The inequality $\overline{\mu}^k(E) > \underline{\mu}^k(E)$ can therefore be true only if $\overline{\mu}^k(E) = +\infty$. If the mapping $\pi^{-1}(E)$ is single-valued, then the uniqueness theorem

[3] When Urysohn conducted his studies (1921), the paper by Hausdorff [4] was unavailable in Moscow.

follows from an integral representation of the measures $\overline{\mu}^k(E)$ and $\underline{\mu}^k(E)$. To demonstrate this fact we consider functions

$$x_i = x_i(y_1, y_2, \dots, y_k), \quad i = 1, 2, \dots, n,$$

that define a mapping in some Cartesian coordinates in R^n and R^k respectively. Since $\pi(E')$ is non-expanding, we obtain

$$|x_i' - x_i''|/|\eta' - \eta''| \leq 1, \tag{1}$$

where as usual, η is the vector with components $y_1, y_2, \dots, y_k$ (and also the point $(y_1, y_2, \dots, y_k)$) and

$$|\eta' - \eta''| = \rho(\eta', \eta'') = \sqrt{\sum (y_i' - y_i'')^2}.$$

Therefore, for the functions x_i the partial derivatives $\partial x_i / \partial y_j$ are defined almost everywhere on E'. We form the expression

$$D(\eta) = \sqrt{\sum \left(\frac{\partial(x_{i_1}, x_{i_2}, \dots, x_{i_k})}{\partial(y_1, y_2, \dots, y_k)} \right)^2},$$

where the sum extends over all possible sequences $(i_1, i_2, \dots, i_k)$.

Theorem on the integral. *Under the above assumptions we have*

$$\overline{\mu}^k(E) = \underline{\mu}^k(E) = \int_{E'} D d\sigma = I^k(E),$$

where the integral is understood in the Lebesgue sense. [4]

Proof. Obviously, it suffices to consider sets E' with $m^k(E') < +\infty$, otherwise E' can be partitioned into a countable sum of such sets.

A function $x_i(y_1, y_2, \dots, y_k) = x_i(\eta)$ is said to be differentiable at a point if

$$\Delta x_i = \sum \frac{\partial x_i}{\partial y_i} \Delta y_j + o(|\Delta \eta|)$$

for $\Delta x_i = x_i' - x_i$ and $\Delta y_j = y_j' - y_j$.

[4] In particular, it follows from our theorem that the integral $I^k(E)$ is independent of the mapping $E = \pi(E')$.

Denoting the matrix $\|\partial x_i/\partial y_j\|$ by $\partial\mathfrak{X}/\partial\eta$ we obtain the vector relation

$$\Delta\mathfrak{X} = \frac{\partial\mathfrak{X}}{\partial\eta}\Delta\eta + o(|\Delta\eta|) \tag{2}$$

at a point where all the functions x_i are differentiable. It follows from Rademacher's results [5] and from (1) that (2) holds almost everywhere. Furthermore, for any $\epsilon > 0$ one can find a subset P of E' such that $m^k(E' - P) \leq \epsilon$, the components of $\partial\mathfrak{X}/\partial\eta$ are continuous on P and (2) holds uniformly.

Since $|D| \leq 1$ and

$$\int_{E'-P} D d\sigma \leq m^k(E' - P) \leq \epsilon,$$

$$\underline{\mu}^k\pi(E' - P) \leq \overline{\mu}^k\pi(E' - P) \leq m^k(E' - P) \leq \epsilon,$$

it suffices to prove our theorem on the integral for P and $Q = \pi(P)$ instead of E' and E. Let η be a point in P and let S be the open ball of radius ρ with centre at η. We also put

$$R = PS, \quad T = \pi(R).$$

Auxiliary theorem. *For $\mu^k = \overline{\mu}^k$ and also for $\mu^k = \underline{\mu}^k$ and any point η in P, we have*

$$\mu^k(T) = \int_R D d\sigma + o(\rho) m^k(S).$$

We assume temporarily that the auxiliary theorem is proved. By Vitali's theorem, for any $\epsilon > 0$ one can find a sequence of balls $S_1, S_2, \ldots, S_m, \ldots$ such that

$$S_i S_j = \varnothing, \quad i \neq j,$$

$$P = \sum R_m + N, \quad m^k(N) = 0,$$

$$\sum m^k(S_m) \leq 2m^k(P),$$

$$\mu^k(T) = \int_R D d\sigma + \epsilon m^k(S_m),$$

therefore

$$\mu^k(Q) = \int_P D d\sigma + 2\epsilon m^k(P).$$

This formula proves our theorem on the integral.

Proof of the auxiliary theorem. Let $D(\eta) \neq 0$. The mapping

$$\zeta' = \Phi(\eta') = \frac{\partial \mathfrak{X}}{\partial \eta}(\eta' - \eta)$$

transforms the ball S into a k-dimensional ellipsoid $E = \Phi(S)$. This is a single-valued continuous mapping. Therefore the ratio

$$|\zeta' - \zeta''|/|\eta' - \eta''|$$

has a positive infimum, and in addition to (2), we have

$$\begin{aligned} \mathfrak{X}' - \mathfrak{X}'' &= \zeta' - \zeta'' + o(\rho)|\zeta' - \zeta''|, \\ |\mathfrak{X}' - \mathfrak{X}''| &= |\zeta' - \zeta''|\{1 + o(\rho)\}. \end{aligned} \tag{3}$$

Let $U = \Phi(R)$. In view of (3), the mapping $T = \pi\Phi^{-1}(U)$ is almost congruent. By Theorem 2 of §4, it follows that

$$\mu^k(T) = m^k(U)\{1 + o(\rho)\}.$$

However, since $m^k(U) = D(\eta)m^k(R)$ and D varies continuously together with η, we finally obtain

$$\mu^k(T) = D(\eta)m^k(R)\{1 + o(\rho)\} = \int_R D d\sigma\{1 + o(\rho)\} = \int_R D d\sigma + o(\rho)m^k(S).$$

It only remains to consider the case $D(\eta) = 0$. Since Φ is continuous, this will imply

$$\int_R D d\sigma = o(\rho)m^k(S).$$

Hence we must show that

$$\underline{\mu}^k(T) \leq \mu^k(T) = o(\rho)m^k(S).$$

The mapping $\zeta' = \Phi(\eta')$ now transforms the ball S into an i-dimensional ellipsoid with $i < k$. Taking an appropriate Cartesian coordinate system we can represent Φ as

$$\begin{aligned} &z_j' = \lambda_j(y_j' - y_j), \quad \lambda \neq 0, \ j \leq i, \\ &z_j' = 0, \quad j \geq i. \end{aligned}$$

We set

$$z_j^* = \lambda_j(y_j' - y_j), \quad j \le i,$$

$$z_j^* = \delta(y_j' - y_j), \quad i \le j \le k,$$

$$z_j^* = 0, \quad k \le j.$$

This is a single-valued mapping $\zeta^* = \Phi_*(\eta')$ from the ball S into a k-dimensinal ellipsoid E^*. It follows that the ratio

$$|\zeta_1^* - \zeta_2^*|/|\eta_1' - \eta_2'|$$

has a positive infimum. Let $U^* = \Phi_*(R)$. The mapping $T = \pi\Phi_*^{-1}(U^*)$ clearly satisfies the condition

$$|\mathfrak{X}_1' - \mathfrak{X}_2| \le |\zeta_1' - \zeta_2'| + o(\rho)|y_1' - y_2'| \le$$

$$\le |\zeta_1^* - \zeta_2^*| + o(\rho)|\eta_1' - \eta_2'| \le |\zeta_1^* - \zeta_2^*|\{1 + o(\rho)\}.$$

By Theorem 2 of §4 we now have

$$\overline{\mu}^k(T) \le m^k(U^*)\{1 + o(\rho)\}.$$

But since δ is arbitrary, the ratio

$$m^k(U^*)/m^k(R) = m^k(E^*)/m^k(S)$$

can be made as small as desired; thus we finally obtain

$$\overline{\mu}^k(T) = o(\rho)m^k(S).$$

Hence, the auxiliary theorem and, along with it, the theorem on the integral is proved.

Proof of the uniqueness theorem. It is obvious that E' can be represented as

$$E' = P_1 + P_2 + \ldots P_m + \ldots + N, \quad P_m \subset P_{m+1},$$

where the P_m are closed and bounded and $m^k(N) = 0$. Let

$$Q_m = \pi(P_m), \quad M = \pi(N).$$

Since $\overline{\mu}^k(M) = 0$, we have

$$\overline{\mu}^k(Q_m) \to \overline{\mu}^k(E) \quad (m \to \infty).$$

Therefore we have to prove the relation

$$\underline{\mu}^k(Q_m) = \overline{\mu}^k(Q_m).$$

However, for closed sets P and Q this relation follows from the theorem on the integral according to the following theorem.

Theorem on uniformization. *If P is a closed bounded set and ϕ is a continuous mapping $Q = \phi(P)$, then there exists an A-subset A of P such that $\phi(A) = Q$, and ϕ is a one-to-one mapping onto A.*

Proof. Let $f, f(C) = P$, be a continuous mapping from a Cantor perfect set (see [1]) onto P. We consider the continuous mapping ψ, $Q = \psi(C) = \phi f(C)$. The pre-image $\psi^{-1}(x')$ of a point $x' \subset Q$ is a closed subset of C. We now denote by $y = \omega(x')$ the infimum

$$y = \inf \psi^{-1}(x')$$

of the set $\psi^{-1}(x')$. The function $y = \omega(x')$, is defined on Q and is semicontinuous; therefore its range A_C is an A-set. Obviously, the mapping $\psi(A_C) = Q$ is one-to-one. Now let $A = f(A_C)$. Thus A is also an A-set and the mapping $Q = \phi(A)$ is one-to-one, as required.

Finally, we note that by the theorem on the integral both functions $\overline{\mu}^k(E)$ and $\underline{\mu}^k(E)$ coincide with the surface measure for Lebesgue squarable surfaces (see [6], p. 315) (we consider only two-dimensional elements, that is, surfaces that can be transformed into a plane square by means of a one-to-one continuous mapping). For surfaces of this class T. Rado proved (see [7], p.445) that the classical integral expression $I^k(E)$ does in fact define the surface Lebesgue measure. The surface measure of sets lying on Lebesgue squarable surfaces, defined by Rademacher (see [8], p. 54 and particularly, §17, p.61), also coincides with $I^k(E) = \overline{\mu}^k(E) = \underline{\mu}^k(E)$.

Appendix. New definition of measure of a linear set

The notion of non-expanding mapping which has been used in this paper to justify a natural notion of measure for sets of a most general type suggests the following simple definition for linear sets. Consider all non-expanding mappings π from a given linear set E onto a closed interval J,

$$\pi(E) = J,$$

and put

$$L(E) = \sup\{L(J)\},$$

where $L(J)$ is the length of the interval J.

Theorem. *For Lebesgue measurable sets E the function $L(E)$ coincides with the Lebesgue measure $m^1(E)$.*

Proof. Let F be a closed bounded subset of E. We define the function $\pi(x)$ by

$$\pi(x) = m^1\{F(x)\},$$

where $F(x)$ denotes the set of all points belonging to F and lying to the left of x. It is easily seen that this function maps the set F, the entire line, and also the set E onto one and the same interval of length $m^1(F)$. Since, in addition, this mapping is non-expanding, we obtain

$$L(E) \geq m^1(F).$$

Furthermore, since F is an arbitrary bounded subset of E, we have

$$L(E) \geq \sup m^1(F) = m^1(E).$$

On the other hand, by Theorem 1 of §1, we have

$$L(E) \leq m^1(E),$$

which completes the proof of our theorem.

1 January 1932

References

1. F. Hausdorff, *Grundzüge der Mengenlehre*, 2nd ed., W. de Gruyter, Berlin-Leipzig, 1927.
2. E. Schmidt, 'Über die Definition des Begriffs der Länge krummer Linien', *Math. Ann.* **55** (1901), 163–176.

3. K. Carathéodory, 'Über das lineare Mass von Punktmengen', *Nachr. Ges. Wiss. Göttingen*, 1914.
4. F. Hausdorff, 'Dimension und aüsseres Mass', *Math. Ann* **79** (1918), 157–172.
5. H. Rademacher, 'Über partielle und totale Differentierbarkeit von Funktionen mehrerer Variabeln und über die Transformationen der Doppelintegrales', *Math. Ann.* **79** (1918), 340–359.
6. H. Lebesgue, Thèse, *Ann. Mat. Pura ed Appl.* (3) **7** (1902), 315.
7. T. Rado, 'Über das Flächenmass rektifizierbarer Flächen', *Math. Ann.* **100** (1928), 445–479.
8. H. Rademacher, 'Über partielle und totale Differentierbarkeit von Funktionen mehrerer Variabeln. II', *Math. Ann.* **81** (1920), 52–63.

22. ON POINTS OF DISCONTINUITY OF FUNCTIONS OF TWO VARIABLES*

(In collaboration with I.Ya. Verchenko)

Let $f(P)$ be a bounded ($|f(P)| \leq M$) single-valued real function defined at each point of a plane region G. Consider a point P_0 in the region G and a ray L issuing from the point P_0 and forming an angle α with the positive direction of the x-axis. We denote by $\Phi(P_0, \alpha)$ the supremum of all upper limits of $f(P)$ as P approaches P_0 along all tangent paths to L. The function $\Phi(P_0, \alpha)$ is called the upper limit of $f(P)$ at P_0 along the α-direction. One can similarly define the lower limit $\Phi(P_0, \alpha)$ of the function $f(P)$ at P_0 along the α-direction. $\Phi(P_0, \alpha)$ considered as a function of α is upper semicontinuous and $\phi(P_0, \alpha)$ is lower semicontinuous. These two functions of α characterize to a great extent the behaviour of $f(P)$ in a neighbourhood of P_0.

1. *The function $f(P)$ is continuous at P_0 if and only if*

$$\Phi(P_0, \alpha) = \phi(P_0, \alpha) = f(P_0)$$

for all α.

2. *P_0 is a point of discontinuity of the first kind if $f(P)$ is not continuous at P_0 and*

$$\Phi(P_0, \alpha) = \phi(P_0, \alpha)$$

for all α.

An arbitrary function $f(P)$ can have at most a countable number of points of discontinuity of the first kind.

Proof. Denote by $\omega(P_0)$ the oscillation of $f(P_0)$ with respect to P_0. Let E_n be the set of those points of discontinuity of the first kind for which the oscillation of $f(P)$ exceeds $1/n$. It can easily be shown that the set E_n does not contain any of its limit points and is therefore at most countable. Consequently, the set of all points of discontinuity of the first kind,

$$E = \sum_n E_n,$$

is also at most countable.

---

* *Dokl. Akad. Nauk SSSR* **1**:3 (1934), 105–106.

3. P_0 *is an ordinary point of the function* $f(P)$ *if, for all* α,

$$\Phi(P_0,\alpha) = \Phi(P_0) \geq f(P_0) \geq \phi(P_0) = \Phi(P_0,\alpha),$$

that is, if $\Phi(P_0,\alpha)$ *and* $\phi(P_0,\alpha)$ *do not depend on* α *and* $f(P_0)$ *lies between them. Each point at which the function* $f(P)$ *is continuous is an ordinary point. An ordinary point at which the function is not continuous is called a point of ordinary discontinuity.*

Theorem. *Given an arbitrary function* $f(P)$, *the set of all its non-ordinary points can be placed on a countable set of rectifiable curves.*

In particular, it follows that the plane measure of the set of non-ordinary points is zero. The proof of this theorem is based on the following auxiliary proposition, which is true when G is a bounded region.

Consider the sector of the circle of radius ρ with centre at P_0 which lies between directions $a-\delta$ and $a+\delta$, and denote by $M(P_0,\alpha,\delta,\rho)$ the supremum of $f(P)$ in the interior of the sector. For any $\alpha,\delta > 0$, $\rho > 0$ and $a > b$, all points P_0 such that

$$\bar{f}(P_0) \geq \alpha, \quad M(P_0,\alpha,\delta,\rho) \leq b$$

can be placed on a finite number of rectifiable curves. [1]

Moscow, 21 December 1933

[1] Here $\bar{f}(P_0)$ denotes the upper limit of $f(P)$ at P_0, that is, the maximum of $f(P_0)$, and $\Phi(P_0) = \sup\Phi(P_0,\alpha)$. Similarly, $\underline{f}(P_0)$ is the lower limit of $f(P)$ at P_0. Then $\omega(P_0) = \bar{f}(P_0) - \underline{f}(P_0)$.

23. ON NORMABILITY OF A GENERAL TOPOLOGICAL LINEAR SPACE *

Our discussion is based on the following definition of topological linear space. A set E is a topological linear space if

1) for the elements of E the operations of addition and multiplication by real numbers are defined which satisfy the axioms for a linear space; [1]

2) for any subset A of E its closure $\overline{A} \subset E$ is defined, which satisfies three axioms for a topological space; [2]

3) the operations of addition and multiplication are continuous.

Open sets of a topological space in the sense of footnote 2, regarded as neighbourhoods of the points of these sets, satisfy in general the first three Hausdorff axioms [3] and the first axiom of separation. However, in our case of topological linear spaces as well as in the general case of topological groups, the second and third axioms of separation are necessarily fulfilled, that is, the space is regular. This fact will be proved in §1. In this connection it seems quite natural that the general theory of linear functionals and operators should be developed in particular for topological linear spaces. A considerable part of this theory, however, has been developed only for the case of normed spaces, that is, spaces where each element is associated with a non-negative number $|x|$ satisfying the following conditions:

$$|ax| = |a|\,|x|, \tag{1}$$

$$|x+y| \leq |x| + |y|; \tag{2}$$

and the topology of the space is determined by the distance between two elements:

$$\rho(x,y) = |x-y|. \tag{3}$$

---

* 'Zur Normierbarkeit eines allgemeinen topologischen linearen Räumes', *Stud. Math.* **5** (1934), 29–33.

[1] See S. Banach, *Théorie des opérations linéaires*, Warszawa, 1931, Chapter 2.

[2] See, for example, P. Alexandroff, 'Über stetige Abbildungen kompakter Räume', *Math. Ann.* **96** (1926), 555. These axioms read as follows.

1. A subset $A \subset E$ consisting of at most one element coincides with its closure.
2. $\overline{\overline{A}} = \overline{A}$ for any set $A \subset E$.
3. $\overline{M \cup N} = \overline{M} \cup \overline{N}$.

[3] See F. Hausdorff, *Grundzüge der Mengenlehre*, Berlin, 1927, 227–229.

The question arises which topological linear spaces can be normed. In other words, what conditions should be imposed so that a topological linear space can be endowed with a norm satisfying conditions (1) and (2) and determining the *a priori* topological relations in this space? In order to answer this question we will use the following definition. [4]

Definition. A set $A \subset E$ is said to be bounded if for any sequence $\{a_n\}$ of real numbers and for any sequence $\{x_n\}$ of elements of A the condition $a_n \to 0$ implies $a_n x_n \to 0$ (where 0 in the latter relation denotes the origin of the space E).

This definition enables us to state the following theorem.

Theorem. *For a space E to be normable it is necessary and sufficient that in E there exist at least one bounded convex neighbourhood of zero.*

Here the set $A \subset E$ is said to be convex if $x \in A$, $y \in A$, $\lambda \geq 0$ and $\mu \geq 0$ imply

$$\frac{\lambda x + \mu y}{\lambda + \mu} \in A. \tag{4}$$

This theorem will be proved in §2.

§1

In this section, E is an arbitrary topological group, that is, an arbitrary topological space on which an operation of addition is defined satisfying all the group axioms, and the addition and subtraction are continuous. We will prove that in this case E is regular or, in other words, that for any neighbourhood $U(x_0)$ of an element $x_0 \in E$ there exists a neighbourhood $V(x_0)$ which belongs to $U(x_0)$ together with its closure. Using the transformation $x' = x - x_0$ we can reduce this problem to the case $x_0 = 0$.

Thus, let a neighbourhood U of zero be given. Since $0 + 0 = 0$, we can, by the continuity of addition, find a neighbourhood of zero V such that $x' \in V$, $x'' \in V$ implies $x' + x'' \in U$. Now let $x \in \overline{V}$. Since $x - x = 0$ and subtraction is continuous, there exists a neighbourhood $W(x)$ of x such that $x' \in W(x)$, $x'' \in W(x)$ implies $x' - x'' \in \overline{V}$. We now find a point x^* that belongs simultaneously to $W(x)$ and V (x^* exists since $x \in \overline{V}$). Since x and

[4] S. Mazur and W. Orlicz, 'Über Folgen linearer Operationen', *Stud. Math.* 4 (1933), 152–157, particularly p. 152.

x^* belong to $W(x)$, it follows that $x - x^* \in \overline{V}$. Further, since $x - x^*$ and x^* belong to V, it follows that $x = (x - x^*) + x^*$ belongs to U, whence

$$\overline{V} \in U, \tag{5}$$

and so on.

§2

The necessity of the conditions of the theorem is obvious since the unit ball, consisting of all points with $|x| < 1$, is a convex bounded neighbourhood of zero. It remains to show that the conditions of the theorem are sufficient. Let U be a convex bounded neighbourhood of zero. We denote by aU the set of points $x = ax'$ such that $x' \in U$. For any $a \neq 0$ the set aU is a convex bounded neighbourhood of zero. We define the norm by

$$|x| = \sup |a|, \quad x \in E \backslash aU, \tag{6}$$

where the supremum extends over all a satisfying the condition $x \in E\backslash aU$.

It can be seen immediately that the norm defined by (6) satisfies condition (1).

We now show that condition (2) also holds. To this end we first note that $x \in aU$ and $y \in aU$ imply

$$\frac{\lambda x + \mu y}{\lambda + \mu} \in aU$$

for $\lambda \geq 0$ and $\mu \geq 0$ and therefore it follows from $|x| \leq \alpha$ and $|y| \leq \alpha$ that

$$\left|\frac{\lambda x + \mu y}{\lambda + \mu}\right| \leq \alpha.$$

Putting $|x| = \lambda,\ |y| = \mu,\ \lambda + \mu = \alpha,\ x' = \frac{\alpha}{\lambda}x,\ y' = \frac{\alpha}{\mu}y$ we obtain

$$|x'| = \frac{\alpha}{\lambda}|x| = \alpha, \quad |y'| = \frac{\alpha}{\mu}|y| = \alpha,$$

$$|x + y| = \left|\frac{\lambda x' + \mu y'}{\lambda + \mu}\right| \leq \alpha = \lambda + \mu = |x| + |y|.$$

It remains to show that the distance $\rho(x, y) = |x - y|$ determines the same limiting relations as in the space E. Let us consider the point 0 and prove that the sets aU, $a > 0$ form a complete system of neighbourhoods of zero. Let W be an arbitrary neighbourhood of zero. We have to find $a > 0$ such that

aU is contained in W. If this were impossible, there would exist sequences $a_n \to 0$ and $x_n \in U$ such that the points $a_n x_n$ would lie outside W for all n and therefore the sequence $a_n x_n$ would not tend to zero, which contradicts the boundedness of U.

Therefore the sets aU form a complete system of neighbourhoods of zero. It can be seen immediately that the ball $|x| < a$ is contained in the set aU. Therefore, if we prove that 0 is an interior point of every ball $|x| < a$, $a > 0$, this will imply that the system of balls $|x| < a$, $a > 0$ is equivalent to the complete system of neighbourhoods aU of zero. But each ball $|x| < a$, $a > 0$ contains a neighbourhood of zero, coinciding with the intersection [5] of $\frac{1}{2}aU$ and $-\frac{1}{2}aU$. Indeed, if $x \in \frac{1}{2}aU$ and $x \in -\frac{1}{2}aU$, then x lies in every neighbourhood bU with $|b| \geq a/2$, whence it follows that $|x| \leq \frac{1}{2}a < a$.

Using the transformation $x' = x - x_0$ we can see that the system of balls with centre at x_0 is equivalent to the complete system of neighbourhoods at this point. Thus, the theorem is proved completely.

26 May 1934

[5] It can easily be seen that the ball $|x| < a$ coincides with the intersection $aU \cap (-aU)$.

24. CONTINUATION OF THE STUDY OF POINTS OF DISCONTINUITY OF FUNCTIONS OF TWO VARIABLES *

(In collaboration with I.Ya. Verchenko)

In the previous paper [1] it was proved that, given an arbitrary bounded function of two variables, all points of its domain G, except for points lying on a countable number of rectifiable curves, are ordinary, that is, satisfy the condition

$$\Phi(P_0, \lambda) = \Phi(P_0) \geq f(P_0) \geq \phi(P_0) = \phi(P_0, \alpha).$$

Among the non-ordinary points the class of so-called *linear points* can be singled out. Namely, we will call P_0 a linear point of a function $f(P)$ if there exists a straight line R passing through P_0 such that both $\Phi(P_0, \alpha)$ and $\phi(P_0, \alpha)$ assume equal values in both directions along R and are constant on the two sides of R. Denoting by ξ the positive direction of R, we can write the following conditions for linearity at the point P_0:

$$\Phi(P_0, \xi) = \Phi(P_0, \xi + \pi) = \Phi_0(P_0),$$
$$\phi(P_0, \xi) = \phi(P_0, \xi + \pi) = \phi_0(P_0),$$
$$\left.\begin{array}{l}\Phi(P_0, \alpha) = \Phi_1(P_0)\\ \phi(P_0, \alpha) = \phi_1(P_0)\end{array}\right\} \quad \xi < \alpha < \xi + \pi,$$
$$\left.\begin{array}{l}\Phi(P_0, \alpha) = \Phi_2(P_0)\\ \phi(P_0, \alpha) = \phi_2(P_0)\end{array}\right\} \quad \xi + \pi < \alpha < \xi + 2\pi,$$

It is easily seen that

$$\Phi_0(P_0) \geq \Phi_1(P_0) \geq \phi_1(P_0) \geq \phi_0(P_0),$$
$$\Phi_0(P_0) \geq \Phi_2(P_0) \geq \phi_2(P_0) \geq \phi_0(P_0).$$

Simple examples show that non-ordinary linear points can fill entire rectifiable curves. By contrast, we have the following theorem.

Theorem. The set of non-linear points of an arbitrary bounded function $f(P)$ has linear measure zero.

Since non-linear points are simultaneously non-ordinary and hence lie on a countable number of rectifiable curves, to prove the theorem it suffices to

* *Dokl. Akad. Nauk SSSR* **4**:7 (1934), 361–362.

[1] *Dokl. Akad. Nauk SSSR* **1**:3 (1934), 104–107. (See No. 22 in this book.)

show that the Lebesgue measure of the set of non-linear points is zero on any rectifiable curve.

In conclusion we show how these results can be applied in the study of plane sets, following Bouligand's approach.

Let $f(P)$ be the characteristic function of a plane point set E, that is, the function equal to one on E and equal to zero outside E. In this case $\Phi(P_0, \alpha)$ can assume only two values, zero and one. The set $C(P_0)$ which is the union of all rays issuing from P_0 such that $\Phi(P_0, \alpha) = 1$ is called (by Bouligand) the contingency set of the set E at the point P_0. The notion of contingency set is a natural generalization of the notion of tangent. It follows immediately from the upper semicontinuity of $\Phi(P_0, \alpha)$ that the contingency set $C(P_0)$ is closed.

At isolated points of E the contingency set is empty also at all points lying at a positive distance from E (the latter are called exterior points of E). By contrast, at all limiting points of E the contingency set contains at least one ray. If a limiting point of E is an ordinary point of the function $f(P)$, then obviously, the contingency set of E at this point contains the whole plane. Along with the case of a contingency set filling the whole plane, two other essentially different cases are possible for linear non-ordinary points of the function $f(P)$:

$$\Phi_0(P_0) = 1, \quad \Phi_1(P_0) = \Phi_2(P_0) = 0$$

and

$$\Phi_0(P_0) = \Phi_1(P_0) = 1, \quad \Phi_2(P_0) = 0.$$

In the former case the contingency set consists of one line (that is, the tangent exists), and in the latter case it consists of a half-plane. Taking into account these distinctions we introduce the following definitions.

Definition 1. A set E is said to have a tangent at P_0 if the contingency set $C(P_0)$ consists of one line.

Definition 2. A point P_0 is said to belong to the regular part of E if $C(P_0)$ consists of a half-plane.

Definition 3. If $C(P_0)$ is the whole plane, then P_0 is said to be an inaccessible point of E.

Definition 4. A point P_0 is said to be exterior to E if it lies at a positive distance from E.

Definition 5. In all other cases P_0 is called an irregular point of E.

Theorem. *The set of points at which the set E has a tangent, the set of points belonging to the regular part of E, and the set of irregular points of E lie on a countable number of rectifiable curves. On each of the curves the set of irregular points has linear measure zero in the sense of Lebesgue. The remaining points in the plane are either inaccessible or exterior to E.*

Moscow, 3 November 1934

25. ON THE CONVERGENCE OF SERIES IN ORTHOGONAL POLYNOMIALS *

In his recent paper [1] I.P. Natanson has proved that for any summable positive weight function $p(x)$ the series in orthogonal polynomials converges almost everywhere to the function $f(x)$, where f admits such a series expansion, provided that the latter function satisfies a Lipschitz condition with exponent α greater than $\frac{1}{2}$. We are going to show that this result remains valid for every function satisfying a Lipschitz condition with arbitrarily small positive exponent α. We establish in fact a somewhat stronger result; namely, the above-mentioned series converges almost everywhere if there exist constants C and $\epsilon > 0$ such that

$$|f(x'') - f(x')| \leq \frac{C}{|\log|x'' - x'||^{1+\epsilon}} \tag{1}$$

for any $x'' \neq x'$.

We recall the basic definitions. Let a positive summable function

$$p(x) > 0$$

be defined on the interval $(0,1)$. Then one can construct a unique sequence of polynomials

$$\omega_0(x),\ \omega_1(x), \ldots,\ \omega_n(x), \ldots,$$

where $\omega_n(x)$ is a polynomial of degree n, such that

$$\begin{aligned} &\int_0^1 p(x)\omega_i(x)\omega_j(x)dx = 0, \quad i \neq j, \\ &\int_0^1 p(x)\omega_i^2(x)dx = 1. \end{aligned} \tag{2}$$

Each function $f(x)$ on $(0,1)$ for which the product $p(x)f(x)$ is summable gives rise to a series

$$\sum_{n=0}^{\infty} c_n\omega_n(x), \tag{3}$$

where

$$c_n = \int_0^1 p(x)f(x)\omega_n(x)dx. \tag{4}$$

---

* *Dokl. Akad. Nauk SSSR* 1:6 (1934), 291–294.

Theorem 1. *If the function $f(x)$ satisfies condition* (1) *on* $(0,1)$, *then the series* (3) *converges to $f(x)$ almost everywhere on* $(0,1)$.

The proof of Theorem 1 is based on the following proposition.

Theorem 2. *The series* (3) *converges almost everywhere, provided that the series*

$$\sum_{n=1}^{\infty} c_n^2 (\log n)^2 \tag{5}$$

is convergent.

Theorem 2 remains valid for any systems of functions ω_i satisfying relations (2), even when they are not polynomials, and for any coefficients c_n, irrespective of whether they are obtained by means of (4) from a function $f(x)$. This theorem is a simple consequence of well-known results of Rademacher and Men'shov. Thus, we set

$$y = \int_0^x p(x)dx. \tag{6}$$

Without loss of generality we can assume that

$$\int_0^1 p(x)dx = 1.$$

Then (6) determines a one-to-one continuous mapping of the interval $(0,1)$ into itself. Note that this mapping transforms sets of positive measure on the x-axis into sets of positive measure on the y-axis. We then set

$$\omega_n(x) = \phi_n(y).$$

The relations (2) readily imply

$$\begin{aligned} &\int_0^1 \phi_i(y)\phi_j(y)dy = 0, \quad i \neq j, \\ &\int_0^1 \phi_i^2(y)dy = 1. \end{aligned} \tag{7}$$

It can be seen that the functions $\phi_n(x)$ form an orthonormal system on $(0,1)$. Rademacher [2] and Men'shov [3] proved that convergence of the series (5) implies convergence of the series

$$\sum_{n=0}^{\infty} c_n \phi_n(y) \tag{8}$$

almost everywhere on $(0,1)$. By virtue of the above, it follows immediately that the series (3) also converges almost everywhere. Indeed, we have

$$\sum_{n=0}^{\infty} c_n \omega_n(x) = \sum_{n=0}^{\infty} c_n \phi_n(y),$$

and if series (3) were to diverge on a set of positive measure on the x-axis, then the series (8) would diverge on a set of positive measure on the y-axis, which, as we have seen, is impossible. Thus, Theorem 2 is proved.

Proof of Theorem 1. If we prove that under the conditions of the theorem the series (3) converges almost everywhere, then it will follow from general properties of orthogonal polynomials that this series converges almost everywhere precisely to the function $f(x)$. Therefore it suffices to prove that under the conditions of Theorem 1 the series (3) converges almost everywhere, and to this end, in view of Theorem 2, it suffices to prove convergence of the series (5).

By condition (1), there exists a constant K such that for any $n > 1$ there is a polynomial $P_n(x)$ of degree n satisfying the inequality

$$|f(x) - P_n(x)| \leq \frac{K}{(\log n)^{1+\epsilon}} \tag{9}$$

on $(0,1)$ (see [4]). Since the sum

$$s_n(x) = \sum_{k=0}^{n} c_k \omega_k(x)$$

gives the best quadratic approximation to the function $f(x)$ on $(0,1)$ with the weight $p(x)$ among all polynomials of degree n, we have

$$R_n = \int_0^1 p(x)\{f(x) - s_n(x)\}^2 dx \leq \int_0^1 p(x)\{f(x) - P_n(x)\}^2 dx \leq \frac{K^2}{(\log n)^{2+2\epsilon}} \int_0^1 p(x) dx = \frac{K^2}{(\log n)^{2+2\epsilon}}. \tag{10}$$

It readily follows from (10) that the series

$$\sum_{n=1}^{\infty} R_n \frac{(\log n)^{1+\epsilon}}{n} \tag{11}$$

is convergent. We now note that

$$R_n = \sum_{k=n+1}^{\infty} c_k^2, \quad R_{n-1} - R_n = c_n^2.$$

Applying the Abel transformation to the series (11) we conclude that the series

$$\sum_{n=2}^{\infty} c_n^2 \lambda_n, \tag{12}$$

where

$$\lambda_n = \sum_{k=1}^{n-1} \frac{(\log k)^{1+\epsilon}}{k},$$

is also convergent. But since λ_n increases more rapidly than

$$(\log n)^2,$$

the convergence of the series (12) implies that the series (5) is convergent. In view of what was said above, this completes the proof of Theorem 1. [1]

21 December 1933

References

1. I.P. Natanson, 'On the problem of expanding functions in series of orthogonal polynomials', *Izv. Akad. Nauk SSSR, Ser. Fiz.-Mat.* **1** (1933), 85–88 (in Russian).
2. H. Rademacher, 'Einige Sätze über Reihen von allgemeinen Orthogonalfunktionen', *Math. Ann.* **87** (1922), 112–138.
3. D. Menchoff, 'Sur les séries de fonctions orthogonales', *Fund. Math.* **4** (1923), 82–105.
4. D. Jackson, *The theory of approximation*, New York, 1930.

[1] The interrelation between the problem posed by Natanson and the Rademacher result which is used in this paper is also indicated in the review of Natanson's paper written by Shokhat (see *Zbl. Math.* **7** (1933), 17). This review contains, however, a number of assertions which are difficult to understand. For example, it is asserted that for a sequence of square-integrable functions $\phi_n(x)$ the relation $\lim_{n\to\infty} \int_a^b \phi_n^2(x)dx = 0$ implies that $\lim_{n\to\infty} \phi_n(x) = 0$ almost everywhere on (a,b). The following example demonstrates a contradiction to this assertion:

$$\phi_n(x) = 0 \text{ if } x < i/2^k \text{ or } x \geq (i+1)/2^k;$$
$$\phi_n(x) = 1 \text{ if } i/2^k \leq x < (i+1)/2^k;$$

where $n = 2^k + 1,\ i = 0, 1, 2, \ldots, 2^k - 1$.

26. LAPLACE TRANSFORMATION IN LINEAR SPACES *

1. Let E be a separable Banach space. Consider a real completely additive set function $F(A)$ defined on all B-measurable sets A in E. For a linear functional f the Lebesgue-Stieltjes integral

$$H(f) = \int_E e^{if(x)} dF(E)$$

always exists. Here x is the variable with respect to which the integration is performed in the space E. The function $H(f)$ depending on the functional f is, by definition, the characteristic function of $F(A)$. It can be shown that the set function $F(A)$ is completely determined by the corresponding function $H(f)$. It would be very interesting to find a simple analytic formula expressing $F(A)$ in terms of $H(f)$.

2. Denoting by $A + x$ the set of points $z = y + x$ where $y \in A$, we can prove that

$$F(A) = \int_E F_1(A - x) dF_2(E)$$

is equivalent to

$$H(f) = H_1(f) H_2(f).$$

This fact can be used in probability theory. In particular, it follows that the fundamental limit theorem of Laplace can be extended to linear spaces. In this case the role of normal distributions is played by distributions $F(A)$ with characteristic functions

$$H(f) = \exp(iM_1(f) - \tfrac{1}{2} M_2(f, f)),$$

where $M_1(f)$ is an arbitrary linear form and $M_2(f, f)$ a symmetric positive definite (or positive semidefinite) quadratic form.

3. Now let

$$M_n(f_1, f_2, \ldots, f_n) = \int_E f_1(x) f_2(x) \ldots f_n(x) dF(E).$$

---

* 'La transformation de Laplace dans les espaces linéaires', *C.R. Acad. Sci. Paris* **200** (1935), 1717–1718.

The multilinear form F_n is called the moment form of order n of the distribution $F(A)$. If the integral

$$R_{n+1} = \int_E ||x||^{n+1} d|F(E)|$$

is finite, we easily obtain the Taylor expansion

$$H(f) = 1 + M_1(f) + \frac{1}{2}M_2(f,f) + \frac{1}{6}M_3(f,f,f) + \ldots + \\ + \frac{1}{n!}M_n(f,f,\ldots,f) + \frac{1}{(n+1)!}\theta R_{n+1}||f||^{n+1} \quad (|\theta| \leq 1).$$

4. Finally, let

$$y = Ux$$

be a linear transformation of E and let

$$G(A) = F\{U(A)\}.$$

For the moments N_n of the distribution $G(A)$ we have

$$N_n(f) = M_n\{V(f)\}.$$

Here $N_n(f) = N_n(f,f,\ldots,f)$, $M_n\{V(f)\} = M_n(V(f),V(f),\ldots,V(f))$ and V is the transformation adjoint to U.

Thus, given a sequence of linear transformations of the space E, it is possible to study the behaviour of moments N_n of various orders for each n separately. Owing to this general fact, in quantum physics one studies moments of the second order (usually considered in the form of linear operators) instead of distributions in linear spaces. If we wanted to develop a non-linear quantum theory, it would be necessary to consider the distributions themselves or their characteristic functions or the set of all moments N_n.

20 May 1935

27. ON THE ORDER OF MAGNITUDE OF THE REMAINDER TERM IN THE FOURIER SERIES OF DIFFERENTIABLE FUNCTIONS *

First statement of the problem. Consider the class of all periodic functions (with period 2π) $f(x)$ that are continuous together with their pth derivative ($p \geq 1$), and for which the absolute value of $f^{(p)}(x)$ is less than or equal to 1. The absolute value of the remainder term

$$R_n(f, x) = f(x) - \tfrac{1}{2}a_0 - \sum_{k=1}^{n}(a_k \cos kx + b_k \sin kx)$$

of the Fourier series of such a function has finite supremum $C_n^{(p)}$, which does not depend on x.

For an arbitrary p times continuously differentiable function $f(x)$ we immediately obtain the inequality

$$|R_n(f, x)| \leq C_n^{(p)} M_p(f), \tag{1}$$

where $M_p(f)$ is the supremum of $|f^{(p)}(x)|$. Clearly, the constant $C_n^{(p)}$ in inequality (1) cannot be made smaller.

Second statement of the problem. We can also consider the broader class of all periodic functions $f(x)$ that are continuous together with their $(p-1)$th derivative, and for which the derivative $f^{(p-1)}(x)$ satisfies the Lipschitz condition

$$|f^{(p-1)}(x) - f^{(p-1)}(y)| \leq |x - y|. \tag{2}$$

Nevertheless, the supremum $C_n^{(p)}$ of $|R_n(f, x)|$ for this broader class of functions remains the same as in the former statement of the problem.

In 1910 Lebesgue [1] showed that the order of magnitude of $C_n^{(1)}$ is $(\log n)/n$. Our aim is to show that, in general,

$$C_n^{(p)} = \frac{4}{\pi^2}\frac{\log n}{n^p} + O\Big(\frac{1}{n^p}\Big). \tag{3}$$

In the case of odd p we will also prove the exact formula

$$C_n^{(p)} = \frac{1}{\pi}\int_0^{2\pi}\Big|\sum_{k=n+1}^{\infty}\frac{\sin kx}{k^p}\Big|dx. \tag{4}$$

---

* 'Zur Grössenordnung des Restgliedes Fourierschen Reihen differenzierbarer Funktionen', *Ann. of Math.* **36** (1935), 521–526.

§1

The presentation below is based on the second statement of the problem. Since the derivative $f^{(p-1)}(x)$ satisfies the Lipschitz condition (2), it follows that the pth derivative also exists almost everywhere, and $|f^{(p)}(x)| \leq 1$. Conversely, an arbitrary measurable function $\phi(x)$ defined everywhere except on a set of measure zero and such that

$$|\phi(x)| \leq 1, \quad \int_0^{2\pi} \phi(x)dx = 0 \tag{5}$$

can be taken as $f^{(p)}(x)$.

We now put

$$D_n^{(0)} = \frac{\sin \frac{2n+1}{2} x}{2\sin(x/2)} = \tfrac{1}{2} + \sum_{k=1}^{n} \cos kx, \tag{6}$$

$$D_n^{(1)} = \frac{x-\pi}{2} + \sum_{k=1}^{n} (\sin kx)/k = - \sum_{k=n+1}^{\infty} (\sin kx)/k, \tag{7}$$

$$D_n^{(2i)}(x) = (-1)^{i+1} \sum_{k=n+1}^{\infty} \frac{\cos kx}{k^{2i}}, \quad i \geq 1, \tag{8}$$

$$D_n^{(2i+1)}(x) = (-1)^{i+1} \sum_{k=n+1}^{\infty} \frac{\sin kx}{k^{2i+1}}, \quad i \geq 1. \tag{9}$$

Obviously,

$$\frac{d}{dx} D_n^{(p)}(x) = D_n^{(p-1)}(x), \ 0 < x < 2\pi, \quad p \geq 1, \tag{10}$$

$$D_n^{(1)}(2\pi) - D_n^{(1)}(0) = \pi, \tag{11}$$

$$D_n^{(p)}(2\pi) - D_n^{(p)}(0) = 0, \quad p \geq 2. \tag{12}$$

From the well-known Dirichlet formula

$$S_n(0) = \frac{1}{\pi} \int_0^{2\pi} f(x) D_n^{(0)}(x)dx$$

we obtain, by integrating by parts and using (10)–(12),

$$S_n(0) = f(0) - \frac{1}{\pi} \int_0^{2\pi} f'(x) D_n^{(1)}(x)dx = f(0) + \frac{1}{\pi} \int_0^{2\pi} f''(x) D_n^{(2)}(x)dx =$$
$$= f(0) + \frac{(-1)^p}{\pi} \int_0^{2\pi} f^{(p)}(x) D_n^{(p)}(x)dx.$$

Consequently,

$$R_n(f,0) = f(0) - S_n(0) = \frac{(-1)^{p+1}}{\pi} \int_0^{2\pi} f^{(p)}(x) D_n^{(p)}(x)dx, \ p \geq 1. \tag{13}$$

Also note that $C_n^{(p)}$ is the supremum of $R_n(f,0)$, so that $C_n^{(p)}$ is equal to the supremum of

$$\frac{1}{\pi}\left|\int_0^{2\pi} \phi(x) D_n^{(p)}(x)dx\right| \tag{14}$$

for all measurable function $\phi(x)$ satisfying the conditions (5). It follows immediately that

$$C_n^{(p)} \leq \frac{1}{\pi}\int_0^{2\pi} |D_n^{(p)}(x)|dx. \tag{15}$$

When p is odd, it follows from (7) and (9) that $D_n^{(p)}$ is an odd function of x. In this case, putting $\phi(x) = +1$ at each point at which $D_n^{(p)} > 0$ and $\phi(x) = -1$ if $D_n^{(p)} < 0$, we obtain a function $\phi(x)$ satisfying the conditions (5). For this function $\phi(x)$ we obtain

$$\frac{1}{\pi}\int_0^{2\pi} \phi(x) D_n^{(p)}(x)dx = \frac{1}{\pi}\int_0^{2\pi} |D_n^{(p)}(x)|dx.$$

Thus, it is seen that for odd p we have

$$C_n^{(p)} = \frac{1}{\pi}\int_0^{2\pi} |D_n^{(p)}(x)|dx. \tag{16}$$

Thus, formula (4) is proved for odd p.

§2. Asymptotic estimation of $C_n^{(p)}$ for $p = 2i + 1$ $(i \geq 0)$

This section is concerned with providing an asymptotic estimate for the integral in (16). We set

$$\begin{aligned} \Phi_k(x) &= -\frac{\cos\frac{2k+1}{2}x}{2\sin(x/2)}, \quad \Psi_k(x) = -\frac{\sin(k+1)x}{4(\sin(x/2))^2}, \\ \Delta(k) &= \frac{1}{k^p} - \frac{1}{(k+1)^p}, \\ \Delta^2(k) &= \Delta(k) - \Delta(k+1). \end{aligned}$$

Here

$$\Phi_k(x) - \Phi_{k-1}(x) = \sin kx, \quad \Psi_k(x) - \Psi_{k-1}(x) = \Phi_k(x).$$

The Abel transformation yields

$$D_n^{(p)}(x) = (-1)^{i+1} \sum_{k=n+1}^{\infty} \frac{\sin kx}{k^p} = (-1)^{i+1}\Big[-\frac{1}{(n+1)^p}\Phi_n(x)+$$

$$+\sum_{k=n+1}^{\infty} \Delta(k)\Phi_k(x)\Big] = (-1)^{i+1}\Big[-\frac{1}{(n+1)^p}\Phi_n(x)-$$

$$-\Delta(n+1)\Psi_n(x) + \sum_{k=n+1}^{\infty} \Delta^2(k)\Psi_k(x)\Big]. \tag{17}$$

We now note that [2]

$$|D_n^{(p)}(x)| \le \sum_{k=n+1}^{\infty} \frac{1}{k^p} = O\Big(\frac{1}{n^{p-1}}\Big),$$

and therefore

$$\int_0^{1/n} |D_n^{(p)}(x)|dx + \int_{2\pi-1/n}^{2\pi} |D_n^{(p)}(x)|dx = O\Big(\frac{1}{n^p}\Big).$$

We also note that

$$\int_{1/n}^{2\pi-1/n} |\Psi(x)|dx = O(n),$$

$$\Delta(k) = O(1/k^{p+1}), \quad \Delta^2(k) = O(1/k^{p+1}),$$

and therefore it follows from (17) that

$$\Big|\int_0^{2\pi} |D_n^{(p)}(x)|dx - \frac{1}{(n+1)^p}\int_{1/n}^{2\pi-1/n} |\Phi_n(x)|dx\Big| \le$$

$$\le \int_0^{1/n} |D_n^{(p)}(x)|dx + \int_{2\pi-1/n}^{2\pi} |D_n^{(p)}(x)|dx+$$

$$+\Delta(n+1)\int_{1/n}^{2\pi-1/n} |\Psi_n(x)|dx + \sum_{k=n+1}^{\infty} \Delta^2(x)\int_{1/k}^{2\pi-1/k} |\Psi_k(x)|dx =$$

$$= O\Big(\frac{1}{n^p}\Big) + O\Big(\frac{1}{n^p}\Big) + O\Big(\frac{1}{n^{p+1}}\Big)O(n)+$$

$$+\sum_{k=n+1}^{\infty} O\Big(\frac{1}{k^{p+2}}\Big)O(k) = O\Big(\frac{1}{n^p}\Big). \tag{18}$$

---

[2] For $p = 1$ the estimate $|D_n^{(1)}(x)| = O(1)$ follows from (7) by virtue of the uniform boundedness of the partial sums of the series $\sum \frac{\sin kx}{k}$.

On the other hand, as in the case of the Lebesgue constants,

$$\frac{1}{\pi}\int_0^{2\pi}\left|\frac{\sin\frac{2n+1}{2}x}{2\sin(x/2)}\right|dx = \frac{1}{\pi}\int_0^{2\pi}|D_n^{(0)}(x)|dx = \frac{4}{\pi^2}\log n + O(1), \tag{19}$$

it can be proved that

$$\frac{1}{\pi}\int_{1/n}^{2\pi-1/n}|\Phi_n(x)|dx = \frac{1}{n}\int_{1/n}^{2\pi-1/n}\left|\frac{\cos\frac{2n+1}{2}}{2\sin(x/2)}\right|dx = \frac{4}{\pi^2}\log n + O(1). \tag{20}$$

Finally, from (18) and (20) we obtain

$$\frac{1}{\pi}\int_0^{2\pi}|D_n^{(p)}(x)|dx = \frac{4}{\pi^2}\frac{\log n}{n^p} + O\Big(\frac{1}{n^p}\Big). \tag{21}$$

Thus, by virtue of (21) and (16), the estimate (3) is proved for odd p.

§3. Asymptotic estimation of $C_n^{(p)}$ for $p = 2i$ ($i \geq 1$)

In this case we put

$$\sigma_n(x) = \tfrac{1}{2}\Big(\frac{\sin((n+1)x/2)}{\sin(x/2)}\Big)^2$$

and note that

$$D_k^{(0)}(x) - D_{k-1}^{(0)}(x) = \cos kx, \quad \sigma_k(x) - \sigma_{k-1}(x) = D_k^{(0)}(x).$$

Using the Abel transformation we obtain

$$\begin{aligned} D_n^{(p)}(x) &= (-1)^{i+1}\sum_{k=n+1}^{\infty}\frac{\cos kx}{k^{2i}} = \\ &= (-1)^{i+1}\Big[-\frac{1}{(n+1)^p}D_n^{(0)}(x) + \sum_{k=n+1}^{\infty}\Delta(k)D_k^{(0)}(x)\Big] = \\ &= (-1)^{i+1}\Big[-\frac{1}{(n+1)^p}D_n^{(0)}(x) - \Delta(n+1)\sigma_n(x) + \\ &+ \sum_{k=n+1}^{\infty}\Delta^2(k)\sigma_k(x)\Big] = \frac{(-1)^i}{(n+1)^p}D_n^{(0)}(x) + Q_n^{(p)}(x). \end{aligned} \tag{22}$$

Further, it is well known that

$$\int_0^{2\pi}|\sigma_n(x)| = O(n).$$

It follows that

$$\int_0^{2\pi} |Q_n^{(p)}(x)dx| \le \Delta(n+1)\int_0^{2\pi} |\sigma_n(x)|dx + \sum_{k=n+1}^{\infty} \Delta^2(k)\int_0^{2\pi} |\sigma_k(x)|dx =$$

$$= O\Big(\frac{1}{n^{p+1}}\Big)O(n) + \sum_{k=n+1}^{\infty} O\Big(\frac{1}{k^{p+2}}\Big)O(k) = O\Big(\frac{1}{n^p}\Big). \tag{23}$$

From (15), (22), (19) and (23) we obtain the following estimate:

$$C_n^{(p)} \le \frac{1}{\pi}\int_0^{2\pi} |D_n^{(p)}(x)|dx \le \frac{1}{\pi(n+1)^p}\int_0^{2\pi} |D_n^{(0)}(x)|dx +$$

$$+\int_0^{2\pi} |Q_n^{(p)}(x)|dx = \frac{4}{\pi^2}\frac{\log n}{n^p} + O\Big(\frac{1}{n^p}\Big). \tag{24}$$

Finally, let $\phi(x) = 0$ on the interval $0 \le x \le 2\pi/(2n+1)$ and $\phi(x) = +1$ or $\phi(x) = -1$ for $x > 2\pi/(2n+1)$, depending on whether $\sin\frac{2n+1}{2}x$ is positive or negative. This function satisfies the conditions (5), and we have

$$\frac{1}{\pi}\int_0^{2\pi} \phi(x)D_n^{(0)}(x)dx = \frac{1}{\pi}\int_{2\pi/(2n+1)}^{2\pi} |D_n^{(0)}(x)|dx = \frac{4}{\pi^2}\log n + O(1),$$

and, by virtue of (22) and (23),

$$\frac{1}{\pi}\Big|\int_0^{2\pi} \phi(x)D_n^{(p)}(x)\Big|dx = \frac{1}{\pi(n+1)^p}\Big|\int_0^{2\pi} \phi(x)D_n^{(0)}(x)\Big|dx + O\Big(\frac{1}{n^p}\Big) =$$

$$= \frac{4}{\pi^2}\frac{\log n}{n^p} + O\Big(\frac{1}{n^p}\Big). \tag{25}$$

By the definition of $C_n^{(p)}$ it follows from (25) that

$$C_n^{(p)} \ge \frac{4}{\pi^2}\frac{\log n}{n^p} + O\Big(\frac{1}{n^p}\Big). \tag{26}$$

Estimates (26) and (24) prove (3) in the case of p even.

Klyaz'ma, near Moscow, 30 November 1934

References

1. H. Lebesgue, 'Sur la représentation trigonométrique approchée des fonctions satisfaisant à une condition de Lipschitz', *Bull. Soc. Math. France* **38** (1910), 184–210.

28. ON THE BEST APPROXIMATION OF FUNCTIONS OF A GIVEN CLASS *

General statement of the problem. It is assumed that a distance is introduced for the functions in question. If we consider the problem of approximating a function f by linear forms

$$\phi = c_1\phi_1 + c_2\phi_2 + \ldots + c_n\phi_n$$

with fixed functions $\phi_1, \phi_2, \ldots, \phi_n$, then the following Chebyshev problem arises: to minimize the distance $\rho(f, \phi)$ by appropriately choosing the coefficients $c_1, c_2, \ldots, c_n$. Putting aside the questions concerning existence and uniqueness, we denote by $E_n(f)$ the infimum of $\rho(f, \phi)$ with respect to ϕ. For a class F of functions f, we further denote by $E_n(F)$ the supremum of $E_n(f)$ over all $f \in F$. The expression $E_n(F)$ is determined by the class F and the functions $\phi_1, \phi_2, \ldots, \phi_n$. We now state a new problem: given F and n, it is required to minimize $E_n(F)$ by choosing the functions $\phi_1, \phi_2, \ldots, \phi_n$. Irrespective of whether this minimum exists, we denote by $D_n(F)$ the infimum of $E_n(F)$.

Thus, to each function class F there corresponds a sequence of completely defined non-negative (possibly infinite) values

$$D_1(F) \geq D_2(F) \geq \ldots \geq D_n(F) \geq \ldots .$$

If the minimum value $D_n(F)$ is actually attained on $E_n(F)$, we can also pose the question of uniqueness in the following way: are the functions $\phi_1, \phi_2, \ldots, \phi_n$ yielding the minimum $D_n(F)$ determined uniquely to within a linear transformation?

2. Specialization. In what follows we will consider only functions of one real variable defined on the interval $[0, 1]$. The distance $\rho(f, \phi)$ is defined as

$$\rho(f, \phi) = \left[\int_0^1 (f - \phi)^2 dx\right]^{\frac{1}{2}}.$$

Under these assumptions we manage to obtain some simple results for the following function classes:

---

* 'Über die beste Annäherung von Funktionen einer gegebenen Funktionenklasse', *Ann. of Math.* **37** (1936), 107–110.

F_p $(p \geq 1)$ consists of all p times differentiable functions f such that

$$\int_0^1 \{f^{(p)}\}^2 dx \leq 1;$$

F_p^* $(p \geq 1)$ consists of all functions belonging to F_p that satisfy the periodicity conditions $f(0) = f(1)$, $f'(0) = f'(1)$, ..., $f^{(p-1)}(0) = f^{p-1}$ (1).

3. Results. In the case of F_p^* we have

$$D_{2m-1}(F_p^*) = D_{2m}(F_p^*) = (1/2\pi m)^p, \quad m = 1, 2, 3, \ldots,$$

and the minimum $D_n(F_p^*)$ is always attained. The best functions $\phi_1, \phi_2, \ldots, \phi_n$ for $n = 2m + 1$ are determined uniquely to within a linear transformation; [1] namely, they are

$$1, \ \sqrt{2} \sin 2\pi kx, \ \sqrt{2} \cos 2\pi kx, \quad k = 1, 2, \ldots, m.$$

In the case of F_1 we have

$$D_n(F_1) = 1/\pi n, \quad n = 1, 2, 3, \ldots,$$

and for each n the best functions $\phi_1, \phi_2, \ldots, \phi_n$ are determined uniquely to within a linear transformation. These functions are

$$1, \ \sqrt{2} \cos \pi kx, \quad k = 1, 2, \ldots, n - 1.$$

In the case of F_p with an arbitrary p the results are no longer this simple, since they depend on an orthonormal system of functions

$$U_{p1}, \ U_{p2}, \ \ldots, \ U_{pr}, \ \ldots,$$

which is specific for each p. This system will be defined in §4. It will be proved that for F_p we have

$$D_n(F_p) = +\infty, \quad n = 1, 2, \ldots, p - 1,$$
$$D_n(F_p) = 1/\sqrt{\lambda_{n-p+1}^{(p)}}, \quad n = p, p + 1, p + 2, \ldots,$$

[1] The assertions on uniqueness as stated in this paper turned out to be false; see the commentary to this paper.

where the $\lambda_k^{(p)}$ are the eigenvalues defined in §4. Asymptotically (when p is fixed), we have

$$D_n(F_p) = (1/\pi n)^p + O(1/n)^{p+1}.$$

In the case of F_p with $n \geq p$, the best functions $\phi_1, \phi_2, \dots, \phi_n$ are determined uniquely to within a linear transformation. These functions are the first n functions U_{pk}, $k = 1, 2, \dots, n$. For $p = 1$ this agrees with the above result, since

$$U_{11} = 1, \ U_{1n} = \sqrt{2} \cos \pi(n-1)x, \quad n = 2, 3, \dots.$$

4. Definition and properties of the functions U_{pn}. For $n = 1, 2, \dots, p$ we put U_{pn} equal to the Legendre polynomial of degree $n-1$ normalized by the condition

$$\int_0^1 U_{pn}^2 \, dx = 1.$$

For $n = p + k, k > 0$, we define U_{pn} as the solution to the system

$$\begin{gathered} (-1)^p y^{(2p)} - \lambda y = 0, \\ y^{(p)}_{x=0,1} = y^{(p+1)}_{x=0,1} = \dots = y^{(2p-1)}_{x=0,1}, \end{gathered} \tag{A}$$

which is normalized in the same fashion and corresponds to the kth non-zero eigenvalue $\lambda_k^{(p)}$ of the system (A) (of course, the eigenvalues $\lambda_k^{(p)}$ are assumed to be arranged in increasing order). In this definition we assume that all non-zero eigenvalues of (A) are positive and simple, which actually is the case. [2]

Among the properties of the system $\{U_{pn}\}$ (for a fixed p) we want to mention the following.

1. It is a complete orthonormal system.

2. The pth derivatives $u_{pn} = d^p U_{pn}/dx^p$ form an orthogonal (although not normalized) system of functions.

3. For each p-times differentiable function f for which

$$\int_0^1 \{f^{(P)}\}^2 dx < +\infty,$$

the expansion $f \sim \sum a_n U_{pn}$ implies $f^{(p)} \sim \sum a_n u_{pn}$.

[2] See the paper by M.G. Krein [1] (then in print). I am grateful to M.G. Krein for informing me about this.

Note that properties 1–3 determine the system $\{U_{pn}\}$ uniquely to within replacement of the Legendre polynomials by any other orthonormal polynomials of degrees less than p.

Finally, we also note that when $n \leq p$, the functions u_{pn} are identically equal to zero, and when $n = p + k$, $k > 0$, the function u_{pn} is the solution of the system

$$\begin{gathered}(-1)^p y^{(2p)} - \lambda y = 0,\\ y_{x=0,1} = y'_{x=0,1} = \ldots = y^{(p-1)}_{x=0,1} = 0,\end{gathered} \tag{B}$$

which corresponds to the kth non-zero eigenvalue $\lambda_k^{(p)}$ (the eigenvalues of problems (A) and (B) obviously coincide). Here the conventional normalization conditions are replaced by

$$\int_0^1 u_{pn}^2 dx = \lambda_k^{(p)}.$$

5. Geometrical interpretation of the proof. We first make some comments on the general formulation of the problem. The set of functions $\phi = c_1\phi_1 + c_2\phi_2 + \ldots + c_n\phi_n$ with fixed functions $\phi_1, \phi_2, \ldots, \phi_n$ forms an n-dimensional linear subspace Φ_n of the space R of all functions in question. $E_n(f)$ is the distance from f to the set Φ_n. The expression $E_n(F)$ is therefore a natural measure of the deviation of the set F from the linear space Φ_n. Hence the infimum $D_n(F)$ of $E_n(F)$ can be called the n-width of the set F.

Our specific discussion relates to the Hilbert space H of all square integrable functions. The sets F_p and F_p^* can be thought of as elliptic cylinders in this space. Geometrically, it is almost obvious that in this case the minimum $D_n(F)$ of the expressions $E_n(F)$ is attained if we consider the n-dimensional linear space Φ_n spanned by the n greatest major axes of the cylinder F. These major axes can easily be found by the classical techniques of variational calculus. In particular, in the case of F_p the directions of the major axes coincide with the above-defined functions U_{pk}. The lengths of the corresponding semi-axes for $k \leq p$ are infinite while for $k = p+i$, $i > 0$, they are equal to $1/\sqrt{\lambda_i^{(p)}}$.

1 May 1935

References

1. M.G. Krein, 'On a special class of differential operators', *Dokl. Akad. Nauk SSSR* **2** (1935), 345–349 (in Russian).

29. ON DUALITY IN COMBINATORIAL TOPOLOGY *

1. The boundary of an r-dimensional algebraic complex is an $(r-1)$-dimensional algebraic complex [2]. Along with the operation of passing to the boundary, we consider the dual operation, which associates with each r-dimensional algebraic complex an $(r+1)$-dimensional complex. The resulting theory is dual to the ordinary homology theory and gives a new understanding of the Poincaré and Alexander classical duality theorems. In addition to ordinary complexes, this theory also covers more general structures which, following Tucker [3], we call cellular spaces. Here we confine ourselves to the study of only finite cellular spaces. By definition, such a space is a finite system of elements (cells). With each cell is associated an integer — its dimension. We distinguish certain pairs of cells whose dimensions differ by 1 and which are said to be mutually incident. Incidence of cells x^r and x^{r-1} of dimensions r and $r-1$ respectively, is denoted $x^r \to x^{r-1}$.

2. A cellular space R is said to be oriented [4] if to each r-dimensional cell x^r of R there corresponds another r-dimensional cell of this space, denoted by $-x^r$, such that (1) $-(-x^r) = x^r$; (2) it follows from $x^r \to x^{r-1}$ that $-x^r \to -x^{r-1}$; and (3) the relations $x^r \to x^{r-1}$ and $x^r \to -x^{r-1}$ are incompatible.

3. An r-dimensional algebraic complex is a function $f(x^r)$ that associates with each cell x^r of an oriented cellular space R an element of a fixed Abelian group J (called the coefficient domain) so that the condition $f(-x^r) = -f(x^r)$ holds. The reader may note that this viewpoint coincides with the conventional interpretation of an algebraic complex as a linear form whose variables are cells and with coefficients taken from a given domain [5]. However, this point of view is logically simpler, which is more convenient particularly in the case of infinite cellular spaces. This advantage will also manifest itself when dimension indices and signs of functions are written. Thus, an r-dimensional algebraic complex (in a given oriented cellular space) can be denoted by $f^r(x^r)$. The totality of r-dimensional algebraic complexes over a given coefficient domain is an Abelian group, denoted by $F^r(R, J)$.

* 'Über die Dualität im Aufbau der Kombinatorischen Topologie', *Mat. Sb.* 1 (1936), 97–102 (see [1]).

4. For an arbitrary $f^r = f^r(x^r)$ we now define

$$g_u f^r = f^{r-1}(x^{r-1}) = \sum_{x^r \to x^{r-1}} f^r(x^r);$$

$$g_0 f^r = f^{r+1}(x^{r+1}) = \sum_{x^{r+1} \to x^r} f^r(x^r);$$

where in both cases the sum extends over the r-dimensional cells incident to x^{r-1} or x^{r+1}. It is clear that $g_u f^r$ is none other than the conventional boundary of the algebraic complex [6].

5. To develop the theory we introduce the following axioms:

AUJ. For all r and all f^r,

$$g_u g_u f^r = 0.$$

AOJ. For all r and all f^r,

$$g_0 g_0 f^r = 0.$$

By means of a simple calculation one can verify that these two axioms are equivalent. Moreover, if one of these axioms is true for the domain of integral coefficients, then both the axioms are true for any coefficient domain.

6. We introduce the following conventional definitions: if $g_u f^r = 0$, then f^r is called a u-cycle; if $g_0 f^r = 0$, then f^r is a 0-cycle. The totality of r-dimensional u-cycles (0-cycles) forms a subgroup $Z_u^r(R,J)$ $(Z_0^0(R,J))$ of the group $F^r(R,J)$. We will say that f_1^r and f_2^r are u-homologous (0-homologous) if there exists f^{r+1} (f^{r-1}) such that $g_u f^{r+1} = f_1^r - f_2^r$ $(g_0 f^{r-1} = f_1^r - f_2^r)$. Homology will be denoted by $f_1^r \overset{u}{\sim} f_2^r$ (*respectively*, $f_1^r \overset{0}{\sim} f_2^r$). When f^r is homologous to 0, we say that f^r is "bounding" or more precisely, "u-bounding", (respectively, "0-bounding"). It follows from AUJ or AOJ that each u-bounding complex is a u-cycle and each 0-bounding complex is a 0-cycle. The bounding cycles will also be called boundaries (u-boundaries and 0-boundaries).

7. The groups $Z_u^r = Z_u^r(R,J)$ $(Z_0^r = Z_0^r(R,J))$ of r-dimensional u-cycles (0-cycles) contain the subgroups H_u^r (H_0^r) of boundaries. The quotient groups $B_u^r = Z_u^r - H_u^r$ and $B_0^r = Z_0^r - H_0^r$ are, by definition, the r-dimensional Betti groups of the given cellular space over the coefficient domain J; more precisely, B_u^r is a u-Betti group and B_0^r is a 0-Betti group; u-groups are, of course, what have up to now been called Betti groups in combinatorial topology. If it is necessary to indicate the cellular space or the coefficient domain, we write in full $B_u^r(R, J)$ (respectively, $B_0^r(R, J)$). One can easily pass to the consideration of pairs of coefficient domains [7] $J' \supset J$ and define the groups $H_u^r(R, J, J')$ and $H_0^r(R, J, J')$ and also

$$B_u^r(R, J, J') = Z_u^r(R, J) - H_u^r(R, J, J')$$

and

$$B_0^r(R, J, J') = Z_0^r(R, J) - H_0^r(R, J, J').$$

8. An oriented cellular space is called a cellular complex if it is isomorphic to (that is, it can be mapped in a one-to-one manner with preservation of the dimension, orientation, and incidence onto) the totality of the oriented cells of a Euclidean cellular complex (that is, a partition of a polyhedron into cells). Here one should bear in mind that the vertices of Euclidean cellular complexes must be taken into account twice, once with the plus sign and once with the minus sign. Accordingly, each one-dimensional element of a Euclidean cellular complex is incident to one positive vertex and one negative vertex.

9. The Betti u-groups, that is, the ordinary Betti groups of a polyhedron partitioned into cells, are known to be topologically invariant (this means that cellular partitions of homeomorphic polyhedra have isomorphic Betti groups). A similar invariance theorem is also true for the Betti 0-groups. This theorem can be deduced [8] from the invariance of u-groups by using the duality theorems in §14.

10. Two oriented cellular spaces R and R' are said to be n-dual if they can be mapped onto each other in a one-to-one manner so that each r-dimensional cell x^r of one space corresponds to an $(n-r)$-dimensional cell $x'^{(n-r)}$ of the other space; pairs of incident cells go into pairs of incident cells, and consequently if two cells x and x' correspond to each other, then $-x$ and $-x'$ are also in correspondence. Clearly, there exists a unique (to within isomorphism) cellular space n-dual to a given cellular space R; this space will be denoted by $D_n R$. In what follows we assume that for a pair of n-dual spaces a definite one-to-one correspondence between them is fixed. This makes it possible to speak of a fixed correspondence between dual cells x and x'.

11. Using the classical definition of an n-dimensional closed manifold M^n, one can easily prove that an oriented cellular space R is isomorphic to a cellular partition of the oriented manifold M^n if and only if R and D_nR are cellular complexes. If R is a cellular partition of M^n, then to within an isomorphism, D_nR is a partition of M^n dual to R in the classical sense.

12. To each algebraic complex $f^r(x^r)$ of a cellular space R there corresponds an algebraic complex $h^{n-r}(x^{n-r})$ of the space D_nR, that is, a function that takes on the values $f^r(x^r)$ on the cells x^{n-r} dual to x^r, while if $f^r = g_u f^{r+1}$, then the functions h^{n-r} and h^{n-r-1} satisfy the duality relation

$$h^{n-r} = g_0 h^{n-r-1}.$$

Similarly, it follows from the relation $f^{r+1} = g_0 f^r$ that

$$h^{n-r-1} = g_u h^{n-r}.$$

This readily implies the isomorphism

$$B_u^r(R, J) \approx B_0^{n-r}(D_nR, J).$$

This isomorphism and §11 imply the first duality theorem for oriented M^n:

$$B_u^r(M^n, J) = B_0^{n-r}(M^n, J).$$

13. In the sequel, which is primarily devoted to the interrelaiton between B_u^r and B_0^r, we must consider the case when the coefficient domain J is an Abelian topological group. Here the group F^r of all algebraic complexes f^r is also topological since it is the topological product of a finite number of factors each of which coincides with J (the number of factors is equal to the number of pairs $(x^r, -x^r)$ in the cellular space R). The topology of the group F^r naturally induces a topology on the subgroups Z_u^r, Z_0^r, H_u^r and H_0^r and also, in the usual way, on the quotient groups B_u^r and B_0^r.

14. Now let A and B be two locally compact Abelian topological groups each of which is the character group of the other (in the sense of Pontryagin [9]). Then the groups $B_u^r(R, A)$ and $B_0^r(R, B)$ are locally bicompact and obey the reciprocity law, according to which $B_u^r(R, A)$ and $B_0^r(R, B)$ are the character groups of each other.

We outline a proof of this assertion. Let a be an arbitrary element of A and b an arbitrary element of B. Then b is a homeomorphism of the group A into the group K of real numbers reduced modulo 1. Thus, $b(a)$ is an element of K, which will be denoted as the product $a \times b$ of the elements a and b. This multiplication is distributive on both sides. The algebraic complex in R relative to A (that is, a function with values in A) will be denoted by $f^r = f^r(x^r)$ and the algebraic complex relative to B will be denoted by $h^r = h^r(x^r)$. We define multiplication of the complex f^r by the complex h^r by the formula

$$f^r \times h^r = \sum_{x^r \in R} f^r(x^r) \times h^r(x^r).$$

This definition allows us to consider f^r as a linear functional on h^r (with values in K). Conversely, h^r can be considered as a linear functional on f^r.

It can be shown by means of calculations that we always have

$$f^r \times g_u h^{r+1} = g_0 f^r \times h^{r+1}, \tag{1}$$

$$f^r \times g_0 h^{r-1} = g_u f^r \times h^{r-1}. \tag{1'}$$

Thus, the operators g_u and g_0 behave like adjoint operators in the theory of linear operators.

The proof of the reciprocity law is as follows. The group $F^r(R, B)$ of all h^r can be regarded as the group of all homeomorphisms of the group $F^r(R, A)$ into K, that is, as the character group of the group $F^r(R, A)$. Furthermore, using (1) one can prove that $Z_0^r(R, B)$ is the annihilator [10] of $H_u^r(R, A)$ in $F^r(R, B)$. It follows that $Z_0^r(R, B)$ is the character group of the quotient group $F^r(R, A) - H_u^r(R, A)$. Further, $Z_u^r(R, A)$ is the annihilator of $H_0^r(R, B)$ in $F^r(R, A)$. It follows that $H_0^r(R, B)$ is the character group of $F^r(R, A) - Z_u^r(R, A)$. Thus, $B_0^r(R, B) = Z_0^r(R, B) - H_0^r(R, B)$ is the character group of the group

$$\{F^r(R, A) - H_u^r(R, A)\} - \{F^r(R, A) - Z_u^r(R, A)\} \approx$$
$$\approx Z_u^r(R, A) - H_u^r(R, A) = B_u^r(R, A).$$

15. We now consider a cellular decomposition of the sphere S^n. The cells of this decomposition will be denoted by x and the cells of the dual decomposition will be denoted by y. Let K be a complex constructed of cells x and let L be the set of cells x that do not belong to K. We denote by K' the set of cells

y dual to the cells in L. It is easily seen that K' is a complex, whereas L is not; however, the latter satisfies the axioms AOJ and AUJ. The group $B^r_u(K')$ is none other than the ordinary Betti group [11] of the complementary space $S^n - K$. We denote this group by $B^r_u(S^n - \overline{K})$.

The formulas in §§16, 17 imply that the groups $B^r_u(L)$, $B^r_0(K')$, and $B^r_0(L)$ are also topological invariants of the complementary space $S^n - \overline{K}$. We denote these groups by $^*B^r_u(S^n - \overline{K})$, $B^r_0(S^n - \overline{K})$, and $^*B^r_0(S^n - \overline{K})$ respectively. Here the group $^*B^r_u(S^n - \overline{K})$ is the Betti group generated by the "open" cycles of $S^n - \overline{K}$ [12].

16. When $1 < r < n$, the formula

$$B^{r-1}_u(K) \approx {}^*B^r_u(S^n - \overline{K}) \approx B^{n-r}_0(S^n - K)$$

holds.

The second isomorphism follows from the duality between L and K' by virtue of §12. We prove the first isomorphism [13].

Any u-cycle z^{r-1} in K is bounding in S^n, that is, $z^{r-1} = g_u f^r$. Obviously, the function f^r, considered on L, is a cycle. The various f^r bounded by the same cycle or by cycles that are homologous (on K) to one another and to the given cycle z^{r-1} are homologous on L.

Thus, each $(r-1)$-dimensional homology class on K corresponds to an r-dimensional homology class on L. Since the correspondence is dual, $B^{r-1}_u(K)$ and $^*B^r_u(S^n - K)$ are isomorphic. Therefore we have proved formula (1). The isomorphism between $B^{r-1}_u(K)$ and $B^{r-1}_0(S^n - K)$ will be called the second duality theorem.

17. Again let A and B be the character groups of each other. Then

$$B^{r-1}_0(K, B) \approx {}^*B^r_0(S^n n - \overline{K}, B) \approx B^{n-r}_u(S^n - \overline{K}, B)$$

is, to within isomorphism, the character group of the group

$$B^{r-1}_u(K, A) \approx {}^*B^r_u(S^n - \overline{K}, A) \approx B^{n-r}_0(S^n - \overline{K}, A).$$

(The proof immediately follows from the reciprocity law in §14.)

This proof contains the Alexander duality theorem, that is, $B^{r-1}_u(K, A)$ and $B^{n-r}_u(S^n - \overline{K}, B)$ are the character groups of each other.

18. In the case $r = 1$ we similarly find that

$$B^0_u(K, A) \approx {}^*B^1_u(S^n - \overline{K}, A) + A \approx B^{n-1}_0(S^n - \overline{K}, A) + A$$

is the character group of the group

$$B^0_0(K, B) \approx {}^*B^1_0(S^n - \overline{K}, B) + B \approx B^{n-1}_u(S^n - \overline{K}, B) + B.$$

Finally, for $r = n$ we see that

$$B^{n-1}_u(K, A) + A \approx {}^*B^n_u(S^n - \overline{K}, A) \approx B^0_0(S^n - \overline{K}, A)$$

is the character group of the group

$$B^{n-1}_0(K, A) + B \approx {}^*B^n_0(S^n - \overline{K}, B) \approx B^0_u(S^n - \overline{K}, B).$$

Remarks

1. The results of this paper as applied to ordinary complexes were established by the author in the spring and summer of 1934 and were partly reported at the International Conference on Tensor Analysis held in Moscow, May 1934. The general theory developed in this paper underlay the report that the author presented at the International Conference on Topology (Moscow, September 1935). In the latter report the author mentioned that a major part of the results had been established by Alexander. See Alexander's papers in Proc. Nat. Acad. Sci. USA **21** (1935), 509–512 published by that time. Alexander also presented his results at the Moscow Conference on Topology. The generalization to the case of closed sets and the construction of homology rings for complexes and closed sets, which were also reported by the author at the Conference on Tensor Analysis in 1934, will be presented in future papers. These constructions were, however, discovered by Alexander and partly published in the above-mentioned paper.

2. For the terminology and notation in combinatorial topology, see P.S. Alexandroff, *Einfachste Grunbegriffe der Topologie*, Springer-Verlag, Berlin, 1932 and particularly, P. Alexandroff and H. Hopf, *Topologie*, Springer-Verlag, Berlin, 1935 Vol. 1:4.

3. A.W. Tucker, 'An abstract approach to manifolds', *Ann. of Math* **34** (1933), 191–243. Cellular spaces are a special case of discrete spaces in the sense

of P.S. Aleksandrov (see his paper 'Sur les espaces discrets', *C.R. Acad. Sci. Paris* **200** (1935), 1649–1650). A considerable part of the results presented here remains valid for discrete spaces more general than cellular spaces, in particular, for partially ordered and finite spaces. Following Tucker we define dimension as a homeomorphism from a given discrete space (regarded as a partially ordered set) into the (ordered) set of integers, which differs from the special definition suggested by Aleksandrov. Also see A. de Rham, 'Sur l'analysis situs des variétés à n-dimensions, Thèse', *J. Math. Pures et Appl., Sér.* 9 **10** (1931), 115–200.

4. See A.W. Tucker, *ibid.*

5. P. Alexandroff and H. Hopf, *Topologie*, Vol. 1, 168–170. It should be noted that if the set of r-dimensional cells belonging to R is empty, then there exists a unique function f^r with an empty domain.

6. In the case of simplicial complexes the definition (using the modern notation of combinatorial topology) reads as follows. First, let a complex $f^r = tx^r$ contain only one simplex x^r. We denote by $x_0^{r-1}, \ldots, x_r^{r-1}$ ($x_0^{r+1}, \ldots, x_r^{r+1}$) the oriented $(r-1)$-dimensional ($(r+1)$-dimensional) simplexes incident to x^r.

We now define

$$g_u f^r = \sum t x_i^{r-1}, \quad g_0 f^r = \sum t x_i^{r+1}.$$

Then, for any algebraic complex $f^r = \sum t^k x^r = \sum f_r^k$,

$$g_u f^r = \sum g_u f_k^r, \quad g_0 f^r = \sum g_0 f_k^r$$

are defined.

7. P. Alexandroff and H. Hopf, *Topologie*, Vol 1, p.205.

8. Here we consider discrete 0-groups. The corresponding character groups are therefore compact.

9. L.S. Pontryagin, 'Topological commutative groups', *Ann. of Math.* **35** (1934), 1–338; E.R. van Kampen, 'Locally bicompact Abelian groups and their character groups', *Ann. of Math.* **36** (1935), 448–463.

10. L.S. Pontryagin, *ibid.*

11. Cf. L. Pontryagin, 'Über den algebraischen Inhalt topologischer Dualitätssätze', *Math. Ann.* **105** (1931), 165–205.

12. Cf. A.W. Tucker, *ibid.*

13. Cf. S. Lefschetz, *Topology*, New York 1930, Chapter 4.

30. HOMOLOGY RINGS OF COMPLEXES AND LOCALLY BICOMPACT SPACES *

§1. The homology ring of a complex

The first section is a continuation of my paper "On Duality in Combinatorial Topology" (see No. 29 of this volume). We retain the notation of that paper, but consider only a special case of the general theory constructed there; namely, the cellular space R is assumed to be a simplicial complex and the additive group of rational numbers is taken as the coefficient domain J. Let two simplexes $x^r = (p_0, p_1, \dots, p_r)$ and $x^s = (q_0, q_1, \dots, q_s)$ be cells in R. If each point q_i differs from all the points p_j and the simplex $(p_0, p_1, \dots, p_r, q_0, q_1, \dots, q_s)$ is a cell of R, then we call x^{r+s+1} the product of x^r by x^s and write

$$x^{r+s+1} = x^r x^s.$$

It is clear that the above-defined multiplication of simplexes is associative; as to commutativity:

$$x^r x^s = (-1)^{(r+1)(s+1)} x^s x^r.$$

The product of two algebraic complexes f^r and f^s is understood to be the algebraic complex f^{r+s+1} given by the formula

$$f^r f^s = f^{r+s+1}(x^{r+s+1}) = \frac{1}{2} \sum_{x^r x^s = x^{r+s+1}} f(x^r) f(x^s). \tag{1}$$

The multiplication on the right-hand side is ordinary multiplication of rational numbers. The sum extends over all possible representations of the simplex x^{r+s+1} in the form of a product $x^r x^s$. An equivalent definition of the product of algebraic complexes was suggested by Alexander [2]. In contrast to the product of simplexes, the product of algebraic complexes is defined in all cases.

If R does not contain $(r+s+1)$-dimensional simplexes, then, as has been proved in my earlier paper, there exists a unique $(r+s+1)$-dimensional algebraic complex, and the product of any f^r and f^s is equal to this complex f^{r+s+1}. It is easily seen that multiplication of algebraic complexes is associative and

* 'Homologierung des Komplexes und des localbikompakten Räumes', *Mat. Sb.* 1 (1936), 701–706. See [1].

distributive with respect to addition. The commutative law is replaced by the relation

$$f^r f^s = (-1)^{(r+1)(s+1)} f^s f^r. \tag{2}$$

For even r it follows that

$$f^r f^s = (-1)^{(r+1)^2} f^r f^r = -f^r f^r = 0. \tag{3}$$

Each vertex of the complex R is defined (according to my earlier paper) by two zero-dimensional cells $+(p)$ and $-(p)$. We introduce a special zero-dimensional complex e^0 such that for any point p we have

$$e^0(+(p)) = +1, \quad e^0(-(p)) = -1.$$

It follows immediately from the definition of the operator g_0 that

$$g_0 f^r = e^0 f^r. \tag{4}$$

In view of (3), $e^0 e^0 = 0$, therefore

$$g_0(g_0 f^r) = e^0 e^0 f^r = 0,$$

which was earlier proved directly.

For an arbitrary s-dimensional simplex x^s we denote by $[x^s]$ the complex (the cellular space) consisting of all faces of the simplex (including the simplex x^s itself). If x^s belongs to R, then clearly each function $f^r(x^r)$ defined for all x^r in R for $r \leq s$ is also defined for all x^r in $[x^s]$. And if f^r is a 0-cycle on R, then f^r remains a 0-cycle if this function is considered as an algebraic complex on $[x^s]$.

Lemma. *Any 0-cycle z^r ($r \leq s$) is 0-homologous to zero on each x^s, that is, there exists a complex f^{r-1} defined on $[x^r]$ such that*

$$g_0 f^{r-1} = z^r \ \textit{on}\ [x^s].$$

To prove the lemma it suffices to choose a vertex p of the simplex x^s; then for each simplex x^{r-1} of $[x^r]$ one can assert that

$f^{r-1}(x^{r-1}) = z^r(+(p)x^{r-1})$ if x^{r-1} does not contain the point p;

$f^{r-1}(x^{r-1}) = 0$ if p is a vertex of x^{r-1}.

The lemma implies that the product of two cycles is always zero. Indeed, let

$$u^{r+s+1} = z^r z^s;$$

then the relations

$$z^r = e^0 f^{r-1}, \quad z^s = e^0 f^{s-1}$$

hold on each $[x^{r+s+1}]$ and therefore

$$u^{r+s+1}(x^{r+s+1}) = e^0 f^{r-1} e^0 f^{s-1} = \pm e^0 e^0 f^{r-1} f^{s-1} = 0.$$

The introduction of a second multiplication leads to new homological invariants. This second multiplication can be defined as:

$$[f^r f^s] = f^{r+s}(x^{r+s}) = \frac{1}{4(r+s+1)} \sum_{x^{r-1}x^0x^{s-1}=x^{r+s}} f^r(x^{r-1}x^0) f^s(x^0 x^{s-1}). \tag{5}$$

The sum on the right-hand side extends over all possible representations of the simplex x^{r+s} in the form of three factors of the indicated dimensions. It is easily seen that the second multiplication is distributive with respect to addition and

$$[f^r, f^s] = (-1)^{rs}[f^s, f^r]. \tag{6}$$

The basic formula relating the first multiplication to the second one is written:

$$(r+s+t+2)[f^r f^s, f^t] = (s+t+1) f^r [f^s, f^t] +$$

$$+(-1)^{(r+1)(s+1)}(r+t+1) f^s [f^r, f^t]. \tag{7}$$

To prove this relation we put $f^r f^s = w^{r+s+1}$. Then

$$(r+s+t+2)[f^r f^s, f^t] = (r+s+t+2)[w^{r+s+1}, f^t] = u^{(r+s+t+1)}(x^{r+s+t+1}) =$$

$$= \frac{1}{4} \sum_{x^{r+s}x^0x^{t-1}=x^{r+s+t+1}} w^{r+s+1}(x^{r+s+1}x^0) f^t(x^0 x^{t-1}) =$$

$$= \frac{1}{8} \sum_{x^{r-1}x^0x^s x^{t-1}=x^{r+s+t+1}} f^r(x^{r-1}x^0) f^s(x^s) f^t(x^0 x^{t-1}) +$$

$$+ \frac{1}{8} \sum_{x^r x^{s-1}x^0x^{t-1}=x^{r+s+t-1}} f^r(x^r) f^s(x^{s-1}x^0) f^t(x^0 x^{t-1}) =$$

$$= (-1)^{(r+1)(s+1)}\frac{1}{2}\sum_{x^s x^{r+t}=x^{r+s+t+1}} f^s(x^s)\times$$

$$\times\frac{1}{4}\sum_{x^{r-1}x^0x^{t-1}=x^{r+s}} f^r(x^{r-1}x^0)f^t(x^0x^{t-1})+\frac{1}{2}\sum_{x^r x^{s+t}=x^{r+s+t+1}} f^r(x^r)\times$$

$$\times\frac{1}{4}\sum_{x^{s-1}x^0x^{t-1}=x^{s+t}} f^s(x^{s-1}x^0)f^t(x^0x^{t-1})=$$

$$= (-1)^{(r+1)(s+1)}(r+t+1)f^s[f^r,f^t]+(s+t+1)f^r[f^s,f^t],$$

as required.

It is easily seen that

$$[e^0,f^r]=f^r. \tag{8}$$

Therefore it follows from (7) that

$$(r+s+2)[e^0f^r,f^s]=(r+s+1)e^0[f^r,f^s]+(-1)^{(r+1)}(s+1)f^rf^s. \tag{9}$$

Applying (9) twice, we obtain

$$[e^0f^r,e^0f^s]=e^0f^rf^s. \tag{10}$$

Formula (10) enables us to introduce a new definition of the second multiplication for 0-cycles. Let z^r and z^s be two 0-cycles and let x^{r+s} be a simplex. By the lemma proved above,

$$z^r=e^0f^{r-1} \text{ and } z^s=e^0f^{s-1}$$

on x^{r+s}. It follows from (10) that

$$[z^r,z^s]=u^{r+s}(x^{r+s})=e^0f^{r-1}f^{s-1}.$$

Clearly, the value of $e^0f^{r-1}f^{s-1}$ on the simplex x^{r+s} does not depend on the choice of f^{r-1} and f^{s-1}. This new definition readily implies a number of facts. First, we show that the product $[z^r,z^s]$ of two 0-cycles is a 0-cycle; indeed, let x^{r+s+1} be an arbitrary simplex and let

$$z^r=e^0f^{r-1} \text{ and } z^s=e^0f^{s-1}$$

for $[x^{r+s+1}]$. It follows that

$$e^0[z^r,z^s]=u^{r+s+1}(x^{r+s+1})=e^0e^0f^{r-1}f^{s-1}=0.$$

Second we prove that the second multiplication is associative for cycles; indeed, let z^r, z^s and z^t be cycles and let x^{r+s+1} be an arbitrary simplex. On $[x^{r+s+t}]$ we have

$$z^r = e^0 f^{r-1}, \quad z^s = e^0 f^{s-1}, \quad z^t = e^0 f^{t-1},$$

and we obtain

$$\Big[[z^r, z^s], z^t\Big] = \Big[z^r, [z^t]\Big] = e^0 f^{r-1} f^{s-1} f^{t-1}$$

for the simplex x^{r+s+t}. We also present the following formula:

$$(r+s+1)e^0[f^r, f^s] = (r+1)[e^0 f^r, f^s] + (-1)^r(s+1)[f^r, e^0 f^s], \qquad (11)$$

which is similar to the differentiation formula in Poincaré rings of skew-symmetric differential forms.

Note that (9) implies

$$(-1)^r(r+s+2)[f^r, e^0 f^s] = (r+s+1)e^0[f^r, f^s] + (-1)^r(r+1)f^r f^s. \quad (9')$$

Multiplying (9) by $(r+1)$ and (9′) by $(s+1)$ and adding together the resulting relations we obtain formula (11).

If z^s is a 0-cycle and z^r is a 0-cycle homologous to zero, then

$$e^0 z^s = 0 \text{ and } z^r = e^0 f^{r-1}.$$

Further, it follows from (11) that

$$(r+s)e^0[f^{r-1}, z^s] = r[z^r, z^s],$$

that is, the cycle $[z^r, z^s]$ is homologous to zero.

It now follows that if the 0-cycles z_1^r and z_2^r as well as z_1^s and z_2^s are pairwise homologous, then the cycles $[z_1^r, z_1^s]$ and $[z_2^r, z_2^s]$ are also homologous. Indeed,

$$[z_1^r, z_1^s] - [z_2^r, z_2^s] = [z_1^r, z_1^s - z_2^s] + [z_1^r - z_2^r, z_2^s] \sim 0$$

since $z_1^s - z_2^s \sim 0$ and $z_1^r - z_2^r \sim 0$. These results imply that the second multiplication of 0-cycles z^r and z^s induces in the usual way the multiplication of elements of the Betti groups $B_0^r(R)$ and $B_0^s(R)$, resulting in an element of the Betti group $B_0^{r+s}(R)$. The elements of all Betti groups $B_0^r(R)$ with this multiplication form the 0-homology ring of the complex R [3].

§2. The homology ring of a locally bicompact space

The results of §1 are valid for an arbitrary collection of vertices. Here R can be any system of simplexes satisfying only the natural requirement that all faces of a simplex x^r belonging to R should also lie in R. We note the following advantage of 0-homology as compared to the ordinary u-homology. The latter homology theory is applicable only to algebraic complexes which have only a finite number of non-zero coefficients on a set of simplexes with a common vertex.

In connection with my paper "The Betti groups of locally bicompact spaces" (see No. 32 of this volume) we now consider a locally bicompact space E and the system R of all simplexes of all possible dimensions whose vertices lie in E. Each function $\bar{f}^r(p_0, p_1, \ldots, p_r)$ defined on E (in the sense of the above-mentioned paper) corresponds to a definite algebraic complex $f^r(x^r)$ of the system R. In this case the products $\bar{f}^r \bar{f}^s = \bar{f}^{r+s+1}$ and $[\bar{f}^r, \bar{f}^s] = \bar{f}^{r+s}$ are defined for two functions $\bar{f}^r$ and $\bar{f}^s$. It is easily seen that to equivalent functions there correspond equivalent products. It is clear that the first and second multiplications of algebraic complexes are also defined in the sense of the above-mentioned paper. Formulas (2)–(4) and (6)–(11) in §1 remain valid. The assertions in §1 can be extended without change to the corresponding Betti groups $B_0^r(E)$, which yields the 0-homology ring of the space E (see [4]).

§3. The case of two-sided manifolds

We now turn to the notation of §1 and assume that the complex R is a two-sided manifold of dimension n. In this case the Betti group $B_0^r(R)$ with any $r \leq n$ is mapped isomorphically onto the group $B_0^{n-r}(R)$ in a well defined manner, as described in the above-mentioned paper (see [1]). In this case rational numbers can also be taken as coefficients. We denote by $\xi_u^r \times \eta_u^s = \zeta^{r+s-n}$ the intersection of elements ξ_u^r and η_u^s of the groups $B_u^r(R)$ and $B_u^s(R)$. The second multiplication of elements ξ_0^r and η_0^s of the groups $B_0^r(R)$ and $B_0^s(R)$ is defined by using intersections of the corresponding elements ξ_u^{n-r} and η_u^{n-s} of the groups $B_u^{n-r}(R)$ and $B_u^{n-s}(R)$, so that $\zeta_0^{r+s} = [\xi_0^r, \eta_0^s]$ is the element corresponding to $\zeta^{n-r-s} = \frac{(r+s)!}{r!s!}\xi_u^{n-r} \times \eta_u^{n-s}$.

Bolshevo-Komarovka, near Moscow, 20 April 1936

Remarks

1. The content of the report presented by the author at the Moscow Conference on Topology was only partly published later. The part of the report omitted here was published earlier (see No. 29 in this volume). The results of §2 of this paper were reported only for the case of compacta. The generalization to arbitrary locally bicompact spaces was developed later (1936).

2. See J.W. Alexander, 'On the ring of a compact metric space', *Proc. Nat. Acad. Sci. USA* **21** (1935), 511–512. In this paper Alexander erroneously assumed that the product of two 0-cycles can be non-zero.

3. See H. Hopf, *J. Reine Angew. Math.* **163** (1930), 73–88; in particular, §1.

4. The first multiplication of functions ($\bar{f}^r \bar{f}^s = \bar{f}^{r+s+1}$) was defined by Alexander in the above-mentioned paper (see [2]) for arbitrary metric spaces. I learned at the Moscow Conference on Topology (September 1935) that Alexander had discovered the second multiplication as well.

31. FINITE COVERINGS OF TOPOLOGICAL SPACES *

(in collaboration with P.S. Aleksandrov)

1. Let F be a topological space. A finite system of $\mathfrak{S}$-closed (open) subsets of the space F is said to be a closed (open) multiplicative covering [1] if , first, $F = \bigcup_{i=1}^{s} A_i$ and, second, each non-empty intersection $A_{i_1} \cap \ldots \cap A_{i_k}$ of elements of the system $\mathfrak{S}$ is also an element of this system. An example of a closed multiplicative covering is a cellular (in particular, simplicial) decomposition of a finite polyhedron.

The greatest among all integers λ such that there exists a decreasing sequence

$$A_{i_1} \supset A_{i_2} \supset \ldots \supset A_{i_\lambda}$$

of different non-empty elements of the multiplicative covering $\mathfrak{S}$ is called the length of this covering. Given an arbitrary finite covering $\mathfrak{U}$ of the space F (that is, an arbitrary finite system of point sets A_i such that $\bigcup_i A_i = F$), we obtain a multiplicative covering $\mathfrak{S} \supset \mathfrak{U}$ by adding to this covering all the possible intersections of its elements. It is obvious that the length of the covering $\mathfrak{S}$ is no greater than the order of the covering $\mathfrak{U}$; however, it may turn out to be less than the order of $\mathfrak{U}$. For example, let F be a square, let $\mathfrak{U}$ be the covering of the square generated by its partition into smaller squares, [2] and let $\mathfrak{S}$ be the multiplicative covering corresponding to $\mathfrak{U}$; then the order of $\mathfrak{U}$ is equal to 4 whereas the length of $\mathfrak{S}$ is equal to 3. The length of any simplicial partition of an n-dimensional polyhedron is equal to n; however, the order of the covering generated by the basic [3] simplexes of such a partition can be arbitrarily large.

The purpose of this paper is to specify the interrelation between the length and order of coverings and, in particular, to prove the following assertion, which seems to be unknown even in the case of an n-dimensional cube and is a strengthening of the Brouwer-Lebesgue Pflasterzatz.

---

* 'Endliche überdeckungen topologischer Räume', *Fund. Math.* **26** (1936), 267–271. (Parts 1 and 2 are due to P.S. Aleksandrov; Parts 3 and 4 are due to A.N. Kolmogorov.)

[1] The terminology is taken from [1].

[2] Partitioning is carried out by drawing segments connecting pairs of opposite sides of the square.

[3] A simplex of a complex K is said to be free if it is a face of a unique basic simplex; a simplex is said to be basic if it is not a face of any other simplex. Each simplex is regarded as being an (improper) face of itself.

Theorem. *If $\epsilon > 0$ is sufficiently small, then the length of each closed and each open multiplicative ϵ-covering of an n-dimensional compactum F is no less than $n+1$.*

Proof. Let $\mathfrak{S}$ be the given multiplicative covering of the compactum F. F can be regarded as a bounded closed subset of the space R^n. We denote by N the nerve and by λ the length of the covering $\mathfrak{S}$.

Lemma. *For $r \geq \lambda$ each free r-dimensional simplex of the nerve N contains a free $(r-1)$-dimensional face.*

Proof of the lemma. Let $x^r = (a_0, \dots, a_r)$ be a free simplex of the nerve N that is a free face of the basic simplex

$$x^u = (a_0, \dots, a_r, a_{r+1}, \dots, a_u), \quad u \geq r.$$

Since $r \geq \lambda$ and, consequently, $r+1 > \lambda$, there exists $p < r$ such that

$$A_0 \cap \dots \cap A_p = A_0 \cap \dots \cap A_p \cap A_{p+1},$$

where A_i is the element of $\mathfrak{S}$ corresponding to the vertex a_i of the nerve N. Then

$$x^{r-1} = (a_0, \dots, a_p, a_{p+2}, \dots, a_r)$$

is the desired face of the simplex x^r. Indeed, let

$$x^v = (a_0, \dots, a_p, a_{p+2}, \dots, a'_r, \dots, a'_v)$$

be the basic simplex of N that has the face x^{r-1}. It follows from the definition of p that

$$x = (a_0, \dots, a_p, a_{p+1}, a_{p+2}, \dots, a_r, a'_r, \dots, a'_v)$$

is also a simplex of N. Since x^r is a face of this simplex, we have $x = x^u$, that is, all the a'_h, $r \leq h \leq v$, are among the a_i, $0 \leq i \leq u$. Hence, x^v is a face of the simplex x^u and, consequently, coincides with x^u since this is a basic simplex. The lemma is proved.

Let N be realized as a Euclidean complex in R^m. Then it follows from the above lemma that $\overline{N}$ can be deformed by compressing [4] consecutively free

[4] See [2], pp. 150–151, or [1], pp. 382–383.

simplexes into a polyhedron of dimension less than λ so that during the deformation each point remains in the basic simplex it belongs to. If $\mathfrak{S}$ is an ϵ-covering and N is realized [5] near $\mathfrak{S}$, then it follows from what has been proved that, using an ϵ-translation, F can be mapped into a polyhedron of dimension less than λ. This proves the theorem.

2. The above proof is true for both closed and open coverings. In the case of open coverings we can also carry out the proof in the following way.

Again let $\mathfrak{S}$ be an ϵ-covering and let N be its nerve realized as a Euclidean complex near $\mathfrak{S}$. Using Kuratowski's method, [6] we map F into $\overline{N}$. This mapping is denoted by κ. The set of points that are mapped into the interior of a simplex $x^r = (a_0, \dots, a_r)$ is

$$A_0 \cap \dots \cap A_r \backslash \bigcup{}' A_i$$

where $\bigcup{}'$ means that the summation extends over all subscripts different from $0, 1, \dots, r$.

Now let $x^n = (a_0, \dots, a_n)$ be an n-dimensional simplex of the nerve N and let $n \geq \lambda$. It follows from

$$A_0 \cap \dots \cap A_p = A_0 \cap \dots \cap A_p \cap A_{p+1}$$

that

$$A_0 \cap \dots \cap A_p \backslash A_{p+1}$$

is empty. Therefore

$$A_0 \cap \dots \cap A_p \backslash \bigcup{}' A_i$$

is also empty. Consequently, the interior of the simplex $(a_0, \dots, a_p)$ does not contain points of the set $\kappa(F)$. In other words, each simplex x^n has a face whose interior does not contain points of the set $\kappa(F)$. Since $\kappa(F)$ is closed, each x^n contains an open set not intersecting $\kappa(F)$. Since this is true for any ϵ, we have $\dim F \leq \lambda - 1$, as required.

3. The case of closed coverings can also be considered using the simple construction below. For our purpose it clearly suffices to prove the following theorem.

---

[5] See [1], pp. 160–161 and Chapter 9, §3, p. 363 ff.

[6] See [3], p. 191 or [1], pp. 366–368.

Theorem. *A compactum F having a closed multiplicative ϵ-covering $\mathfrak{S}$ of length λ also has a closed ϵ-covering whose order is less than or equal to λ.*

Proof. Let $A_1, \ldots, A_s$ be elements of the covering $\mathfrak{S}$. By the rank of an element A_k of $\mathfrak{S}$ is meant the length of the subcovering $\mathfrak{S}_h$ of $\mathfrak{S}$ consisting of the elements contained in A_h.[7] The elements of rank r of the covering $\mathfrak{S}$ will be denoted by A_{ri}, $i = 1, \ldots, s_r$. Two distinct elements of rank r intersect at an element of rank less than r.

We now put

$$B_{ri} = U(A_{ri}, d_r),$$

where the $d_r > 0$ are determined consecutively for $r = 1, \ldots, \lambda$ in a manner which will be described later. The number d_1 must be sufficiently small so that the B_{1i} are pairwise disjoint. If $d_{r'}$, $r' < t$, have already been determined, then we take d_r so small that for any A_{ri} and A_{rj}, $A_{ri} \cap A_{rj} = A_{ik}$, the inclusions $\overline{B}_{ri} \cap \overline{B}_{rj} \subset B_{r'k}$ hold. Finally, let

$$C_{ri} = \overline{B_{ri} \setminus \bigcup_{r'<r} B_{r'k}} \subset \overline{B}_{ri} \bigcup_{r'<r} B_{r'k}.$$

For $i \neq j$ we have

$$C_{ri} \cap C_{rj} = \varnothing,$$

since if

$$A_{ri} \cap A_{rj} = A_{r'k},$$

then

$$C_{ri} \cap C_{rj} \subset \overline{B}_{ri} \cap \overline{B}_{rj} \setminus B_{r'k} \subset B_{r'k} \setminus B_{r'k} = \varnothing.$$

Thus, if a point belongs to several sets C, then all these sets have different subscripts r. However, r is the rank of an element of $\mathfrak{S}$ and therefore it can assume only λ different values, so that a point of F can belong to at most λ different sets C, that is, the order of the covering generated by these sets does not exceed λ.

4. Our theorem can be extended *mutatis mutandis*, to the case of an arbitrary normal space R (without the assumptions that this space is compact and metrizable). Namely, the following fact is true.

---

[7] A_h is also an element of $\mathfrak{S}_h$.

Let $\mathfrak{U} = \{G_1, \ldots, G_u\}$ be a finite open covering and $\mathfrak{S} = \{A_1, \ldots, A_s\}$ a finite closed covering of a normal space R such that $\mathfrak{S}$ is a refinement of $\mathfrak{U}$ (that is, each A_i is contained in at least one G_h). Then there exists a closed covering $\mathfrak{C} = \{C_1, \ldots, C_\nu\}$ that is a refinement of $\mathfrak{U}$ and whose order does not exceed the length of the covering $\mathfrak{S}$.

The proof is the same as in §3, except that $U(A_{ri}, d_r)$ should be replaced by open sets $U_{ri} \supset A_{ri}$ satisfying the conditions in §3 stated in relation to $U(A_{ri}, d_r)$. In particular, if $A_{ri} \subset G_h$, then we also have $U_{ri} \subset G_h$.

Bolshevo-Komarovka, 1935

References

1. P. Alexandroff and H. Hopf, *Topologie*, Vol. I, Berlin 1935.
2. P.S. Alexandroff, 'Untersuchungen über Gestalt und Lage abgeschlossener Mengen beliebiger Dimension', *Ann. of Math.* **30** (1928–1929), 101–187.
3. K. Kuratowski, 'Sur un théorème fondamental concernant le nerf d'un système d'ensembles', *Fund. Math.* **20** (1933), 191–196.

32. THE BETTI GROUPS OF LOCALLY BICOMPACT SPACES *

Let [1] J be a discrete Abelian group and let θ be a bicompact Abelian group. [2] Also, let R be a locally bicompact Hausdorff space. Here, for each non-negative integer r, I am going to define two Betti groups $B_0^r(R, J)$ and $B_u^r(R, \theta)$, which are called the 0-group and the u-group of the space R relative to r (the dimension) and the coefficient domains J and θ, respectively. In the case when R is a compactum (a compact metric space) the u-groups are isomorphic, in the sense of Vietoris, to the Betti groups with respect to θ; in this case the 0-groups are isomorphic to the groups defined by Alexander in 1935. [3]

1. Definition of the groups $B_u^r(R, \theta)$. A function $\phi^r(\epsilon_0, \ldots \epsilon_r)$ is called a complex (more precisely, an r-dimensional complex) if it satisfies the following conditions:

α) $\phi^r(\epsilon_0, \ldots, \epsilon_r)$ is single-valued and is defined for all systems of $r+1$ subsets $\epsilon_0, \ldots, \epsilon_r$ of the space R that are bicompact in R;

β) the values of $\phi^r(\epsilon_0, \ldots, \epsilon_r)$ are elements of the group θ;

γ) ϕ^r is skew-symmetric in all its arguments;

δ) ϕ^r is additive in all arguments, that is,

$$\phi^r(\epsilon_0, \ldots, \epsilon_i' + \epsilon_i'', \ldots, \epsilon_r) = \phi^r(\epsilon_0, \ldots, \epsilon_i', \ldots, \epsilon_r) + \phi^r(\epsilon_0, \ldots, \epsilon_i'', \ldots, \epsilon_r);$$

ϵ) $\phi^r(\epsilon_0, \ldots, \epsilon_r)$ is equal to the zero element of the group θ provided that the closures $\bar{\epsilon}_o, \ldots, \bar{\epsilon}_r$ of the sets $\epsilon_0, \ldots, \epsilon_r$ contain no common point.

The complexes ϕ^r form an additive Abelian group Φ^r. This group can be mapped homomorphically into the group Φ^{r-1} by means of an operator g_u in the following way. When R is bicompact, we define g_u by

$$g_u \phi^r \equiv \phi^{r-1}(\epsilon_0, \ldots, \epsilon_{r-1}) = \phi^r(R, \epsilon_0, \ldots, \epsilon_{r-1}).$$

If R is not bicompact, we replace R in the foregoing formula by an open set $G \supset \bar{\epsilon}_0 + \ldots + \bar{\epsilon}_{r-1}$ bicompact in R; it follows from ϵ) that this definition does

* 'Les groupes de Betti des espaces localement bicompacts', *C.R. Acad. Sci. Paris* **202** (1936), 1144–1147.

[1] For all notions in general topology see P. Alexandroff and H. Hopf, *Topologie*, Vol. 1, Berlin, 1935. The notion of a locally bicompact space was introduced by P.S. Aleksandrov (see *Math. Ann.* **92** (1924), 294).

[2] See E.R. van Kampen, *Ann. of Math.* **36** (1935), 448–463.

[3] See J.W. Alexander, *Proc. Nat. Acad. Sci. USA* **21** (1935), 511–512.

not depend on the choice of the set G. The complex $\phi^{r-1} = g_u\phi^r$ is called the boundary of ϕ^r. Complexes whose boundaries are equal to zero are called cycles (r-dimensional u-cycles). Each boundary is a cycle since

$$g_u g_u \phi^r(\epsilon_0, \ldots \epsilon_{r-2}) = \phi^r(G, G, \epsilon_0, \ldots, \epsilon_{r-2}) = 0.$$

The group Z^r of all r-dimensional cycles contains the subgroup Γ^r of all cycles homologous to zero (that is, the cycles for which there exists ϕ^{r+1} such that $g_u\phi^{r+1} = \phi^r$). By definition, the quotient group $Z^r - \Gamma^r$ is the Betti group $B^r_u(R, \theta)$.

We shall define neighbourhoods in Φ^r in such a way that Φ^r becomes bicompact. This implies that the groups Z^r, Γ^r, and $B^r_u(R, \theta)$ will also be bicompact. [4] Neighbourhoods of each element ϕ^r of the group Φ^r correspond to certain finite systems of subsets in the space R, bicompact in R, which are defined as follows. Let $M_1, \ldots, M_n$ be a system in question. With each system of $r+1$ subscripts $i_0, \ldots, i_r$ none of which exceeds n we associate a neighbourhood $U_{i_0 \ldots i_r}$ of the element $\phi^r(M_{i_0}, \ldots, M_{i_r})$ of the group θ. By definition, the neighbourhood of the complex ϕ^r consists of all complexes ψ^r such that for an arbitrary subset $i_0, \ldots, i_r$ of the numbers $1, \ldots, n$ the element $\psi^r(M_{i_0}, \ldots, M_{i_r})$ of the group θ does not belong to $U_{i_0, \ldots, i_r}$.

2. Definition of the groups $B^r_0(R, J)$. Consider the system $\overline{F}^r$ of all functions $\bar{f}^r(p_0, \ldots, p_r)$ satisfying the following conditions:

a) $\bar{f}^r(p_0, \ldots, p_2)$ is a single-valued function defined for all systems of $r+1$ points $p_0, \ldots, p_r$ of the space R;

b) the values of $\bar{f}^r$ are elements of the group J;

c) $\bar{f}^r$ is skew-symmetric in all its arguments;

d) for each function $\bar{f}^r$ there exists a finite system $S_{\bar{f}^r}$ of pairwise disjoint subsets, bicompact in R, such that

1°. $\bar{f}^r(p_0, \ldots, p_r) = \bar{f}^r(p'_0, \ldots, p'_r)$ if for any i the two points p_i and p'_i belong to the same element of $S_{\bar{f}^r}$;

⟨2°. $\bar{f}r(p_0, \ldots, p_r) = 0$ if at least one of the arguments $p_0, \ldots, p_r$ does not belong to any of the elements of $S_{\bar{f}r}$.⟩

Two functions $\bar{f}^r_1$ and $\bar{f}^r_2$ are said to be equivalent if for each point p of the space R there exists a neighbourhood $U(p)$ such that $\bar{f}^r_1(p_0, \ldots, p_r) =$

[4] See van Kampen's memoir mentioned above.

$\bar{f}_2^r(p_0, \ldots, p_r)$ provided that all points p_i belong to $U(p)$. An equivalence class of functions $\bar{f}^r$ will be called a 0-complex of dimension r, and will be denoted by f^r. The complexes form a group F^r, which is none other than the quotient group $F^r = \overline{F}^r - O^r$. We now define the 0-boundary of $\bar{f}^r$ as the function

$$g_0 \bar{f}^r \equiv \bar{f}^{r+1}(p_0, \ldots, p_{r+1}) = \sum_{i=0}^{r+1} (-1)^i \bar{f}^r(p_0, \ldots, p_{i-1}, p_{i+1}, \ldots, p_{r+1}).$$

The boundaries of the two equivalent functions are equivalent; therefore the boundary $g_0 f^r$ of the complex f^r is defined as the complex of the boundaries of functions belonging to f^r. The complex f^r is called a cycle (0-cycle of dimension r) if its boundary is equal to zero (that is, the boundary is equal to the complex O^{r+1} containing the zero function $\overline{O}^{r+1}$). The complexes that are boundaries (that is, those homologous to zero) form the subgroup H^r of the group Z^r consisting of all r-dimensional cycles (since $g_0 g_0 f^r = O^{r+2}$). By definition, the quotient group $Z^r - H^r$ is the Betti group $B_0^r(R, J)$.

16 March 1936

33. PROPERTIES OF THE BETTI GROUPS OF LOCALLY BICOMPACT SPACES *

1. The 0- and u-groups of an arbitrary locally bicompact space R are interrelated in the following way. [1]

Let J be an arbitrary discrete abelian group and let θ be its character group in the sense of Pontryagin; [2] then θ is a bicompact abelian group with discrete character group isomorphic to the group J. Generally, two abelian groups one of which is discrete and the other bicompact are said to be *reciprocal* [3] if one of them is isomorphic to the character group of the other.

The reciprocity theorem. *For each $r \geq 0$ the groups $B_u^r(R,\theta)$ and $B_0^r(R,J)$ are reciprocal.*

Proof. Since θ is the character group of the group J, the product $a \times \alpha$ of an arbitrary element a of J with any element α of the group θ is an element $\alpha(a)$ of the continuous cyclic group K (the group of real numbers reduced modulo 1). For two functions $\bar{\phi}^r$ and $\bar{f}^r$ satisfying conditions (α)–(δ) and (a)–(d) in (B1), respectively, we define the product $\bar{f}^r \times \bar{\phi}^r$ in the following manner: let $M_1, \ldots, M_n$ be elements of the system of sets $S_{\bar{f}}^r$ defined in §2 in (B1) by condition (d); we set

$$\bar{f}^r \times \bar{\phi}^r = \sum \bar{f}^r(p_{i_0}, \ldots, p_{i_r}) \times \bar{\phi}^r(M_{i_0}, \ldots, M_{i_r}), \tag{1}$$

where p_i is an arbitrary point in M_i and the subscripts i_k independently run over the set $1, \ldots, n$. Obviously, the product $\bar{f}^r \times \bar{\phi}^r$ is none other than an n-fold integral in the sense of Radon-Stieltjes:

$$\bar{f}^r \times \bar{\phi}^r = \int \ldots \int \bar{f}^r(p_0, \ldots, p_r)\bar{\phi}^r(d_{p_0}R, \ldots, d_{p_r}R). \tag{1'}$$

Thus, each function $\bar{\phi}^r$ determines a homeomorphic mapping of the group $\overline{F}^r$ into the group K. It is easily seen that, conversely, each homeomorphism

---

* 'Propriétés des groupes de Betti des espaces localement bicompacts', *C.R. Acad. Sci. Paris* **202** (1936), 1325–1327.

[1] This paper is essentially based on my previous paper (see No. 32 in this book) which will be referred to as (B1).

[2] For references to results related to topological groups and, particularly, to the theory of their characters, see my paper (B1) and Pontryagin's memoir 'Topological commutative groups', *Ann. of Math.* **35** (1934), 361–388.

[3] Dual in the sense of Pontryagin's character theory.

from $\overline{F}^r$ into K can be defined in this way by choosing $\bar{\phi}^r$ in an appropriate manner. It follows that the group $\overline{\Phi}^r$ of all functions $\bar{\phi}^r$ satisfying conditions (α)–(δ) and with the topology introduced as in (B1) is reciprocal to $\overline{F}^r$. However, the subgroup Φ^r of the group $\overline{\Phi}^r$ is precisely the annihilator in $\overline{F}^r$ of the group O^r of functions $\bar{f}^r$ equivalent to zero. Hence, the groups Φ^r and F^r of the complexes ϕ^r and f^r are also reciprocal. This allows us to define the product $f^r \times \phi^r$ of elements of the groups F^r and Φ^r. Further, we prove the basic relation

$$f^r \times g_u \phi^{r+1} = g_0 f^r \times \phi^{r+1}. \tag{2}$$

It follows that Γ^r is the annihilator of the group Z^r in Φ^r, and consequently, $\Phi^r - \Gamma^r$ is reciprocal to Z^r. It also follows that Z^r is the annihilator of the group H^r in Φ^r, and hence $\Phi^r - Z^r$ is reciprocal to H^r. Therefore the group $B_0^r(R, J) = Z^r - H^r$ is reciprocal to $Z^r - \Gamma^r = B_u^r(R, \theta)$.

Fundamental systems. We denote by $\sum$ a finite or infinite partition $R = \sum B_\alpha$ of the space R, where B_α are disjoint subsets. The partition $\sum$ is said to be locally finite if each bicompact set in R intersects only a finite number of sets B_α. A system S of locally finite partitions is called a fundamental system if it possesses the following properties: 1) given two partitions $\sum'$ and $\sum''$ belonging to the system S, the partition $\sum'\sum'' = \sum$ whose elements are intersections of arbitrary elements of $\sum'$ and $\sum''$ also belongs to S; 2) for any finite system $U_1, \dots, U_n$ of open sets covering a bicompact subset A in R, among the partitions of R belonging to S there exists at least one such that any element of it having a common point with A is contained in some U_i.

Finite unions of sets taken from partitions $\sum$ belonging to a given fundamental system S form a field of sets, [4] which we denote by $T(S)$.

First reduction theorem. *For an arbitrary fundamental system S and a 0-cycle z^r there exists a cycle y^r homologous to z^r and constant[5] in each of its arguments on the elements of a partition $\sum$ belonging to S.*

---

[4] A system of sets T is called a field if the union, intersection, and difference of any two sets belonging to T is again a set belonging to T.

[5] A complex f^r is said to be constant on the elements of a partition if it contains a function $\bar{f}^r$ possessing this property.

Second reduction theorem. *Let* $\phi^r(\epsilon_0, \dots, \epsilon_r)$ *be a u-cycle. If* $\phi^r(\epsilon_0, \dots, \epsilon_r)$ *is equal to* 0 *whenever* $\epsilon_0, \dots, \epsilon_r$ *belong to the field* $T(S)$ *of a fundamental system* S*, then the cycle* ϕ^r *is homologous to zero.*

We denote by $\bar{f}^r_S$ the functions satisfying conditions (a)–(d) in §2 in (B1) and constant on the elements of at least one partition $\sum$ belonging to the fundamental system S. Further, we denote by ϕ^r_S the functions satisfying conditions (β)–(ϵ) and defined only on the sets belonging to $T(S)$. The functions f^r_S form a group $\widetilde{F}^r_S$. In this case all considerations in (B1) remain valid and lead to the definition of the groups $B^r_u(R, \theta, S)$ and $B^r_0(R, J, S)$. Thus, by the reduction theorems, these two groups are isomorphic to the groups $B^r_u(R, \theta)$ and $B^r_0(R, J)$, respectively.

30 March 1936

34. THE BETTI GROUPS OF METRIC SPACES *

In this paper [1] we study the general theory of Betti groups in two special cases of metric spaces, namely the case of compacta and the case of open manifolds.

1. Let R be an arbitrary compactum. Our primary aim in this case is to prove the following

Theorem. *The group $B_u^r(R,\theta)$ is isomorphic to the ordinary Betti group $B^r(R,\theta)$ of the space R (in the sense of Vietoris).* [2]

In order to prove this theorem we first note that in the case of compacta general fundamental systems can be replaced by fundamental sequences $S = \{\Sigma_1, \ldots, \Sigma_k, \ldots\}$ of successive subpartitions of R into disjoint sets:

$$R = B_1^k + \ldots + B_{s_k}^k, \quad B_i^k B_j^k = 0 \ (i \neq j), \ (\Sigma_k)$$

where the diameters of B_i^k tend to zero as k tends to infinity. Then it suffices to show that $B_u^r(R,\theta,S)$ is isomorphic to $B^r(R,\theta)$. Thus, associated with the system S is the projection spectrum Π_S of the space R in the sense of Aleksandrov; [3] to this end it suffices to select a point a_i^k in each B_i^k and to regard these points as the vertices of the nerve N_k of the system of sets $\overline{B}_1^k, \ldots, \overline{B}_{s_k}^k$. Therefore we arrive at the following proposition:

Lemma. *There is an isomorphism preserving the homology between $Z^r(R,\theta,S)$ and the group of all true projection cycles c^r of the spectrum Π_S.*

Indeed, let $\phi^r(E_0, \ldots, E_r)$ be a u-cycle. We set

$$c_k^r(a_{i_0}, \ldots, a_{i_r}) = \phi^r(B_{i_0}, \ldots, B_{i_r})$$

for each r-dimensional simplex $(a_{i_0}, \ldots, a_{i_r})$ belonging to N_k.

The resulting algebraic subcomplexes c_k^r of the nerve N_k are cycles and the sequence $\{c_k^r\}$ is a true projection cycle c^r. It is easily seen that the correspondence between ϕ^r and c^r is the desired isomorphism. On obtaining a true Vietoris cycle we only have to show that, conversely, an arbitrary Vietoris

---

* 'Les groupes de Betti des espaces métriques', *C.R. Acad. Sci. Paris* **202** (1936), 1558–1560.

[1] It is based on my papers Nos. 32 and 33 in this book, which will be referred to as (BI) and (BII).

[2] L. Vietoris gave the definitions only when the coefficient domains are the group of integers and a group of order 2 (*Math. Ann.* **97** (1927), 454–472). General coefficient domains were introduced by P.S. Aleksandrov (see *Fund. Math.* **20** (1933), 140–151).

[3] See *Ann. of Math.* **30** (1928), 101–187, particularly p. 107.

cycle is homologous to a true projection cycle. The proof of this fact (carried out by Aleksandrov [4] in the case when θ is a finite group) is based on the compactness of the coefficient domain.

2. Now let R be an open manifold of dimension n. We consider an arbitrary cellular decomposition M^n of the space R. Then, with minor changes, the earlier developed theory [5] can be applied. To this end we consider odd functions $\phi^r(x^r)$ of the oriented cells x^r in M^n whose values are elements of θ, that is, classical infinite algebraic subcomplexes. The operators g_u and g_0 are defined as in the above-mentioned memoir; the former operator yields the boundary of the algebraic complex $\phi^r(x^r)$ in the classical sense. One can automatically define the groups Z_u^r and Z_0^r of u-cycles and 0-cycles as well as the subgroups Γ_u^r and Γ_0^r of the cycles serving as boundaries. The Betti groups $B_u^r(M^n,\theta) = Z_u^r - \Gamma_u^r$ are isomorphic to the groups $B_u^r(R,\theta)$ obtained in (BI). Further, it can be shown that for any two cellular decompositions M_1^n and M_2^n of R, the groups $B_0^r(M_1^n,\theta)$ and $B_0^r(M_2^n,\theta)$ are also isomorphic. Consequently, we may write $B_0^r(R,\theta)$ instead of $B_0^r(M^n,\theta)$.

On the other hand, consider finite algebraic subcomplexes M^n, that is, odd functions $f^r(x^r)$ with values in J that are non-zero only for a finite number of x^r. As usual, these functions lead to the Betti groups $B_u^r(M^n,J)$ and $B_0^r(M^n,J)$. It can be easily proved that $B_0^r(M^n,J)$ is isomorphic to the group $B_0^r(R,J)$ defined in (BI) and that, to within isomorphism, $B_u^r(M^n,J)$ is independent of the choice of the cellular decomposition M^n, and therefore the group $B_u^r(M^n,J)$ can be regarded as the group $B_u^r(R,J)$; the latter is, however, the ordinary Betti group of M^n in the sense of classical combinatorial Analysis Situs. Using two dual cellular decompositions we can finally prove the Poincaré duality theorem that is, the isomorphisms

$$B_u^r(R,\theta) \approx B_0^{n-r}(R,\theta), \tag{1}$$

$$B_u^r(R,J) \approx B_0^{n-r}(R,J). \tag{2}$$

It should also be noted that, by the duality theorem in (BII), the groups (1) and (2) are dual.

20 April 1936

[4] *Ibid.*, p.162.

[5] See *Mat. Sb.* 1:1 (1936), 97–102; (see No. 29 in this volume).

35. RELATIVE CYCLES. THE ALEXANDER DUALITY THEOREM *

1. Let [1] R be a locally bicompact space and let θ be a closed set in R. A complex $\phi^r(E_0, \ldots, E_r)$ which is a u-complex in the terminology of (BI) is said to lie on θ if for any $E_0, \ldots, E_r$ we have

$$\phi^r(E_0, \ldots, E_r) = \phi^r(E_0 Q, \ldots, E_r Q).$$

Following Lefschetz, [2] a complex is called a relative cycle with respect to Q, or simply, a cycle modulo Q, it its boundary lies on Q. Let ζ^r be a cycle modulo Q. We will say that ζ^r is a boundary modulo Q or that ζ^r is homologous to zero modulo Q if there exists a complex ϕ^{r+1} such that $g_u \phi^{r+1} - \zeta^r$ lies on Q. The cycles/boundaries modulo Q form a subgroup Γ_Q^r of the group Z_Q^r of all r-dimensional cycles modulo Q. The group

$$B_u^r(R, Q, \theta) = Z_Q^r - \Gamma_Q^r$$

is the Betti u-group of the space R.

The annihilator of the group Γ_Q^r in F^r is the group Z_Q^r of all r-dimensional 0-cycles that are equal to zero on Q (in the sense that $z^r(p_0, \ldots, p_r) = 0$ when all the points p_i belong to Q). The group H_Q^r of all 0-cycles/boundaries that are zero on Q is the annihilator of Z_Q^r in F^r. Using an argument similar to that in (BII) one can see that the groups $Z_Q^r - H_Q^r = B_0^r(R, Q, J)$ and $B_u^r(R, Q, \theta)$ are dual.

2. **Theorem**. *The groups* $B_0^r(R, Q, J)$ *and* $B_0^r(R - Q, J)$ *are isomorphic. So are the groups* $B_u^r(R, Q, \theta)$ *and* $B_u^r(R - Q, \theta)$.

The latter isomorphism immediately follows from the former isomorphism, which can be proved by means of the following lemma.

Lemma. *An arbitrary* 0*-cycle that is equal to zero on* Q *is homologous to a cycle that is non-zero only on a bicompact set in* $R - Q$.

---

* 'Cycles relatifs. Théorème de dualité de M. Alexander', *C.R. Acad. Sci. Paris* **202** (1936), 1641–1643.

[1] This paper is based on my previous papers (Nos. 32–34 in this book) which will be referred to as (BI), (BII), and (BIII).

[2] Naturally, S. Lefschetz stated this definition for relative cycles in the ordinary sense of Vietoris. He also studied the notions of relative homology and Betti groups for this case (see S. Lefschetz, *Topology*, New York 1930).

3. The preceding discussion can be used in the proof of the following theorems.

First duality theorem. *Let R be a locally bicompact space with zero groups $B_0^r(R, J)$ and $B_0^{r-1}(R, J)$. Then for any set $Q \subset R$ closed in R the groups $B_0^{r-1}(Q, J)$ and $B_0^r(R - Q, J)$ are isomorphic.*

Second duality theorem. *Let R be a locally bicompact space with zero groups $B_u^r(R, \theta)$ and $B_u^{r-1}(R, \theta)$. Then for any set $Q \subset R$ closed in R the groups $B_u^{r-1}(Q, \theta)$ and $B_u^r(R - Q, \theta)$ are isomorphic.*

Now let $R = R^n$ be a Euclidean space of dimension n. In the case $1 < r < n$ the hypotheses of the duality theorems hold. Consequently, it follows from formulas (1) and (2) in (BIII) that in this case the isomorphic groups

$$B_0^{r-1}(Q, J) \approx B_0^r(R^n - Q, J) \approx B_u^{n-r}(R^n - Q, J)$$

are dual to the groups

$$B_u^{r-1}(Q, \theta) \approx B_u^r(R^n - Q, \theta) \approx B_0^{n-r}(R^n - Q, \theta).$$

In particular, it follows that $B_u^{r-1}(Q, \theta)$ and $B_u^{n-r}(R^n - Q, J)$ are dual, which is precisely the Alexander duality theorem in the general form suggested recently by Pontryagin.

4 May 1936

36. ON OPEN MAPPINGS *

In this paper we present an example of an open mapping[1] from a one-dimensional continuum onto a two-dimensional continuum.

1. We first make a remark concerning passage to a topological limit in the sense of Aleksandrov-Freudenthal (see [1, 2]). Let

$$X_1, X_2, \ldots, X_n, \ldots \tag{1}$$

be a sequence of topological spaces and suppose that for each n a continuous mapping ϕ_n from X_n into X_{n-1} ($n = 2, 3, \ldots$) is defined. Then, by definition, the points of the limit space X are the sequences

$$x = (x_1, x_2, \ldots, x_n, \ldots) \tag{2}$$

where $x_n \in X_n$ and $x_{n-1} = \phi_n(x_n)$. For each point (2) in X the nth "coordinate" x_n is denoted by $\Phi_n(x)$ so that Φ_n maps X into X_n. We introduce a topology in X by calling $x \in X$ a point of adherence of the set $M \subset X$ if and only if for each n the point $\Phi_n(x)$ is a point of adherence for $\Phi_n(M)$ in X_n.

Along with (1), we now consider another sequence

$$Y_1, Y_2, \ldots, Y_n, \ldots \tag{3}$$

of topological spaces together with mappings ψ_n from Y_n into Y_{n-1} and with limit space Y. By analogy with Φ_n, we construct mappings Ψ_n from Y into Y_n.

Finally, we assume that each X_n is mapped onto Y_n by an open f_n. We suppose further that for each $x_n \in X_n$ the relation

$$\psi_n f_n(x_n) = f_{n-1}\phi_n(x_n) \tag{4}$$

holds. Clearly, (4) implies that the mapping f from X into Y generated by the mappings f_n:

$$f(x) = f\{x_n\} = \{f_n(x)\},$$

* 'Über offene Abbildungen', *Ann. of Math.* **38** (1937), 36–38.

[1] A continuous mapping of a space X into a space Y is said to be open if the image of each open set in X is open in Y.

is an open mapping from X onto Y. Indeed, if $G \subset X$ is open in X and $x \in G$, then there exists n such that $x_n = \Phi_n(x)$ is not a point of adherence for $\Phi_n(X - G)$. Since the mapping $f_n : X_n \to Y_n$ is open, it follows that $y_n = f_n(x_n)$ is not a point of adherence for $f_n\Phi_n(X - G) = \Psi_n f(X - G)$. Consequently, the point $y = f(x)$, whose nth coordinate is y_n, is not a point of adherence for $f(X - G) \supset Y - f(G)$, whence it follows that the set $f(G)$ is open and thus the mapping f is open.

2. We now proceed to the construction of the desired example. Let

$$\epsilon_1, \epsilon_2, \ldots, \epsilon_n, \ldots$$

be a decreasing sequence of positive numbers convergent to zero. We construct two-dimensional Euclidean polyhedra Y_n recursively (the polyhedra can be, for instance, thought of as being located in R^4). Let Y_1 be a square. Assume that a triangulation of Y_{n-1} has already been constructed and that the diameter of each triangle in this triangulation is less than ϵ_n. We cut out a square hole in each triangle and glue up the hole with a Möbius strip. The Möbius strips must be realized as triangulated surfaces.

The mapping ψ_n leaves fixed each point of Y_n that is simultaneously a point of Y_{n-1}, and each of the added Möbius strips is mapped continuously onto the interior of the corresponding hole so that the points on the edges remain fixed. The limit space Y resulting from this construction is precisely the well-known Pontryagin surface and, hence, it is a two-dimensional continuum (in the sense of Brouwer's definition of dimension) [3].

3. Before proceeding to the construction of the one-dimensional continuum X and an open mapping from this continuum onto Y we make a brief preliminary remark on the following fact.

A torus (we mean a toroidal surface) can be mapped onto the Möbius strip by an open mapping so that a certain tame closed curve which is non-bounding on the torus (for instance, the equator) is mapped onto the edge of the strip.

Indeed, let u and v be geographical coordinates: $-\pi \leq u < +\pi$, $-\pi \leq v < +\pi$. By identifying the point (u, v) with $(u + v, -v)$ we obtain the desired mapping of the toroidal surface onto the Möbius strip (the curve $v = 0$ goes into the edge of the strip). Obviously, a mapping with the above properties can also be defined as a simplicial mapping.

Again let X_1 be a square and let f_1 be the identity mapping from X_1 into Y_1. Assume that the triangulated surface of X_{n-1} and the open simplicial mapping $f_{n-1} : X_{n-1} \to Y_{n-1}$ have already been constructed. Each hole cut out in a triangle in Y_{n-1} (in order to pass to Y_n) is the image of one or several holes in X_{n-1} under the mapping f_{n-1}. We cut out all these holes and glue the triangulated toroidal surface to the edge of each hole along the equator. Thus, X_{n-1} goes into X_n. Using an open simplicial mapping we map each torus onto the Möbius strip, gluing the corresponding hole in Y_{n-1}. The resulting open simplicial mapping from X_n onto Y_n will be denoted by f_n. Further, we define a mapping ϕ_n from X_n onto Y_{n-1} such that

$$f_{n-1}\phi_n(x_n) = \psi_n f_n(x_n),$$

which, obviously, can be done.

By construction, the f_n are open mappings and satisfy conditions (4), and therefore they determine an open mapping from X onto Y. Finally, it is easily seen (perhaps the simplest way is to apply the Aleksandrov theorem on ϵ-mappings) that X is one-dimensional.

We also note that Y can be embedded topologically in a four-dimensional Euclidean space and X (like any curve) can be embedded in three-dimensional space.

Bolshevo-Komarovka, near Moscow

References

1. P. Alexandroff, 'Sur les suites des espaces topologiques', *C. R. Acad. Sci. Paris* **200** (1935), 1708–1711.
2. H. Freudenthal, 'Die R_n-adische Entwicklung von Räumen und Gruppen', *Proc. K. Akad. Wetensch. Amsterdam* **38** (1935), 414–417.
3. L. Pontrjagin, 'Sur une hypothèse fondamentale de la théorie de la dimension', *C.R. Acad. Sci. Paris* **190** (1930), 1105–1107.

37. SKEW-SYMMETRIC FORMS AND TOPOLOGICAL INVARIANTS *

The author's aim is to develop a kind of finite-difference calculus which, on the one hand, can lead to differential operations on skew-symmetric tensors (multivectors) by means of a passage to the limit and, on the other hand, are closely related to notions of combinatorial topology. In particular, using this finite-difference calculus one can find new invariants of complexes and closed sets and prove some generalizations of the duality theorems.

1. We will consider skew-symmetric functions[1]

$$f_n(x_0, x_1, x_2, \dots, x_n)$$

of $(n+1)$ arguments $x_0, x_1, \dots, x_n$. Since such a function does not change its value under any even permutation of $x_0, x_1, \dots, x_n$ we say that f_n is a function of an oriented simplex

$$x^n = (x_0, x_1, \dots, x_n).$$

When the simplex changes its orientation, the function changes its sign:

$$f_n(-X^n) = -f_n(X^n).$$

In what follows the functions $f_n(X^n)$ are assumed to be defined for all n-dimensional simplexes X^n contained in a set K of simplexes of different dimensions. The set K is assumed to satisfy the following conditions:

(1) if X^n belongs to K, so does $-X^n$;

(2) if a simplex X^n belongs to K, then all its $(n-1)$-dimensional faces also belong to K.

Conditions (1) and (2) are generalizations of the requirements imposed on complexes in combinatorial topology. For K to be a complex it is also necessary that each vertex x belong to only a finite number of simplexes in K. In what follows the latter requirement is not introduced.

Multiplication of simplexes is defined as

$$(x_0, x_1, \dots, x_n)(y_0, y_1, \dots, y_m) = X^n Y^m = Z^{n+m+1} =$$
$$= (x_0, x_1, \dots, x_n, y_0, y_1, \dots, y_m).$$

---

* In: Proc. Seminar on Vector and Tensor Analysis and Applications in Geometry, Mechanics and Physics, Moscow, Leningrad: GONTI 1 (1937), 345–347 (in Russian).

[1] For simplicity, we assume that the functions in question take real values, although certain other cases are also significant in applications to topology.

Multiplication of functions $f_n(X^n)$ is determined by the formula

$$f_n f_m = f_{n+m+1}(X^{n+m+1}) = \tfrac{1}{2} \sum_{X^n X^m = X^{n+m+1}} f_n(X^n) f_m(X^m),$$

where the sum extends over all pairs of simplexes X^n and X^m whose product is X^{n+m+1}.

It is easily seen that multiplication of functions is associative and distributive relative to addition. Instead of the commutative law we have the relation

$$f_n f_m = (-1)^{(n+1)(m+1)} f_m f_n.$$

In particular, when n is even, we obtain

$$f_n f_n = (-1)^{(n+1)^2} f_n f_n = -f_n f_n = 0.$$

In order that our theory be applicable to simplexes of dimension 0, we must assume that to each element x_0 there correspond two simplexes $+(x_0)$ and $-(x_0)$ of dimension 0. With this stipulation, we set

$$e_0(+(x_0)) = 1, \quad e_0(-(x_0)) = -1$$

for any element x_0.

The function $e_0(X^0)$ of zero-dimensional simplexes satisfies the above conditions.

We now define the operation rot as

$$\operatorname{rot} f_n = e_0 f_n.$$

The operation rot on f_n is clearly a function f_{n+1} of simplexes of dimension $n+1$. By the second condition imposed on K, the operation is defined for all simplexes x^{n+1} in K. It follows directly from the relation

$$e_0 e_0 = 0$$

that

$$\operatorname{rot}(\operatorname{rot} f_n) = 0.$$

Moreover, the operation rot is obviously distributive with respect to addition.

A function f_n will be called a cycle if

$$\operatorname{rot} f_n = 0.$$

Further, a cycle f_n is homologous to zero if there exists a function f_{n-1} such that $\operatorname{rot} f_{n-1} = f_n$.

The cycles, as well as the cycles homologous to zero, form a group with respect to addition. The quotient group of the group of all cycles relative to the group of cycles homologous to zero is called the Betti group of K. The interrelation between the Betti groups defined in this way and the ordinary Betti groups considered in combinatorial topology will be studied in one of my future papers.

2. The product of two cycles is always equal to zero. It is useful to introduce a second multiplication for cycles. The second multiplication, denoted by

$$[f_n, f_m],$$

can be defined for any functions f_n and f_m of the type in question; however, I have managed to prove the associative law for this multiplication only in the case of cycles. The general definition of the second multiplication is

$$[f_n, f_m] = f_{n+m}(X^{n+m}) =$$

$$= \frac{1}{4(n+m+1)} \sum_{X^{n-1}X^0X^{m-1}=X^{n+m}} f_n(X^{n-1}X^0) f_m(X^0X^{m-1}).$$

The second multiplication is distributive relative to addition. When f_n, f_m and f_k are cycles, the associative law

$$[[f_n, f_m], f_k] = [f_n, [f_m, f_k]]$$

holds.

Instead of the commutative law we have

$$[f_n, f_m] = (-1)^{nm}[f_m, f_n].$$

The operation rot is related to this second multiplication by the formula

$$(n+m+1)\operatorname{rot}[f_n, f_m] = (n+1)[\operatorname{rot} f_n, f_m] + (-1)^n(m+1)[f_n, \operatorname{rot} f_m].$$

The product (resulting from the second multiplication) of a cycle homologous to zero by any cycle is a cycle homologous to zero. This makes it possible to define the homology ring of the system of K in the same manner as it is usually defined for manifolds.

38. A STUDY OF THE DIFFUSION EQUATION WITH INCREASE IN THE AMOUNT OF SUBSTANCE, AND ITS APPLICATION TO A BIOLOGICAL PROBLEM *

In collaboration with I.G. Petrovskii and N.S. Piskunov

Introduction

For the sake of simplicity we consider the two-dimensional diffusion equation

$$\frac{\partial v}{dt} = k\left(\frac{\partial^2 v}{\partial x^2} + \frac{\partial^2 v}{\partial y^2}\right), \quad k > 0, \tag{1}$$

where x and y are the coordinates of a point in the plane, t is time and v is the density of substance at the point (x, y) at time t. We now assume that diffusion is accompanied by increase in the amount of substance at a rate which depends on the density at the given point and time. We then obtain the equation

$$\frac{\partial v}{\partial t} = k\left(\frac{\partial^2 v}{\partial x^2} + \frac{\partial^2 v}{\partial y^2}\right) + F(v). \tag{2}$$

It is natural that we are interested only in the values of $F(v)$ for $v \geq 0$. Assume that $F(v)$ is a function of v which is continuous, differentiable the required number of times and satisfies the conditions

$$F(0) = F(1) = 1; \tag{3}$$

$$F(v) > 0, \ 0 < v < 1; \tag{4}$$

$$F'(0) = \alpha > 0, \ F'(v) < \alpha, \ 0 < v \leq 1. \tag{5}$$

We thus assume that for very small v the rate $F(v)$ of increase in v is proportional to v (with proportionality factor α), and as v approaches 1, there arises a state of "saturation" when v no longer increases. Accordingly we will consider only solutions of equation (2) satisfying the condition

$$0 \leq v \leq 1. \tag{6}$$

Arbitrary initial values of v for $t = 0$ satisfying condition (6) determine one and only one solution [1] of equation (2) for $t > 0$ subject to the same condition (6).

---

* *Bull. Moscow Univ., Math. Mech.* 1:6 (1937), 1–26.

[1] This fact will be proved in §3.

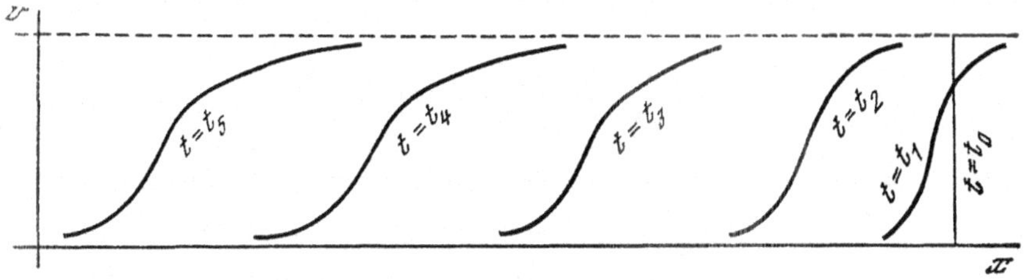

Fig. 1

Further, we will assume that the density v does not depend on the coordinate y. In this case the basic equation (2) is written as

$$\frac{dv}{dx} = k\frac{d^2v}{dx^2} + F(v). \tag{7}$$

We now assume that at the initial time $t = 0$ we have $v = 0$ for $x < a$ and $v = 1$ (that is, the density takes its maximum value) for $x > b \geq a$. Naturally, the region of densities close to 1 will expand, with increasing t, from right to left, displacing thereby the region of small densities to the left. In the special case $a = b$ the pattern looks approximately as shown in Figure 1. The segment of the density curve (as a function of x) on which the major part of density drop from 1 to 0 occurs, moves to the left with increasing time. In its shape the density curve approaches a definite limiting curve for $t \to \infty$. The problem is to find the limiting shape of the density curve and the limiting rate of its motion from right to left. It turns out that the desired limiting rate is

$$\lambda_0 = 2\sqrt{k\alpha}, \tag{8}$$

and the limiting shape of the density curve is determined by the solution of the equation

$$\lambda_0\frac{dv}{dx} = k\frac{d^2v}{dx^2} + F(v) \tag{9}$$

that vanishes when $x = -\infty$ and is equal to one for $x = +\infty$. Such a solution always exists and is unique, to within a transformation $x' = x + c$, which does not change the shape of the curve.

Note that the equation (9) can be obtained in the following manner. We seek a solution of equation (7) such that the curve representing the dependence of v on x does not change its shape with varying t and the curve itself moves from right to left at a rate λ. This solution has the form

$$v(x,t) = v(x + \lambda t). \tag{10}$$

Regarding v as a function of one variable $z = x+\lambda t$, we obtain the equation

$$\lambda \frac{dv}{dz} = k\frac{d^2v}{dz^2} + F(v).$$

This equation turns out to have solutions satisfying the above conditions for equation (9) with any $\lambda \geq \lambda_0$. However, only for $\lambda = \lambda_0$ do we obtain the desired limiting shape of the density curve under the above initial conditions. In order to understand the fact that there are solutions of the form (10) to equation (7) for $\lambda > \lambda_0$, that is, such that the region of large (close to 1) densities moves at a rate greater than λ_0, which at first glance seems astonishing, we examine the limiting case $k = 0$. In this case there is no diffusion and equation (7) can easily be integrated. Under our initial conditions, at points $x < a$ where the density was initially equal to zero it remains zero for any $t > 0$. However, it can easily be shown by calculation that for any $\lambda > \lambda_0$ there exist solutions (10) to equation (7) satisfying all the above conditions. Here the apparent motion of substance from right to left is due to the increase in density at each point occurring independently of the course of the process at other points.

In §1 the facts presented in the Introduction will be applied to the study of biological problems; in §§2, 3 these facts will be proved.

§1

Consider an area populated by a species. We first assume that a dominant gene A is distributed over the area with constant concentration p $(0 \leq p \leq 1)$. Further, we assume that in the struggle for existence, individuals with the character A (that is, those belonging to the genotypes AA and Aa) have an advantage over individuals not possessing this character (that is, those belonging to the genotype aa); more precisely, we assume that the ratio of the probability that an individual with the character A survives to the corresponding probability for an individual without the character is equal to

$$1+\alpha$$

where α is a small positive number. Then, up to terms of order α^2, we obtain for the increment of the concentration p per one generation the formula (see [1])

$$\Delta p = \alpha p(1-p)^2. \tag{11}$$

Now let the concentration p be different at different points in the area occupied by the species, that is, let p depend on the coordinates of the point in the plane (x, y). If the individuals of the species under consideration were fixed at the points of the area, (11) would still be valid. Assume, however, that during the time between birth and reproduction, each individual moves at random (all the directions of motion being equiprobable) and travels some distance. Let $f(r)dr$ be the probability that an individual passes a distance lying between r and $r + dr$; then

$$\rho = \sqrt{\int_0^\infty r^2 f(r)dr}$$

is the root-mean-square path. Therefore, instead of (11) we obtain

$$\Delta p(x,y) = \int_{-\infty}^{\infty}\int_{-\infty}^{\infty} p(\xi,\eta)\frac{f(r)}{2\pi r}d\xi\, d\eta - p(x,y) +$$

$$+\alpha p(x,y)\{1 - p(x,y)\}^2, \tag{12}$$

where

$$r = \sqrt{(x-\xi)^2 + (y-\eta)^2}.$$

We now assume that p is differentiable with respect to x, y and time t (measured in generations), α and ρ are small, and the third moment

$$\rho^3 = \int_0^\infty |r^3| f(r)dr$$

is small as compared to ρ^2. In this case, expanding $p(\xi,\eta)$ in (12) into a Taylor series in $(\xi - x)$ and $(\eta - y)$ and confining ourselves to the terms of second order (the terms of first order disappear), we obtain [2] an approximate differential equation for p:

$$\frac{\partial p}{\partial t} = \frac{\rho^2}{4}\left(\frac{\partial^2 p}{\partial x^2} + \frac{\partial^2 p}{\partial y^2}\right) + \alpha p(1-p)^2. \tag{13}$$

All considerations concerning the general equation (2) are applicable to (13).

---

[2] As to the passage from (12) to (13), see, for instance, similar considerations in A. Ya. Khinchin [2].

Let us stress again the stipulated assumptions. We have assumed that the concentration p changes smoothly, depending on place and time (differentiability with respect to x, y, t), the changes result from the selection in favour of the dominant character A in the ratio $(1+\alpha) : 1$ and from the random motion of individuals such that the root-mean-square path of an individual during the time between birth and reproduction is ρ, and finally, α and ρ are small (ρ is small as compared to the paths on which substantial changes in the concentration p occur). In this case, taking one generation as a time unit we obtain equation (13).

We now consider the case when a large area has already been occupied by the gene A with concentration p close to 1. It is natural to assume that there is an intermediate concentration region along the boundary of the area. Beyond this region p is assumed to be close to zero. In view of the positive selection, the area occupied by the gene A expands, in other words, its boundary moves towards places that have not yet been occupied by the gene A, and along this boundary there always remains the intermediate concentration region. Our first problem is to find the *rate of advance of the gene A*, that is, the rate at which the boundary of the area occupied by the gene A moves along the normal to this boundary. Formula (8) readily yields an answer to this question: since in this case k is equal to $\rho^2/4$, it follows that the desired rate is

$$\lambda = \rho\sqrt{\alpha}. \tag{14}$$

The second problem which naturally arises is to find the width of the intermediate region. By formula (9), the concentration p along the normal to the boundary satisfies the equation

$$\lambda \frac{dp}{dn} = \frac{\rho^2}{4} \frac{d^2p}{dn^2} + \alpha p(1-p)^2,$$

whence, on dividing by α and substituting λ from (14), we obtain

$$\frac{\rho}{\sqrt{\alpha}} \frac{dp}{dn} = \frac{1}{4} \frac{\rho^2 d^2p}{\alpha dn^2} + p(1-p)^2.$$

Introducing the new variable ν by means of the formula

$$n = (\rho/\sqrt{\alpha})\nu \tag{15}$$

we obtain the equation

$$\frac{dp}{d\nu} = \frac{1}{4}\frac{d^2p}{d\nu^2} + p(1-p)^2, \tag{16}$$

which contains neither α nor ρ. The boundary conditions for this equation are the same as in the case of (9):

$$p(-\infty) = 0, \quad p(+\infty) = 1.$$

It follows from (15) that the width of the intermediate region is proportional to

$$L = \rho/\sqrt{\alpha}. \tag{17}$$

For example, let $\rho = 1$, $\alpha = 0.0001$; then $\lambda = 0.01$ and $L = 100$.

§2

In this section we consider the equation

$$\lambda\frac{dv}{dx} = k\frac{d^2v}{dx^2} + F(v), \tag{18}$$

where λ and k are assumed to be positive and the function $F(v)$ satisfies the conditions in the introduction.

We are going to establish the relations between λ, k and $\alpha = F'(0)$ for which this equation has a solution satisfying the conditions

$$0 \le v(x) \le 1,$$
$$v(x) \to 1 \text{ as } x \to +\infty \text{ and } v(x) \to 0 \text{ as } x \to -\infty.$$

Let $dv/dx = p$. Then

$$\frac{d^2v}{dx^2} = \frac{dp}{dv}\frac{dv}{dx} = \frac{dp}{dv}p.$$

On substituting this into equation (18) we obtain

$$\frac{dp}{dv} = \frac{\lambda p - F(v)}{kp}. \tag{19}$$

We are interested in those integral curves of this equation that pass between the straight lines $v = 0$ and $v = 1$ in the plane (p, v). Generally, these can include curves of the following types:

1. integral curves that are separated from the lines $v = 0$ and $v = 1$ by at least a distance $\epsilon > 0$;

2. integral curves that go infinitely far away from the axis v and asymptotically approach one of the lines $v = 0$ or $v = 1$;

3. integral curves that intersect one of these lines at a finite point lying on the v-axis;

4. integral curves that approach the points $v = 0,\ p = 0$ and $v = 1,\ p = 0$ and do not belong to any of the former types.

However, it can be easily seen that integral curves of the first type cannot correspond to solutions of equation (18) satisfying the stated conditions, since for such solutions v cannot be arbitrarily close to 0 or 1.

Integral curves of the second type do not exist at all, since curves of this type must have points such that for very large $|p|$ the value of $|dp/dv|$ is very large. However, the ratio $(\lambda p - F(v))/kp$ is approximately equal to λ/k for large $|p|$ (in view of the boundedness of $F(v)$ on the interval $(0, 1)$).

Corresponding to the integral curves of the third type are solutions of equation (18) that do not necessarily remain between 0 and 1. Indeed, assume, for example, that a curve of this type approaches a point $v = 1,\ p = p_1 \neq 0$. In the vicinity of the line $v = 1$ we have

$$\frac{dp}{dv} \approx \frac{\lambda}{k} \neq 0,$$

therefore p can be regarded here as a function of v. Let $p = \phi(v)$. Since $\phi(1) = p_1 \neq 0$, it follows that $|\phi(v)|$ is greater than a positive constant C on a small interval $(1-\epsilon, 1+\epsilon)$. We denote by x_0 the value of x for which $v = 1-\epsilon$. Then, integrating the equation $dv/dx = \phi(x)$, we find

$$\int_{x_0}^{x} dx = x - x_0 = \int_{1-\epsilon}^{v} \frac{dv}{\phi(v)}.$$

It follows that x does not exceed $2\epsilon/C$ in absolute value when v varies from $1-\epsilon$ to $1+\epsilon$. Therefore, when x changes from x_0 to $x_0 + 2\epsilon/C$, v necessarily passes through 1.

It remains to consider integral curves of the fourth type. Each of the points $v = 0,\ p = 0$ and $v = 1,\ p = 0$ is a singular point of the differential equation (19). An integral curve of the fourth type must approach each of these points without intersecting the lines $v = 0$ and $v = 1$ and therefore it does not twist.

Thus, in order that such curves can exist, the characteristic equation for each of these points must have real roots. We write $F(v)$ in the form

$$F(v) = \alpha v + \phi_1(v).$$

Then, clearly, $\phi_1(v) = o(v)$. Therefore the characteristic equation at the point $v = 0,\ p = 0$ is given by

$$\begin{vmatrix} \lambda - \rho & -\alpha \\ k & -\rho \end{vmatrix} = 0,$$

whence

$$\rho^2 - \lambda\rho + \alpha k = 0. \tag{20}$$

This equation has real roots if

$$\lambda^2 \geq 4\alpha k.$$

To obtain the characteristic equation at the point $v = 1,\ p = 0$ we make a change of variables, putting $v = 1 - u$. This results in

$$\frac{dp}{du} = \frac{-\lambda p + \Phi(u)}{kp},$$

where $\Phi(u) = F(1 - u)$.

Obviously, $F'(1) \leq 0$ and $\Phi'(0) = -F'(1) = A \geq 0$. Consequently,

$$\Phi(u) = Au + o(u),$$

and the characteristic equation at the point $v = 1,\ p = 0$ takes the form

$$\begin{vmatrix} -\lambda - \rho & A \\ k & -\rho \end{vmatrix} = 0,$$

whence

$$\rho^2 + \lambda\rho - Ak = 0. \tag{21}$$

This equation has real roots when

$$\lambda^2 \geq -4Ak.$$

Since $\alpha > 0$, it follows that equation (20) has real roots of the same sign. Therefore the point $(0, 0)$ is a node. All integral curves that fall in a sufficiently small neighbourhood of this point pass through it. As to equation (21), it has roots of different signs when $A > 0$. Therefore if $A > 0$, then only two integral curves pass through the point (1, 0), along well-defined directions. Let these directions be specified by the equations

$$m_1 u + n_1 p = 0, \quad m_2 u + n_2 p = 0. \tag{22}$$

The coefficients m_1, n_1, m_2, n_2 are known [3] to be determined from the equations

$$k m_1 - \rho_1 n_1 = 0, \quad k m_2 - \rho_2 n_2 = 0, \tag{23}$$

where ρ_1 and ρ_2 are the roots of the characteristic equation (21). Since the roots are of different sign, it follows that the slopes of the lines (22) have different signs as well. [4] Therefore each of the angles formed at the intersection of the lines $v = 1$ and $p = 0$ contains only one integral curve of equation (19) passing through the point $v = 1, p = 0$. Figure 2 shows an approximate configuration of these curves. The curve II intersects the p-axis below the origin since equation (19) implies that $dp/dv > 0$ in the part of the strip between $v = 0$ and $v = 1$ that lies below the v-axis. Therefore the curve II may be excluded from further consideration. It remains to examine the curve I. [5]

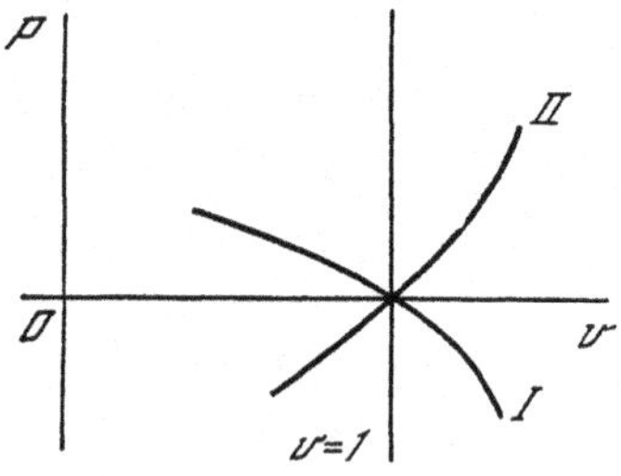

Fig. 2

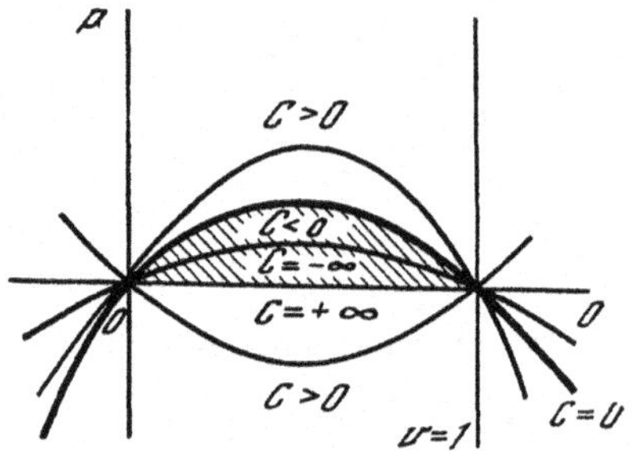

Fig. 3

We are going to prove that each curve of type I intersects the p-axis at the origin. We first of all prove that this curve cannot intersect the p-axis below

[3] See, for example, Bendixson [3] or Stepanov [4].

[4] Using this technique one can show in just the same way that both slopes of the tangents to the integral curves of (19) at the origin are positive.

[5] If $A = 0$, then one can only assert that there exists at least one integral curve of the type I approaching the point (1, 0) along a definite direction whose slope is negative (see [5]).

the origin. To this end we consider the isoclines of equation (19). The equation of the family of these lines has the form

$$\frac{\lambda p = F(v)}{kp} = C. \tag{24}$$

Here C is the value of dp/dv at the point v, p. Hence

$$p = \frac{F(v)}{\lambda - Ck}. \tag{24'}$$

Equation (24) specifies a family of curves passing through the points (0, 0) and (1, 0). Figure 3 schematically demonstrates this family. Each curve is supplied with the corresponding value of C. The heavy line represents the curve corresponding to $C = 0$. As the vertices of the curves go up, C increases accordingly and approaches the value λ/k which corresponds to the lines $v = 0$ and $v = 1$. In the region lying between the curve $C = 0$ and the v-axis (shaded in Figure 3) we have $C < 0$, and C becomes very large in absolute value at points near the v-axis. Below the v-axis we have $C > 0$, and C decreases from $+\infty$ to λ/k as the vertex of the curve goes down from the v-axis to $-\infty$.

It is now easily seen that an integral curve I (see Figure 2) cannot intersect the axis Op below the origin. Indeed, in this case the curve I would intersect the v-axis. Since $dp/dv = -\infty$ on the upper side of this axis and $dp/dv = +\infty$ on the lower side, the integral curve I is convex toward the line $v = 1$ at the point where it intersects the v-axis. Therefore in order that this curve pass through the point (1, 0) it is necessary that dp/dv become infinite above the v-axis, which is impossible. For the same reason, an integral curve I cannot intersect the line $v = 1$ above the v-axis.

We now prove that an integral curve I cannot intersect the p-axis above the origin. To this end it suffices to prove that there exists a ray passing through the origin and lying in the first quadrant that does not intersect any integral curve intersecting the positive semi-axis p. From equation (24) we obtain [6]

$$\left(\frac{\overline{dp}}{dv}\right)_{v=0} = \frac{\alpha}{\lambda - Ck}.$$

We now find C for which $(\overline{dp}/dv)_{v=0} = C$. To this end we have to solve the equation

$$\frac{\alpha}{\lambda - Ck} = C,$$

[6] $\frac{\overline{dp}}{dv}$ denotes the derivative of the function $p = p(v)$ determined by equation (24′).

that is,

$$kC^2 - C\lambda + \alpha = 0,$$

whence

$$C = \frac{\lambda \pm \sqrt{\lambda^2 - 4\alpha k}}{2k}. \tag{25}$$

Since we are assuming that

$$\lambda^2 \geq 4\alpha k,$$

it follows that the two values of C in (25) are real and positive. Denote by C_0 one of these values and draw the line

$$p = C_0 v. \tag{26}$$

It is easily seen that for all points of the strip between $v = 0$ and $v = 1$ that lie above the line (26), or even on the line itself (except for the origin),[7] we have

$$\frac{dp}{dv} > C_0.$$

Therefore none of the integral curves passing through a point on the p-axis above the origin can cross the part of the line (26) located above the v-axis. We have thus proved that each integral curve of type I (see Figure 2) passes through the origin.

We now prove that there exists only one integral curve of type I. (Of course, we have to prove this only for $A = 0$.) Indeed, we have proved that all curves of type I pass through the origin. On the other hand, it follows from (19) that for $p > 0$ and fixed v the derivative dp/dv increases with p. It follows that two integral curves issuing from the origin cannot pass through the point (1, 0).

We now prove that the curve I corresponds to the solution of equation (18) satisfying the conditions stated at the beginning. First we note that any perpendicular to the v-axis intersects the integral curve I of equation (19) only at one point, since otherwise dp/dv would take the value ∞ above the v-axis. Therefore p is a function of v, that is, $p = \phi(v)$, along this curve. Also recall that the curve I intersects the v-axis at the point (1, 0) at an angle whose tangent is negative and at the origin at an angle with positive tangent. Therefore for small values of v we have

$$p = k_1 v + o(v), \tag{27}$$

[7] Here p is the function of v defined by equation (18).

while for small values of $1 - v$ we obtain

$$p = k_2(1 - v) + o(1 - v), \tag{28}$$

where k_1 and k_2 are positive.

Recall now that $p = dv/dx$. Therefore $dv/dx = \phi(v)$, whence $dx = dv/\phi(v)$. Integrating the latter equation we obtain

$$x - x_0 = \int_{v_0}^{v} \frac{dv}{(v)}, \quad 0 < v_0 < 1.$$

By virtue of (27) and (28), it follows that $x \to -\infty$ as $v \to 0$ and $x \to \infty$ for $v \to 1$, as required.

§3

Instead of equation (7), which was discussed in the Introduction, we here consider the equation

$$\frac{dv}{dt} - \frac{d^2v}{dx^2} = F(v), \tag{29}$$

where the function $F(v)$ satisfies the following conditions:

$$F(0) = F(1) = 0; \tag{30}$$

$$F(v) > 0, \ 0 < v < 1; \tag{31}$$

$$F'(0) = 1; \tag{32}$$

$$F'(v) < 1, \ 0 < v \leq 1; \tag{33}$$

$F'(v)$ is bounded and continuous on (0, 1). Moreover, we assume that $F(v)$ is differentiable the required number of times. The general equation (7) presented in the Introduction can always be reduced to the form (29) by means of the change of variables

$$x = \sqrt{k/\alpha}\bar{x} \text{ and } t = \bar{t}/\alpha.$$

In this section our primary aim is to prove that as $t \to \infty$ the part of the density curve $v(t, x)$ (as a function of x) corresponding to the major portion of density drop from 1 to 0 moves to the left with increasing time at a rate approaching 2 (from below) and the shape of the curve itself approaches that of the graph of the solution $u(x)$ of the equation

$$\frac{d^2u}{dx^2} - 2\frac{du}{dx} + F(v) = 0 \tag{34}$$

that vanishes as $x \to -\infty$ and tends to 1 as $x \to \infty$. The existence of this solution has been proved in §2.

Before proving the basic assertions in this section we consider the equation

$$\frac{dv}{dt} - \frac{d^2v}{dx^2} = F(x,t,v),$$

a special case of which is equation (29). We will prove the existence of a solution taking prescribed values at $t = 0$, and study some properties of the solution.

Theorem 1. *Consider the equation*

$$\frac{dv}{dt} - \frac{d^2v}{dx^2} = F(x,t,v), \tag{35}$$

where the continuous bounded function $F(x,t,v)$ satisfies the Lipschitz condition with respect to v and x, that is,

$$|F(x_2,t,v_2) - F(x_1,t,v_1)| < k|v_2 - v_1| + k|x_2 - x_1| \tag{36}$$

(where k is a constant not depending on x, t, v). Let $f(x)$ be a bounded function defined for all values of x. For simplicity, we assume that $f(x)$ has only a finite number of points of discontinuity. Then there exists a unique function $v(x,t)$, bounded for bounded values of t, which for $t > 0$ satisfies equation (35) and for $t = 0$ is equal to $f(x)$ at each point at which this function is continuous. In what follows, when saying that $v(x,t)$ is equal to $f(x)$ for $t = 0$ we will always mean only the points of continuity of $f(x)$.

Proof. Let $v_0(x,t)$ be a bounded function satisfying for $t > 0$ the equation

$$\frac{\partial v}{\partial t} - \frac{\partial^2 v}{\partial x^2} = 0 \tag{37}$$

and equal to $f(x)$ for $t = 0$. Substituting this function for v into the right-hand side of (35) we obtain the solution of this equation that vanishes on the x-axis (see [6]), using the formula

$$\tilde{v}_1(x,t) = \frac{1}{2\sqrt{\pi}} \int_0^t d\eta \int_{-\infty}^{\infty} \frac{\exp(-(x-\xi)^2/4(t-\eta))}{\sqrt{t-\eta}} F(\xi,\eta,v_0(\xi,\eta))d\xi. \tag{38}$$

The function

$$v_1(x,t) = v_0(x,t) + \tilde{v}_1(x,t)$$

is equal to $f(x)$ for $t = 1$, and for $t > 0$ it satisfies the equation

$$\frac{\partial v_1}{\partial t} - \frac{\partial^2 v_1}{\partial x^2} = F[x, t, v_0(x,t)].$$

More generally, using the formula

$$v_{i+1}(x,t) = v_0(x,t) + \frac{1}{2\sqrt{\pi}} \int_0^t d\eta \int_{-\infty}^{+\infty} \frac{\exp(-(x-\xi)^2/4(t-\eta))}{\sqrt{t-\eta}} F(\xi, \eta, v_i) d\xi \tag{39}$$

we find the function $v_{i+1}(x,t)$ satisfying the equation

$$\frac{\partial v_{i+1}}{\partial t} - \frac{\partial^2 v_{i+1}}{\partial x^2} = F(x, t, v_i) \tag{40}$$

for $t > 0$ and equal to $f(x)$ for $t = 0$.

We prove that the sequence of functions $v_i(x,t)$ is uniformly convergent. Indeed, taking into account (36) we find from (39) that

$$M_{i+1}(t) = \sup_{\eta \le t} |v_{i+1}(x,\eta) - v_i(x,\eta)| \le$$

$$\le \frac{1}{2\sqrt{\pi}} \int_0^t d\eta \int_{-\infty}^{+\infty} \frac{\exp(-(x-\xi)^2/4(t-\eta))}{\sqrt{t-\eta}} |F(\xi, \eta, v_i(\xi,\eta)) -$$

$$-F(\xi, \eta, v_{i-1}(\xi,\eta))| d\xi \le \int_0^t k M_i(\eta) d\eta, \tag{41}$$

since

$$\int_{-\infty}^{\infty} \frac{\exp(-(x-\xi)^2/4(t-\eta))}{\sqrt{t-\eta}} d\xi = 2\sqrt{\pi}.$$

However, denoting by M_0 an upper bound of the values of $|f(x)|$ and $F(x,t,0)$ we obtain

$$|v_0(x,t)| \le M_0,$$

and, by virtue of (38), it follows that

$$M_1 \le \int_0^t (k+1) M_0 dt = (k+1) M_0 t = Mt.$$

Hence, using inequality (41) we easily obtain

$$M_i \le \frac{M k^{i-1} t^i}{i!},$$

which readily implies the uniform convergence of the sequence v_i.

We set

$$v(x,t) = \lim_{i\to\infty} v_i(x,t).$$

When $t = 0$, the function $v(x,t)$ is equal to $f(x)$. Moreover, as is easily seen, this function satisfies the equation

$$v(x,t) = v_0(x,t) + \frac{1}{2\sqrt{\pi}} \int_0^t d\eta \int_{-\infty}^{+\infty} \frac{\exp(-(x-\xi)^2/4(t-\eta))}{\sqrt{t-\eta}} F(\xi,\eta,v(\xi,\eta))d\xi. \tag{42}$$

Hence, it is clear that $v(x,t)$ is a continuous function of x and t for $t > 0$. It is proved in the above-cited memoir of Gevrey [6], pp. 343–344, that for any bounded function F the second term on the right-hand side of (42) has a bounded derivative with respect to x. By condition (36), it follows that for $t > 0$ the function $F(x,t,v(x,t))$ has bounded derivatives of any order with respect to x, and therefore $v(x,t)$ satisfies equation (35) (see [6], p. 351).

The uniqueness of the bounded solution is proved in the following way. Assume that there exist two bounded functions $v_1(x,t)$ and $v_2(x,t)$ taking the same value at $t = 0$. Then they satisfy the equation

$$v_2(x,t) - v_1(x,t) =$$

$$= \frac{1}{2\sqrt{\pi}} \int_0^t d\eta \frac{\exp(-(x-\xi)^2/4(t-\eta))}{\sqrt{t-\eta}} [F(\xi,\eta,v_2) - F(\xi,\eta,v_1)]d\xi. \tag{43}$$

We set

$$M(t) = \sup_{t\geq\eta} |v_2(x,\eta) - v_1(x,\eta)|.$$

Using (36) we obtain from (43) the inequality

$$M(t) \leq k \int_0^t M(\eta)d\eta,$$

which is impossible.

Remark. For regions bounded by lines $x = \phi_1(t)$ and $x = \phi_2(t)$ on the left and on the right and by straight lines $t = t_0$ and $t = t_1 > t_0$ from above and below, respectively, the above-mentioned memoir by Gevrey [6] contains a proof of the existence and uniqueness of a bounded function satisfying equation (35) in the interior of the region and taking prescribed bounded continuous values on the lines $x = \phi_1(t)$, $x = \phi_2(t)$, and $t = t_0$. It can similarly be proved that there exists a unique bounded function satisfying equation (35) in the interior of a region G bounded only on one side by a line $x = \phi(t)$ and by straight lines $t = t_0$ and $t = t_1 > t_0$ from above and below and taking prescribed bounded continuous values on the lines $x = \phi(t)$ and $t = t_0$.

Theorem 2. *If the function $F(x,t,v)$ is replaced by another function $F_1(x,t,v)$ such that always*

$$F_1(x,t,v) \geq F(x,t,v),$$

then the function $v(x,t)$ does not decrease, provided that the initial conditions remain unchanged.

Remark. If (35) is interpreted as the heat equation, then the function $F(x,t,v)$ characterizes the heat source intensity, and physically Theorem 2 becomes quite clear.

Proof. Let the functions $v(x,t)$ and $v_1(x,t)$ satisfy equation (35) and the equation

$$\frac{\partial v_1}{\partial t} - \frac{\partial^2 v_1}{\partial x^2} = F_1(x,t,v_1),$$

respectively.

After term-by-term subtraction we find that the function

$$w(x,t) = v_1(x,t) - v(x,t)$$

satisfies the equation

$$\frac{\partial w}{\partial t} - \frac{\partial^2 w}{\partial x^2} = F_1(x,t,v_1) - F(x,t,v).$$

We set

$$w(x,t) = \bar{w}(x,t)\ \exp(-kt),$$

where k is the same as in inequality (36). Then

$$\frac{\partial \bar{w}}{\partial t} - \frac{\partial^2 \bar{w}}{\partial x^2} = k\bar{w} + \exp(kt)[F_1(x,t,v_1) - F(x,t,v)].$$

Hence,

$$\bar{w}(x,t) = \frac{1}{2\sqrt{\pi}} \int_0^t d\eta \int_{-\infty}^{\infty} \frac{\exp(-(x-\xi)^2/4(t-\eta))}{\sqrt{t-\eta}} \{k\bar{w} + \exp(k\eta)[F_1(\xi,\eta,v_1) -$$

$$-F(\xi,\eta,v)]\}d\xi = \frac{1}{2\sqrt{\pi}}\int_0^t d\eta \int_{-\infty}^{\infty} \frac{\exp(-(x-\xi)^2/4(t-\eta))}{\sqrt{t-\eta}}\{k\bar{w}+$$

$$+\exp(k\eta)[F_1(\xi,\eta,v_1) - F(\xi,\eta,v_1) + F(\xi,\eta,v_1)-$$

$$-F(\xi,\eta,v)]d\xi \geq \frac{1}{2\sqrt{\pi}}\int_0^t d\eta \int_{-\infty}^{\infty} \frac{\exp(-(x-\xi)^2/4(t-\eta))}{\sqrt{t-\eta}}\{k\bar{w}+$$

$$+\exp(k\eta)[F(\xi,\eta,v_1) - F(\xi,\eta,v)]\}d\xi \geq$$

$$\geq \frac{1}{2\sqrt{\pi}}\int_0^t d\eta \int_{-\infty}^{\infty} \frac{\exp(-(x-\xi)^2/4(t-\eta))}{\sqrt{t-\eta}}(k\bar{w} - k|\bar{w}|)d\xi. \tag{44}$$

The expression in the square brackets is equal to zero if $\bar{w} \geq 0$ and to $-2k\bar{w}$ if $\bar{w} \leq 0$. We denote by $-m(t)$ the infimum of $\bar{w}(\xi,\eta) - |\bar{w}(\xi,\eta)|$ for $\eta \leq t$. To prove the theorem it obviously suffices to show that $m(t) \equiv 0$. To this end we note that (44) implies the inequality

$$\bar{w}(x,t) \geq -k\int_0^t m(\eta)d\eta,$$

and therefore

$$m(t) \leq k\int_0^t m(\eta)d\eta,$$

which is only possible if $m(t) \equiv 0$.

Theorem 3. *When $f(x)$ increases, the value of $v(x,t)$ does not decrease.*

The physical meaning of this theorem is as clear as that of the foregoing theorem, provided that (35) is interpreted as the heat equation for a bar. The function $f(x)$ represents the initial temperature of the bar. When this temperature increases, the subsequent temperature also increases.

Proof of Theorem 3. Let $v_1(x,t)$ and $v_2(x,t)$ satisfy equation (35), let these functions be equal to $f_1(x)$ and $f_2(x)$ respectively, for $t = 0$, and let $f_2(x) \geq f_1(x)$. We prove that $v_2 \geq v_1$.

The function $w = v_2 - v_1$ satisfies the equation

$$\frac{\partial w}{\partial t} - \frac{\partial^2 w}{\partial x^2} = F(x,t,v_2) - F(x,t,v_1).$$

By condition (36),

$$F(x,t,v_2) - F(x,t,v_1) \geq -k|w|.$$

Therefore by Theorem 2, the function $w(x,t)$ is not less than the function $v^*(x,t)$, that is equal to $f_2(x) - f_1(x)$ when $t = 0$, while for $t > 0$ it satisfies the equation

$$\frac{\partial v^*}{\partial t} - \frac{\partial^2 v^*}{\partial x^2} = -k|v^*|.$$

If t is bounded, then $\exp(-kt)v^{**}$ is a (unique, by Theorem 1) bounded solution of this equation, where $v^{**}(x,t)$ satisfies equation (37) with the initial condition $v^{**}(x,0) = f_2(x) - f_1(x)$ (this function is clearly non-negative). Consequently,

$$w = v_2 - v_1 \geq 0,$$

as required.

Theorem 4. *If everywhere $f(x) \geq 0$ and $F(x,t,0) = 0$, then we also have*

$$v(x,t) \geq 0.$$

Proof. By Theorem 3, when $f(x)$ decreases, the function $v(x,t)$ does not increase. When $f(x) \equiv 0$, we have $v(x,t) \equiv 0$. Consequently, $f(x) \geq 0$ implies that $v(x,t) \geq 0$, as required.

Theorem 5. *If, in addition to the hypothesis of Theorem 4, $f(x) > 0$ on an interval of positive length, then for $t > 0$ we have*

$$v(x,t) > 0.$$

Proof. The proof follows from that of Theorem 3 by setting $v_2 = v$ and $v_1 = 0$ and taking into account the fact that the function $v^{**}(x,t)$ can be represented by a Poisson integral and is therefore necessarily positive for $t > 0$.

Theorem 6. *If $F(x,t,1) \equiv 0$ and $f(x) \leq 1$, then $v(x,t) \leq 1$.*

Proof. By Theorem 3, when $f(x)$ increases, the function $v(x,t)$ does not decrease. When $f(x) \equiv 1$, we have $v(x,t) \equiv 1$. The theorem now follows.

Theorem 7. *If for $t = 0$ the function $v(x,t)$ is equal to a monotone increasing differentiable function $f(x)$ and, for $t > 0$, satisfies the equation*

$$\frac{\partial v}{\partial t} - \frac{\partial^2 v}{\partial x^2} = F(t,v), \tag{45}$$

then $v(x,t)$ is a non-decreasing function of x for any $t > 0$.

Proof. By Theorem 1,

$$v(x,t) = v_0(x,t) + \frac{1}{2\sqrt{\pi}} \int_0^t d\eta \int_{-\infty}^{\infty} \frac{F(\eta, v(\xi,\eta)) \exp(-(x-\xi)^2/4(t-\eta))}{\sqrt{t-\eta}} d\xi, \tag{46}$$

where for $t > 0$ $v_0(x,t)$ satisfies the equation

$$\frac{\partial v}{\partial t} - \frac{\partial^2 v}{\partial x^2} = 0 \tag{47}$$

and equals $f(x)$ when $t = 0$. If $f(x)$ is differentiable, then $\partial v_0(x,t)/\partial x$ tends to $f'(x)$ as (x,t) tends to $(x,0)$ (see [6], pp. 330–331). On the other hand, the partial derivative with respect to x of the second term on the right-hand side of (46) does not exceed $(4/\sqrt{\pi})Mt^{1/2}$ in absolute value, provided that $|F| \leq M$ (see [6], p. 344). Therefore $\partial v(x,t)/\partial x$ tends to $f'(x)$ as $t \to 0$. If we also assume that the function $v(x,t)$ has derivatives $\partial^2 v/\partial t \partial x$ and $\partial^3 v/\partial x^3$, which is true when $F(t,v)$ is three times differentiable with respect to v, then the function $w(x,t) = \partial v(x,t)/\partial x$ satisfies the equation

$$\frac{\partial w}{\partial t} - \frac{\partial^2 w}{\partial x^2} = \frac{\partial F}{\partial v} w. \tag{48}$$

Whence, applying Theorem 4, we find that $w(x,t) \geq 0$, which is the desired result.

Theorem 8. *If*

$$f^{(\epsilon)}(x) \to f^{(0)}(x) \text{ as } \epsilon \to 0,$$

so that

$$\int_{-\infty}^{+\infty} |f^{(\epsilon)} - f^{(0)}| dx \to 0,$$

then for any $t > 0$ the function $v^{(\epsilon)}(x,t)$ satisfying (35) when $t > 0$ and equal to $f^{(\epsilon)}(x)$ when $t = 0$, tends to the function $v^{(0)}(x,t)$ (also) satisfying equation (35) when $t > 0$ and equal to $f^{(0)}(x)$ when $t = 0$.

Proof. We will look for $v^{(\epsilon)}(x,t)$ and $v^{(0)}(x,t)$ by the technique of successive approximation, as in the proof of Theorem 1. The functions $v_0^{(\epsilon)}$ and $v_0^{(0)}$ are determined by the formulas

$$v_0^{(\epsilon)}(x,t) = \frac{1}{2\sqrt{\pi}} \int_{-\infty}^{\infty} f^{(\epsilon)}(\xi) \frac{\exp(-(x-\xi)^2/4t)}{\sqrt{t}} d\xi,$$

$$v_0(x,t) = \frac{1}{2\sqrt{\pi}} \int_{-\infty}^{\infty} f^{(0)}(\xi) \frac{\exp(-(x-\xi)^2/4t)}{\sqrt{t}} d\xi.$$

It follows immediately that

$$v_0^{(\epsilon)}(x,t) \to v_0(x,t) \text{ as } \epsilon \to 0$$

when $t > 0$.

The difference $\tilde{v}_1^{(\epsilon)}(x,t) - \tilde{v}_1^{(0)}(x,t)$ (we use the same notation as in the proof of Theorem 1) is given by the formula

$$\frac{1}{2\sqrt{\pi}} \int_0^t d\eta \int_{-\infty}^{+\infty} \frac{\exp(-(x-\xi)^2/4(t-\eta))}{\sqrt{t-\eta}} [F(\xi,\eta,v_0^{(\epsilon)}) - F(\xi,\eta,v_0^{(0)})] d\xi,$$

whence

$$v_1^*(x,t) = |\tilde{v}_1^{(\epsilon)}(x,t) - \tilde{v}_1^{(0)}(x,t)| \leq$$

$$\leq \frac{k}{2\sqrt{\pi}} \int_0^t d\eta \int_{-\infty}^{+\infty} \frac{\exp(-(x-\xi)^2/4(t-\eta))}{\sqrt{t-\eta}} |v_0^{(\epsilon)}(\xi,\eta) - v_0^{(0)}(\xi,\eta)| d\xi.$$

We set

$$v_0^*(x,t) = \frac{1}{2\sqrt{\pi}} \int_{-\infty}^{\infty} |f^{(\epsilon)}(\xi) - f^{(0)}(\xi)| \frac{\exp(-(x-\xi)^2/4(t-\eta))}{\sqrt{t}} d\xi.$$

Obviously,

$$v_0^*(x,t) \geq |v_0^{(\epsilon)}(x,t) - v^{(0)}(x,t)|,$$

and therefore

$$v_1^*(x,t) \leq \frac{k}{2\sqrt{\pi}} \int_0^t d\eta \int_{-\infty}^{+\infty} \frac{\exp(-(x-\xi)^2/4(t-\eta))}{\sqrt{t-\eta}} v_0^*(\xi,\eta) d\xi =$$

$$= k \int_0^t v_0^*(x,t) d\eta = ktv_0^*(x,t).$$

The latter inequality follows from the fact that $v_0^*(x,t)$ satisfies equation (47). Thus,

$$v_1^*(x,t) \leq ktv_0^*(x,t).$$

Using the same technique we find that

$$v_i^*(x,t) = |v_i^{(\epsilon)}(x,t) - v_i^{(0)}(x,t)| \leq \frac{(kt)^i}{i!} v_0^*(x,t).$$

It follows that, by selecting a sufficiently small ϵ, we can make the sum $\sum_{i=0}^{\infty} v_i^*(x,t)$ and hence the expression $|v^{(\epsilon)}(x,t) - v^{(0)}(x,t)|$ arbitrarily small, as required.

Theorem 9. *The function $v(x,t)$ that satisfies equation (45) for $t > 0$ is zero for $t = 0$ and $x < 0$ and is equal to 1 for $t = 0$ and $x < 0$, is a non-decreasing function of x for any $t > 0$, and $\partial v(x,t)/\partial x > 0$ for $t > 0$.*

Proof. According to the above lemma, the function $v(x,t)$ can be regarded as the limit of the functions $v^{(\epsilon)}(x,t)$ as $\epsilon \to 0$ which assume the same values as $v(x,t)$ on the x-axis for $|x| \geq \epsilon$, are continuous together with their derivatives with respect to x throughout the x-axis, and are monotone. However, it has just been proved (Theorem 7) that for $t > 0$ the function $v^{(\epsilon)}(x,t)$ is monotone increasing in x. Therefore this is also true for $v(x,t)$.

We now prove that for $t > 0$ the derivative $\partial v(x,t)/\partial x$ is positive. To this end we have only to show that for $t > 0$ the relation $\partial v(x,t)/\partial x = 0$ is impossible. This can be done in the following way. For $t > 0$ the derivative $\partial v(x,t)/\partial x$ satisfies equation (48). Therefore the expression $\bar{w}(x,t) = \exp(Mt)\partial v(x,t)/\partial x$ satisfies the equation

$$\frac{\partial \bar{w}}{\partial t} - \frac{\partial^2 \bar{w}}{\partial x^2} = \left[\frac{\partial F}{\partial v} + M\right]\bar{w},$$

where M is an upper bound of $|\partial F/\partial v|$.

Since

$$\partial F/\partial v + M \geq 0,$$

it follows from Theorem 2 that for $t > t_0 > 0$ the function $\bar{w}(x,t)$ is not less than the function $\bar{\bar{w}}(x,t)$ equal to $\bar{w}(x,t)$ for $t = t_0$ and satisfying the equation

$$\frac{\partial \bar{\bar{w}}}{\partial t} - \frac{\partial^2 \bar{\bar{w}}}{\partial x^2} = 0$$

for $t > t_0$.

The function $\bar{\bar{w}}(x,t)$ is positive for all $t > t_0$ since for $t = t_0$ it is not identically equal to zero if t_0 is sufficiently small.

In what follows we will denote by $v(x,t)$ the function satisfying equation (29) for $t > 0$, equal to 0 for $t = 0$ and $x < 0$, and equal to 1 for $t = 0$ and $x > 0$.

Theorem 10. *For any fixed $x < 0$ we have*

$$v(x - 2t, t) \to 0 \text{ as } t \to +\infty.$$

Proof. The function $\bar{v}(x,t) = v(x-2t,t)$ satisfies the equation

$$\frac{\partial \bar{v}}{\partial t} - \frac{\partial^2 \bar{v}}{\partial x^2} = -2\frac{\partial \bar{v}}{\partial x} + F(\bar{v}).$$

On the other hand, the function $v^*(x,t) = v(x-2t,t)\exp(-x)$ satisfies the equation [8]

$$\frac{\partial v^*}{\partial t} - \frac{\partial^2 v^*}{\partial x^2} = [F(\bar{v}) - \bar{v}]\exp(-x).$$

By the conditions (32) and (33) imposed on $F(v)$, the function $F(v) - v$ is non-positive. Therefore $v^*(x,t)$ is less than the function satisfying equation (37) for $t > 0$ and, for $t = 0$, equal to zero for $x < 0$ and to $\exp(-x)$ for $x > 0$. The latter function tends to zero uniformly with respect to x as $t \to \infty$.

Theorem 11. *For a fixed t we regard the derivative $\partial v(x,t)/\partial x$ as a function of v. This is possible in view of Theorem 9. Let*

$$\partial v(x,t)/\partial x = \psi(v,t). \tag{49}$$

Then as t increases and v is fixed, the function ψ does not increase.

Proof. Consider the functions $v(x,t)$ and $v(x+c,t+t_0) = v_{t_0}(x,t)$, where c is a constant and $t_0 > 0$. We put

$$w(x,t) = v(x,t) - v_{t_0}(x,t).$$

Let $\mathfrak{M}$ be the set of points (x,t) such that $w(x,t) > 0$. First we prove that this set is bounded only on the left by a line emanating from the origin and along which the coordinate t is nowhere decreasing. To prove this we note that $w(x,t)$ satisfies the equation

$$\frac{\partial w}{\partial t} - \frac{\partial^2 w}{\partial x^2} = k(x,t), \tag{50}$$

where $k(x,t)$ is a bounded function, namely

$$k(x,t) = F'(\bar{u}(x,t)),$$

[8] It is easily seen that the function $v^*(x,t)$ remains bounded when $t > 0$ is bounded.

and $\bar{u}(x,t)$ is a number lying between $v(x,t)$ and $v_{t_0}(x,t)$. Therefore the set $\mathfrak{M}$ cannot contain isolated portions[9] not adjoining the x-axis. Thus, the set consists of a single portion adjoining the positive semi-axis x, as is clear. In order to prove that the set $\mathfrak{M}$ is bounded on the left by a line along which t is non-decreasing we assume that, on the contrary, this line contains a portion of the form shown in Figure 4. For example, suppose that, starting from the point A, this line goes downward. Then the function $w(x,t)$ assumes negative values to the right of the line OA whereas on the line OA it is equal to 0, while for $x > 0$ it takes on positive values on the x-axis. However, using the same methods as in the proof of Theorem 4 we can show that this is impossible.

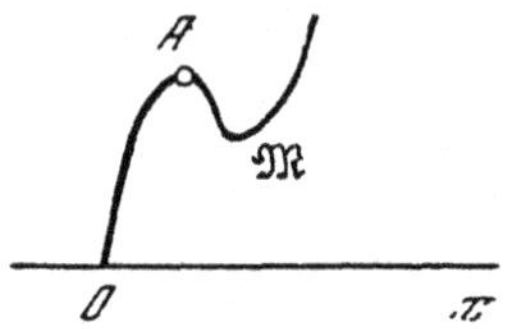

Fig. 4

In just the same way it can be proved that the set $\mathfrak{M}$ is unbounded from the right.

The above remarks allow us to prove the stated theorem quite simply. Indeed, in view of the arbitrariness in the choice of c, $v(x_0,t)$ and $v_{t_0}(x_0,t)$ can be made equal for any pre-assigned t and some $x = x_0$. Then, by the above argument,

$$v(x,t) \geq v_{t_0}(x,t)$$

for all $x > x_0$, and consequently

$$\frac{\partial v}{\partial x}(x_0,t) \geq \frac{\partial v_{t_0}}{\partial x}(x_0,t),$$

which is the desired result.

Theorem 12. *For any t we have*

$$\frac{\partial v(x,t)}{\partial x} \geq u'(x)$$

[9] The proof of a similar assertion for the case of finite portions can be found in [7], pp. 386–387. It can be shown that the same is true for infinite portions as well. Cf. the Remark to Theorem 1.

provided that $v(x,t) = u(x)$. *Here* $u(x)$ *is the solution to equation* (34) *which was discussed at the beginning of this section.*

The proof is completely similar to that of the foregoing theorem. It is only necessary to replace the function $v_{t_0}(x,t)$ by $u(x+c)$ and the function $w(x,t)$ by the difference $v(x,t) - u(x+c)$.

Theorem 13. *Let*

$$v^*(x,t) = v(x + \phi(t), t),$$

where the function $\phi(t)$ *is chosen in such a way that*

$$v^*(0,t) = c = \text{const.}$$

Then

$$v^*(x,t) \to v^*(x) \text{ uniformly with respect to } x \text{ as } t \to \infty.$$

Proof. It follows from (49) that

$$x = \int_c^{v^*} \frac{dv}{\psi(v,t)}. \tag{51}$$

By Theorem 11, the integrand increases monotonically as $t \to \infty$. Moreover, by Theorem 12, the integral $\int_c^{v^*} \frac{\partial v}{\psi(v,t)}$ cannot increase indefinitely. Therefore we can pass to the limit under the integral sign in (51). Let

$$\psi(v,t) \to \psi(v) \text{ as } t \to \infty.$$

Then the limit of (51) is

$$x = \int_c^{v^*} \frac{dv}{\psi(v)}.$$

Since, by Theorem 12,

$$\psi(v) > 0,$$

it follows that this condition determines a function v^* of x. It remains to show that $v^*(x,t)$ converges to $v^*(x)$ uniformly. To this end we note that (51) implies that $x(v^*,t)$ converges uniformly to $x(v)$ on any interval $\epsilon < v^* < 1 - \epsilon$. If we now take into account the fact that, by Theorem 11, the function $\psi(v^*,t)$ is bounded on each such interval, it follows that the function $v^*(x,t)$ converges uniformly to $v^*(x)$ for the values of x such that $v(x)$ is contained between ϵ and $1-\epsilon$ (where ϵ is arbitrarily small). But outside this interval of values of x, $v^*(x,t) \to v^*(x)$ uniformly since for sufficiently large t the function $v^*(x,t)$ assumes values that differ slightly from 0 or 1.

Theorem 14. *As $t_0 \to +\infty$ the sequence of functions*

$$v_{t_0}(x,t) = v[x + \phi(t_0),\ t + t_0]$$

converges uniformly to a solution $\bar{v}(x,t)$ of equation (29) *in the region $t \leq T =$ const. The function $\phi(t_0)$ is defined so that*

$$v_{t_0}(0,0) = c = \text{const}$$

for all t_0.

Proof. The function

$$w(x,t) = v_{t_0}(x,t) - v_{t_0+T}(x,t)$$

satisfies the equation

$$\frac{\partial w}{\partial t} - \frac{\partial^2 w}{\partial x^2} = F'(\bar{v})w, \tag{52}$$

where $\bar{v}(x,t)$ is contained between $v_{t_0}(x,t)$ and $v_{t_0+T}(x,t)$. By Theorem 13, for sufficiently large t_0 we have

$$|w(x,0)| < \epsilon,$$

where $\epsilon > 0$ is arbitrarily small. By Theorems 2 and 3, the function $w(x,t)$ is less than the function $\bar{w}(x,t) = \epsilon \exp(kt)$, where k is an upper bound for the values of $|F'(u)|$, since for $t=0$ the latter function assumes a value that is not less than $w(x,0)$ and for $t>0$ satisfies the equation

$$\frac{\partial \bar{w}}{\partial t} - \frac{\partial^2 w}{\partial x^2} = k|\bar{w}|,$$

the right-hand side of which is not less than the right-hand side of (52) for $w = \bar{w}$. It can similarly be proved that

$$\bar{w}(x,t) > -\epsilon \exp(kt).$$

Thus, we have shown that the sequence of functions $v_{t_0}(x,t)$ converges uniformly in a region $t < T$ to a function $\bar{v}(x,t)$ as $t_0 \to +\infty$. Let us show that $\bar{v}(x,t)$ satisfies equation (29).

Using (42) we write

$$v_{t_0}(x,t) = v_{t_0,0}(x,t) + \\ + \frac{1}{2\sqrt{\pi}} \int_0^t d\eta \int_{-\infty}^{+\infty} \frac{\exp(-(x-\xi)^2/4(t-\eta))}{\sqrt{t-\eta}} F(v_{t_0}(\xi,\eta))d\xi. \tag{53}$$

We can pass to the limit in this relation by substituting $\bar{v}$ for v_{t_0}. However, as was shown in the proof of Theorem 1, a function satisfying equation (53) satisfies equation (29) as well.

Theorem 15. *As $t_0 \to +\infty$ the first partial derivatives of $v_{t_0}(x,t)$ with respect to x and t tend to the corresponding partial derivatives of $\bar{v}(x,t)$, uniformly in a region $\epsilon < t < T$ where ϵ and T are positive constants.*

Proof. The uniform convergence of $\partial v_{t_0}/\partial x$ can be proved by means of relation (53). Indeed, for $t > \epsilon$ the partial derivative of the first term on the right-hand side with respect to x converges uniformly since this term can be represented as a Poisson integral. In order to show that this is also true for the second term for $t < T$ we consider the difference between the two values of it corresponding to $t_0 = t_0'$ and $t_0 = t_0''$. This difference is equal to

$$\frac{1}{2\sqrt{\pi}} \int_0^t d\eta \int_{-\infty}^{+\infty} \frac{\exp(-(x-\xi)^2/4(t-\eta))}{\sqrt{t-\eta}} \left[F(v_{t_0'}) - F(v_{t_0''})\right] d\xi. \tag{54}$$

By Theorem 14, the difference

$$F(v_{t_0'}(\xi,t)) - F(v_{t_0''}(\xi,t))$$

is arbitrarily small for sufficiently large t_0' and t_0''. By applying to this case the above-mentioned result of Gevrey, we see that for sufficiently large t_0' and t_0'' the partial derivative of (54) with respect to x tends uniformly to zero, provided that $t < T$.

The function $w_{t_0}(x,t) = \partial v_{t_0}(x,t)/\partial x$ satisfies the equation

$$\frac{\partial w_{t_0}}{\partial t} - \frac{\partial^2 w_{t_0}}{\partial x^2} = F'(v_{t_0})w_{t_0}.$$

We have already proved that for $\epsilon < t < T$ the right-hand side of this equation converges uniformly as $t_0 \to +\infty$. Therefore the argument used for proving the uniform convergence of $\partial v_{t_0}/\partial x$ can be applied in the proof of the uniform convergence of $\partial w_{t_0}/\partial x = \partial^2 v_{t_0}/\partial x^2$. And since the function v_{t_0} satisfies equation (29), the uniform convergence of $\partial v_{t_0}/\partial t$ is also proved.

Theorem 16. *Let the function $v_{t_0}(x,t)$ $(\bar{v}(x,t))$ be equal to a constant c on the line $x = \phi_{t_0}(t)$ $(x = \phi(t))$. Then*

$$\phi_{t_0}(t) \to \phi'(t) \text{ as } t_0 \to \infty$$

uniformly with respect to t for $\epsilon < t < T$.

Proof. The value of $\phi'_{t_0}(t)$ $(\phi'(t))$ at the point $(\phi_{t_0}(t), t)$ $(\phi(t), t))$ is equal to

$$-\frac{\partial v_{t_0}/\partial t}{\partial v_{t_0}/\partial x} \quad \left(-\frac{\partial \bar{v}/\partial t}{\partial \bar{v}/\partial x}\right).$$

By Theorems 12 and 14,

$$[\phi_{t_0}(t) - \phi(t)] < \epsilon_1$$

throughout the region $\overline{G}$ $(\epsilon < t < T)$, provided that t_0 is sufficiently large and ϵ_1 is arbitrarily small. By Theorem 15, for the same values of the arguments the difference between the numerators and the denominators of the fractions

$$\frac{\partial v_{t_0}/\partial t}{\partial v_{t_0}/\partial x} \text{ and } \frac{\partial \bar{v}/\partial t}{\partial \bar{v}/\partial x} \tag{55}$$

is arbitrarily small, uniformly in the region $\overline{G}$. Moreover, the derivative $\partial \bar{v}/\partial x$ does not exceed a positive constant in the strip $\phi(t) - \epsilon_2 < x < \phi(t) + \epsilon_2$. Consequently, the fractions in (55), for the same values of the arguments and sufficiently large t_0, differ by less than ϵ_3 in the strip

$$\epsilon < t < T, \quad \phi(t) - \epsilon_2 < x < \phi(t) + \epsilon_2.$$

Also, taking into account the fact that the expression $\frac{\partial \bar{v}/\partial t}{\partial \bar{v}/\partial x}$ is uniformly continuous in this strip and therefore the difference between its values at points in this strip with the same t is arbitrarily small for sufficiently small ϵ_3, we complete the proof of the theorem.

Theorem 17. *For any t we have*

$$\bar{v}(x,t) = u(x+2t) \quad \textit{and} \quad d\phi/dt \to -2 \textit{ as } t \to \infty$$

(the notation is the same as in Theorem 14).

Proof. Consider the function

$$v^*(x,t) = \bar{v}(x + c_1(t), t),$$

where the function $c_1(t)$ is chosen so that

$$v^*(0,t) \equiv c = \text{const}.$$

Then

$$\frac{\partial v^*}{\partial t} = \frac{\partial^2 v^*}{\partial x^2} + c_1(t)\frac{\partial v^*}{\partial x} + F(v^*).$$

On the other hand, by the definition of $\bar{v}(x,t)$ the value of $v^*(x,t)$ does not depend on t for any x. Therefore

$$\partial v^*/\partial t = 0 \text{ and } c_1'(t) = \text{const}.$$

It follows from §2 that this constant cannot be greater than -2 and, by Theorem 10, the constant cannot be less than -2. Consequently , it is equal to -2, and, by Theorem 16, we have

$$d\phi/dt \to -2 \text{ as } t \to \infty,$$

as required.

Remark. Assume that the initial values of $v(x,t)$ differ from those considered up to now; namely, let

1) $v(x,0) = 1$ for $x \geq c_1$;

2) $v(x,0) = 0$ for $x \leq c_2 < c_1$;

3) $v(x,0)$ assumes arbitrary values between 0 and 1 for $c_2 < x < c_1$.

Then it is easily seen that in this case the rate at which the region in which the major part of the drop of v from 1 to 0 occurs moves to the left, nevertheless tends to 2 since

$$v(x - c_1, t) \leq \bar{v}(x,t) \leq v(x - c_2, t),$$

where $\bar{v}(x,t)$ denotes the solution of equation (29) satisfying the new initial conditions.

References

1. R.A. Fisher, *The genetical theory of natural selection*, Oxford Univ. Press, 1930.
2. A.Ya. Khinchin, *Asymptotic laws of probability theory*, ONTI, Moscow, 1936 (in Russian).
3. I. Bendixson, 'Sur les courbes définies par les équations différentielles', *Acta Math.* **24** (1901), 1–88.
4. V.V. Stepanov, *A course of differential equations*, Fizmatgiz, Moscow, 1958 (in Russian).
5. I. Petrowsky, 'Über das Verhalten der Integralkurven eines Systems gewöhnlichen Differentialgleichungen in der Nähe eines singularen Punktes', *Mat. Sb.* **41** (1934), 107–156.
6. M. Gevrey, 'Sur les équations aux dérivées partielles du type parabolique', *J. Math. Pures et Appl. Ser.* 6 **9** (1913), 305–471.
7. I. Petrowsky, 'Zur ersten Randwertaufgabe der Warmeleitungsgleichung', *Comp. Math.* **1** (1935), 383–419.

39. A SIMPLIFIED PROOF OF THE BIRKHOFF-KHINCHIN ERGODIC THEOREM *

§1. Introduction

The theorem in question was stated by Birkhoff as a theorem of mechanics or, as one can say, a theorem related to the evolution of an arbitrary system whose state is completely determined by a finite number of parameters and whose behaviour is governed by differential equations admitting an integral invariant.

Assume that the state of the system under consideration can be represented by a point P on a closed n-dimensional differential-geometric manifold M^n. Further, let the coordinates $x_1, x_2, \ldots, x_n$ of the point P satisfy the differential equations

$$dx_i/dt = X_i(x_1, x_2, \ldots, x_n) \tag{1}$$

whose right-hand sides do not involve t explicitly. Assume that the solution of (1) is unique. Then for each t equations (1) determine a transformation of any subset E on the manifold M^n:

$$T_t E = E',$$

where E' is the set of points P' into which the points P of E go in time t. By an integral invariant is meant a set function $I(E)$ such that

$$I(T_t E) = I(E). \tag{2}$$

We assume that there exists an integral invariant $I(E)$ on all measurable subsets of the manifold M^n, satisfying the condition

$$I(M^n) = 1. \tag{3}$$

The theorem which Birkhoff himself proved under somewhat narrower assumptions reads: *for any real-valued function $f(P)$ defined on M^n and summable relative to $I(E)$* [1] *the limit*

$$\lim_{C\to+\infty} \frac{1}{C}\int_0^C f(T_t P)dt = \psi(P)$$

* *Uspekhi Mat. Nauk* **5** (1938), 52–56.

[1] This means that the integral

$$\int_{M^n} |f(P)| I(dM^n)$$

exists and is finite.

exists for all P except, possibly, on at most a set U for which $I(U) = 0$.

This theorem can be regarded as a special case of the following more general theorem. *Given a stationary stochastic process in the sense of Khinchin,* [2] *that is, a set of random variables x_t depending on the parameter t, $-\infty < t < +\infty$, such that the distributions of the systems*

$$(x_{t_1}, x_{t_2}, \dots, x_{t_n}) \text{ and } (x_{t_1+\tau}, x_{t_2+\tau}, \dots, x_{t_n+\tau})$$

coincide for any $n, t_1, t_2, \dots, t_n$, and τ; suppose that the variables x_t have finite mathematical expectation $\mathbf{E}|x_t|$ (which are the same for all t since the process is stationary) and are continuous in t with probability 1. [3] *Then with probability 1 the integral*

$$\lim_{C \to +\infty} \frac{1}{C} \int_0^C x_t dt \tag{4}$$

exists.

This is the general Birkhoff-Khinchin theorem. The above Birkhoff theorem is in fact a special case, since if the probability that a point P lies in the set E is assumed to be equal to $I(E)$, then the random variables

$$x_t = f(T_t P)$$

satisfy all the conditions of the general theorem.

§2. Reduction to the discrete case

In this section we will show that the existence of the limit in (4) with probability 1 is implied by the existence of the limit

$$\lim \frac{1}{n} \int_0^n x_t dt \tag{5}$$

with the same probability, where n runs only over the positive integers. Introducing the expression

$$\bar{x}_i = \int_i^{i+1} x_t dt$$

and noting that

$$\mathbf{E}|\bar{x}_i| \le \mathbf{E} \int_i^{i+1} |x_t| dt = \mathbf{E}|x_0|,$$

[2] A.Ya. Khinchin, 'Correlation theory of stationary stochastic processes', *Uspekhi Mat. Nauk* **5** (1938), 42–51 (in Russian)

[3] This assumption can be weakened.

we can see that the proof of the general Birkhoff-Khinchin theorem thus reduces to the proof of the following theorem, related to discrete sequences of random variables.

Main Theorem. *Let* x_i $(-\infty < i < +\infty)$ *be a stationary sequence*[4] *of random variables such that the mathematical expectation* $\mathbf{E}|x_i|$ *is finite. Then the limit*

$$\lim_{n\to\infty} \frac{x_0 + x_1 + \ldots + x_{n-1}}{n}$$

exists with probability 1.

Thus, we assume that the limit (5) exists with probability 1 and put

$$y_n = \int_n^{n+1} |x_t|dt.$$

Clearly,

$$\mathbf{E}(y_0) = \int_0^1 \mathbf{E}|x_t|dt = \mathbf{E}|x_0| < +\infty.$$

Therefore the series of probabilities

$$\sum_{n=1}^{\infty} \mathbf{P}\Big\{\frac{1}{n}y_n > \epsilon\Big\} = \sum_{n=1}^{\infty} \mathbf{P}\Big\{\frac{1}{n}y_0 > \epsilon\Big\} = \sum_{n=1}^{\infty} \mathbf{P}\{y_0 > n\epsilon\} =$$
$$= \sum_{n=1}^{\infty}\sum_{m=n}^{\infty} \mathbf{P}\{(m+1)\epsilon \geq y_0 > m\epsilon\} =$$
$$= \sum_{m=1}^{\infty} m\mathbf{P}\{(m+1)\epsilon > y_0 \geq m\epsilon\} \leq \frac{1}{\epsilon}\mathbf{E}(y_0)$$

converges for any $\epsilon > 0$. It follows that

$$\lim_{n\to\infty} \frac{1}{n} y_n = \lim_{n\to\infty} \frac{1}{n} \int_n^{n+1} |x_t|dt = 0 \tag{6}$$

with probability 1.

The existence of the limit (5) implies that the limit

$$\lim_{C\to+\infty} \frac{1}{C} \int_0^{[C]} x_t dt$$

[4] The definition of stationarity remains literally the same; only t should be replaced by an index i assuming only integral values.

exists where $[C]$ denotes the integral part of C and C runs over a continuous range of values. In the case of (6) we obtain

$$\lim_{C\to+\infty}\sup\left|\frac{1}{C}\int_0^C x_t dt - \frac{1}{C}\int_0^{[C]} x_t dt\right| =$$

$$= \lim_{C\to+\infty}\sup\left|\frac{1}{C}\int_{[C]}^C x_t dt\right| \leq \lim_{n\to\infty}\sup\frac{1}{n}\int_n^{n+1}|x_t|dt = 0.$$

Therefore if, simultaneously, the limit (5) exists and (6) holds, then the limit (4) exists with probability 1.

§3. Proof of the Main Theorem

We set

$$h_{ab} = \frac{x_a + x_{a+1} + \ldots + x_{b-1}}{b-a}.$$

It is required to prove that h_{0b} tends to a definite limit with probability 1 as $b \to +\infty$.

Lemma. *If the probability that h_{0b} tends to a limit as $b \to +\infty$ is less than 1, then there exist two numbers α and β, $\alpha < \beta$, such that the probability that the inequalities*

$$\lim_{b\to+\infty}\sup h_{0b} > \beta, \quad \lim_{b\to+\infty}\inf h_{0b} < \alpha \tag{7}$$

hold simultaneously is positive.

Proof. Let

$$(\alpha_n, \beta_n), \ n = 1, 2, \ldots,$$

be the sequence of all intervals $\alpha_n < \beta_n$ with rational end points. If $\lim h_{0b}$ does not exist, then the system of intervals (α_n, β_n) contains a first interval such that $\lim\sup h_{0b} < \alpha_n$ and $\lim\inf h_{0b} < \alpha_n$. Thus, the event K when there is no $\lim h_{0b}$ is subdivided into a countable number of events K_n. But if the probability $P(K)$ is positive, then there exists n such that $P(K_n)$ is also positive. This proves the lemma.

Proof of the Main Theorem. We denote by K the event when the inequalities (7) hold. We assume that all the x_i take definite values and proceed to a detailed study of the behaviour of the means h_{ab}. An interval (a, b) is said to be singular (relative to β) if $h_{ab} > \beta$ and $h_{ab'} \leq \beta$ for all b', $a < b' < b$. It is easily shown that two singular intervals (a, b) and (a_1, b_1) cannot overlap, that is, the case when $a < a_1 < b < b_1$ is impossible. Indeed, if the intervals overlap, then

$$h_{ab} = \frac{(a_1 - a)h_{aa_1} + (b - a_1)h_{a_1 b}}{b - a},$$

and $h_{ab} > \beta$ would imply that either $h_{aa_1} > \beta$, which is impossible since (a, b) is a singular interval, or $h_{a_1 b} > \beta$, which is impossible since the interval (a_1, b_1) is singular.

The difference $b - a$ is called the rank of the interval (a, b). Further, an interval (a, b) is said to be s-singular if it is singular, has rank less than or equal to s and is not contained in any greater singular interval whose rank does not exceed s. Obviously, each singular interval of rank not exceeding s is contained in a unique s-singular interval. Indeed, among the singular intervals containing (a, b) whose rank is no higher than s there must be at least one of smallest length; if there were two such intervals, then they would overlap, which, by what has been proved, is impossible. As to s-singular intervals, the only possibility is that any two such intervals lie outside each other.

Let K_s denote the event when inequalities (7) hold and jointly there exists $t \leq s$ such that $h_{0t} > \beta$. Clearly, $\lim K_s = K$; therefore

$$\lim P(K_s) = P(K). \tag{8}$$

Our aim is to obtain a contradiction to the assumption that $P(K) > 0$. If $P(K) > 0$, then clearly, starting from some s, all the $P(K_s)$ are also positive. In what follows we consider only these s.

In the case of K_s there exists an s-singular interval (a, b) with $a \leq 0 < b$. Indeed, among the values $t \leq s$ such that $h_{0t} > \beta$ in the case of K_s, there is a smallest one (denoted by t'). The interval $(0, t')$ is singular. Consequently, it is contained in an s-singular interval (a, b). Conversely, if there exists an s-singular interval (a, b) with $a \leq 0 < b$, then there is $t \leq s$ such that $h_{0t} > \beta$. Indeed, if $a = b$, then one can set $t = b$, and if $a < 0$, then it follows from

$$h_{ab} = \frac{-ah_{a0} + bh_{0b}}{b - a},$$

$h_{ab} > \beta$, and $h_{a0} \leq \beta$ that $h_{0b} > \beta$ and consequently, one can also set $t = b$. Thus, the event K_s is completely determined by the inequalities (7) and the existence of an s-singular interval (a, b) with $a \leq 0 < b$. Moreover, there can exist only one s-singular interval (a, b) satisfying the inequality $a \leq 0 < b$ in each specific case, since any two such intervals would overlap at the point 0, which is impossible. Denoting by p the value $-a$ and by q the value $b-a$ we see that the event K_s is the union of pairwise incompatible events K_{pq} corresponding to the existence of s-singular intervals:

$$K_s = \sum K_{pq}; \; q = 1, \ldots, s; \; p = 0, 1, \ldots, q-1. \tag{9}$$

The change $i' = i+p$ transforms the event K_{0q} into K_{pq}; therefore, in view of stationarity, we have $P(K_{pq}) = P(K_{0q})$ and $\mathbf{E}K_{pq}(x_0) = \mathbf{E}K_{0q}(x_p)$. Taking into account the fact that the inequality $h_{0q} > \beta$ holds for the event K_{0q} we thus obtain

$$\begin{aligned} P(K_s)\mathbf{E}K_s(x_0) &= \sum_{p,q} P(K_{pq})\mathbf{E}K_{pq}(x_0) = \sum_q P(K_{0q}) \sum_p \mathbf{E}K_{0q}(x_p) = \\ &= \sum_q P(K_{0q})\mathbf{E}K_{0q}(qh_{0q}) > \sum_q (K_{0q})q\beta = \sum_{p,q} P(K_{pq})\beta = P(K_s)\beta, \end{aligned}$$

whence

$$\mathbf{E}K_s(x_0) > \beta.$$

Since $K_s \to K$, it follows that

$$\mathbf{E}K(x_0) \geq \beta.$$

On the other hand, it can similarly be proved that

$$\mathbf{E}K(x_0) \leq \alpha,$$

which yields the desired contradiction. Consequently, $P(K) = 0$, which proves the assertion of the main theorem.

40. ON INEQUALITIES FOR SUPREMA OF CONSECUTIVE DERIVATIVES OF AN ARBITRARY FUNCTION ON AN INFINITE INTERVAL *

§1. Statement of the problem and results

Consider a function $f(x)$ whose first n derivatives are bounded on the entire real line. Here the boundedness of the nth derivative is understood in the sense that the $(n-1)$st derivative has bounded derived numbers. [1] We denote by $M_k(f)$ the supremum of the absolute value of the kth derivative of the function $f(x)$:

$$M_k(f) = \sup |f^{(k)}(x)|, \quad k = 0, 1, 2, \ldots, n.$$

We note that here $M_n(f)$ is the supremum of the absolute values of the derived numbers of the $(n-1)$st derivative.

The aim of this paper is to prove the following theorem which I stated in [1]:

Theorem I. *Given three positive numbers M_0, M_k, M_n $(0 < k < n)$; then in order that a function $f(x)$ such that*

$$M_0 = M_0(f), \quad M_k = M_k(f), \quad M_n = M_n(f),$$

exist, it is necessary and sufficient that the condition

$$M_k \leq C_{nk} M_0^{(n-k)/n} M_n^{k/n}, \tag{1}$$

should hold, where

$$C_{nk} = K_{n-k} : K_n^{(n-k)/n},$$

$$K_i = \begin{cases} \frac{4}{\pi}\left(1 - \frac{1}{3^{i+1}} + \frac{1}{5^{i+1}} - \frac{1}{7^{i+1}} + \ldots\right) & \text{for even } i, \\ \frac{4}{\pi}\left(1 + \frac{1}{3^{i+1}} + \frac{1}{5^{i+1}} - \frac{1}{7^{i+1}} + \ldots\right) & \text{for odd } i, \end{cases} \tag{2}$$

Inequality (1) can be written in the form

$$M_k^n \leq C_{nk}^n M_0^{n-k} M_n^k. \tag{1*}$$

* *Uchen. Zap. MGU* Vol. 30, No. 3 (1939), 3–16.

[1] The results do not alter if we consider only functions $f(x)$ possessing a continuous nth derivative, or even analytic functions: in these cases condition (1) of Theorem I remains necessary and sufficient.

The advantage of this form is that all the coefficients C^n_{nk} are rational. We present the values of C^n_{nk} for $n \leq 5$:

n	$k=1$	$k=2$	$k=3$	$k=4$
2	2			
3	$\frac{9}{8}$	3		
4	$\frac{512}{375}$	$\frac{36}{25}$	$\frac{24}{5}$	
5	$\frac{1953125}{1572864}$	$\frac{125}{72}$	$\frac{225}{128}$	$\frac{15}{2}$

We also give approximate values of the constants C_{nk} for $n \leq 7$:

n	$k=1$	$k=2$	$k=3$	$k=4$	$k=5$	$k=6$
2	1.41421					
3	1.04004	1.44225				
4	1.08096	1.09545	1.48017			
5	1.04426	1.11665	1.11942	1.49631		
6	1.04298	1.08001	1.14520	1.14280	1.50892	
7	1.03451	1.07289	1.10472	1.16471	1.15137	1.51748

We make some additional remarks on the asymptotic behaviour of the coefficients C_{nk}. Since, obviously,

$$K_i \to 4/\pi \text{ as } i \to \infty,$$

the following properties hold:

1) as $n \to \infty$ and for bounded $n-k$ we have

$$C_{nk} = K_{n-k} + \epsilon_{nk}, \; \epsilon_{nk} \to 0;$$

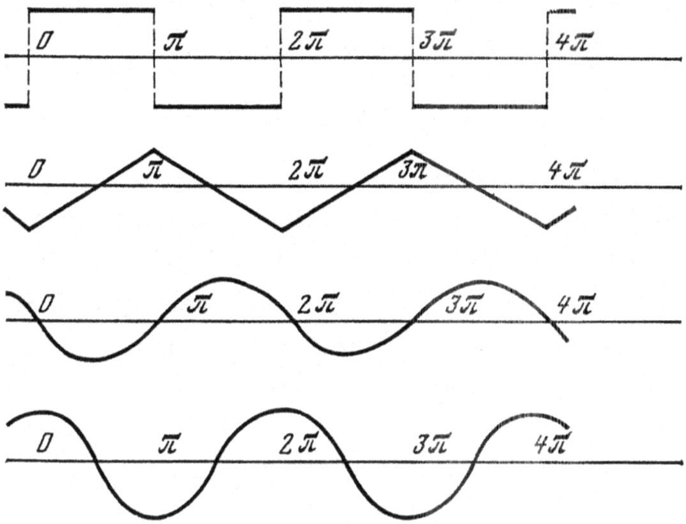

Fig. 1

2) as $n \to \infty$ and $n - k \to \infty$,

$$C_{nk} = \left(\frac{4}{\pi}\right)^{k/n} + \epsilon'_{nk}, \quad \epsilon'_{nk} \to 0.$$

In particular,

$$C_{n,n-1} \to K_1 = \pi/2 \text{ as } n \to \infty,$$

$$C_{n,1} \to 1 \qquad \text{as } n \to \infty.$$

The last two relations are of special interest, since for all n and k $(0 < k < n)$ the inequalities [2]

$$1 < C_{nk} < \pi/2 \tag{3}$$

hold.

For the case $n = 2$, $k = 1$ Theorem I was proved in 1914 in [2], while for all $n < 5$ and $n = 5$, $k = 2$ it was proved by G.E. Shilov [3]. In the proof of Theorem I we use the fact that, in particular, the extremal relation

$$M_k(f) = C_{nk} M_0^{(n-k)/n}(f) M_n^{k/n}(f) \tag{4}$$

is fulfilled, for a given n and an arbitrary k $(0 < k < n)$, for the function

$$f_n(x) = \frac{4}{\pi} \sum_{m=0}^{\infty} \frac{\sin\left((2m+1)x - \frac{\pi}{2}n\right)}{(2m+1)^{n+1}}. \tag{5}$$

The functions $f_n(x)$, whose graphs are given in Figure 1, were considered earlier in another context by N.I. Akhiezer and M.G. Krein [4].

---

[2] The proof is elementary and we do not present it here.

It can be easily shown that the validity of relation (4) for the function $f_n(x)$ implies its validity for any function of the form

$$\phi_n(x) = af_n(bx + c), \tag{6}$$

where $a > 0$, $b > 0$, and c are arbitrary constants. The constants a and b can be selected so that $M_0(\phi_n)$ and $M_n(\phi_n)$ assume preassigned arbitrary positive values. [3] The conjecture that the functions of the form (6) give the extremum of $M_k(f)$ for given $M_0(f)$ and $M_n(f)$ was put forward by E.G. Shilov. The content of this paper is, in essence, a proof of this conjecture.

The functions $f_n(x)$ are periodic with period 2π and, moreover, satisfy the relation

$$f_n\left(\frac{\pi}{2}m + x\right) = (-1)^{m+n+1} f_n\left(\frac{\pi}{2}m - x\right).$$

By virtue of this relation, the behaviour of the function $f_n(x)$ on the whole real line is uniquely determined by its values on a single closed interval of the form

$$m\pi/2 \le x \le (m+1)\pi/2.$$

For $n > 0$ the function $f_n(x)$ varies monotonically on each of these intervals from zero at one of its ends to $\pm K_n$ at the other end. Analogous periodicity and symmetry properties are also characteristic for functions of the form (6). These periodicity and symmetry properties for functions $\phi_n(x)$ of the form (6) readily imply the following: if, along with $M_0(\phi_n)$ and $M_n(\phi_n)$, the value ξ of the function $\phi_n(x)$ at a point x_0 satisfying the condition

$$|\xi| \le M_0(\phi_n)$$

is also fixed, then these three parameters specify exactly two functions among all functions of the form (5) when $|\xi| < M_0(\phi_n)$ and exactly one function when $|\xi| = M_0(\phi_n)$. Denoting these two functions (which coincide in the case $|\xi| = M_0(\phi_n)$) by $\phi_{n,1}(x)$ and $\phi_{n,2}(x)$, we have

$$\phi'_{n,1}(x_0) = -\phi'_{n,2}(x_0)$$

and, consequently,

$$|\phi'_{n,1}(x_0)| = |\phi'_{n,2}(x_0)|.$$

[3] As to the constant c, varying it clearly does not affect the values of $M_i(\phi_n)$.

Now consider a function $f(x)$ with finite $M_i(f)$ for $i \leq n$. By a *comparison function* of order n for the function $f(x)$ at a point x_0 we will mean a function of the form (6) such that

$$M_0(\phi_n) = M_0(f), \quad M_n(\phi_n) = M_n(f), \quad \phi_n(x_0) = f(x_0).$$

Using this definition we can prove the following theorem.

Theorem II. *If $\phi_n(x)$ is a comparison function of order n for a function $f(x)$ at a point x_0, then*

$$|f'(x_0)| \leq |\phi_n'(x_0)|. \tag{7}$$

We note that for $n = 1$ the derivatives $f'(x)$ and $\phi_n'(x)$ do not necessarily exist; however, inequality (7) remains valid in this case as well for the derived numbers. Theorem II is a strengthening of the inequality (1) for $k = 1$. Indeed, inequality (7) implies that for any point x_0 we have

$$|f'(x_0)| \leq M_1(\phi_n) = C_{n,1} M_0^{\frac{n-1}{n}}(\phi_n) M_n^{\frac{1}{n}}(\phi_n) = C_{n,1} M_0^{\frac{n-1}{n}}(f) M_n^{\frac{1}{n}}(f),$$

whence

$$M_1(f) \leq C_{n,1} M_0^{\frac{n-1}{n}}(f) M_n^{\frac{1}{n}}(f),$$

which is the inequality (1) for the case $k = 1$.

The proof of Theorems I and II is carried out in the following way: in §2 the sufficiency of condition (1) of Theorem I is proved by considering functions of the form

$$\chi_n(x) = a f_n(bx) + d;$$

in §3, for the case $k = 1$ the necessity of condition (1) (in Theorem I) and Theorem II are proved simultaneously by induction on n; in §4 induction on k is used to prove the necessity of condition (1) in the general case.

§2. Sufficiency of condition (1)

For any positive M_0, M_k and M_n the relation

$$M_k^n = \gamma_{nk}^n M_0^{n-k} M_n^k$$

uniquely determines for $0 < k < n$ the factor

$$\gamma_{nk} = \gamma_{nk}(M_0, M_k, M_n) > 0.$$

For an arbitrary function $f(x)$ with finite $M_i(f)$, $i \leq n$, we put

$$\gamma_{nk}(f) = \gamma_{nk}\{M_0(f),\ M_k(f),\ M_n(f)\}.$$

We say that a triple of positive numbers (M_0, M_k, M_n) is "feasible" if there is a function f with

$$M_0(f) = M_0,\ M_k(f) = M_k,\ M_n(f) = M_n.$$

We prove the following lemma.

Lemma. *If a triple (M_0, M_k, M_n) is "feasible", then all triples (M_0', M_k', M_n') such that*

$$\gamma_{nk}(M_0', M_k', M_n') \leq \gamma_{nk}(M_0, M_k, M_n)$$

are also feasible.

Proof. By the hypothesis of the lemma, there is a function $f(x)$ such that

$$M_0(f) = M_0,\ M_k(f) = M_k,\ M_n(f) = M_n.$$

It can easily be shown that for any $a > 0$ and $b > 0$ the function

$$\phi(x) = af(bx)$$

satisfies the relation

$$\gamma_{nk}(\phi) = \gamma_{nk}(f).$$

The constants $a > 0$ and $b > 0$ can be selected so that

$$M_k(\phi) = M_k',\ M_n(\phi) = M_n'.$$

These relations and the inequality

$$\gamma_{nk}(M_0', M_k', M_n') \leq \gamma_{nk}(M_0, M_k, M_n) = \gamma_{nk}(f) = \gamma_{nk}(\phi)$$

imply that

$$M_0' \geq M_0(\phi).$$

We now put

$$\chi(x) = \phi(x) + d,$$

where d is a constant. Clearly,

$$M_k(\chi) = M_k(\phi) = M'_k, \quad M_n(\chi) = M_n(\phi) = M'_n.$$

Varying the constant d we can arbitrarily increase $M_0(\chi)$ in relation to $M_0(\phi)$ and, in particular, we can arrange matters so that

$$M_0(\chi) = M'_0.$$

We thus see that the triple (M'_0, M'_k, M'_n) is in fact "feasible".

By virtue of the above lemma, it is clear that the sufficiency of condition (1) is proved if for any given k and n $(0 < k < n)$ there is at least one function $f(x)$ such that

$$\gamma_{nk}(f) = C_{nk}.$$

However, this condition is certainly satisfied by the function $f_n(x)$ determined by formula (5). Indeed, it can readily be shown that for $k = 1, 2, \dots, n$ we have the relations

$$f_n^{(k)}(x) = f_{n-k}(x), \tag{8}$$

$$M_k(f_n) = \sup |f_{n-k}(x)| = K_{n-k}. \tag{9}$$

Relations (8), (9) and (2) imply that $\gamma_{nk}(f_n) = C_{nk}$.

§3. The case $k = 1$

In this section we prove Theorem II and the necessity of condition (1) of Theorem I for the case $k = 1$. To prove the two propositions it suffices to show the following.

A) *Theorem* II *is true for* $n = 1$.

B) *If Theorem* II *is true for* $n = m$, *then condition* (1) *of Theorem* I *is necessary for* $n = m + 1$.

C) *If Theorem* II *is true for* $n = m$ *and condition* (1) *of Theorem* I *is necessary for* $n = m + 1$, *then Theorem* II *is true for* $n = m + 1$.

To prove proposition A it suffices to note that for any x we have

$$|f_1'(x)| = 1,$$

and, consequently, in the case $n = 1$ the absolute value of the derivative of the comparison function

$$\phi_1(x) = af_1(bx) + c$$

in Theorem II is constant:

$$|\phi_1'(x)| = M_1(\phi_1) = M_1(\phi).$$

Thus, for $n = 1$ the whole content of Theorem II reduces to the trivial inequality

$$|f'(x_0) \leq M_1(f).$$

We now turn to the proof of proposition B. By an *upper comparison function* of order n for a function $f(x)$ at a point x_0 we will mean a function of the form

$$\bar{\phi}_n(x) = af_n(bx + c)$$

such that

$$M_0(\bar{\phi}_n) > M_0(f), \quad M_n(\bar{\phi}_n) = M_n(f), \quad \bar{\phi}_n(x_0) = f(x_0).$$

It can easily be seen that for an upper comparison function the absolute value of the derivative $|\bar{\phi}_n'(x_0)|$ is always greater than that of the corresponding comparison function ϕ_n' in the proper sense. It is also clear that an upper comparison function $\bar{\phi}_n(x)$ can always be selected so that in any given finite interval containing the point x_0 the function $\bar{\phi}_n(x)$ itself is arbitrarily close to $\phi_n(x)$ and its derivative is arbitrarily close to the derivative $\phi_n'(x)$. Therefore, for each n the assertion of Theorem II is equivalent to the following:

Theorem II*. *If $\bar{\phi}_n(x)$ is an upper comparison function of order n for a function $f(x)$ at a point x_0, then*

$$|f'(x_0)| < |\bar{\phi}_n'(x_0)|. \tag{10}$$

We now assume that Theorems II and II* have been proved for $n = m$ and consider a function $f(x)$ with finite $M_i(f)$ up to $i = m + 1$.

For an arbitrarily small $\epsilon > 0$ there is a point x_0 at which

$$|f'(x_0)| > M_1(f) - \epsilon.$$

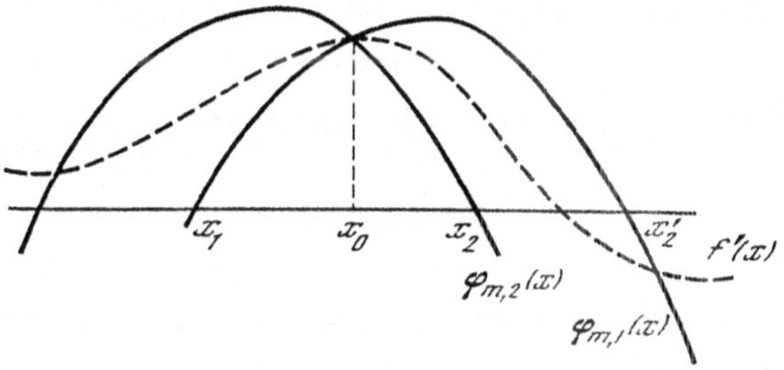

Fig. 2

Without loss of generality, we can assume that $f'(x_0) > 0$; otherwise, we consider the function $-f(x)$ instead of $f(x)$. For the function $f'(x)$ we construct two comparison functions $\phi_{m,1}(x)$ and $\phi_{m,2}(x)$ at the point x_0 satisfying the conditions

$$\phi'_{m,1}(x_0) \geq 0, \quad \phi'_{m,2}(x_0) \leq 0.$$

Denote by x_1 the zero of the function $\phi_{m,1}(x)$ nearest to x_0 on the left and by x_2 the zero of the function $\phi_{m,2}(x)$ nearest to x_0 on the right (see Figure 2). We now prove that on the closed intervals $[x_1; x_0]$ and $[x_0; x_2]$ the inequalities

$$f'(x) \geq \phi_{m,1}(x) \tag{11a}$$

and

$$f'(x) \geq \phi_{m,2}(x) \tag{11b}$$

hold respectively. We prove the first of these inequalities. To this end, instead of the function $\phi_{m,1}(x)$ itself we consider an upper comparison function $\bar{\phi}_{m,1}(x)$ of order n for $f(x)$ at the same point x_0. We denote by $\bar{x}_1$ the zero of the function $\bar{\phi}_{m,1}$ nearest to x_0 on the left and show that in the interval $(x_1; x_0)$ the inequality

$$f'(x) > \bar{\phi}_{m,1}(x) \tag{12a}$$

holds. Indeed, if the inequality were violated somewhere in the interval $(\bar{x}_1; x_0)$, there would exist a first point ξ on the left of x_0 belonging to the interval $(\bar{x}_1; x_0)$ on which the inequality is violated. At the point ξ we obviously have the relations

$$f'(\xi) = \bar{\phi}_{m,1}(\xi), \tag{13}$$

$$f''(\xi) \geq \bar{\phi}'_{m,1}(\xi) > 0. \tag{14}$$

Relation (13) shows that $\bar{\phi}_{m,1}(x)$ can be regarded as an upper comparison function of order m for $f'(x)$ at the point ξ. Consequently, by Theorem II (which is assumed to have been proved for $n = m$),

$$|f''(\xi)| < |\bar{\phi}'_{m,1}(\xi)|. \tag{15}$$

The contradiction between (14) and (15) proves that inequality (12a) cannot be violated in the interval $(\bar{x}_1; x_0)$.

If $\bar{\phi}_{m,1}(x)$ is selected so that it is sufficiently close to $\phi_{m,1}(x)$, then the point $\bar{x}_1$ is arbitrarily close to x_1. Therefore, by passing to the limit in (12a) we obtain inequality (11a) on the closed interval $[x_1; x_0]$. Inequality (11b) is proved in a similar way.

Inequalities (11a) and (11b) imply that

$$f(x_2) - f(x_1) \geq \int_{x_1}^{x_0} \phi_{m,1}(x)dx + \int_{x_0}^{x_2} \phi_{m,2}(x)dx. \tag{16}$$

If ϵ is sufficiently small, then $\phi_{m,2}(x)$ is arbitrarily close to $\phi_{m,1}(x)$ on $[x_0; x_2]$ and the point x_2 is arbitrarily close to the zero x_2' of the function $\phi_{m,1}(x)$ nearest to x_0 on the right (see Figure 2). Consequently, the right-hand side of inequality (16) can be made arbitrarily close to

$$\int_{x_1}^{x_2'} \phi_{m,1}(x)dx. \tag{17}$$

We now note that

$$\phi_{m,1}(x) = af_m(bx + c)$$

is the derivative of the function

$$\phi_{m+1}(x) = (a/b)f_{m+1}(bx + c).$$

It can easily be seen that the integral (17), extended over the interval between two neighbouring zeros of the function $\phi_{m,1}(x)$, over which it is positive, is exactly equal to

$$2M_0(\phi_{m+1}).$$

On the other hand, since

$$f(x_2) - f(x_1) \geq 2M_0(f),$$

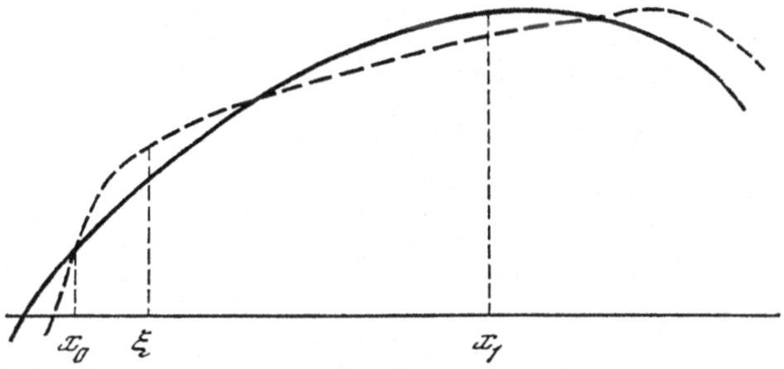

Fig. 3

by virtue of (16) and the fact that the right-hand side of (16) is arbitrarily close to the integral (17), we obtain the inequality

$$M_0(f) \leq M_0(\phi_{m+1}).$$

Finally, this inequality, together with the relations

$$M_1(\phi_{m+1}) = C_{m+1,1} M_0^{\frac{m}{m+1}}(\phi_{m+1}) M_{m+1}^{\frac{1}{m+1}}(\phi_{m+1}),$$
$$M_1(\phi_{m+1}) = M_0(\phi_{m,1}) = M_0(f') = M_1(f),$$
$$M_{m+1}(\phi_{m+1}) = M_m(\phi_{m,1}) = M_m(f') = M_{m+1}(f)$$

result in

$$M_1(f) \leq C_{m+1,1} M_0^{\frac{m}{m+1}}(f) M_{m+1}^{\frac{1}{m+1}}(f),$$

which completes the proof of proposition B.

We now prove proposition C. Assume that the necessity of condition (1) for $n = m + 1$ and $k = 1$ has already been proved and that Theorems II and II* have been proved for $n = m$. Assume that, in contrast to proposition C, Theorem II* is false for $n = m + 1$. Then there is a function $f(x)$ having an upper comparison function $\bar{\phi}_{m+1}(x)$ of order $m+1$ at a point x_0 such that the inequality

$$|f'(x_0)| \geq |\bar{\phi}'_{m+1}(x_0)|$$

holds. Since $M_0(\bar{\phi}_{m+1}) > M_0(f)$, the point x_0 cannot be a maximum point of $\bar{\phi}_{m+1}$, and hence $\bar{\phi}_{m+1}(x_0) \neq 0$ and $f'(x_0) \neq 0$ at x_0. Without loss of generality, we can assume that $f(x_0) \geq 0$ and $f'(x_0) > 0$. The other cases can be reduced to this situation by replacing, if necessary, $f(x)$ by $-f(x)$ and x by $-x$. The comparison function can also be selected so that $\bar{\phi}'_{m+1}(x_0) > 0$ (see Figure 3).

Now let x_1 be the maximum point of $\bar{\phi}_{m+1}(x)$ nearest to x_0 on the right. We have

$$f(x_0) = \bar{\phi}_{m+1}(x_0),$$
$$f(x_1) \leq M_0(f) < M_0(\bar{\phi}_{m+1}) = \bar{\phi}_{m+1}(x).$$

Consequently, on the closed interval $[x_0; x_1]$ the difference

$$f(x) - \bar{\phi}_{m+1}(x)$$

attains its maximum at a point ξ distinct from x_1. At the point ξ we have

$$f'(\xi) = \bar{\phi}_{m+1}(\xi), \quad f''(\xi) \leq \bar{\phi}''_{m+1}(\xi). \tag{18}$$

Differentiating the function

$$\bar{\phi}_{m+1}(x) = af_{m+1}(bx + c)$$

we obtain the function

$$\bar{\phi}_m(x) = \bar{\phi}'_{m+1} = abf_m(bx + c).$$

We now note that, by virtue of the definition of an upper comparison function, $\bar{\phi}_{m+1}(x)$ satisfies

$$M_0(f) < M_0(\bar{\phi}_{m+1}), \tag{19}$$
$$M_m(f') = M_{m+1}(f) = M_{m+1}(\bar{\phi}_{m+1}) = M_m(\phi_m). \tag{20}$$

Inequality (1), which is assumed to have been proved for the case $n = m + 1$ and $k = 1$, together with (19) yields

$$M_0(f') = M_1(f) \leq C_{m+1,1} M_0^{\frac{m}{m+1}}(f) M_{m+1}^{\frac{1}{m+1}}(f) <$$
$$< C_{m+1,1} M_0^{\frac{m}{m+1}}(\bar{\phi}_{m+1}) M_{m+1}^{\frac{1}{m+1}}(\bar{\phi}_{m+1}) = M_1(\bar{\phi}_{m+1}) = M_0(\bar{\phi}_m). \tag{21}$$

From (20), (21) and the relation

$$\bar{\phi}_m(\xi) = \bar{\phi}'_{m+1}(\xi) = f'(\xi)$$

we conclude that $\bar{\phi}_m(x)$ is an upper comparison function of order m for $f'(x)$ at the point ξ. By Theorem II*, which is assumed to have been proved for $n = m$, it follows that

$$|f''(\xi) < |\bar{\phi}'_m(\xi)| = |\bar{\phi}''_{m+1}(\xi)|$$

or, since $\bar{\phi}''_{m+1}(\xi) \leq 0$,

$$f''(\xi) > \bar{\phi}''_{m+1}(\xi). \tag{22}$$

The contradiction between inequalities (18) and (22) proves proposition C.

§4. Necessity of condition (1) for any n

Assume that the necessity of condition (1) has already been proved for all n and $k \leq m$ $(m \geq 1)$. We will prove that in this case its necessity holds for $k = m+1$ as well. To this end we consider a function $f(x)$ with finite $M_0(f)$, $M_{m+1}(f)$ and $M_n(f)$ $(m+1 < n)$. Clearly,

$$M_0(f') = M_1(f), \quad M_m(f') = M_{m+1}(f), \quad M_{n-1}(f') = M_n(f).$$

Since the necessity of condition (1) for $k = m$ has already been proved, we have

$$M_m(f') \leq C_{n-1,m} M_0^{\frac{n-1-m}{n-1}}(f') M_{n-1}^{\frac{m}{m-1}}(f')$$

or, what is the same,

$$M_{m+1}(f) \leq C_{n-1,m} M_1^{\frac{n-1-m}{n-1}}(f) M_n^{\frac{m}{n-1}}(f). \tag{23}$$

Moreover, since for the case $k = 1$ the necessity of condition (1) has been proved for any n, we have

$$M_1(f) \leq C_{n,1} M_0^{(n-1)/n}(f) M_n^{1/n}(f). \tag{24}$$

The substitution of (24) into (23) results in

$$M_{n+1}(f) \leq C_{n-1,m} C_{n,1}^{\frac{n-1}{n}} M_0^{\frac{n-1-m}{n}}(f) M_n^{\frac{m+1}{n}}(f),$$

and, since

$$C_{n-1,m} C_{n,1}^{(n-1)/n} = C_{n,m+1},$$

we finally obtain

$$M_{m+1}(f) \leq C_{n,m+1} M_0^{\frac{n-m-1}{n}}(f) M_n^{\frac{m+1}{n}}(f),$$

which proves the desired proposition.

References

1. A.N. Kolmogorov, 'Une généralisation de l'inégalité de M.J. Hadamard entre les bornes supérieures des dérivées successives d'une fonction', *C.R. Acad. Sci. Paris* **207** (1938), 764–765.
2. J. Hadamard, 'Sur le module maximum d'une fonction et de ses dérivées', *C.R. Acad. Sci. Paris* **41** (1914), 68–72.
3. Yu.G. Bosse (G.E. Shilov), 'On inequalities between derivatives', *Sb. Rabot. Stud. Nauch. Kruzhkov MGU* **1** (1937), 17–27.
4. N.I. Akhiezer and M.G. Krein, 'On best approximation of trigonometric sums by differentiable periodic functions', *Dokl. Akad. Nauk SSSR* **15** (1937), 107–112 (in Russian).

41. ON RINGS OF CONTINUOUS FUNCTIONS ON TOPOLOGICAL SPACES *

(in collaboration with I.M. Gel'fand)

This paper is related to studies by M.H. Stone [2] and to the above paper by G.E. Shilov.[1] In contrast to the latter, we consider the ring of continuous functions on a topological space as a purely algebraic object without defining any topological relations in it. It turns out that in the case of bicompact spaces, considered by M.H. Stone, and also in some much more general cases, even the purely algebraic structure of the ring of continuous functions determines the topological space to within a homeomorphism.

For an arbitrary topological space S we consider the following two rings: the ring $C(S)$ of all real continuous functions on S and the ring $C'(S)$ of all bounded functions belonging to $C(S)$.

When studying these rings it is natural to confine ourselves to the case of completely regular spaces S. This is due to the fact that the rings $C(S)$ and $C'(S)$ of an arbitrary topological space S are isomorphic, respectively, to the rings $C(\rho S)$ and $C'(\rho S)$ of a certain completely regular space ρS, introduced by E. Čech [1]. In the paper by Čech, along with the space ρS, a continuous mapping $y = \rho(x)$ of the space S on the space ρS is defined possessing the following property: the real continuous functions on S coincide with the functions of the form

$$f(x) = \phi[\rho(x)],$$

where $\phi(y)$ is a continuous real function on the sapce ρS. This fact readily implies that the rings $C(S)$ and $C'(S)$ are isomorphic to the rings $C(\rho S)$ and $C'(\rho S)$ respectively.

In view of what has been said, *in what follows we will always assume that the original space S itself is completely regular* (in this case ρS coincides with S).

An ideal of the ring C is said to be *maximal* if it does not coincide with the whole ring and is not contained in any larger ideal different from the ring C itself. We now consider the set γ of maximal ideals of the ring C; they

* *Dokl. Akad. Nauk SSSR* **22**:1 (1939), 11–15 (in Russian).

[1] A reference to a paper by G.E. Shilov, 'Ideals and subrings of the ring of continuous functions', published in the same issue of the journal (pp. 7–10).

form a topological space using the following definition: a maximal ideal α is an adherence point of a set of maximal ideals $\mathfrak{M}$ if it contains the intersection of all maximal ideals contained in $\mathfrak{M}$. It can easily be verified that for an arbitrary ring C this definition of adherence points transforms the set γ into a T_1 space (in the terminology of P.S. Alexandrov and E. Hopf, *Topologie* I). We note that this method of introducing a topology in the set of maximal ideals was used earlier by M.H. Stone [2].

The spaces γ corresponding to the rings $C(S)$ and $C'(S)$ will be denoted by $\gamma(S)$ and $\gamma'(S)$ respectively.

§1. The case of a bicompact space

If the space S is bicompact, then the rings $C(S)$ and $C'(S)$ coincide (since all continuous functions on S are bounded) and there is no need to consider them separately.

Theorem I. *If the space S is bicompact, then it is homeomorphic to the space $\gamma(S)$.*

To prove the theorem we first prove the following lemma.

Lemma. *For any ideal A of the ring $C(S)$ not coinciding with the whole ring there is a point a in the space S at which all functions belonging to A vanish.*

Proof of the lemma. Assume the contrary. Then for each point ξ of the space S there is a function $f_\xi(x)$ belonging to A and non-zero at the point ξ. The function $f_\xi(x)$ is non-zero in a neighbourhood $u(\xi)$ of the point ξ. By the Borel-Lebesgue theorem, there is a finite set of points

$$\xi_1, \xi_2, \ldots, \xi_n$$

such that the neighbourhoods

$$u(\xi_1), u(\xi_2), \ldots, u(\xi_n)$$

cover the whole space S. The function

$$\phi(x) = f_{\xi_1}^2(x) + f_{\xi_2}^2(x) + \ldots + f_{\xi_n}^2(x)$$

belongs to the ideal A and is non-zero everywhere. Therefore any function $f(x)$ belonging to the ring $C(S)$ can be represented in the form

$$f(x) = \psi(x)\phi(x)$$

and, consequently, in contrast to the assumption, it also belongs to A. This contradiction proves the lemma.

Proof of Theorem I. The above lemma implies that any ideal A not coinciding with the whole ring $C(S)$ is contained in the ideal $I(a)$ consisting of all functions vanishing at a point a. Hence, only ideals of this type can be maximal. It is also clear that each ideal of type $I(a)$ is in fact maximal. By virtue of the complete regularity of the space S, the maximal ideals $I(a)$ and $I(a')$ corresponding to two different points a and a' are different, since there is a function belonging to $C(S)$, vanishing at the point a, and non-zero at the point a'. Thus, by associating the maximal ideal $I(a)$ with each point a of the space S we establish a one-to-one correspondence between the space S and the set $\gamma(S)$ of maximal ideals of the ring $C(S)$.

It remains to prove that this correspondence between S and $\gamma(S)$ is a homeomorphism.

Let $\mathfrak{M}$ be the set of maximal ideals corresponding to a set M of points in the space S. If a is an adherence point of the set M, then all functions contained in the intersection of all ideals belonging to $\mathfrak{M}$, that is, vanishing at all points of M, vanish at the point a as well, that is, are contained in the ideal $I(a)$, which, by definition, means that $I(a)$ is an adherence point of the set $\mathfrak{M}$. Conversely, if a is not an adherence point of M, then, by the complete regularity of S, there is a function $f(x)$ belonging to $C(S)$, vanishing on M, and non-zero at the point a; this function is contained in the intersection of all ideals belonging to $\mathfrak{M}$ and does not belong to $I(a)$; hence, in this case $I(a)$ is not an adherence point of $\mathfrak{M}$. Thus, the homeomorphism of the spaces S and $\gamma(S)$ is proved.

Theorem II. *For two bicompact spaces S and S_1 to be homeomorphic it is necessary and sufficient that the rings $C(S)$ and $C(S_1)$ be algebraically isomorphic.*

Proof. The necessity of the condition of the theorem is obvious, since any homeomorphism between S and S_1 automatically induces an isomorphic mapping of the ring $C(S)$ onto the ring $C(S_1)$. The sufficiency of the condition follows from Theorem I, since isomorphism of $C(S)$ and $C(S_1)$ obviously implies homeomorphism of the spaces $\gamma(S)$ and $\gamma(S_1)$; by Theorem I, these are homeomorphic to the spaces S and S_1.

§2. The ring $C'(S)$ in the general case

The investigation of the properties of the ring $C'(S)$ for an arbitrary completely regular space S reduces to the case of a bicompact space. To this end we must make use of the space βS introduced by E. Čech [1]. Namely, according to Čech, each completely regular space S is associated with a certain space βS; it is uniquely determined to within a homeomorphism and is characterized by the following conditions:

1) βS is bicompact;

2) βS contains S;

3) S is dense in βS;

4) each bounded continuous real function on S can be extended to βS with preservation of continuity.

The fact that S is dense in βS implies that the extension of continuous functions from S to βS can be performed in a unique manner. Consequently, this gives rise to a one-to-one correspondence between bounded continuous functions on S and βS. It can easily be seen that the correspondence determines an isomorphic mapping of the ring $C'(S)$ onto the ring $C'(\beta S)$. We have thus proved the following theorem:

Theorem III′. *The ring $C'(S)$ is isomorphic to the ring $C'(\beta S)$* or, what is the same, *to the ring $C(\beta S)$.*

Theorems I and III′ imply

Theorem IV′. *The space $\gamma(S)$ is homeomorphic to the space βS.*

By virtue of Theorem III′, it is clear that two non-homoeomorphic spaces S and S_1 can have isomorphic rings $C'(S)$ and $C'(S_1)$. To show this it suffices to take an arbitrary non-bicompact space as S and put $S_1 = \beta S$; then, by Theorem III′, the ring $C'(S)$ is isomorphic to the ring $C'(S_1)$. However, the following theorem holds:

Theorem V′. *For two spaces S and S_1 satisfying the first axiom of countability to be homeomorphic it is necessary and sufficient that the rings $C'(S)$ and $C'(S_1)$ be algebraically isomorphic.*

Proof. The sufficiency of the condition of the theorem is proved as in the case of Theorem I. The necessity of the condition follows from the following fact established by Čech [1]: for two spaces S and S_1 satisfying the first axiom of countability, homeomorphism of βS and βS_1 implies homeomorphism of the orginal spaces S and S_1. Indeed, isomorphism of $C'(S)$ and $C'(S_1)$ implies homeomorphism of $\gamma'(S)$ and $\gamma'(S_1)$, that is, by Theorem IV$'$, homeomorphism of βS and βS_1 and consequently, by virtue of the above-mentioned result of Čech, it also implies homeomorphism of S and S_1.

§3. The ring $C(S)$ in the general case

For the ring $C(S)$ a theorem analogous to Theorem III$'$ related to the ring $C'(S)$ does not hold. For example, if S is the space of integers (that is, a countable set of isolated points) and $S_1 = \beta S$, then $C(S)$ and $C(S_1)$ are non-isomorphic (here we do not present a proof of this fact). However, theorems analogous to Theorems IV$'$ and V$'$ remain valid.

Theorem IV. *The space $\gamma(S)$ is homeomorphic to the space $\beta(S)$.*

Theorem V. *For two spaces S and S_1 satisfying the first axiom of countability to be homeomorphic it is necessary and sufficient that the rings $C(S)$ and $C(S_1)$ be algebraically isomorphic.*

Theorem V can be derived from Theorem IV in exactly the same way as Theorem V$'$ has been derived from Theorem IV$'$. Therefore it only remains to prove Theorem IV.

Proof of Theorem IV. We associate with each function $f(x)$ belonging to $C(S)$ the function

$$f_1(x) = (2/\pi)\arctan[f(x)].$$

The function $f_1(x)$ is continuous and bounded on S. Consequently, it can be extended to the whole space βS with preservation of continuity. In a certain sense, this makes it possible to extend the function $f(x)$ to the whole space βS. Namely, at each point of the set $\beta S \backslash S$ we put

$$f(x) = \begin{cases} \tan[(\pi/2)f_1(x)], & \text{if } |f_1(x)| \neq 1; \\ +\infty, & \text{if } f_1(x) = +1; \\ -\infty, & \text{if } f_1(x) = -1. \end{cases}$$

After the functions $f(x)$ are extended to the whole space βS, the proof of Theorem IV can be carried out, with certain complications, by analogy with

Theorem I. We confine ourselves to the presentation of the main points of the proof.

We first prove the following lemma.

Lemma 1. *For any ideal A of the ring $C(S)$ not coinciding with the whole ring there is a point a in the space βS at which all functions belonging to A vanish.*

Any point a of the space βS determines an ideal $I(a)$ of the ring $C(S)$, consisting of all functions belonging to the ring such that all products

$$\phi(x) = \psi(x)f(x)$$

with factors $\psi(x)$ belonging to $C(S)$ vanish at a (if a belongs to $\beta S \backslash S$, then the value $\phi(a)$ is understood in the above-mentioned sense). On the basis of Lemma 1, it can be proved that all ideals of type $I(a)$ are maximal and that there are no other maximal ideals in the ring $C(S)$.

Lemma 2. *If the point a belongs to S, then the ideal $I(a)$ consists of all functions $f(x)$ in the ring $C(S)$ that vanish at a. If a belongs to $\beta S \backslash S$, then the ideal $I(a)$ consists of all functions $f(x)$ in the ring $C(S)$ that vanish on the set of points N_f of the original space S for which a is an adherence point.*

We do not give a proof of Lemma 2. This lemma implies that the ideals $I(a)$ and $I(a')$ corresponding to two different points a and a' are different. Hence, a one-to-one correspondence between the points a of the space βS and the maximal ideals $I(a)$ is established. By analogy with what has been done in the case of Theorem I, it can be proved that this correspondence between βS and $\gamma(S)$ is a homeomorphism.

Remark. Isomorphism of the rings $C(S)$ and $C(S_1)$ implies (by Theorem IV) homeomorphism of the spaces βS and βS_1 and, consequently (by Theorem III′), homeomorphism of the rings $C'(S)$ and $C'(S_1)$ as well. As to the converse, isomorphism of the rings $C'(S)$ and $C(S_1)$ does not, in general, imply isomorphism of the rings $C(S)$ and $C(S_1)$. For instance, if S is the space of integers and $S_1 = \beta S$, then the rings $C'(S)$ and $C'(S_1)$ are isomorphic, whereas the rings $C(S)$ and $C(S_1)$ are not isomorphic. Thus, the ring $C(S)$ provides more adaptable means for studying topological properties of the space S than does the ring $C'(S)$. However, there are nevertheless non-homeomorphic spaces

S and S_1 with isomorphic rings $C(S)$ and $C(S_1)$. For example, such are the space S of transfinite numbers $< \Omega$ and the space S_1 of transfinite numbers considered with the usual limit relations defined in them. [2]

17 November 1938

References

1. E. Čech, *Ann. of Math.* **38** (1937), 823–844.
2. M.H. Stone, *Trans. Amer. Math. Soc.* **41** (1937), 375–381.

[2] Here Ω denotes the first uncountable transfinite number.

42. CURVES IN A HILBERT SPACE INVARIANT WITH RESPECT TO A ONE-PARAMETER GROUP OF MOTIONS *

By a *motion* in a Hilbert space H we will mean any transformation

$$y = Kx$$

of the space H into itself that can be represented in the form

$$Kx = a + Ux,$$

where a is a fixed element of the space H and U is a unitary operator.

We define a continuous one-parameter group (COG) of motions as a set $\{K_t\}$ of motions K_t depending on a real parameter t running through all values between $-\infty$ and $+\infty$, which satisfies the following conditions:

1) for any s and t,

$$K_{s+t} = K_s K_t;$$

2) for any element x of the space H the condition $t_n \to t$ implies

$$\|K_{t_n} x - K_t x\| \to 0$$

as $n \to \infty$.

Let K_t be a COG of motions and let x_0 be a fixed element of the space H. Consider the function

$$\xi(t) = K_t x_0$$

of the real argument t.

Obviously, this function is continuous (in the sense of strong convergence in H).

From the standpoint of geometry, the point

$$x = \xi(t)$$

describes a curve in the space. An arbitrary motion K transforms the function $\xi(t)$ into a new function

$$\xi_1(t) = K\xi(t).$$

---

* *Dokl. Akad. Nauk SSSR* **26** (1940), 6–9 (in Russian).

If the motion $K = K_s$ belongs to the original COG of motions, then

$$\xi_1(t) = \xi(t-s),$$

that is, the curve $\xi(t)$ is mapped onto itself under the transformation $K = K_s$. In other words, the curve $\xi(t)$ is invariant with respect to the transformations K_s belonging to the original COG. If K does not belong to the original COG, then in general, under the transformation K the curve $\xi(t)$ goes into a new curve congruent to the former. Below we will not discuss the geometrical terminology related to the notion of a curve and state the problem under investigation in analytical form.

Definition. A function $\xi(t)$ with values belonging to H and defined for all real t is said to be of class $\mathfrak{K}$ if it can be represented as $\xi(t) = K_t x_0$, where x_0 is an element of H and $\{K_t\}$ is a COG of motions of the space H.

The aim of this paper is to study functions of class $\mathfrak{K}$ and, in particular, properties of them that are invariant with respect to motions, that is, the properties that remain invariable under transition from the function $\xi(t)$ to a function

$$\xi_1(t) = K\xi(t),$$

whatever the motion K may be.

We consider a function $\xi(t)$ of class $\mathfrak{K}$ and put

$$B_\xi(\tau_1, \tau_2) = (\xi(t+\tau_1) - \xi(t),\ \xi(t+\tau_2) - \xi(t)). \tag{1}$$

(Obviously, $B_\xi\ (\tau_1, \tau_2)$ does not depend on t.) Denote, respectively, by H_ξ, G_ξ, and α_ξ the smallest closed linear subspace of the space H containing all differences $\xi(t+\tau) - \xi(t)$, the space $H - H_\xi$ (consisting of all elements x belonging to H and orthogonal to H_ξ), and the dimension of G_ξ (which can be equal to $0, 1, 2, \ldots,$ or ∞).

Using the above notation we state the following theorems.

Theorem 1. *If $\xi_1(t)$ and $\xi_2(t)$ are of class $\mathfrak{K}$, then for a motion K transforming ξ_1 into ξ_2, that is a motion such that*

$$\xi_2(t) = K\xi_1(t),$$

to exist it is necessary and sufficient that the following two conditions hold simultaneously: 1) $\alpha_{\xi_1} = \alpha_{\xi_2}$*; and* 2) $B_{\xi_1}(\tau_1, \tau_2) = B_{\xi_2}(\tau_1, \tau_2)$ *for any* τ_1, τ_2.

Theorem 2. *Given α ($\alpha = 0, 1, 2, \ldots$ or ∞) and a function $B(\tau_1, \tau_2)$, then in order that at least one function $\xi(t)$ of class $\mathfrak{K}$ such that*

$$\alpha_\xi = \alpha, \;\; B_\xi(\tau_1, \tau_2) = B(\tau_1, \tau_2)$$

should exist, it is necessary and sufficient that $B(\tau_1, \tau_2)$ be representable in the form

$$B(\tau_1, \tau_2) = \oint_{-\infty}^{\infty} (e^{i\lambda\tau_1} - 1)(e^{-i\lambda\tau_2} - 1)dF(\Delta_\lambda) + \theta\tau_1\tau_2, \tag{2}$$

where the integral is understood in the sense

$$\oint_{-\infty}^{\infty} = \lim_{\epsilon\to 0}\int_{-\infty}^{-\epsilon} + \lim_{\epsilon\to 0}\int_{\epsilon}^{+\infty},$$

θ is an arbitrary constant, and the interval function $F(\Delta_\lambda)$ satisfies the following conditions:

A) *it is defined for all intervals Δ_λ not having zero inside or as end point and it is additive and right continuous with respect to the end points of Δ_λ;*

B) *$F(\Delta_\lambda) \geq 0$;*

C) *the integrals $\int_{-\infty}^{-1} dF(\Delta_\lambda)$, $\oint_{-1}^{1} \lambda^2 dF(\Delta_\lambda)$, $\int_1^\infty dF(\Delta_\lambda)$ are finite.*

Theorem 3. *A function $B(\tau_1, \tau_2)$ representable as (2) can, under conditions* A), B), *and* C, *be represented in this form in a unique manner.*

This uniquely determined function $F(\Delta_\lambda)$ and the constant θ corresponding to $B_\xi(\tau_1, \tau_2)$ for a given function $\xi(t)$ of class $\mathfrak{K}$ will be denoted by $F_\xi(\Delta_\lambda)$ and θ_ξ, respectively.

Theorems 1–3 imply that α_ξ, 0_ξ and $F_\xi(\Delta_\lambda)$ form a complete system of invariants relative to motions for a curve $\xi(t)$ of class $\mathfrak{K}$. And for a curve of class $\mathfrak{K}$ with given α ($\alpha = 0, 1, 2, \ldots$, or ∞), θ (which can be an arbitrary complex number), and $F(\Delta_\lambda)$ to exist it is necessary and sufficient that the interval function $F(\Delta_\lambda)$ satisfy conditions A), B), and C) of Theorem 2.

The spectral representation (2) of the function $B_\xi(\tau_1, \tau_2)$ is related to the spectral representation of the function $\xi(t)$ itself described in the following theorem.

Theorem 4. *Each function $\xi(t)$ of class $\mathfrak{K}$ can be represented uniquely in the form*

$$\xi(t) = \int_{-\infty}^{\infty} (e^{i\lambda t} - 1)d\Phi(\Delta_\lambda) + ut + v, \tag{3}$$

where u and v are elements of H and $\Phi(\Delta_\lambda)$ is a function of Δ_λ with values in H possessing the following properties:

1°. *The function $\Phi(\Delta_\lambda)$ is defined for all intervals Δ_λ not having zero as an interior or end point and is additive and right continuous with respect to the end points of Δ_λ.*

2°. *If Δ'_λ and Δ''_λ do not intersect, then*

$$(\Phi(\Delta'_\lambda),\ \Phi(\Delta''_\lambda)) = 0.$$

3°. *The integrals*

$$\int_{-\infty}^{-1} dF(\Delta_\lambda),\ \oint_{-1}^{1} \lambda^2 dF(\Delta_\lambda),\ \int_{1}^{\infty} dF(\Delta_\lambda),$$

where

$$F(\Delta_\lambda) = \|\Phi(\Delta_\lambda)\|^2, \tag{4}$$

are finite.

Here the integrals $\oint$ are understood in the sense indicated in Theorem 2, and

$$F(\Delta_\lambda) = F_\xi(\Delta_\lambda),\ \|u\|^2 = \theta_\xi,$$

where $F_\xi(\Delta_\lambda)$ and θ were defined earlier.

The function $\Phi(\Delta_\lambda)$ and the elements u and v of the space H corresponding, by Theorem 4, to a function $\xi(t)$ of class $\mathfrak{K}$ will be denoted by $\Phi_\xi(\Delta_\lambda)$, u_ξ, v_ξ, respectively.

Theorem 5. *For given u and v belonging to H and an interval function $\Phi(\Delta_\lambda)$ with values in H to determine a function $\xi(t)$ of class $\mathfrak{K}$ by means of formula* (3) *it is necessary and sufficient that $\Phi(\Delta_\lambda)$ satisfy conditions* 1°, 2°, *and* 3° *of Theorem* 4.

In conclusion we note that a function of class $\mathfrak{K}$ can be represented in the form

$$\xi(t) = U_t x_0, \tag{5}$$

where $\{U_t\}$ is a COG of unitary operators, when the norm

$$\|\xi(t)\| = r$$

is constant for all t, that is, if and only if the curve $\xi(t)$ lies on the sphere

$$\|x\| = r.$$

The class of functions representable in the form (5) will be denoted by $\mathfrak{K}_0$. For functions of class $\mathfrak{K}_0$, propositions analogous to Theorems 1–5 are derived directly from the spectral representation of the corresponding group $\{U_t\}$ and are well known, sometimes in a somewhat different form.

The investigation of an arbitrary function of class $\mathfrak{K}$ can be reduced to the consideration of functions of class $\mathfrak{K}_0$ based on the following fact: if $\xi(t)$ belongs to $\mathfrak{K}$, then

$$\zeta_h(t) = [\xi(t+h) - \xi(t)]/h$$

belongs to $\mathfrak{K}_0$. The function $\zeta_h(t)$ can be represented in the form

$$\zeta_h(t) = U_t\zeta_h(0),$$

where the operators

$$U_t = \int e^{i\lambda t} dE(\Delta_\lambda)$$

do not depend on h. Therefore

$$\zeta_h(t) = \int_{-\infty}^{\infty} e^{i\lambda t} d\Psi_h(\Delta_\lambda),$$

where

$$\Psi_h(\Delta_\lambda) = E(\Delta_\lambda)\zeta_h(0).$$

The existence of the limit

$$\lim_{h\to 0} \Psi_h(\Delta_\lambda) = \Psi(\Delta_\lambda)$$

can be proved and Φ is defined as

$$\Phi(\Delta_\lambda) = \int_{\Delta_\lambda} \frac{d\Psi(\Delta_\lambda)}{i\lambda}.$$

This function Φ is none other than the function Φ_ξ in Theorem 2. Complete proofs will be published in another paper.

23 November 1939

43. WIENER SPIRALS AND SOME OTHER INTERESTING CURVES IN A HILBERT SPACE *

Here we consider some special cases of curves studied in the previous paper [1].

By a *similarity transformation* in a Hilbert space H we shall mean an arbitrary transformation A of H into itself that is representable in the form

$$A = a + qUx,$$

where a is a fixed element of H, U is a unitary operator, and q is a positive real number.

Definition. A function $\xi(t)$ of class $\mathfrak{K}$ is said to belong to class $\mathfrak{U}$ if for any real $k \neq 0$ there is a similarity transformation A_k such that

$$\xi(kt) = A_k \xi(t)$$

for all t.

If the Hilbert space in our definition is replaced by a finite-dimensional unitary space, then the only functions of class $\mathfrak{U}$ are linear functions of the form

$$\xi(t) = ut + v$$

(where u and v are fixed elements of the space). Therefore it is of particular interest that in a Hilbert space there are some other types of functions of class $\mathfrak{U}$. Geometrically, each function of class $\mathfrak{U}$ determines a curve in the space H that is invariant with respect to the two-parameter group of similarity transformations under which the curve can be mapped onto itself so that any given pair of its points x and $y \neq x$ is transformed into another given pair of points x' and $y' \neq x'$ lying on the same curve.

Theorem 6. *The function $B_\xi(\tau_1, \tau_2)$ corresponding to a function $\xi(t)$ of class $\mathfrak{U}$ can be represented as*

$$B_\xi(\tau_1, \tau_2) = c\big[|\tau_1|^\gamma + |\tau_2|^\gamma - |\tau_1 - \tau_2|^\gamma\big],$$

where c and γ are real constants satisfying the inequalities

$$c \geq 0, \ 0 \leq \gamma \leq 2.$$

---

* *Dokl. Akad. Nauk SSSR* **26** (1940), 115–118 (in Russian).

It is obvious that when the function $B_\xi(\tau_1, \tau_2)$ is not identically equal to zero, the constants

$$c = c_\xi, \quad \gamma = \gamma_\xi$$

are determined uniquely from $B_\xi(\tau_1, \tau_2)$ and, consequently, from the function $\xi(t)$ itself. It is also clear that $B_\xi(\tau_1, \tau_2)$ can be equal to zero identically only when $\xi(t)$ is constant. In what follows we do not consider this exceptional case and assume that

$$\gamma_\xi > 0, \quad c_\xi > 0.$$

We thus see that a function of class $\mathfrak{U}$ is characterized, to within motions in the space H, by the invariants α_ξ, c_ξ, and γ_ξ. As to the corresponding curves, they are determined, to within congruence, by the invariants α_ξ and γ_ξ (whereas variation of c_ξ does not affect the shape of the curve represented by the function $\xi(t)$ and is related only to a change in the choice of the parameter t).

Theorem 7. *For any α ($= 0, 1, 2, \ldots,$ or ∞), c ($c > 0$) and γ ($0 < \gamma \leq 2$) there is at least one function $\xi(t)$ of class $\mathfrak{U}$ with*

$$\alpha_\xi = \alpha, \quad c_\xi = c, \quad \gamma_\xi = \gamma.$$

For the functions of class $\mathfrak{U}$ corresponding to given α, c and γ we have, for $\gamma > 2$, the relations

$$\theta_\xi = 0, \quad F_\xi(\Delta_\lambda) = \frac{c}{D} \int_{\Delta_\lambda} \frac{d\lambda}{|\lambda|^{\gamma+1}},$$

where

$$D = 4 \int_0^\infty \frac{(\sin(\lambda/2))^2}{\lambda^{\gamma+1}} d\lambda.$$

In the case $\gamma = 2$ we have

$$\theta_\xi = 2c, \quad F_\xi(\Delta_\lambda) = 0.$$

In this case ($\gamma = 2$) the function $\xi(t)$ itself is linear (that is, geometrically it represents a straight line in H).

We now consider the special class $\mathfrak{M}$ of functions $\xi(t)$ of class $\mathfrak{U}$ such that

$$\gamma_\xi = 1.$$

The curves corresponding to this class of functions will be called *Wiener spirals.* According to the above, a Wiener spiral is determined to within congruence by the single invariant α_ξ. If we are interested only in the location of a curve in the corresponding space H_ξ, we can say that all Wiener spirals are congruent. The exact meaning of this is the following: for any two Wiener spirals determined by functions ξ_1 and ξ_2, there is a one-to-one correspondence between H_{ξ_1} and H_{ξ_2} of the form

$$y = a + Ux,$$

where a is a fixed element of H_{ξ_1} and U is an iosmetric linear operator transforming the former spiral into the latter.

Theorem 8. *For a function of class $\mathfrak{K}$ to belong to the class $\mathfrak{M}$ it is necessary and sufficient that for any two disjoint intervals*

$$s_1 < t < t_1, \quad s_2 < t < t_2$$

of the t-axis the relation

$$[\xi(t_1) - \xi(s_1),\ \xi(t_2) - \xi(s_2)] = 0$$

hold.

In terms of geometry, Theorem 8 can be stated thus: Wiener spirals are characterized completely by the following two properties:

1) they are invariant with respect to a COG of motions;

2) their chords corresponding to two non-overlapping arcs are orthogonal.

Proofs of the above theorems will be published in another paper. Here we present some additional remarks and examples.

1. For $\gamma < 2$ a function $\xi(t)$ of class $\mathfrak{U}$ is non-differentiable (both in the sense of strong convergence and in the sense of weak convergence).

2. If $\xi(t)$ belongs to the class $\mathfrak{U}$, then the values of ξ corresponding to different values of t are different. This property does not necessarily hold for functions of class $\mathfrak{K}$, which includes, in particular, periodic functions.

3. The values of a function $\xi(t)$ of class $\mathfrak{U}$ are unbounded in norm. Since the values of a function of class $\mathfrak{K}_0$ have constant norm, it follows that the classes $\mathfrak{U}$ and $\mathfrak{K}_0$ are disjoint.

4. *Example of a realization of a Wiener spiral.* Consider the Hilbert space H_0 of complex functions $f(z)$ of a real variable z $(-\infty < z < \infty)$ with finite integral

$$\int_{-\infty}^{\infty} |f(z)|^2 dz,$$

where the scalar product is defined by the usual formula

$$(f, g) = \int_{-\infty}^{\infty} f(z)\bar{g}(z) dz.$$

For $t \geq 0$ we assume $\xi(t)$ to be equal to the function $f(z)$ equal to 1 for $0 \leq z \leq t$ and equal to 0 for the other values of z.

It can readily be verified that $f(z)$ satisfies the conditions of Theorem 8.

5. *Application of functions of classes* $\mathfrak{K}$, $\mathfrak{K}_0$, $\mathfrak{U}$ *and* $\mathfrak{M}$ *to probability theory.* Consider a probability field (for the definitions see [2]). The complex random variables x of the field with finite mathematical expectation

$$\mathbf{E}|x|^2$$

form a finite- or infinite-dimensional unitary space R if their scalar product is defined by the formula

$$(x, y) = \mathbf{E}(x\bar{y}).$$

Of particular interest is the case of an infinite-dimensional space with a countable basis. In this case R satisfies all the axioms of a Hilbert space. If a random variable $\xi(t)$ is associated with each t, then $\xi(t)$ is called a random function. We assume that $\xi(t)$ belongs to R for each t and is continuous in the sense of convergence in norm in R. Then

a) the fact that $\xi(t)$ belongs to the class $\mathfrak{K}_0$ is equivalent to the *stationarity* of the random function $\xi(t)$ *in the broad sense* (in probability theory an important role is also played by a different notion of "stationarity in the narrow sense", which we do not discuss here). Random functions stationary in the broad sense have been studied thoroughly by A. Ya. Khinchin [3];

b) if a random function $\xi(t)$ belongs to the class $\mathfrak{K}$, then it is natural to call it a *random function with stationary* (in the broad sense) *increments.* A detailed study of these functions is an urgent problem of probability theory which can be carried out on the basis of the results of my paper [1];

c) in terms of probability theory, the condition

$$[\xi(t_1) - \xi(s_1), \quad \xi(t_2) - \xi(s_2)] = 0$$

of Theorem 8 means that the correlation coefficient of the increments of $\xi(t)$ on two disjoint intervals of the t-axis is equal to zero. Thus, random functions of class $\mathfrak{M}$ are functions with stationary (in the broad sense) and uncorrelated increments.

It is a special case of this kind of random functions, encountered when studying Brownian motion, that led N. Wiener as early as 1923 (see [4]) to considerations which, in terms of geometry, give rise to the above-mentioned Wiener spirals.

28 November 1939

References

1. A.N. Kolmogorov, 'Curves in a Hilbert space invariant with respect to a one-parameter group of motions', *Dokl. Akad. Nauk SSSR* **26** (1940), 6–9 (in Russian). (No. 42)
2. A.N. Kolmogorov, *Basic notions of probability theory*, ONTI, Moscow, Leningrad, 1936; 2nd ed., Nauka, Moscow, 1974 (in Russian).
3. A.Ya. Khinchin, 'Correlation theory of stationary stochastic processes', *Uspekhi Mat. Nauk* **5** (1938), 42–51 (in Russian).
4. N. Wiener, 'Differential-space', *J. Math. and Phys.* **11**:3 (1923), 122–174.

44. POINTS OF LOCAL TOPOLOGICAL CHARACTER OF COUNTABLY-MULTIPLE OPEN MAPPINGS OF COMPACTA *

Consider a continuous single-value mapping

$$y = f(x)$$

from a compactum X onto a compactum Y. A mapping f is said to be *open* if the image of any open set in the space X is an open set in the space Y. A mapping f is said to be *countably multiple* if the full inverse image $f^{-1}(y)$ of any point y in Y is at most countable. A point x in X is called a *point of local topological character* of the mapping f if there is a neighbourhood U of x that is mapped topologically onto its image $f(U)$ under the mapping f. Below, in addition to the results by P.S. Aleksandrov [1], we prove the following theorem.

Theorem. *If a mapping f is open and countably multiple, then the set T_f of its points of local topological character is everywhere dense in X.*

To prove the theorem it clearly suffices to show that the existence of a non-empty open set $U \subset X$ containing no points of local topological character of an open mapping f implies the existence in

$$V = f(U)$$

of a point y^* with an at most countable complete inverse image $f^{-1}(y^*)$. To find such a point y^* we construct points

$$x^{(n)}_{i_1 i_2 \dots i_n} \in U, \quad y^{(n)} \in V \ (n = 1, 2, \dots; \ i_k = 0, 1)$$

and open sets

$$U^{(n)}_{i_1 i_2 \dots i_n} \subset U, \quad V^{(n)} \subset V \ (n = 1, 2, \dots; \ i_k = 0, 1)$$

possessing the following properties:

1) $f(x^{(n)}_{i_1 i_2 \dots i_n}) = y^{(n)}$,

2) $x^{(n)}_{i_1 i_2 \dots i_n} \in U^{(n)}_{i_1 i_2 \dots i_n}$,

* *Dokl. Akad. Nauk SSSR* **30** (1941), 477–479.

3) $y^{(n)} \in V^{(n)}$,

4) $\overline{U^{(n+1)}_{i_1 i_2 \ldots i_n i_{n+1}}} \subset U^{(n)}_{i_1 i_2 \ldots i_n}$,

5) $\overline{V^{n+1}} \subset \overline{V^{(n)}}$,

6) $V^{(n)} \subset f(U^{(n)}_{i_1 i_2 \ldots i_n})$,

7) $\overline{U^{(n+1)}_{i_1 i_2 \ldots i_n 0}} \cdot \overline{U^{(n+1)}_{i_1 i_2 \ldots i_n 1}} = 0$,

8) the diameters of the sets $V^{(n)}$ and $U^{n}_{i_1 i_2 \ldots i_n}$ are less than $1/n$.

Conditions 3), 5) and 8) imply that the points $y^{(n)}$ converge to a limit point y^* for $n \to \infty$. Conditions 2), 4) and 8) imply that for any sequence

$$i_1, i_2, \ldots, i_n, \ldots$$

consisting of zeros and ones the points $x^{(n)}_{i_1 i_2 \ldots i_n}$ converge to a limit point

$$x_{i_1 i_2 \ldots i_n \ldots}$$

as $n \to \infty$.

Condition 7) implies that the points $x_{i_1 i_2 \ldots i_n}$ corresponding to different sequences $i_1, i_2, \ldots, i_n$ are different. It follows from the continuity of the mapping f that for any $i_1, i_2, \ldots, i_n \ldots$ we have

$$f(x_{i_1 i_2 \ldots i_n \ldots}) = y^*,$$

that is, the complete inverse image $f^{-1}(y^*)$ of the point y^* has the power of the continuum.

We now proceed to the construction of the points $x^{(n)}_{i_1 i_2 \ldots i_n}$ and the open sets $U^{(n)}_{i_1 i_2 \ldots i_n}$ and $V^{(n)}$.

a) For $n = 1$ we take as $y^{(1)}$ an arbitrary point belonging to V and having at least two different inverse images $x^{(1)}_0$ and $x^{(1)}_1$ in U. By virtue of the absence of points of local topological character in U, such a point $y^{(1)}$ exists. For $U^{(1)}_0$ and $U^{(1)}_1$ we take arbitrary neighbourhoods of the points $x^{(1)}_0$ and $x^{(0)}_1$ satisfying the conditions

$$\overline{U^{(0)}_0} \subset U, \quad \overline{U^{(1)}} \subset U_1 \text{ and } \overline{U^{(1)}_0} \cdot \overline{U^{(1)}_1} = 0.$$

Finally, for $V^{(1)}$ we take a neighbourhood of the point $y^{(1)}$ whose closure is contained in V, $f(U_0^{(1)})$ and $f(U_1^{(0)})$. By virtue of the openness of the mapping f, such a neighbourhood of $y^{(1)}$ exists.

b) Assume that the points $x^{(n)}_{i_1i_2\dots i_n}$ and $y^{(n)}$ and the open sets $U^{(n)}_{i_1i_2\dots i_n}$ and $V^{(n)}$ have already been constructed for $n = N$. We now construct them for $n = N+1$. We set

$$\nu = i_1 + 2i_2 + 2^2 i_3 + \ldots + 2^{N-1} i_N;$$

we denote the system of indices $i_1i_2\dots i_n$ by $S(\nu)$. The systems of indices $i_1i_2\dots i_n0$ and $i_1i_2\dots i_n1$ are denoted by $S(\nu)0$ and $S(\nu)1$, respectively. In the new notation the sets $U^{(N+1)}_{i_1i_2\dots i_N i_{N+1}}$ we are about to construct are designated by the symbols $U^{(N+1)}_{S(\nu)0}$ and $U^{(N+1)}_{S(\nu)1}$, where

$$\nu = 0,\ 1,\ 2,\ \ldots,\ 2^N - 1.$$

When constructing these sets we also have to introduce points η_ν, ξ'_ν, and ξ''_ν and open sets W possessing the following properties:

1) $\xi'_\nu \in U^{(N+1)}_{S(\nu)0}$, $\quad \xi''_\nu \in U^{N+1}_{S(\nu)1}$,

2) $\eta_\nu \in W_\nu$,

3) $f(\xi'_\nu) = f(\xi''_\nu) = \eta_\nu$,

4) $W_\nu \subset f(U^{(N+1)}_{S(\nu)0})$, $\quad W_\nu \subset f(U^{(N+1)}_{S(\nu)1})$,

5) $W_{\nu+1} \subset W_\nu$,

6) $W_0 \subset V^{(N)}$.

The construction of the points η_ν, ξ'_ν, and ξ''_ν and the open sets $U^{(N+1)}_{S(\nu)0}$, $U^{(N+1)}_{S(\nu)1}$, and W_ν is carried out by induction by passing from ν to $\nu+1$.

b$_1$) For $\nu = 0$ we take a point η_0 in $V^{(N)}$ having at least two different inverse images ξ'_0 and ξ''_0. For $U^{(N+1)}_{S(0)0}$ and $U^{(N+1)}_{S(0)1}$ we take neighbourhoods of the points ξ'_0 and ξ''_0 satisfying the conditions

$$\overline{U^{N+1}_{S(0)0}} \subset U^{(N)}_{S(0)}, \quad \overline{U^{(N+1)}_{S(0)1}} \subset U^{(N)}_{(S(0)} \text{ and } \overline{U^{(N+1)}_{S(0)0}} \cdot \overline{U^{(N+1)}_{S(0)1}} = 0.$$

By the openness of f, there is a neighbourhood W_0 of the point η_0 contained in $V^{(N)}$, $f(U^{(N+1)}_{S(0)0})$, and $f(U^{N+1}_{S(0)1})$.

b$_2$) We now assume that η_ν, ξ'_ν, ξ''_ν, $U^{(N+1)}_{S(\nu)0}$, $U^{(N+1)}_{S(\nu)1}$ and W_ν have already been constructed for $\nu = \beta$ and then construct them for $\nu = \beta+1$. For $\eta_{\beta+1}$ we take a point belonging to W_β and having at least two different inverse images $\xi'_{\beta+1}$ and $\xi''_{\beta+1}$ in $U^{(N)}_{S(\beta+1)}$. For $U^{(N+1)}_{S(\beta+1)0}$ and $U^{(N+1)}_{S(\beta+1)1}$ we take neighbourhoods of the points $\xi'_{\beta+1}$ and $\xi''_{\beta+1}$ satisfying the conditions

$$U^{(N+1)}_{S(\beta+1)0} \subset U^{(N)}_{S(\beta+1)}, \quad \overline{U^{(N+1)}_{S(\beta+1)0}} \subset U^{(N)}_{S(\beta+1)}$$

$$\text{and} \quad \overline{U^{(N+1)}_{S(\beta+1)0}} \cdot \overline{U^{(N+1)}_{S(\beta+1)1}} = 0.$$

By the openness of f, there is a neighbourhood $W_{\beta+1}$ of the point $\eta_{\beta+1}$ contained in

$$W_\beta, \; f(U^{(N+1)}_{S(\beta+1)0}) \text{ and } f(U^{(N+1)}_{S(\beta+1)1}).$$

When $\nu = \nu_0 = 2^N - 1$ in our construction, we put $y^{(N+1)} = \eta_\nu$. By virtue of the properties 4), 5), and 6) of the sets W_ν, the set W_{ν_0} is contained in both $f(U^{(N+1)}_{S(\nu)0})$ and $f(U^{(N+1)}_{S(\nu)1})$.

Therefore the point $y^{(N+1)}$ has inverse images

$$x^{(N+1)}_{S(\nu)0} \text{ and } x^{(N+1)}_{S(\nu)1}$$

belonging to each of the sets $U^{(N+1)}_{S(\nu)0}$ and $U^{(N+1)}_{S(\nu)1}$ respectively. By virtue of the openness of f, the point $y^{(N+1)} \in W_{\nu_0} \subset V^{(N)}$ has a neighbourhood $V^{(N+1)}$ whose closure is contained in $V^{(N)}$ and in both $f(U^{(N+1)}_{S(\nu)0})$ and $f(U^{(N+1)}_{S(\nu)1})$.

It can easily be verified that the points $x^{(n)}_{i_1 i_2 \dots i_n}$ and open sets $U^{(n)}_{i_1 i_2 \dots i_N}$ and $V^{(n)}$ that we have constructed satisfy conditions 1)–8).

20 December 1940

References

1. P.S. Aleksandrov, 'On countably-multiple open maps', *Dokl. Akad. Nauk SSSR* **4**:7 (1936), 283–287 (in Russian).

45. LOCAL STRUCTURE OF TURBULENCE IN AN INCOMPRESSIBLE VISCOUS FLUID AT VERY LARGE REYNOLDS NUMBERS *

§1. We denote by

$$u_\alpha(P) = u_\alpha(x_1, x_2, x_3, t), \quad \alpha = 1, 2, 3,$$

the velocity components at time t at a point with rectangular Cartesian coordinates x_1, x_2, x_3. When studying turbulence it is natural to regard the velocity components $u_\alpha(P)$ at each point $P = (x_1, x_2, x_3, t)$ of the region G under consideration belonging to the four-dimensional space (x_1, x_2, x_3, t) as *random variables* in the sense of probability theory (for this approach, see the paper by Millionshchikov [1]).

Denoting the mathematical expectation of a random variable A by $\overline{A}$, we assume that

$$\overline{u_\alpha^2} \text{ and } \overline{(\partial u_\alpha / \partial x_\beta)^2}$$

are finite and bounded in each bounded subregion of the region G.

We introduce new coordinates in the four-dimensional space (x_1, x_2, x_3, t):

$$y_\alpha = x_\alpha - x_\alpha^{(0)} - u_\alpha(P^{(0)})(t - t^{(0)}), \quad s = t - t^{(0)}, \tag{1}$$

where

$$P^{(0)} = (x_1^{(0)}, x_2^{(0)}, x_3^{(0)}, t^{(0)})$$

is a fixed point belonging to G. Note that the coordinates y_α of a point P depend on the random variables $u_\alpha(P^{(0)})$ and therefore they are themselves random variables. In the new coordinates the velocity components are

$$w_\alpha(P) = u_\alpha(P) - u_\alpha(P^{(0)}). \tag{2}$$

Suppose that for fixed values of the components $u_\alpha(P^{(0)})$ the points $P^{(k)}$ $(k = 1, 2, \ldots, n)$ with coordinates $y_\alpha^{(k)}$ and $s^{(k)}$ in the coordinate system (1) lie in the region G. Then we can consider the $3n$-dimensional conditional probability distribution F_n of the variables

$$w_\alpha^{(k)} = w_\alpha(P^{(k)}), \quad \alpha = 1, 2, 3; \ k = 1, 2, \ldots, n,$$

---

* *Dokl. Akad. Nauk SSSR* **30**:4 (1941), 299–301.

for given

$$u_\alpha^{(0)} = u_\alpha(P^{(0)}).$$

In general, the distribution F_n depends on the parameters $x_\alpha^{(0)}, t^{(0)}, u_\alpha^{(0)}, y_\alpha^{(k)}$ and $s^{(k)}$.

Definition 1. A turbulence is said to be *locally homogeneous* in a region G if for any fixed n, $y_\alpha^{(k)}$ and $s^{(k)}$ the distribution F_n does not depend on $x_\alpha^{(0)}, t^{(0)}$, or $u_\alpha^{(0)}$ as long as all the points $P^{(k)}$ lie in G.

Definition 2. A turbulence is said to be *locally isotropic* in a region G if it is locally homogeneous and the distributions indicated in Definition 1 are invariant with respect to rotations and reflections of the original coordinate axes x_1, x_2, x_3.

As compared to the notion of *isotropic turbulence* introduced by Taylor, the above definition of locally isotropic turbulence is *narrower* in the sense that it requires that the distributions F_n be independent of $t^{(0)}$, that is, stationary in time, and is *broader* in the sense that the constraints are imposed only on the distributions of the velocity differences and not on the distribution of the velocities themselves.

§**2.** The isotropy hypothesis in Taylor's sense is in good agreement with experiment in the case of turbulence generated by a flow passing through a grid (see [3]). In the majority of other situations of practical significance, this hypothesis can only be regarded as a rather crude approximation to reality, even for small regions G and at very large Reynolds numbers.

By contrast, in our opinion, it seems very likely that in the case of an arbitrary turbulent flow with sufficiently large Reynolds number [1]

$$\mathrm{Re} = LU/\nu$$

in sufficiently small regions G of the four-dimensional space (x_1, x_2, x_3, t) not lying close to the boundaries of the flow or other singularities of it, the hypothesis of local isotropy holds with high accuracy. Here, by "small regions" are meant those whose linear dimensions and time scales are small relative to L and

$$T = U/L,$$

[1] Here L and U are the typical length and velocity scales for the flow as a whole.

respectively.

Naturally, this hypothesis, which we have put forward in such a general and somewhat ambiguous form, cannot be proved rigorously.[2]

To make its experimental verification possible for individual special cases, we present below a number of consequences of the hypothesis of local isotropy.

§3. We denote by y a vector with components y_1, y_2, y_3 and consider the random variables

$$w_\alpha(y) = w_\alpha(y_1, y_2, y_3) = u_\alpha(x_1 + y_1, x_2 + y_2, x_3 + y_3, t) - u_\alpha(x_1, x_2, x_3, t). \quad (3)$$

By virtue of the hypothesis of local isotropy, the distributions of (3) do not depend on x_1, x_2, x_3 or t. For the first moments of the variables $w_\alpha(y)$,

[2] Here we present only some general consideration in favour of the suggested hypothesis. For very large Re a turbulent flow can be represented in the following way. The mean flow (characterized by the mathematical expectations $\bar{u}_\alpha$) is accompanied by "first-order fluctuations" imposed on it, consisting of chaotic motions of individual fluid volumes with diameters of the order of $l^{(1)} = l$ (where l is the Prandtl mixing length) relative to one another; we denote by $v^{(1)}$ the order of the velocities of these relative motions. In turn, the first-order fluctuations become unstable at very large Re and there appear "second-order fluctuations" imposed on the former, the latter having mixing length $l^{(2)} < l^{(1)}$ and relative velocities $v^{(2)} < v^{(1)}$. This process of successive refinement of turbulent fluctuations goes on until for fluctuations of sufficiently high order n the Reynolds number $\mathrm{Re}^{(n)} = l^{(n)} v^{(n)}/\nu$ turns out to be sufficiently small, so that the effect of viscosity on the nth-order fluctuations is appreciable and prevents the formation of $(N+1)$th-order fluctuations imposed on the former.

From the energetic standpoint, the process of turbulent mixing can be thought of in the following manner: the first-order fluctuations absorb the energy of the mean motion and transfer it subsequently to higher-order fluctuations; as to the energy of the smallest fluctuations, it is dissipated and transformed into thermal energy due to viscosity.

By virtue of the chaotic mechanism of transfer of motion from lower-order fluctuations to higher-order ones, it is natural to assume that, within the limits of space regions which are small as compared to $l^{(1)}$, small-scale higher-order fluctuations are subject to an approximately spatially isotropic statistical regime. Within short time intervals this regime can naturally be regarded as being stationary, even when the flow as a whole is non-stationary.

Since at very large Re the velocity component differences

$$w_\alpha(P) = u_\alpha(P) - u_\alpha(P^{(0)})$$

at nearby points P and $P^{(0)}$ of the four-dimensional space (x_1, x_2, x_3, t) are determined almost exclusively by higher-order fluctuations, the presented scheme leads to the hypothesis of local isotropy in small regions in the sense of the definitions in §§1, 2.

local isotropy implies that

$$\overline{w_\alpha(y)} = 0. \tag{4}$$

Therefore we proceed to the study of the second moments[3]

$$B_{\alpha\beta}(y^{(1)}, y^{(2)}) = \overline{w_\alpha(y^{(1)}) w_\beta(y^{(2)})}. \tag{5}$$

Local isotropy implies that

$$B_{\alpha\beta}(y^{(1)}, y^{(2)}) = \tfrac{1}{2}[B_{\alpha\beta}(y^{(1)}, y^{(1)}) + B_{\alpha\beta}(y^{(2)}, y^{(2)}) - \\ - B_{\alpha\beta}(y^{(2)} - y^{(1)}, y^{(2)} - y^{(1)})] \tag{6}$$

This formula allows us to confine ourselves to second moments of the form $B_\alpha(y, y)$. For these we have

$$B_{\alpha\beta}(y, y) = \overline{B}(r) \cos\theta_\alpha \cos\theta_\beta + \delta_{\alpha\beta} B_{nn}(r), \tag{7}$$

where $r^2 = y_1^2 + y_2^2 + y_3^2$, $y_\alpha = r\cos\theta_\alpha$, $\delta_{\alpha\beta} = 0$ for $\alpha \neq \beta$, and $\delta_{\alpha\beta} = 1$ for $\alpha = \beta$;

$$\overline{B}(r) = B_{dd}(r) - B_{nn}(r), \tag{8}$$

$$B_{dd}(r) = \overline{[w_1(r, 0, 0)]^2}, \quad B_{nn}(r) = \overline{[w_2(r, 0, 0)]^2}. \tag{9}$$

For $r = 0$ we have

$$B_{dd}(0) = B_{nn}(0) = \frac{\partial}{\partial r} B_{dd}(0) = \frac{\partial}{\partial r} B_{nn}(0) = 0, \tag{10}$$

$$\frac{\partial^2}{\partial r^2} B_{dd}(0) = 2\overline{\left(\frac{\partial w_1}{\partial y_1}\right)^2} = 2a^2, \quad \frac{\partial^2}{\partial r^2} B_{nn}(0) = 2\overline{\left(\frac{\partial w_2}{\partial y_1}\right)^2} = 2a_n^2. \tag{11}$$

Formulas (6)–(11) have been derived without using the assumption that the fluid is incompressible. This assumption implies the equation

$$r\partial B_{dd}/\partial r = -2\overline{B}, \tag{12}$$

making it possible to express B_{nn} in terms of B_{dd}. Formulas (12) and (11) imply that

$$a_n^2 = 2a^2. \tag{13}$$

[3] All results in §3 are completely analogous to those obtained in [1, 2, 4] for the case of isotropic turbulence in Taylor's sense.

Further, it is easy to show that (under the assumption of incompressibility) the average energy dissipation rate per unit mass is given by

$$\bar{\epsilon} = \nu\left\{2\overline{\left(\frac{\partial w_1}{\partial y_1}\right)^2} + 2\overline{\left(\frac{\partial w_2}{\partial y_2}\right)^2} + 2\overline{\left(\frac{\partial w_3}{\partial y_3}\right)^2} + \overline{\left(\frac{\partial w_2}{\partial y_1} + \frac{\partial w_1}{\partial y_2}\right)^2} + \right.$$
$$\left. + \overline{\left(\frac{\partial w_3}{\partial y_2} + \frac{\partial w_2}{\partial y_3}\right)^2} + \overline{\left(\frac{\partial w_1}{\partial y_3} + \frac{\partial w_3}{\partial y_1}\right)^2}\right\} = 15\nu a^2. \tag{14}$$

§**4.** Consider the transformation of coordinates

$$y'_\alpha = y_\alpha/\eta, \quad s' = s/\sigma. \tag{15}$$

In the new coordinates the velocities, the kinetic viscosity, and the average energy dissipation rate per unit mass are expressed by the following formulas:

$$w'_\alpha = w_\alpha \sigma/\eta, \quad \nu' = \nu\sigma/\eta^2, \quad \bar{\epsilon}' = \bar{\epsilon}\sigma^3/\eta^2. \tag{16}$$

We now state the following hypothesis.

The first similarity hypothesis. *The distributions F_n for locally isotropic turbulence are uniquely determined by the parameters ν and $\bar{\epsilon}$.*

For

$$\eta = \lambda = \sqrt{\nu/a} = \nu^{3/4}/\bar{\epsilon}^{\,1/4}, \tag{17}$$
$$\sigma = 1/a = \sqrt{\nu/\bar{\epsilon}} \tag{18}$$

the transformation of coordinates (15) results in the parameters $\nu' = 1$, $\bar{\epsilon}' = 1$.

Therefore by virtue of the above similarity hypothesis, the corresponding function

$$B'_{dd}(r') = \beta_{dd}(r') \tag{19}$$

must be the same in all cases of locally isotropic turbulence. The formula

$$B_{dd}(r) = \sqrt{\nu\bar{\epsilon}}\,\beta_{dd}(r/\lambda) \tag{20}$$

in combination with the earlier derived expressions shows that in the case of locally isotropic turbulence the second moments $B_{\alpha\beta}(y^{(1)}, y^{(2)})$ can be uniquely expressed in terms of ν, $\bar{\epsilon}$ and the universal function β_{dd}.

§5. To determine the behaviour of the function $\beta_{dd}(r')$ for large r' we introduce one more hypothesis.

The second similarity hypothesis.[4] *If the absolute values of the vectors $y^{(k)}$ and their differences $y^{(k)} - y^{(k')}$ (where $k' \neq k$) are large relative to λ, the distributions F_n are uniquely determined by the parameter $\bar{\epsilon}$ and do not depend on ν.*

We put

$$y''_\alpha = y'_\alpha / k^3, \quad s'' = s'/k^2, \tag{21}$$

where y'_α and s' are determined according to formulas (15), (17) and (18). Since

$$\bar{\epsilon}' = \bar{\epsilon}'' = 1,$$

for any k, by virtue of the above hypothesis, for r' large relative to $\lambda' = 1$ we have

$$B''_{dd}(r'') \approx B'_{dd}(r'') = \beta_{dd}(r'/k^3).$$

On the other hand, formula (20) implies that

$$B''_{dd}(r'') = (1/k^2) B'_{dd}(r') = (1/k^2)\beta_{dd}(r').$$

Hence, for large r' we have

$$\beta_{dd}(r'/k^3) \approx (1/k^2)\beta_{dd}(r),$$

whence

$$\beta_{dd}(r') \approx C (r')^{2/3}, \tag{22}$$

where C is an absolute constant. By virtue of (17), (20) and (22) for r large relative to λ we have

$$B_{dd}(r) \approx C \bar{\epsilon}^{2/3} r^{2/3}. \tag{23}$$

From (23) and (12) it can easily be derived that for r large relative to λ,

$$B_{nn}(r) \approx 4/3 B_{dd}(r). \tag{24}$$

---

[4] In terms of the schematic description of turbulence presented in Footnote 2, λ is the scale of the smallest fluctuations, whose energy is dissipated and transformed directly into thermal energy due to viscosity. The meaning of the second similarity hypothesis is that for fluctuations of intermediate orders for which $l^{(k)}$ is greater than λ the mechanism of energy transfer from larger fluctuations to smaller ones does not depend on the viscosity.

We note that instead of formula (24), by virtue of (13) for r small as compared to λ the relation

$$B_{nn}(r) \approx 2B_{dd}(r) \tag{25}$$

holds.

28 December 1940

References

1. M.D. Millionshchikov, 'Degeneration of homogeneous isotropic turbulence in viscous incompressible fluids', *Dokl. Akad. Nauk SSSR* **22**:5 (1939) 236–240 (in Russian).
2. G.I. Taylor, 'Statistical theory of turbulence' I–IV, *Proc. Roy. Soc. London* **A 151** (1935), No. 874, 421–478.
3. S. Goldstein, ed., *Modern developments in fluid mechanics*, Vol. 1, §95, Oxford Univ. Press, 1938.
4. Th. von Karman, 'The fundamentals of the statistical theory of turbulence', *J. Aeronaut. Sci.* **4**:4 (1937), 131–138.

46. ON THE DEGENERATION OF ISOTROPIC TURBULENCE IN AN INCOMPRESSIBLE VISCOUS FLUID *

As in [1, 2], we assume that the velocity components

$$u_\alpha(P,t) = u_\alpha(x_1, x_2, x_3, t)$$

at the point $P = (x_1, x_2, x_3)$ at time t are random variables and denote by $\overline{A}$ the mathematical expectation of a random variable A.

In the case of isotropic turbulence in Taylor's sense (see [3, 4]) we have

$$\bar{u}_\alpha = 0, \tag{1}$$

and the second moments

$$b_{\alpha\beta} = \overline{u_\alpha(x_1^{(1)}, x_2^{(1)}, x_3^{(1)}, t) u_\beta(x_1^{(2)}, x_2^{(2)}, x_3^{(2)}, t)} \tag{2}$$

are given by the formulas (see [1, 5])

$$b_{\alpha\beta} = \bar{b}(r,t)\cos\theta_\alpha \cos\theta_\beta + \delta_{\alpha\beta} b_{nn}(r,t), \tag{3}$$

where

$$r^2 = (x_1^{(2)} - x_1^{(1)})^2 + (x_2^{(2)} - x_2^{(1)})^2 + (x_3^{(2)} - x_3^{(1)})^2,$$

$$x_\alpha^{(2)} - x_\alpha^{(1)} = r\cos\theta_\alpha, \quad \delta_{\alpha\beta} = 0 \text{ for } \alpha \neq \beta, \quad \delta_{\alpha\beta} = 1 \text{ for } \alpha = \beta;$$

$$\bar{b}(r,t) = b_{dd}(r,t) - b_{nn}(r,t); \tag{4}$$

$$b_{dd}(r,t) = \overline{u_1(x_1, x_2, x_3, t) u_1(x_1 + r, x_2, x_3, t)}; \tag{5}$$

$$b_{nn}(r,t) = \overline{u_2(x_1, x_2, x_3, t) u_2(x_1 + r, x_2, x_3, t)}; \tag{6}$$

$$\partial b_{dd}/\partial r = -(2/r)\bar{b}. \tag{7}$$

Formula (7), together with

$$b_{dd}(0,t) = b_{nn}(0,t) = \overline{[u_\alpha(x_1, x_2, x_3, t)]^2} = b(t), \tag{8}$$

---

* *Dokl. Akad. Nauk SSSR* **31**:6 (1941), 538–541.

makes it possible to express b_{nn} in terms of b_{dd}. Thus, all the second moments can be found from a single function $b_{dd}(r,t)$.

Von Karman and Howarth [5] derived the equation

$$\frac{\partial b_{dd}}{\partial t}+2\left(\frac{\partial d_{nnd}}{\partial r}+\frac{4}{r}b_{nnd}\right)=2\nu\left(\frac{\partial^2 b_{dd}}{\partial r^2}+\frac{4}{r}\frac{\partial b_{dd}}{\partial r}\right) \tag{9}$$

for $b_{dd}(r,t)$, where

$$b_{nnd}(r,t)=\overline{u_2^2(x_1,x_2,x_3,t)u_1(x_1+r,x_2,x_3,t)}. \tag{10}$$

Loitsyanskii [6] showed that (9) implies that

$$\Lambda=\int_0^\infty b_{dd}(r,t)r^4dr \tag{11}$$

does not depend on time.

As the "turbulence scale" we take the length

$$L=(\Lambda/b)^{1/5}. \tag{12}$$

Using L and the average value of the velocity components

$$\mathbf{u}=\sqrt{b}, \tag{13}$$

we introduce the Reynolds number

$$\mathrm{Re}=L\mathbf{u}/\nu=(\Lambda^{1/5}/\nu)\mathbf{u}^{3/5}. \tag{14}$$

Since $\mathbf{u}\to 0$ as $t\to\infty$, (14) implies that the Reynolds number Re is small for large t. At this terminal stage of degeneration of isotropic turbulence it is legitimate to apply the theory neglecting the third moments. As was shown in [1], in this case

$$b_{dd}(r,t)=\frac{k}{(4\nu t)^{5/2}}\exp\left(-\frac{r^2}{8\nu t}\right), \tag{15}$$

$$b_{nn}(r,t)=\frac{k}{(4\nu t)^{5/2}}\left(1-\frac{r^2}{8\nu t}\right)\exp\left(-\frac{r^2}{8\nu t}\right) \tag{16}$$

and consequently,

$$\mathbf{u}=At^{-5/4}, \tag{17}$$

$$L=B\sqrt{t}. \tag{18}$$

It is easy to express k, A and B in terms of ν and Λ.

For *large* Re, assuming that $b_{dd}(r,t)$ has the form

$$b_{dd}(r,t) = b\psi(r/L), \tag{19}$$

von Karman found (see [5]) that [1]

$$\mathbf{u} = \mathbf{u}_0(t-t_0)^{-p}, \quad L = L_0(t-t_0)^{1-p}, \tag{20}$$

but he did not determine the exponent p.

Below we present a new derivation of formula (20), which implies that

$$p = 5/7. \tag{21}$$

We first show that for sufficiently large Re the assumption (19) implies

$$db/dt = -Kb^{3/2}L^{-1} \tag{22}$$

where K is an absolute constant. Formula (22) can be justified in different ways. If the assumption made in my paper [2] is adopted, then the following argument can be used. In isotropic turbulence the average energy per unit mass is equal to $(3/2)b$. Therefore the average energy dissipation rate per unit mass is

$$\bar{\epsilon} = -\tfrac{3}{2}db/dt. \tag{23}$$

Since using the notation $B_{dd}(r)$ in [2] we can write

$$b_{dd}(r) = b - \tfrac{1}{2}B_{dd}(r), \tag{24}$$

formula (23) in [2] implies that for r small relative to L and large relative to λ we have

$$b_{dd}(r) \approx b\left(1 - \frac{C}{2b}\bar{\epsilon}^{2/3}r^{2/3}\right). \tag{25}$$

For small ρ we derive the expression [2]

$$\psi(\rho) \approx 1 - g\rho^{2/3} \tag{26}$$

---

[1] Equation (77) in [6] was integrated incorrectly. Actually, the exponent in formula (78) must be equal to $AB/[5+AB]$ and that in formula (81) must be equal to $-5/[5+AB]$. The value $p = 5/7$ corresponds to $AB = 2$.

[2] We assume that relation (26) remains valid for arbitrarily small ρ. However, it has a real meaning only for ρ large relative to λ/L, since the assumption (19) itself can naturally be adopted only for r large relative to λ.

for the function $\psi(\rho)$, where

$$g = (C/2)\bar{\epsilon}^{2/3}L^{2/3}b^{-1}.$$

Consequently,

$$\bar{\epsilon} = (2gb/C)^{3/2}L^{-1}. \tag{27}$$

Putting

$$K = \tfrac{2}{3}(2g/C)^{3/2} \tag{28}$$

we obtain formula (22) from (23) and (27). Formula (28) shows how the coefficient K is expressed in terms of the absolute constant C introduced in [2] and the parameter g determined from (26) taking into account the form of the function $\psi(\rho)$.

From (12) and (22) we find that

$$db/dt = -K\Lambda^{-1/5}b^{17/10}. \tag{29}$$

Integration of equation (29) results in

$$b = (10/7K)^{10/7}\Lambda^{2/7}(t-t_0)^{-10/7}. \tag{30}$$

Relations (30), (13) and (12) imply

$$u = (10/7K)^{5/7}\Lambda^{1/7}(t-t_0)^{-5/7}, \tag{31}$$

$$L = (7K/10)^{2/7}\Lambda^{1/7}(t-t_0)^{2/7}. \tag{32}$$

A comparison of (31) and (32) with (20) shows that we actually have $p = 5/7$.

4 March 1941

References

1. M.D. Millionshchikov, 'Degeneration of homogeneous isotropic turbulence in viscous incompressible fluids', *Dokl. Akad. Nauk SSSR* **22**:5 (1939), 236–240 (in Russian).
2. A.N. Kolmogorov, 'Local structure of turbulence in an incompressible viscous fluid at very large Reynolds numbers', *Dokl. Akad. Nauk SSSR* **30**:4 (1941), 299–303 (in Russian). (See No. 45.)

3. G.I. Taylor, 'Statistical theory of turbulence', I–IV, *Proc. Roy. Soc. London* **A151** (1935), No. 874, 421–478.
4. S. Goldstein, ed. *Modern developments in fluid mechanics*, Vol. 1, §91, Oxford Univ. Press, 1938.
5. Th. von Karman and L. Howarth, 'On the statistical theory of isotropic turbulence', *Proc. Roy. Soc. London* **A164** (1938), No. 917, 192–245.
6. L.G. Loitsyanskii, 'Some basic regularities of an isotropic turbulent flow', *Trudy TsAGI* **440** (1939), 3–23.

47. DISSIPATION OF ENERGY IN ISOTROPIC TURBULENCE *

In my paper [1] the definition of the notion of local isotropy was stated and the functions

$$B_{dd}(r) = \overline{[u_d(M') - u_d(M)]^2}, \quad B_{nn}(r) = \overline{[u_n(M') - u_n(M)]^2} \tag{1}$$

were introduced, where r denotes the distance between the points M and M', $u_d(M')$ and $u_d(M)$ are the velocity components at the points M and M' in the direction of $\overline{MM'}$, and $u_n(M)$ and $u_n(M')$ are the velocity components at M and M' respectively, in a direction perpendicular to $\overline{MM'}$.

In what follows we also need the third order moments

$$B_{ddd}(r) = \overline{[u_d(M') - u_d(M)]^3}. \tag{2}$$

Locally isotropic turbulence in an incompressible fluid is described by the equation

$$4\bar{\epsilon} + \left(\frac{dB_{ddd}}{dr} + \frac{4}{r}B_{ddd}\right) = 6\nu\left(\frac{d^2B_{dd}}{dr^2} + \frac{4}{r}\frac{dB_{dd}}{dr}\right), \tag{3}$$

which is analogous to the well-known von Karman equation for isotropic turbulence in Taylor's sense. Here $\bar{\epsilon}$ denotes the average energy dissipation rate per unit mass. Equation (3) can be written as

$$\left(\frac{d}{dr} + \frac{4}{r}\right)\left(6\nu\frac{dB_{dd}}{dr} - B_{ddd}\right) = 4\epsilon \tag{4}$$

and, by virtue of the condition $dB_{dd}(r)/dr\big|_{r=0} = B_{ddd}(0) = 0$, it yields

$$6\nu dB_{dd}/dr - B_{ddd} = (4/5)\bar{\epsilon}r. \tag{5}$$

As is well known, for small r we have

$$B_{dd} \approx (1/15v)\bar{\epsilon}r^2, \tag{6}$$

that is, $$6\nu dB_{dd}/dr \approx (4/5)\bar{\epsilon}r.$$

Thus, for small r the second term on the left-hand side of equation (5) is infinitesimally small relative to the first term. Conversely, for large r the first term is negligibly small as compared to the second term, that is, we can assume that

$$B_{ddd} \approx -(4/5)\bar{\epsilon}r. \tag{7}$$

It is natural to assume that for large r the ratio

$$S = B_{ddd} : B_{dd}^{3/2}, \tag{8}$$

* *Dokl. Akad. Nauk SSSR* **32**:1 (1941), 19–21.

that is, the *skewness* of the distribution of the difference

$$\Delta u_d = u_d(M') - u_d(M),$$

remains constant. Under this assumption, for large r we have

$$B_{dd} \approx C\bar{\epsilon}^{2/3} r^{2/3}, \tag{9}$$

where

$$C = (-4/5S)^{2/3}. \tag{10}$$

Relation (9) was derived in [1] on the basis of somewhat different considerations. [1]

In [1] we introduced the *local scale* of turbulence

$$\lambda = (\nu^3/\bar{\epsilon})^{1/4} \tag{11}$$

and justified the assumption that

$$B_{dd}(r) = \sqrt{\nu\bar{\epsilon}}\beta_{dd}(r/\lambda), \quad B_{nn}(r) = \sqrt{\nu\bar{\epsilon}}\beta_{nn}(r/\lambda), \tag{12}$$

where β_{dd} and β_{nn} are universal functions such that

$$\beta_{dd}(\rho) \approx (1/15)\rho^2, \quad \beta_{nn}(\rho) \approx (2/15)\rho^2 \tag{13}$$

for small ρ and

$$\beta_{dd}(\rho) \approx C\rho^{2/3}, \quad \beta_{nn}(\rho) \approx (4/3)C\rho^{2/3} \tag{14}$$

for large ρ. In isotropic turbulence in Taylor's sense the laws of locally isotropic turbulence must hold at distances much less than the *integral scale* of turbulence L (for a precise definition see my paper [3]). In this case the correlation coefficients

$$R_{dd}(r) = \overline{(u_d(M')u_d(M))} : b, \quad R_{nn}(r) = \overline{(u_n(M')u_n(M))} : b, \tag{15}$$

where b is the average value of the square of the velocity components, are related to $B_{dd}(r)$ and $B_{nn}(r)$ by

$$B_{dd} = 2b(1 - R_{dd}), \quad B_{nn} = 2b(1 - R_{nn}). \tag{16}$$

By (16) and (12), for r small relative to L we must have

$$1 - R_{dd} \approx \frac{\sqrt{\nu\bar{\epsilon}}}{2b}\beta_{dd}\left(\frac{r}{\lambda}\right), \quad 1 - R_{nn} \approx \frac{\sqrt{\nu\bar{\epsilon}}}{2b}\beta_{nn}\left(\frac{r}{\lambda}\right). \tag{17}$$

---

[1] A.M. Obukhov obtained relation (9) independently, by calculating the balance in the spectral distribution of the fluctuation energy (see [2]).

If r is small relative to L and large relative to λ, then by (14) and (11),

$$1 - R_{dd} \approx (1/2)C\bar{\epsilon}^{2/3}b^{-1}r^{2/3}, \tag{18^1}$$

$$1 - R_{nn} \approx (2/3)C\bar{\epsilon}^{2/3}b^{-1}r^{2/3}. \tag{18^2}$$

Formulas (18) make it possible to determine the constant C from experimental data The correlation coefficients R_{dd} and R_{nn} were measured most thoroughly by Dryden, Schubauer, Mock, and Scramstad [4]. Representing formula (18^2) in the form

$$1 - R_{nn} \approx 2C(kr)^{2/3}, \quad k = \bar{\epsilon} : (3b)^{3/2}, \tag{19}$$

we have calculated the values of the coefficient k corresponding to turbulence at a distance 40M from a grid with a mesh size M equal to 1″, 3.25″, and 5″ using the empirical formula (17) in [4] and putting $b = \sqrt{\bar{u}^2}$ and $\bar{\epsilon} = (3/2)Ud\sqrt{\bar{u}^2}/dx$ in the notation of [4]:

M (inch)	1	3.25	5
k (cm^{-1})	0.197	0.065	0.042

With these values of k, the k curves in Figure 5 in [4], taking into account the correction for the wire length, are in good agreement with formula (19) for values of r not very large relative to L and

$$C = 3/2. \tag{20}$$

The local scale λ is so small that under the conditions of the experiments described in [4] the deviations from relations (18) for small r cannot be detected.[2]

[2] In this connection we note that the attempt of Millionshchikov [5] to apply the theory of isotropic turbulence neglecting third order moments to experimental data in [4] is based on a misunderstanding. It can easily be seen that under the conditions of the experiments in [4] the terms involving second order moments in the equations relating second order moments to third order ones (for example, in equation (3)) are *much less* than those involving third order moments.

In Figure 3 in [5], the comparison with the theoretical curve of R_{nn} (obtained by neglecting third order moments) using the experimental data in [4] was performed incorrectly since: 1) the data under consideration relate to a definite spectral component of fluctuations and not to the total fluctuations; and 2) the experimental data do not correspond to the condition $8\nu t = 7.56$ which determines the theoretical curve.

In his later studies [6] (see also M.D. Millionshchikov, *Izv. Akad. Nauk SSSR Ser. Geogr. i Geofiz.* 5:4/5 (1941), 433–446 (in Russian)) Millionshchikov himself estimated the errors due to the neglect of third order moments, using subtle considerations involving third and fourth order moments.

The curves in Figure 28 in [4] cannot be used directly to determine C, since they do not include the correction for the wire length. However, for r small relative to L they confirm with sufficient accuracy the relation

$$(1 - R_{nn})/(1 - R_{dd}) = 4/3 \tag{21}$$

implied by (18). Including the correction for the wire length, the authors of [4] estimated the ratio on the left-hand side of (21) as 1.28 (see [4, p. 29]), which appears to be sufficiently close to the theoretical value 4/3 if we take into account the restricted accuracy of an experiment.

References

1. A.N. Kolmogorov, 'Local structure of turbulence in an incompressible viscous fluid at large Reynolds numbers', *Dokl. Akad. Nauk SSSR* **30**:4 (1941), 299–303 (in Russian). (See No. 45.)
2. A.M. Obukhov, 'On the energy distribution in the spectrum of a turbulent flow', *Dokl. Akad. Nauk SSSR* **32**:1 (1941), 22–24 (in Russian).
3. A.N. Kolmogorov, 'On the degeneration of isotropic turbulence in an incompressible viscous fluid', *Dokl. Akad. Nauk SSSR* **31**:6 (1941), 538–541 (in Russian). (See No. 46.)
4. H.L. Dryden, G.B. Schubauer, W.C. Mock and H.K. Scramstad, 'Measurements of intensity and scale of wind-tunnel turbulence and their relation to the initial Reynolds number of spheres', *Nat. Adv. Com. Aeronaut.* **581** (1937).
5. M.D. Millionshchikov, 'Damping of fluctuation velocity in aerodynamical tunnels', *Dokl. Akad. Nauk SSSR* **22**:5 (1939), 241–242 (in Russian).
6. M.D. Millionshchikov, 'On the influence of third order moments in isotropic turbulence', *Dokl. Akad. Nauk SSSR* **32**:9 (1941), 615–617 (in Russian).

48. EQUATIONS OF TURBULENT MOTION IN AN INCOMPRESSIBLE FLUID *

The general pattern of turbulent motion can be described (according to Taylor and Richardson) in the following way. The mean flow is accompanied by turbulent fluctuations imposed on it and having different scales, beginning with maximal scales of the order of the "external scale" of turbulence L (the "mixing length") to smallest scales of the order of the distances λ at which the effect of viscosity becomes appreciable (the "internal scale" of turbulence). In regions whose dimensions are small relative to L the velocity field is smooth. Most large-scale fluctuations receive energy from the mean flow and transfer it to fluctuations of smaller scales. Thus, there appears a flux of energy transferred continuously from fluctuations of larger scales to those of smaller scales. Dissipation of energy, that is, transformation of energy into heat, occurs mainly in fluctuations of scale λ. The amount of energy w dissipated in unit time per unit volume is the basic characteristic of turbulent motion for all scales.

The most characteristic properties of turbulence are those relating to motion in regions whose dimensions are small as compared to L. The local structure of turbulence (that is, its structure at scales $\ll L$) can be regarded as being spatially homogeneous and isotropic. Its study is carried out by analogy with Taylor's and von Karman's study of isotropic turbulence. However, in this case one should consider moments of various orders, defined as average values of products of components of vector velocity differences at two different points in space. By virtue of the isotropy, these moments are functions depending only on the distance r between the points (for $r \ll L$). For example, the second order moments B_{dd} and B_{nn} are the average values of the squares of differences of the projections of the velocities at two points on the line joining the points and the direction perpendicular to the line, respectively. For $r \ll L$ these moments turn out to be proportional to r^2, while for $r \gg \lambda$ (and $r \ll L$) B_{dd} and B_{nn} are proportional to $r^{2/3}$. [1] The internal scale of turbulence λ itself is proportional to $\nu^{3/4}(\rho/w)^{1/4}$, where ν is the viscosity of the fluid and

* *Izv. Akad. Nauk SSSR Ser. Fiz.* **6**:1–2 (1942), 56–58. Summary of the paper presented at the General Meeting of the Division of Physics and Mathematics of the USSR Academy of Sciences; 26–28 January 1942, Kazan'.

[1] *Dokl. Akad. Nauk SSSR* **30**:4 (1941), 299–303 (in Russian). (No. 45 in the present edition).

ρ is its density. These results are in good agreement with measurements of the velocity correlation performed by Dryden et al.

A somewhat less thorough mathematical investigation can be carried out for a homogeneous and isotropic (at all scales) turbulent motion in the case when there is no mean flow; such a motion must damp out with time. In addition to Taylor's and von Karman's results, the conservation law found by L.G. Loitsyanskii[2] can be applied to show[3] that in this case the fluctuation velocity of motion decays inversely proportionally to $t^{5/7}$. The scale of motion L increases proportionally as $t^{2/7}$.

Based on the above local properties of turbulence, we can construct (using a cruder approximate assumption) a complete system of equations of turbulent motion. These equations are written as

$$\frac{D\bar{v}_i}{Dt} = F_i - \frac{\partial}{\partial x_i}\left(\frac{\bar{p}}{\rho} + b\right) + A \sum_j \frac{\partial}{\partial x_j}\left[\frac{b}{\omega}\left(\frac{\partial \bar{v}_i}{\partial x_j} + \frac{\partial \bar{v}_j}{\partial x_i}\right)\right], \tag{1}$$

$$\frac{D\omega}{Dt} = -\frac{7}{11}\omega^2 + A' \sum_j \frac{\partial}{\partial x_j}\left[\frac{b}{\omega}\frac{\partial \omega}{\partial x_j}\right], \tag{2}$$

$$\frac{Db}{Dt} = -b\omega + \frac{A}{3}\frac{b}{\omega}\bar{\epsilon} + A'' \sum_j \frac{\partial}{\partial x_j}\left[\frac{b}{\omega}\frac{\partial b}{\partial x_j}\right]. \tag{3}$$

Here D/Dt is the symbol of the total time derivative, F_i is an external force, $\bar{v}_i$ is the velocity of the mean motion, and $\bar{p}$ is the average value of the pressure; $b = (1/3)\sum_j \overline{v_j'^2}$ is equal to one third of the mean square of the fluctuation velocity; ω is a certain average "frequency", defined as $\omega = c\sqrt{b}/L$ where c is a constant, and $\epsilon = \sum_{j,i}\left(\frac{\partial \bar{v}_i}{\partial x_j} + \frac{\partial \bar{v}_j}{\partial x_i}\right)^2$. Finally, A, A' and A'' are numerical constants which must be determined once and for all by comparing an individual solution of the equations with experimental data. The ratio b/ω in equations (1), (2) and (3) plays the role of a "diffusion coefficient" for the average velocity, the "frequency" ω, and the fluctuation velocity b, respectively. The first term in (3) is equal to two thirds of the turbulent dissipation of energy related to a unit mass of the fluid $(2w/3\rho)$, and the second term is equal to two thirds of the energy transferred by turbulent fluctuations from the mean motion. The ratio of the coefficients in the first terms in (1) and (2) is determined by

---

[2] *Trudy TsAGI* **440** (1939), 3 (in Russian).

[3] *Dokl. Akad. Nauk SSSR* **31**:6 (1941), 538–541 (in Russian) (No. 46 in the present edition).

the above-mentioned Loitsyanskii relation. Equations (1)–(3) determine the average velocity vector $\bar{v}_i$, the fluctuation velocity b, and the scale of motion L.

The solution of these equations meets with severe difficulties. They can be integrated in finite form for fluid motion between movable parallel planes, the result being in agreement with well-known results of von Karman.

In conclusion, some preliminary results on numerical integration of equations (1)–(3) for a turbulent flow in a tube of a circular cross-section were presented.

Discussion. L.D. Landau noted that A.N. Kolmogorov had, for the first time, suggested a correct interpretation of the local structure of a turbulent flow. As to the equations of turbulent motion, L.D. Landau said that, in his opinion, they must necessarily take into account the fact that the presence of the curl of the velocity in a turbulent flow is confined to a finite region in space; qualitatively, correct equations must result in this very kind of vortex distribution.

P.L. Kapitsa also took part in the discussion.

26 January 1942

49. A REMARK ON CHEBYSHEV POLYNOMIALS LEAST DEVIATING FROM A GIVEN FUNCTION *

In 1918 A. Haar [1] managed to state extremely general uniqueness conditions for a polynomial of best approximation. However, for thirty years since its publication, Haar's result has not given rise to any new specific studies of analytical character. This seems to be due to the fact that no sufficiently simple and analytically natural cases of applicability of Haar's conditions have been found outside the framework of "Chebyshev systems" in the sense of S.N. Bernshtein (see [2], §§1, 2 in Chapter 1). The work by Haar, along with the whole classical theory of best approximation, deals with real functions. In this paper we show that Haar's theorem can be readily extended to complex functions. This makes it possible to apply it, for instance, to the problem of best approximation of a continuous complex function on any closed bounded set in the complex plane by ordinary polynomials (see [4]).

Besides the extension of Haar's theorem to the complex case (see Theorem 3 below) we establish a complex analogue of a characteristic property of polynomials of best approximation (see Theorem 1 below), found by Chebyshev himself. Since the proofs of all subsequent results, which are directly related to Chebyshev's work, are quite elementary, we present all the proofs in full, although some of them are merely slight modifications of §§44, 45 in the book by N.I. Akhiezer [3]. It is the publication of this book that has aroused our interest in this group of problems.

We consider polynomials of the form

$$P_\alpha(z) = \alpha_1\phi_1(z) + \alpha_2\phi_2(z) + \ldots + \alpha_n\phi_n(z),$$

where ϕ_k are continuous complex functions selected once and for all, which are assumed to be defined and continuous on a compactum K.

For an arbitrary continuous complex function $f(z)$ on K and any system

$$\alpha = (\alpha_1, \alpha_2, \ldots, \alpha_n)$$

of n complex numbers we set

$$D_{\alpha,f}(z) = P_\alpha(z) - f(z).$$

* *Uspekhi Mat. Nauk* **3**:1 (1948), 216–221.

As in the real case, the maximum

$$L(\alpha, f) = \max_{z \in K} |D_{\alpha,f}(z)|$$

exists and the minimum

$$\mathcal{E}(f) = \min_{\alpha} L(\alpha, f)$$

is attained for at least one polynomial $P_\alpha(z)$.

The set of $z \in K$ such that

$$|D_{\alpha,f}(z)| = L(\alpha, f)$$

will be denoted as $M_{\alpha,f}$.

Theorem 1. *For the relation*

$$L(\alpha, f) = \mathcal{E}(f)$$

to hold it is necessary and sufficient that the following condition be fulfilled:

(T) *whatever the polynomial $P_\beta(z)$ may be,*

$$\min_{z \in M_{\alpha,f}} \operatorname{Re}\{P_\beta(z)\overline{D_{\alpha,f}(z)}\} \leq 0.$$

Proof of the sufficiency. Assume that the condition (T) is fulfilled and consider a polynomial $P_{\alpha'}(z)$. Let

$$P_\beta(z) = P_\alpha(z) - P_{\alpha'}(z).$$

By virtue of condition (T), there is a point $z_0 \in M_{\alpha,f}$ such that

$$\operatorname{Re}\{P_\beta(z_0)\overline{D_{\alpha,f}(z_0)}\} \leq 0.$$

At this point we have

$$|D_{\alpha',f}(z_0)|^2 = \{D_{\alpha,f}(z_0) - P_\beta(z_0)\}\{\overline{D_{\alpha,f}(z_0) - P_\beta(z_0)}\} =$$

$$= D_{\alpha,f}(z_0)\overline{D_{\alpha,f}(z_0)} - P_\beta(z_0)\overline{D_{\alpha,f}(z_0)} - \overline{P_\beta(z_0)}D_{\alpha,f}(z_0) +$$

$$+ P_\beta(z_0)\overline{P_\beta(z_0)} = |D_{\alpha,f}(z_0)|^2 - 2\operatorname{Re}\{P_\beta(z_0)\overline{D_{\alpha,f}(z_0)}\} + |P_\beta(z_0)|^2 \geq L^2(\alpha, f).$$

Thus,

$$L(\alpha', f) \geq L(\alpha, f)$$

for any α', as required.

Proof of the necessity. Assume that condition (T) is violated, that is, there is a polynomial $P_\beta(z)$ such that

$$\operatorname{Re}\{P_\beta(z)\overline{D_{\alpha,f}(z)}\} > 0$$

everywhere on $M_{\alpha,f}$.

Since the set $M_{\alpha,f}$ is closed, there exists $\epsilon > 0$ and an open set $G \supseteq M_{\alpha,f}$ such that

$$\operatorname{Re}\{P_\beta(z)\overline{D_{\alpha,f}(z)}\} \geq \epsilon$$

for all $z \in G$.

We set

$$H = \max_{z \in K} |P_\beta(z)|,$$
$$h = L(\alpha, f) - \max_{z \in K} |D_{\alpha,f}(z)|,$$

where

$$F = K \backslash G.$$

Since the set F is closed and $|D_{\alpha,f}(z)|$ does not attain $L(\alpha, \beta)$ on it, we have $h > 0$.

We take $\lambda > 0$ such that

$$\lambda \leq \epsilon / H^2, \quad \lambda \leq h/2H,$$

and set

$$P_{\alpha'}(z) = P_\alpha(z) - \lambda P_\beta(z).$$

It can readily be shown that

$$|D_{\alpha',f}(z)|^2 = |D_{\alpha,f}(z)|^2 - 2\lambda \operatorname{Re}\{P_\beta(z)\overline{D_{\alpha,f}(z)}\} + \lambda^2 |P_\beta(z)|^2 \leq$$
$$\leq L^2(\alpha, f) - 2\lambda\epsilon + \lambda \frac{\epsilon}{H^2} H^2 = L^2(\alpha, f) - \lambda\epsilon$$

on G. On the other hand, we have

$$|D_{\alpha',f}(z)| \leq |D_{\alpha,f}(z)| + \lambda |P_\beta(z)| \leq L(\alpha, f) - h + \frac{h}{2H} H = L(\alpha, f) - \frac{h}{2}$$

on F. These two equations show that

$$L(\alpha', f) < L(\alpha, f),$$

that is, $P_\alpha(z)$ is not a polynomial of best approximation, as required. Theorem 1 is proved completely.

We now introduce the following assumptions:

(A) the functions $\phi_1(z)$, $\phi_2(z)$, $\dots$, $\phi_n(z)$ are linearly independent on K;

(B) K contains at least $n+1$ points;

(H) any polynomial $P_\beta(z)$ whose coefficients β_k are not all equal to zero vanishes in K at at most $n-1$ points.

Conditions (A) and (B) only serve to exclude trivial cases not of interest for our consideration. On the contrary, condition (H) substantially restricts the class of admissible systems of functions $\phi_k(z)$.

Theorem 2. *Under the conditions* (B) *and* (H), *the relation*

$$L(\alpha, f) = \mathcal{E}(f)$$

implies that

$$|D_{\alpha,f}(z)| = \mathcal{E}(f)$$

at at least $n+1$ *points.*

Proof. Assume that, contrary to the assertion of the theorem, the set $M_{\alpha,f}$ consists only of points

$$z_1, z_2, \dots, z_q; \quad q \le n.$$

Since, by (B), the set $M_{\alpha,f}$ does not coincide with K, for $k = 1, 2, \dots, q$ we have

$$|D_{\alpha,f}(z_k)| = L(\alpha, f) > 0.$$

We extend in an arbitrary manner the system of points $z_1, z_2, \dots, z_q$ (in the case $q < n$) to a system of n distinct points

$$z_1, z_2, \dots, z_n.$$

It follows from (H) that the determinant

$$|\phi_k(z_m)|, \quad k, m = 1, 2, \dots, n,$$

is non-zero. Therefore the system of equations

$$\beta_1\phi_1(z_k) + \beta_2\phi_2(z_k) + \ldots + \beta_n\phi_n(z_k) = D_{\alpha,f}(z_k), \quad k = 1, 2, \ldots, q,$$

is consistent. Now any solution of the system corresponds to a polynomial $P_\beta(z)$ that contradicts condition (T) of Theorem 1, as can easily be shown. Hence $P_\alpha(z)$ cannot be a polynomial of best approximation. This contradiction proves the theorem.

As was indicated at the beginning of this paper, we are interested in conditions under which the *uniqueness theorem* holds:

(U) *For any continuous complex function $f(z)$ on K there is only one polynomial*

$$P_\alpha(z) = T_f(z)$$

possessing the property

$$L(\alpha, f) = \mathcal{E}(f).$$

Theorem 3. *If condition* (A) *holds, then propositions* (U) *and* (H) *are equivalent* (*that is,* (H) *is a necessary and sufficient condition for* (U)).

Proof of the implication (H) $\to$ (U). It follows from (A) that K contains at least n points. If the number of points in K is exactly equal to n, then any function $f(z)$ is a polynomial and $f(z)$ is itself the unique best approximation with $\mathcal{E}(f) = 0$. It remains to consider the case when condition (B) is fulfilled and Theorem 2 can be applied.

Assume, contrary to proposition (U), that there are two different polynomials $P_{\alpha'}(z)$ and $P_{\alpha''}(z)$ such that

$$L(\alpha', f) = L(\alpha'', f) = \mathcal{E}(f).$$

We construct the polynomial $P_\alpha(z)$ corresponding to

$$\alpha_k = (\alpha'_k + \alpha''_k)/2; \quad k = 1, 2, \ldots, n.$$

It is readily seen that

$$L(\alpha, f) = \mathcal{E}(f).$$

By Theorem 2, there are $n + 1$ points

$$z_1, z_2, \ldots, z_{n+1},$$

at which

$$|D_{\alpha,f}(z_k)| = \mathcal{E}(f).$$

It is proved in the usual way that

$$D_{\alpha',f}(z_k) = D_{\alpha'',f}(z_k) = D_{\alpha,f}(z_k)$$

at all these points, which, together with assumption (H), leads to a contradiction. This proves the implication (H)→(U).

Proof of the implication (U)→(H). Assume that, contrary to the assertion, condition (H) is violated, that is, there is a system of n different points $z_1, z_2, \ldots, z_n$ such that the determinant

$$|\phi_k(z_m)|, \quad k, m = 1, 2, \ldots, n,$$

is equal to zero, that is, there are

$$\gamma_1, \gamma_2, \ldots, \gamma_n$$

not equal identically to zero and satisfying the condition

$$\gamma_1\phi_k(z_1) + \gamma_2\phi_k(z_2) + \ldots + \gamma_n\phi_k(z_n) = 0$$

for any $k = 1, 2, \ldots, n$.

It is easy to construct a continuous complex function $g(z)$ on K possessing the following properties:

(a) $g(z_m) = \overline{\gamma}_m/|\gamma_m|$ at those points z_m where $\gamma_m \neq 0$;

(b) $|g(z)| \leq 1$ everywhere on K.

Taking a function $g(z)$ of this kind we put

$$f(z) = g(z)\Big\{1 - \frac{1}{H}|P_\gamma(z)|\Big\},$$

where

$$H = \max_{z \in K} |P_\gamma(z)|.$$

Lemma. $\mathcal{E}(f) = 1$.

To prove the lemma we assume that, contrary to the assertion, there is a polynomial $P_\alpha(z)$ satisfying the inequality

$$L(\alpha, f) < 1.$$

Then at those points z_m where $\gamma_m \neq 0$ we have

$$f(z_m) = \overline{\gamma}_m/|\gamma_m|,$$

$$\operatorname{Re}\{\gamma_m P_\alpha(z_m)\} = \operatorname{Re}\{\gamma_m f(z_m) + \gamma_m D_{\alpha,f}(z_m)\} =$$

$$= \operatorname{Re}\{|\gamma_m| + \gamma_m D_{\alpha,f}(z_m)\} > 0,$$

$$\operatorname{Re}\sum_{m=1}^{n} \gamma_m P_\alpha(z_m) > 0,$$

which contradicts the relation

$$\sum_{m=1}^{n} \gamma_m P_\alpha(z_m) = \sum_{k=1}^{n} \alpha_k \sum_{m=1}^{n} \gamma_m \phi_k(z_m) = 0.$$

This contradiction proves the lemma.

We now prove that for

$$|\epsilon| \leq 1/H$$

any polynomial

$$P_\beta(z) = \frac{\epsilon}{H} P_\nu(z)$$

satisfies the relation

$$L(\beta, f) = 1 = \mathcal{E}(f).$$

This follows from the estimate

$$|D_{\beta,f}(z)| \leq |f(z)| + |P_\beta(z)| = \left|1 - \frac{1}{H} P_\gamma(z)\right| + \left|\frac{\epsilon}{H} P_\gamma(z)\right| =$$

$$= 1 - \frac{1}{H}\left|P_\gamma(z)\right| + \frac{|\epsilon|}{H}\left|P_\gamma(z)\right| \leq 1.$$

We thus arrive at a contradiction to condition (U), which proves the implication (U)→(H). Theorem 3 is proved completely.

We can return to the theory of best approximation of real functions by real polynomials using the following proposition.

Theorem 4. *If the functions $\phi_k(z)$ $(k = 1, 2, \ldots, n)$ and $f(z)$ are real, then the relation*

$$L(\alpha, f) = \mathcal{E}(f)$$

(where $\mathcal{E}(f)$ is defined as the minimum of $L(\alpha, f)$ over complex polynomials $P_\alpha(z)$) is attained for a real polynomial $P_\alpha(z)$.

The proof of Theorem 4 is implied directly by the simple remark that, under the assumption of the theorem, the polynomial

$$P_{\alpha'}(z) = \tfrac{1}{2}\{P_\alpha(z) + \overline{P_\alpha(z)}\}$$

is real and satisfies the inequality

$$L(\alpha', f) \leq L(\alpha, f)$$

for any complex polynomial $P_\alpha(z)$.

References

1. A. Haar, *Math. Ann.* **78** (1918), 294–311.
2. S.N. Bernshtein, *Extremal properties of polynomials* Part 1, ONTI, Moscow, Leningrad, 1937 (in Russian).
3. N.I. Akhiezer, *Lectures on approximation theory*, Gostekhteorizdat, Moscow, Leningrad, 1947 (in Russian).
4. P. Montel, *Leçons sur les séries de polynomes à une variable complexe*, Paris, 1910, pp. 66–71.

50. ON THE BREAKAGE OF DROPS IN A TURBULENT FLOW *

The paper by M.K. Baranaeva, E.N. Teverovskii and E.L. Tregubova [1] published immediately above this paper contains interesting experimental data on the breakage of drops of fluids insoluble in water in a turbulent flow of water. The authors of [1] believe that, besides the fact that this phenomenon is of interest in its own right, it can serve as an additional means for verifying our concepts of the local structure of turbulent fluctuations.

The theory of local structure of turbulent fluctuations was elaborated by A.M. Obukhov and me. It was verified using data of direct measurements of fluctuation velocities and found convincing evidence and a number of applications in the study of scattering of sound, ultrashort radio waves and light in a turbulent atmosphere. According to the theory, the average value v_d^2 of the square of the velocity difference at two points lying at a distance d is given by the formula

$$v_d^2 = (\nu/\lambda_0)^2 f(d/\lambda_0), \qquad (\alpha)$$

where ν is the kinematic viscosity and f is a universal function with asymptotic behaviour

$$f(k) \sim k^2 \quad \text{for small } k, \qquad (\beta)$$

$$f(k) \sim k^{2/3} \quad \text{for large } k. \qquad (\gamma)$$

(We follow, when possible, the notation in [1] and use the symbol $\sim$ in the sense of approximate proportionality without writing the proportionality coefficient.)

The parameter λ_0 in formula (α) is the "internal scale" of turbulence. The components in the spectral representation of turbulent fluctuations with scales much less than λ_0 must be negligibly small. As to the velocity differences at distances d much less than λ_0, according to (β) their root mean square value decreases proportionally with d. For many physical processes these differences can also turn out to be "sufficiently large" at distances d much less than the internal scale λ_0. Therefore the discovery of physical processes generated by these differences does not at all mean that the theory put forward by Obukhov and me must be revised. [1]

* *Dokl. Akad. Nauk SSSR* **66**:5 (1949), 825–828 (in Russian).

[1] As to the attempt to justify the necessity of applying the "two thirds law" (γ) for d much less than λ_0 made in the paper [2] by E.N. Teverovskii, it cannot be recognized as sufficiently convincing. The phenomena described in [2] can be given a quite different explanation.

We now consider a turbulent flow of a "first" fluid (it was water in experiments in [1]) and a thin jet of a "second" fluid having the same density (this requirement was fulfilled to a high accuracy in the experiments in [1]), with kinematic viscosity ν' and surface tension σ at the interface between the fluids.

If σ were equal to zero and ν' were equal to ν, then from the standpoint of mechanics, the jet of the second fluid would not differ in any way from a jet of the first fluid singled out purely geometrically. The behaviour of such a jet was not wholly adequately described in [1]. Actually, this jet must not disintegrate into drops of diameter of the order of λ_0, but must be deformed to become a winding thread thinning and branching indefinitely. No limit to the mixing of the jet in the flow would be obtained in such an idealized scheme.

The limit to this mixing is due only to surface tension. Owing to the surface tension, the jet disintegrates into drops, the drops are broken to a certain limit, and the drops of sufficiently small diameter d are preserved, since for small d the breaking forces acting on them due to the velocity differences, which are of the order of v_d, are small for small d and can no longer overcome the surface tension.

The hypothesis of universality of the dimensionless local structure of turbulent flows implies that the behaviour of a drop of diameter d must depend only on the dimensionless ratios d/λ_0 and ν'/ν and the Weber number

$$\mathrm{We}_d = \sigma / v_d^2 d\rho.$$

Instead of the ratios d/λ_0 and ν'/ν we can introduce two Reynolds numbers

$$\mathrm{Re}_d = v_d d/\nu, \quad \mathrm{Re}'_d = v_d d/\nu'.$$

It seems that in the region of diameters of drops of the order of the internal scale λ_0, the number of dimensionless parameters characterizing the process cannot be reduced and there must be three such parameters:

$$(\mathrm{We}_d,\ d/\lambda_0,\ \nu'/\nu)$$

or, equivalently

$$(\mathrm{We}_d,\ \mathrm{Re}_d,\ \mathrm{Re}'_d).$$

However, the values of d much smaller than λ_0 correspond to a region where viscous forces substantially dominate the inertia forces, and there remain only two dimensionless characteristics:

$$\mathrm{We}_d \text{ and } \nu'/\nu.$$

In the region where d is much larger than λ_0, the role of the viscosity of the first fluid must be insignificant. If the viscosity ν' of the second fluid is less than or of the same order as that of the first fluid, then for d much larger than λ_0 we can neglect the viscosity of both fluids. If the viscosity of the second fluid is much larger than the viscosity of the first fluid, then the former can be neglected only for d exceeding the scale

$$\lambda_0' = (\nu'/\nu)^{3/4}\lambda_0.$$

Thus we obtain a table of dimensionless characteristics:

	$\nu' \ll \nu$ or $\nu' \approx \nu$	$\nu' \gg \nu$
$d \ll \lambda_0$	We_d and ν'/ν	
$d \approx \lambda_0$	We_d, d/λ_0 and ν'/ν	
$d \gg \lambda_0$	We_d	We_d and Re_d'
$d \gg \lambda_0'$	—	We_d

In the experiments described in [1], water was used as the first fluid and two different mixtures were used as the second fluid. The surface tension σ was almost the same in the two cases under consideration. The ratio ν'/ν was different in the two versions of the experiment but did not differ from unity by very much. As is seen from Figure 2 in [1], the effect of the difference between the values of ν' in the two versions of the experiment turned out to be within the limits of the spread of points for each of the versions separately. Therefore, for now, we do not discuss formula (3) suggested in [1], whose first term must take into account the effect of variation of the ratio ν'/ν (in [1] the symbol η denotes the viscosity $\eta = \nu'\rho$ of the *second* fluid) and consider the case of a constant ratio ν'/ν less than or exceeding unity by a small amount.

In this case, following the authors of [1], it is natural to assume that the maximum diameter d_0 of drops possessing sufficient stability to be preserved for a long time must correspond to a definite "critical" Weber number W_{d_0}. Strictly speaking, by virtue of the above, this assumption can be recognized as being justified only outside the limits of the region of the values of d_0 of the same order as λ_0, and the critical Weber number may turn out to be different in the regions $d_0 \ll \lambda_0$ and $d_0 \gg \lambda_0$. Using formulas (β) and (γ) we finally obtain:

$$d_0 = \left(\frac{\sigma}{\mathrm{We}_{\mathrm{cr}}\nu^2\rho}\right)^{1/3}\lambda_0^{4/3} \quad \text{in the case } d_0 \ll \lambda_0,$$

$$d_0 = \left(\frac{\sigma}{\mathrm{We}^*_{\mathrm{cr}}\nu^2\rho}\right)^{3/5}\lambda_0^{8/5} \quad \text{in the case } d_0 \gg \lambda_0.$$

For a tube of diameter D, outside the laminar boundary layer we have

$$\lambda_0 = g(r/D)(\nu^3 D/u_*^3)^{1/4},$$

where u_* is the so-called dynamic velocity, r is the distance from the axis of the tube, and g is a universal function. Substituting this expression for λ_0 into the above formulas for d_0 we find

$$d_0 \sim g^{4/3}\left(\frac{r}{D}\right)\left(\frac{\sigma\nu D}{\mathrm{We}_{\mathrm{cr}}\rho}\right)^{1/3} u_*^{-1} \qquad \text{in the case } d_0 \ll \lambda_0, \qquad \text{(I)}$$

$$d_0 \sim g^{8/5}\left(\frac{r}{D}\right)\left(\frac{\sigma}{\mathrm{We}^*_{\mathrm{cr}}\rho}\right)^{3/5} D^{2/5} u_*^{-6/5} \qquad \text{in the case } d_0 \gg \lambda_0, \qquad \text{(II)}$$

Since the ratio u/u_* of the average velocity in the tube to the dynamic velocity varies slowly, formula (II) corresponds to formulas (7) and (8) in [1]. Probably, when explaining the experimental results presented in [1], it would be more advisable to apply formula (I) instead of (II). However, given the accuracy of experiments attainable at present, the difference btween the exponent -1 in formula (I) and the exponent $-6/5$ in formula (II) can hardly be detected. Judging from Figure 2 in [1], direct data processing results in an exponent somewhat less than one. However, the authors of [1] believe reasonably that this may be due to the fact that at large velocities the time the drops stay in the fluid is insufficient for the breakage process to manage to terminate.

Further, the authors raise a very interesting problem: since v_d is only the root mean square value of the velocity difference at a distance d, is it possible that there can appear, however rarely, much greater differences at the same distance?

To answer this question it is necessary to know the *probability distribution* for the velocity differences at a given distance d.

If it were possible to set up experiments on the breakage of drops so that each drop could stay sufficiently long in a region with constant characteristics of the local structure, then these experiments could prove important for the further development of our concepts of the local structure of turbulence.

Unfortunately, the internal scale λ_0 in a tube varies appreciably over the cross section. Those drops that go rather far away from the axis of the tube must undergo breakage to notably less diameters than those remaining all the time near the axis.

This makes it necessary to interpret with extreme care the data in [1] on the dependence of the minimum size of the drops on the time of their stay in the flow.

In our opinion, the development of further studies in the topic under discussion must go in the following two directions.

1. The above concept of the existence of a definite lower bound for the diameters of the drops d_0 below which no further breakage occurs within the limits of a region with given characteristics of the local structure of the flow must be developed taking into account the effect produced on the diameter d_0 by the viscosity of the second fluid ν'. It is necessary to try to apply this theory as a first approximation to the calculation of various specific problems involving inhomogeneous flows.

2. Experiments must be staged to investigate the dependence of the distribution of the sizes of drops on the time of their stay in a flow with strictly constant local characteristics.

References

1. M.K. Baranaev, E.N. Teverovskii and E.L. Tregubova, 'On the measure of minimal fluctuations in a turbulent flow', *Dokl. Akad. Nauk SSSR* **66**:5 (1949), 821–824 (in Russian).
2. E.N. Teverovskii, 'On coagulation of aerosol particles in a turbulent atmosphere', *Izv. Akad. Nauk SSSR Ser. Geogr. i Geofiz.* **12**:1 (1948), 7–9 (in Russian).

51. ON DYNAMICAL SYSTEMS WITH AN INTEGRAL INVARIANT ON A TORUS *

Below we consider coordinate systems on a torus T^2 that associate with each point P a pair of real numbers (x, y) defined modulo 2π.

Analyticity of a coordinate function$f(x, y)$ is understood as its analyticity for all real x and y. Obviously, a coordinate function $f(x, y)$ is a single-valued function $f(P)$ of the point P only when it is periodic with period 2π in each of the variables x and y. We consider a dynamical system on T^2 defined in some coordinate system by a system of equations

$$dx/dt = A(x, y), \quad dy/dt = B(x, y) \tag{1}$$

and possessing an integral invariant

$$I(g) = \iint_g U(x, y)dx\,dy. \tag{1'}$$

Naturally, the functions A, B and U are assumed to be single-valued functions of the point P. Moreover, we assume that they are analytic and satisfy the conditions

$$A^2 + B^2 > 0, \quad U > 0$$

everywhere on T^2.

Under these assumptions the well-known general results of Poincaré, Denjoy and Kneser (for example, see [1], Ch.2, §2) can be supplemented in the following way.

Theorem 1. *There is an analytic transformation of coordinates that reduces the system* (1) *to the form*

$$\frac{dx}{dt} = \frac{1}{F(x, y)}, \quad \frac{dy}{dt} = \frac{\gamma}{F(x, y)}, \tag{2}$$

and the integral invariant $I(g)$ *to the form*

$$I(g) = \iint_g F(x, y)dx\,dy, \tag{2'}$$

---

* *Dokl. Akad. Nauk SSSR* **93**:5 (1953), 763–766.

where γ is a constant.

The case of a rational number γ is related to closed trajectories and, qualitatively, is quite elementary. If γ is irrational, then, as is known, all trajectories are everywhere dense in T^2 and the dynamical system is transitive (see [1]). The above assumptions of analyticity of the functions A, B and U make it possible to show that here, in a sense, the "general" case is the case of the dynamical system written as

$$du/dt = \lambda_1, \quad dv/dt = \lambda_2 \tag{3}$$

in an appropriate coordinate system, where λ_1 and λ_2 are constants. Namely, the following theorem holds.

Theorem 2. *If there exist $C > 0$ and $h > 0$ such that*

$$|m - n\gamma| \geq Ch^n \tag{4}$$

for all integers m and $n > 0$, then there exists an analytic transformation of coordinates under which the system (2) *goes into system* (3), *where*

$$\lambda_2 = \gamma\lambda_1,$$

and the integral invariant $I(g)$ takes the form

$$I(g) = K \iint_g du\, dv, \tag{3'}$$

where K is a constant.

Condition (4) is fulfilled (for an appropriate choice of the parameters C and h depending on γ) for all γ except for a set of Lebesgue measure zero (see, for example, [2]).

Obviously, when Theorem 2 is applicable, the dynamical system has a discrete spectrum with eigenfrequencies

$$\lambda_{rs} = r\lambda_1 + s\lambda_2,$$

where r and s are integers, and corresponding eigenfunctions ϕ_{rs} which, after the system is reduced to the form (3), are written as

$$\phi_{rs}(u, v) = e^{i(ru+sv)}. \tag{5}$$

By the analyticity of the coordinate transformation that reduces the system to the form (3), the eigenfunctions are analytic in the original coordinate system as well.

For irrational γ not satisfying condition (4) there are a number of other possibilities, an exhaustive discussion of which requires going over to the standpoint of the general metrical theory of dynamical systems. Here it is natural to include in the consideration not only non-analytic but also discontinuous transformations of coordinates, with the only requirement that the mapping of a coordinate torus (x, y) onto a coordinate torus (u, v) as well as its inverse mapping, be Lebesgue measurable and transform sets of measure zero to sets of measure zero (metrical absolute continuity). It is natural to carry out the consideration to within a set of measure zero and, for example, to regard a transformation as being analytic, differentiable a given number of times, or continuous when it coincides, to within a set of measure zero, with a transformation that is analytic, differentiable the given number of times, or continuous.

Theorem 3 *stated below is concerned with just these absolutely continuous (in the metrical sense) transformations considered to within a set of measure zero.*

For a correct understanding of it, one should note that two transformations of this kind reducing a dynamical system with γ irrational to the form (3) can differ only by a linear transformation

$$u' = au + bv, \quad v' = cu + dv$$

with integral coefficients and of determinant equal to ± 1.

Theorem 3. *For an appropriate choice of the irrational number γ and the analytic function $F(x, y)$, each of the following cases is possible:*

1) *system* (2) *reduces to the form* (3) *by a transformation of coordinates which is infinitely differentiable but not analytic;*

2) *system* (2) *reduces to the form* (3) *by a transformation that is differentiable k times but is not differentiable $(k+1)$ times;* 3) *reduction of the system* (2) *to the form* (3) *is possible only by means of a transformation discontinuous everywhere;*

4) *reduction of the system* (2) *to the form* (3) *is impossible.*

Obviously, in cases 1), 2), and 3) the system has a discrete spectrum with system of eigenfunctions of the form (5), and the eigenfunctions turn out to be non-analytic, not differentiable $(k+1)$ times, and discontinuous everywhere, respectively.

It is highly probable that in case 4) the spectrum must be continuous, but for now we have proved only the following theorem.

Theorem 4. *For an appropriate choice of the irrational number γ and the analytic function $F(x,y)$, the spectrum of the dynamical system is continuous.* [1]

The transformation of coordinates whose existence is stated in Theorem 2 and in assertions 1), 2) and 3) of Theorem 3 can be obtained in the following way:

$$u = \frac{2\pi}{a_0}\{\tau(x,y) + R(y-\gamma x)\}, \quad v = y + \gamma(u-x),$$

where

$$\tau(x,y) = \int_0^x F(\xi, y+\gamma(\xi - x))d\xi, \quad S(y) = \tau(2\pi, y),$$

$$a_0 = \frac{1}{2\pi}\int_0^{2\pi} S(y)dy = \frac{1}{2\pi}\int_0^{2\pi}\int_0^{2\pi} F(x,y)dx\,dy,$$

and $R(y)$ is a 2π-periodic solution of the functional equation

$$R(y - 2\pi\gamma) - R(y) = S(y) - a_0 \tag{6}$$

satisfying the condition

$$\int_0^{2\pi} R(y)dy = 0.$$

If $S(y)$ is represented in the form of a trigonometric series

$$S(y) = a_0 + \sum_{n\neq 0} a_n e^{iny},$$

then $R(y)$ can be formally written as

$$R(y) = \sum_{n\neq 0} \frac{a_n}{e^{i2\pi n\gamma} - 1} e^{iny}. \tag{7}$$

Condition (4) guarantees the convergence of the series (7) to an analytic function. The divisors involved in formula (7), which can be very small for an

---

[1] That is, there is no non-constant measurable function $\phi(P)$ (to within a set of measure zero) that varies along the trajectories according to the law

$$\phi(P_t) = e^{i\lambda t}\phi(P_0).$$

appropriate choice of the irrational number γ, account for the appearance of the anomalies mentioned in Theorem 3.

If the series

$$\sum_{n\neq 0}\left|\frac{a_n}{e^{i2\pi n\gamma}-1}\right|^2 \tag{8}$$

converges, then the spectrum is discrete and the eigenfunctions have the form (5). Under the assumption of analyticity of the function $F(x,y)$ made at the very beginning, the coefficients a_n decrease rapidly with increasing $|n|$, whereas for any irrational number γ small divisors in the series (7) may correspond only to a rarefied sequence of values of n. Therefore, when the series (8) is divergent, it should be expected that the properties of the series (7) are analogous to those of lacunary series with a divergent sum of squares of coefficients, that is, no measurable function $R(y)$ can be associated in a natural way with the series (7). It is therefore probable that in the case of a divergent series (8), reduction of the dynamical system to the form (3) is impossible. It is likely that in this case the spectrum is necessarily continuous, that is, for the dynamical systems under consideration the case of a mixed spectrum is excluded and the discrete spectrum always has exactly two independent frequencies.

In conclusion, we note that the role of "small divisors" preventing the convergence of series is well-known in celestial mechanics.

13 November 1953

References

1. V.V. Nemytskii and V.V. Stepanov, *Qualitative theory of differential equations*, 2nd ed., Fizmatgiz, Moscow, 1959 (in Russian).
2. A.Ya. Khinchin, 'Metrical properties in the theory of irrational numbers', *Uspekhi Mat. Nauk* **1** (1936), 7–32 (in Russian).

52. ON THE PRESERVATION OF CONDITIONALLY PERIODIC MOTIONS UNDER SMALL VARIATIONS OF THE HAMILTON FUNCTION *

We consider in the $2s$-dimensional phase space of a dynamical system with s degrees of freedom a region G, represented as the product of an s-dimensional torus T by a region S in an s-dimensional Euclidean space. The points of the torus will be characterized by circular coordinates $q_1, \dots, q_s$ (the replacement of q_α by $q_\alpha + 2\pi$ does not change the position of the point q), and the coordinates of a point p belonging to S will be denoted by $p_1, \dots, p_s$. Assume that in the region G the equations of motion in the coordinates $(q_1, \dots, q_s, p_1, \dots, p_s)$ have the canonical form

$$\frac{dq_\alpha}{dt} = \frac{\partial}{\partial p_\alpha} H(q,p), \quad \frac{dp_\alpha}{dt} = -\frac{\partial}{\partial q_\alpha} H(q,p). \tag{1}$$

In what follows, the Hamilton function H is assumed to depend on a parameter θ, defined for all $(q,p) \in G$, $\theta \in (-c;+c)$, and to be independent of time. In essence, the consideration below is related to real functions, but imposes rather strong conditions on the smoothness of the function $H(q,p,\theta)$, stronger than the condition of infinite differentiability. For simplicity, in what follows we assume that *the function $H(p,q,\theta)$ is analytic in the variables (q,p,θ) jointly.*

Below the summation over Greek subscripts extends from 1 to s. Ordinary vector notation is used: $(x,y) = \sum_\alpha x_\alpha y_\alpha$ and $|x| = +\sqrt{(x,x)}$. By an integral vector is meant a vector all components of which are integers. A set of points $(q,p) \in G$ with $p = c$ is denoted by T_c. In Theorem 1 it is assumed that S contains the point $p = 0$, that is, $T_0 \subseteq S$.

Theorem 1. *Let*

$$H(q,p,0) = m + \sum_\alpha \lambda_\alpha p_\alpha + \tfrac{1}{2} \sum_{\alpha\beta} \Phi_{\alpha\beta}(q) p_\alpha p_\beta + O(|p|^3), \tag{2}$$

where m and λ_α are constants, and let the inequality

$$|(n,\lambda)| \geq c/|n|^2 \tag{3}$$

* *Dokl. Akad. Nauk SSSR* **98:4** (1954), 527–530.

be fulfilled for a certain choice of the constants $c > 0$ and $\eta > 0$ and all integral vectors n. Moreover, let the determinant formed from the average values

$$\phi_{\alpha\beta}(0) = \frac{1}{(2\pi)^s} \int_0^{2\pi} \int_0^{2\pi} \Phi_{\alpha\beta}(q) dq_1 \dots dq_s$$

of the functions

$$\Phi_{\alpha\beta}(q) = \frac{\partial^2 H}{\partial p_\alpha \partial p_\beta}(q, 0, 0)$$

be non-zero:

$$|\phi_{\alpha\beta}(0)| \neq 0. \tag{4}$$

Then there exist analytic functions $F_\alpha(Q, P, \theta)$ and $G_\alpha(Q, P, \theta)$ defined for all sufficiently small θ and all points (Q, P) belonging to a neighbourhood V of the set T_0 that determine a contact transformation

$$q_\alpha = Q_\alpha + \theta F_\alpha(Q, P, \theta), \quad p_\alpha = P_\alpha + \theta G_\alpha(Q, P, \theta)$$

of V into $V' \subseteq G$ reducing H to the form

$$H = M(\theta) + \sum_\alpha \lambda_\alpha P_\alpha + O(P^2) \tag{5}$$

($M(\theta)$ does not depend on Q or P).

The significance of Theorem 1 in mechanics can easily be understood. It shows that, under conditions (2) and (3), an s-parameter family of conditionally periodic motions

$$q_\alpha = \lambda_\alpha t + q_\alpha^{(0)}, \ p_\alpha = 0,$$

existing at $\theta = 0$ cannot disappear under a small variation of the Hamilton function H; namely, the variation results only in a displacement of the s-dimensional torus T_0, along the trajectories of the motions: it is transformed into a torus $P = 0$, which is filled with trajectories of conditionally periodic motions with the same frequencies $\lambda_1, \dots, \lambda_s$.

The transformation

$$(Q, P) = K_\theta(q, p)$$

whose existence is asserted in Theorem 1 can be constructed as the limit of transformations

$$(Q^{(k)}, P^{(k)}) = K_\theta^{(k)}(q, p),$$

where the transformations

$$(Q^{(1)}, P^{(1)}) = L^{(1)}(q, p), \quad (Q^{(k+1)}, P^{(k+1)}) = L_\theta^{(k+1)}(Q^{(k)}, P^{(k)})$$

are found by a "generalized Newton's method" (see [1]). In this paper we confine ourselves to the construction of the transformation $K_\theta^{(1)} = L_\theta^{(1)}$, which makes it possible to understand the role of conditions (3) and (4) of Theorem 1. We determine the transformation $L_\theta^{(1)}$ by means of the formulas

$$\begin{aligned} Q_\alpha^{(1)} &= q_\alpha + \theta Y_\alpha(q), \\ p_\alpha &= P_\alpha^{(1)} + \theta\Big\{\sum_\beta P_\beta^{(1)} \frac{\partial Y_\beta}{\partial q_\alpha} + \xi_\alpha + \frac{\partial}{\partial q_\alpha} X(q)\Big\} \end{aligned} \tag{6}$$

(it can be readily verified that this is a contact transformation) and find the constants ξ_α and ζ and the functions $X(q)$ and $Y_\beta(q)$ by proceeding from the requirement that

$$\begin{aligned} H = m + \sum_\alpha \lambda_\alpha p_\alpha + \tfrac{1}{2} \sum_{\alpha,\beta} \Phi_{\alpha\beta}(q) p_\alpha p_\beta + \\ + \theta\{A(q) + \sum_\alpha B_\alpha(q) p_\alpha\} + O(|p|^3 + \theta|p|^2 + \theta^2) \end{aligned} \tag{7}$$

should have the form

$$H = m + \theta\zeta + \sum \lambda_\alpha P_\alpha^{(1)} + O(|P^{(1)}|^2 + \theta^2). \tag{8}$$

Substituting (6) into (7) we find

$$\begin{aligned} H = m + \sum_\alpha \lambda_\alpha P_\alpha^{(1)} + \theta\Big\{A + \sum_\alpha \lambda_\alpha \Big(\xi_\alpha + \frac{\partial X}{\partial q_\alpha}\Big)\Big\} + \\ + \theta \sum_\alpha P_\alpha^{(1)} \Big\{B_\alpha + \sum_\beta \Phi_{\alpha\beta}(q)\Big(\xi_\beta + \frac{\partial X}{\partial q_\beta}\Big) + \sum_\beta \lambda_\beta \frac{\partial Y_\alpha}{\partial q_\beta}\Big\} + \\ + O(|P^{(1)}|^2 + \theta^2). \end{aligned}$$

Thus, condition (8) reduces to the requirement that the equations

$$A + \sum \lambda_\alpha \Big(\xi_\alpha + \frac{\partial X}{\partial q_\alpha}\Big) = \zeta, \tag{9}$$

$$B_\alpha + \sum_\beta \Phi_{\alpha\beta}\Big(\xi_\beta + \frac{\partial X}{\partial q_\beta}\Big) + \sum_\beta \lambda_\beta \frac{\partial Y_\alpha}{\partial q_\beta} = 0 \tag{10}$$

hold.

We introduce the function

$$Z_\alpha(q) = \sum_\beta \Phi_{\alpha\beta}(q) \frac{\partial}{\partial q_\beta} X(q). \tag{11}$$

Expanding the functions $\Phi_{\alpha\beta}$, A, B_α, X, Y_α, and Z_α into Fourier series of the form

$$X(q) = \sum x(n) e^{i(n,q)}$$

and putting, for the sake of definiteness,

$$x(0) = 0, \quad y_\alpha(0) = 0, \tag{12}$$

we obtain the following equations for the unknown Fourier coefficients $x(n)$, $y_\alpha(n)$, $z_\alpha(n)$ and the constants ξ_α and ζ:

$$a(0) + \sum \lambda_\alpha \xi_\alpha = \zeta, \tag{13}$$

$$a(n) + i(n, \lambda) x(n) = 0 \text{ for } n \neq 0, \tag{14}$$

$$b_\alpha(0) + \sum_\beta \phi_{\alpha\beta}(0) \xi_\beta + z_\alpha(0) = 0, \tag{15}$$

$$b_\alpha(n) + \sum_\beta \phi_{\alpha\beta}(n) \xi_\beta + z_\alpha(n) + i(n, \lambda) y_\alpha(n) = 0 \text{ for } n \neq 0. \tag{16}$$

It is easy to see that the system of equations (11)–(16) possesses a unique solution under conditions (3) and (4). Condition (3) is important when finding $x(n)$ from (14) and $y_\alpha(n)$ from (16). Condition (4) is important for the determination of ξ_β from (15). Since the coefficients of the Fourier series of the analytic functions $\Phi_{\alpha\beta}$, A, and B_α decrease at least as rapidly as ρ^h ($\rho < 1$) with increasing $|n|$, condition (3) implies not only the formal solvability of equations (13)–(16) but also the convergence of the Fourier series of the functions X, Y_α and Z_α and the analyticity of these functions. The construction of further approximations encounters no new difficulties. Only the application of condition (3) in the proof of the convergence of the mappings $K_\theta^{(k)}$ to an analytic limit mapping K_θ is somewhat more intricate.

The condition of absence of "small divisors" (3) should be presumed to be fulfilled "in general" since for any $\eta > s - 1$ there exists $c(\lambda)$ such that

$$|(n, \lambda)| \geq c(\lambda)/|n|^\eta$$

at all points of the s-dimensional space $\lambda = (\lambda_1, \ldots, \lambda_s)$, except at a set of Lebesgue measure zero, for any integers $n_1, n_2, \ldots, n_s$ (see [2]). It is natural to assume that, "in general", condition (4) is also fulfilled. Since

$$\phi_{\alpha\beta}(0) = \frac{\partial \lambda_\beta}{\partial p_\alpha}(0),$$

where

$$\lambda_\beta(p) = \frac{1}{(2\pi)^s} \int_0^{2\pi} \cdots \int_0^{2\pi} \frac{dq_\beta}{dt} dq_1 \ldots dq_s,$$

is the "average frequency" corresponding to the coordinate q_β for fixed momenta $p_1, \ldots, p_s$, condition (3) means that the Jacobian of the average frequencies with respect to the momenta is non-zero.

We now proceed to the consideration of the special case when $H(p, q, 0)$ depends only on p, that is, $H(q, p, 0) = W(p)$. In this case, for $\theta = 0$ each torus T_p consists of entire trajectories of conditionally periodic motions with frequencies

$$\lambda_\alpha(p) = \partial W / \partial p_\alpha.$$

If the Jacobian

$$J = \left| \frac{\partial \lambda_\alpha}{\partial p_\beta} \right| = \left| \frac{\partial^2 W}{\partial p_\alpha \partial p_\beta} \right| \tag{17}$$

is non-zero, then Theorem 1 can be applied to almost all tori T_p. There arises the natural conjecture that for small θ the "displaced tori" appearing in accordance with Theorem 1 occupy a major part of the region G. This fact is confirmed by Theorem 2 below. When stating the latter theorem, we assume that the region S is bounded and consider the set M_θ of those points $(q^{(0)}, p^{(0)}) \in G$ for which the solution

$$q_\alpha(t) = f_\alpha(t; q^{(0)}, p^{(0)}, \theta), \quad p_\alpha(t) = g_\alpha(t; q^{(0)}, p^{(0)}, \theta)$$

of the system of equations (1) with initial conditions

$$q_\alpha(0) = q_\alpha^{(0)}, \quad p_\alpha(0) = p_\alpha^{(0)}$$

has trajectories not falling outside the region G for t varying from $-\infty$ to $+\infty$ and is conditionally periodic with periods $\lambda_\alpha = \lambda_\alpha(q^{(0)}, p^{(0)}, \theta)$, that is, has the form

$$f_\alpha(t) = \phi_\alpha(e^{i\lambda_1 t}, \ldots, e^{i\lambda_s t}), \quad g_\alpha(t) = \psi_\alpha(e^{i\lambda_1 t}, \ldots, e^{i\lambda_s t}).$$

Theorem 2. *If $H(q,p,0) = W(p)$ and the determinant* (17) *is non-zero in the region S, then for $\theta \to 0$ the Lebesgue measure of the set M_θ tends to the total measure of the region G.*

It seems that, in a sense, the "general case" is the case when the set M_θ has an everywhere dense complement for all positive θ. Complications of this kind appearing in the theory of analytic dynamical systems were indicated in my paper [3] in connection with a more specific situation.

31 August 1954

References

1. L.V. Kantorovich, 'Functional analysis and applied mathematics', *Uspekhi Mat. Nauk* **3**:6 (1948), 163 (in Russian).
2. J.F. Koksma, *Diophantische Approximationen*, Springer, Berlin, 1936.
3. A.N. Kolmogorov, 'On dynamical systems with an integral invariant on a torus', *Dokl. Akad. Nauk SSSR* **93**:5 (1953), 763–766 (in Russian). (See No. 51.)

53. THE GENERAL THEORY OF DYNAMICAL SYSTEMS AND CLASSICAL MECHANICS *

Introduction

It was a surprise to me that I would have to present a paper at the final session of the Congress in this large hall, which had been known to me rather as a place for the performance of great musical compositions of the world conducted by Mengelberg. The paper which I have prepared, without taking into account that it would occupy such an honourable position in the programme of the Congress, is devoted to a rather special range of problems. My aim is to elucidate ways of applying basic concepts and results in the modern general metrical and spectral theory of dynamical systems to the study of conservative dynamical systems in classical mechanics. However, it seems to me that the subject I have chosen may also be of broader interest, as one of the examples of the appearance of new, unexpected and profound relationships between different branches of classical and modern mathematics.

In his famous address at the Congress in 1900, D. Hilbert said that the unity of mathematics and the impossibility of its division into independent branches stem from the very nature of the science of mathematics. The most convincing evidence for the correctness of this idea is the appearance of new focal points at each stage in the development of mathematics, where, in the solution of quite specific problems, notions and methods of quite different mathematical disciplines become necessary and are involved in new interrelations. For the mathematics of the 19th century one of these focal points was the problem of integration of systems of differential equations of classical mechanics, where problems of mechanics, the theory of differential equations, problems of the calculus of variations, multidimensional differential geometry, the theory of analytic functions, and the theory of continuous groups were organically interwoven.

After the work of H. Poincaré, the fundamental role of topology for this range of problems became clear. On the other hand, the Poincaré-Carathéodory recurrence theorem initiated the "metrical" theory of dynamical systems in the

* In: *Proc. Intern. Congr. Math.* 1954, Vol. 1, 315–333; also in: *Trudy Mezhdunar. Mat. Kongres. Amsterdam*, 1954, USSR Academy of Sciences Press, Moscow, 1961, 187–208 (in Russian).

sense of the study of properties of motions holding for "almost all" initial states of the system. This gave rise to the "ergodic theory", which was generalized in different ways and became an independent centre of attraction and a point of interlacing for methods and problems of various most recent branches of mathematics (abstract measure theory, the theory of groups of linear operators in Hilbert and other infinite-dimensional spaces, the theory of random processes, etc.). At the preceding International Congress in 1950 the extensive paper by Kakutani [23] was devoted to general problems of ergodic theory.

As is well known, topological methods have found significant applications in the theory of oscillations, in particular, in the solution of quite specific problems arising in the study of automatic control systems, electrical engineering, etc. However, these real physical and technical applications deal mainly with non-conservative systems. Here the problem usually reduces to finding individual asymptotically stable motions (in particular, stable stationary points and stable limiting cycles) and studying pencils of integral curves attracted to these asymptotically stable motions.

In conservative systems, asymptotically stable motions are impossible. Therefore, for instance, the determination of individual periodic motions, however interesting it may be from the viewpoint of mathematics, has only a rather restricted real physical significance in the case of conservative systems. For conservative systems, the metrical approach is of basic importance making it possible to study properties of a major part of motions. For this purpose, contemporary general ergodic theory has elaborated a system of notions whose conception is highly convincing from the viewpoint of physics. However, up to now the progress made towards the application of these modern approaches to the analysis of specific problems of classical mechanics has been more than limited.

First we consider the following problem. Assume that a motion on an s-dimensional analytic manifold V^s is determined by a canonical system of differential equations with analytic Hamilton function $H(q_1, \ldots, q_s, p_1, \ldots, p_s)$. Let there be k single-valued analytic first integrals $I_1, \ldots, I_k$ and suppose that the conditions

$$I_1 = C_1; \ \ldots; \ I_k = C_k$$

single out an analytic manifold M^{2s-k} in the phase space Ω^{2s}. As we know, for almost all values of $C_1, \ldots, C_k$ there appears in a natural way an analytic invariant density on M^{2s-k}, which makes it possible to apply the general principles of the metrical theory of dynamical systems to motions on M^{2s-k}. It is reasonable to resort to these more modern means when, apart from the integrals $I_1, \ldots, I_k$, there are no single-valued analytic first integrals independent of the former or when their determination encounters severe difficulties and other classical methods for completing the integration of the system also prove inapplicable. In such cases it is necessary to use a qualitative approach in order to find out whether the motion on M^{2s-k} is transitive (that is, whether almost the entire manifold M^{2s-k} consists of a single ergodic set) and then, in the transitive case, to determine the nature of the spectrum or, in the absence of transitivity, to study, to within a set of measure zero (or at least to within a set of small measure), the decomposition of M^{2s-k} into ergodic sets and the nature of the spectrum on these ergodic sets.

There are only two specific problems of classical mechanics known to me where this programme has been realized to a certain degree.

1. For inertial motion on a closed surface V^2 of everywhere negative curvature,[1] Hopf found in 1939 that a motion on a three-dimensional manifold L_h^3 singled out by the condition of constancy of energy $H = h$, is transitive and the spectrum is continuous (see [8]).

2. As will be shown below, for inertial motion on analytic surfaces that are sufficiently close to a triaxial ellipsoid, the motion on L_h^3 is non-transitive and, to within a set of small measure, decomposes into two-dimensional tori T^2 on each of which the motion is transitive and the spectrum is discrete (see the end of §2).

However, I believe that the time has now come when considerably more rapid progress can be made.

§1. Analytic dynamical systems and their stable properties

Dynamical systems of classical mechanics are a special case of analytic dynamical systems with an integral invariant. The medium of such a dynamical system

---

[1] Perhaps it is useful to note that it is possible to specify in ordinary Euclidean space a closed surface V^2 of genus two and to place a finite number of centres of attraction or repulsion in its vicinity so that they generate on V^2 a potential of forces such that the motion of a material point on V^2 under the action of these external forces is mathematically equivalent to inertial motion in a metric possessing an everywhere negative curvature.

is an n-dimensional analytic manifold Ω^n (the phase space of the system). Accordingly, the admissible transformations of the coordinates $x_1, \dots, x_n$ of a point $x \in \Omega^n$ are always assumed to be analytic.

The right-hand sides of the differential equations

$$dx_\alpha/dt = F_\alpha(x_1, \dots, x_n) \tag{1}$$

determining the motion and the invariant density $M(x)$ generating an invariant measure

$$m(A) = \int_A M(x)dx_1 \dots dx_n$$

are assumed to be analytic coordinate functions. [2]

In accordance with what was said in the Introduction, we are primarily interested in canonical systems, that is, systems with $n = 2s$, where the coordinates of a point $(q,p) \in \Omega^{2s}$ are partitioned into two groups $q_1, \dots, q_s$ and $p_1, \dots, p_s$, the admissible transformations of coordinates are contact transformations, the equations of motion are canonical equations of the form

$$\frac{dq_\alpha}{dt} = \frac{\partial H}{\partial p_\alpha}, \quad \frac{dp_\alpha}{dt} = -\frac{\partial H}{\partial q_\alpha}, \tag{2}$$

and the invariant density is

$$M(q,p) = 1.$$

Particular attention will be paid to finding which of the properties of dynamical systems are "typical" for "arbitrary" F_α and M (or an "arbitrary" function $H(q,p)$ in the case of canonical systems) and which of them can manifest themselves only by way of an "exception". However, this is quite an intricate problem. The approach from the standpoint of the category of corresponding sets in the spaces of systems of functions $\{F_\alpha, M\}$ (or functions H), despite the well-known achievements in this direction obtained in the general theory of abstract dynamical systems, is of interest rather as a means for proving existence than as a direct way for solving actual problems set by researchers in physics and mechanics, however stylized and idealized their statement may be. By contrast, the approach from the standpoint of measure theory appears to be quite reasonable and natural as viewed from physics (for instance, as it

[2] In what follows, when speaking simply of a "measure" without any further stipulation, we always mean the measure m.

was set forth forcibly by von Neumann [1]), but its application is hampered by the absence of a natural measure in function spaces.

We will follow two routes. First, to obtain positive results establishing that a certain type of dynamical systems should be recognized as being essential, not "exceptional", and from any reasonable point of view, should not be "neglected" (in the way that sets of measure zero are neglected), we will use the notion of stability in the sense of preservation of a certain type of behaviour of a dynamical system under small variation of the functions F_α and M or of the function H. From this standpoint, any type of behaviour of a dynamical system for which there exists at least one example of its stable realization should be recognized as being important and not negligible. According to the accepted approach from the viewpoint of analytic functions, "smallness" of variation of a function $f_0(x)$ will be understood in the sense that it may be replaced by a function

$$f(x) = f_0(x) + \theta\phi(x,\theta)$$

with a small value of the parameter θ, where the function ϕ is analytic with respect to the variables $x_1, \ldots, x_n$, θ jointly. This approach to the problem is open to criticism, but it allows for some interesting results. When it is allowable to confine ourselves to nearness of the functions f_0 and f in the sense of the nearness of their derivatives of a bounded order, a special stipulation will be made.

To obtain negative results, establishing the non-essential, exceptional nature of a certain phenomenon, we will use a somewhat artificial technique. Namely, if a class K of functions $f(x)$ admits of the introduction of a finite number of functionals

$$F_1(f),\ F_2(f),\ \ldots,\ F_r(f)$$

on it, which in a sense can naturally be recognized as assuming, "generally speaking, arbitrary" values

$$F_1(f) = C_1, \ldots, F_r(f) = C_r$$

belonging to a region in the r-dimensional space of points $C = (C_1, \ldots, C_r)$, we will regard any phenomenon that can occur only for C belonging to a set of r-dimensional Lebesgue measure zero as being exceptional and "negligible".

We begin the review of specific results with an application of the above idea to the study of dynamical systems whose phase space is a two-dimensional torus.

§2. Dynamical systems on a two-dimensional torus and some canonical systems with two degrees of freedom

Here and in what follows, the points of the torus T^2 are assumed to be determined by circular coordinates x_1, x_2 (the point x does not change when x_α is replaced by $x_\alpha + 2\pi$). According to what was said earlier, the functions F_α on the right-hand sides of the equations

$$\frac{dx_1}{dt} = F_1(x_1, x_2), \quad \frac{dx_2}{dt} = F_2(x_1, x_2)$$

and the invariant density $M(x_1, x_2)$ are assumed to be analytic, to satisfy the additional conditions

$$F_1^2 + F_2^2 > 0, \; M > 0, \tag{1}$$

and for simplicity, the normalization condition $m(T^2) = 1$. We introduce the average rotation frequencies

$$\lambda_1 = \int_{T^2} F_1(x)dm, \quad \lambda_2 = \int_{T^2} F_2(x)dm.$$

In the case under consideration, a slight strengthening of results by Poincaré, Denjoy and Kneser leads to the conclusion that the equations of motion can be reduced to the form

$$\frac{dx_1}{dt} = \lambda_1 M(x_1, x_2), \quad \frac{dx_2}{dt} = \lambda_2 M(x_1, x_2)$$

by means of an analytic transformation of coordinates.

It is well-known that in the case of an irrational value of the quotient

$$\gamma = \lambda_1/\lambda_2,$$

all the trajectories are everywhere dense and the measure m is transitive. Moreover, following Markov [2], it can easily be proved that when γ is irrational, the dynamical system is strictly ergodic, that is, contains exactly a single ergodic set E whose points have their own measure equal to

$$\mu_E = cm,$$

where c is a constant. The natural assertion that, "in general", under conditions (1) a motion on a two-dimensional torus possesses all these properties is an example of the application of the above-mentioned principle of neglecting those cases in which a finite system of functionals (λ_1 and λ_2 in the case under consideration) assumes values belonging to a set of measure zero (the set of points (λ_1, λ_2) with a rational quotient γ in the case under study).

In [3] I managed to make some further progress. Namely, I proved that if there exist $c > 0$ and $h > 0$ such that the inequality

$$|r - s\gamma| \geq ch^s \tag{2}$$

holds for all integers r and s, then the equations of motion can be brought to the form

$$dx_1/dt = \lambda_1, \quad dx_2/dt = \lambda_2 \tag{3}$$

using an analytic transformation of coordinates.

As is known from the theory of Diophantine approximations, condition (2) is fulfilled (for appropriate c and h) for almost all irrational numbers γ. Hence, except for the cases when γ is approximated "abnormally well" by fractions r/s, under conditions (1) an analytic dynamical system with an integral invariant on a torus T^2 necessarily admits of only almost periodic or, even more specifically, "conditionally periodic" motions with two independent frequencies λ_1 and λ_2.

We know many problems of classical mechanics with two degrees of freedom ($s = 2$ and $n = 4$) where, owing to the existence of two first integrals I_1 and I_2 which are single-valued on the entire manifold Ω^4, the four-dimensional manifold Ω^4 is decomposed, excluding some exceptional at most three-dimensional manifolds, into two-dimensional manifolds

$$L^2_{C_1C_2} = L^2 \quad (I_1 = C_1, \; I_2 = C_2).$$

Since at stationary points the four equations

$$\frac{\partial H}{\partial q_1} = \frac{\partial H}{\partial q_2} = \frac{\partial H}{\partial p_1} = \frac{\partial H}{\partial p_2} = 0$$

hold, in the case of analytic function H the set of these points on Ω^4 is at most countable. Therefore they may fall on a manifold L^2 only by way of exception. It follows that almost all compact manifolds L^2 must be tori (as orientable compact two-dimensional manifolds admitting of a vector field without zero vectors).

As we know, problems of classical mechanics of the type in question are always integrable. It is the qualitative investigation of special problems of this type (motion under the action of gravity on a surface of revolution, inertial motion on the surface of a triaxial ellipsoid, etc., the motion of a point in a plane under the Newtonian attraction of two immobile centres, etc.) that leads to a great number of examples of the decomposition of the space Ω^4, mainly into tori T with windings of trajectories of conditionally periodic motions with two independent frequencies λ_1 and λ_2, filling them everywhere densely. Generally speaking, between these tori there lies an everywhere dense set of tori decomposing into closed trajectories, owing to the commensurability of the frequencies, and in a discrete manner, decomposing into at most three-dimensional singular manifolds on which, in particular, rest points and so-called asymptotic motions are located. The consideration of these integrable problems provides many interesting examples of rather complicated decompositions of the phase space Ω into ergodic sets and a remainder of "irregular points", lying on the trajectories of asymptotic motions. [3]

In my paper [3] mentioned above, it is pointed out that for exceptional irrational values of γ (not satisfying condition (2)) there are in fact a number of new possibilities which sometimes seem rather unexpected for analytic systems (this will be discussed later). However, in the above-mentioned problems of classical mechanics these exceptional cases do not appear, for the following quite simple reason: in these problems the transformation to circular coordinates ξ_1, ξ_2 on the tori T^2 and to the parameters C_1 and C_2 of the tori is performed by way of contact transformations. Therefore the equations preserve the canonical form

$$\frac{d\xi_\alpha}{dt} = \frac{\partial}{\partial C_\alpha} H, \quad \frac{dC_\alpha}{dt} = -\frac{\partial}{\partial \xi_\alpha} H,$$

and since the invariance of the tori T^2 takes place only when

$$dC_1/dt = dC_2/dt = 0,$$

[3] In this connection we note that the very instructive qualitative analysis of the problem of attraction by two immobile centres carried out in the well-known treatise by Charlier has proved to be incomplete and partly incorrect and has twice been corrected [4], [5].

the function H depends only on C_1 and C_2, which results in equations (3) with constant λ_1 and λ_2 on each of the tori T^2 without any exception.

Therefore the real significance for classical mechanics of the above analysis of dynamical systems on T^2 depends on whether there are sufficiently important examples of canonical systems with two degrees of freedom, not integrable by classical methods, and in which an essential role is played by invariant (with respect to the transformations S^t) two-dimensional tori.

To show that such examples exist we consider, following the investigation of the neighbourhood of an elliptic periodic motion carried out by Birkhoff [6], a system with circular coordinates q_1, q_2 and momenta p_1, p_2 for which

$$H(q,p) = W(p).$$

The equations of motion have the form

$$\frac{dq_\alpha}{dt} = \frac{\partial W}{\partial p_\alpha}, \quad \frac{dp_\alpha}{dt} = 0.$$

It is obvious that the tori T_c^2 defined by the conditions

$$p_1 = c_1, \quad p_2 = c_2$$

are invariant, and on each of them there occurs a conditionally periodic motion

$$\frac{dq_\alpha}{dt} = \lambda_\alpha(c) = \frac{\partial}{\partial c_\alpha} W(c_1, c_2)$$

with two frequencies depending on c. Assume that the Jacobian of the frequencies λ_α with respect to the momenta p_α is non-zero:

$$\left|\frac{\partial \lambda_\alpha}{\partial p_\beta}\right| = \left|\frac{\partial^2 W}{\partial p_\alpha \partial p_\beta}\right| \neq 0.$$

It turns out that in this case the decomposition of the region under consideration, lying in the four-dimensional space Ω^4, into two-dimensional tori T^2 is basically stable with respect to small variations in H of the form

$$H(q,p,\theta) = W(p) + \theta S(q,p,\theta).$$

To give a precise statement, we consider a region $G \subseteq \Omega^4$ determined by the condition $p \in B$, where B is a bounded region in the p-plane. Under the assumption that the functions W and S are analytic and condition (4) holds, it is possible to prove the following: for any $\epsilon > 0$ there exists $\delta > 0$ such that for $|\theta| < \delta$, in the dynamical system

$$\frac{dq_\alpha}{dt} = \frac{\partial}{\partial p_\alpha} H(q,p,\theta), \quad \frac{dp_\alpha}{dt} = -\frac{\partial}{\partial q_\alpha} H(q,p,\theta)$$

the entire region G, except for a set of measure less than ϵ, consists of invariant two-dimensional tori T^2 on each of which the motion is determined by the equations

$$d\xi_1/dt = \lambda_1, \quad d\xi_2/dt = \lambda_2$$

in appropriate circular coordinates (depending analytically on q, p), where λ_1 and λ_2 are constant on each T^2, that is, the motion is conditionally periodic with two periods.

The method of proof consists in studying the behaviour of the original tori T_c^2 with frequencies $\lambda_\alpha(c)$ satisfying condition (2) under variation of θ and establishing that for sufficiently small ϵ each of the tori is not destroyed and is merely displaced in Ω with preservation of trajectories of conditionally periodic motions with constant frequencies λ_α on it.

Probably many of you will already have guessed that, in essence, what we are talking about is a certain modification of the idea of the possibility of avoiding the appearance of abnormal "small divisors" when calculating disturbed orbits, which has been extensively discussed in the literature on celestial mechanics. However, in contrast to ordinary perturbation theory, we obtain exact results instead of the conclusion that the series of some approximation of finite order (relative to θ) are convergent. This is achieved because instead of calculating the disturbed motion for fixed initial conditions, we change the initial conditions themselves so that, with varying θ, we always deal with motions having normal frequencies λ_α (in the sense of condition (2)).

We make the following three additional remarks on what has been said.

1. The theorem on the reducibility of motions on T^2 to the form (3) can also be proved under the conditions that the functions F_α and M are differentiable a sufficiently large finite number of times (naturally, with a corresponding weakening of the conclusion). By contrast, the theorem on the preservation of tori in Ω^4 seems to require that the functions $W(p)$ and $S(q,p,\theta)$, if not analytic, have an infinite number of derivatives obeying some constraints on their order of growth.

2. The exceptional set of measure $< \epsilon$ mentioned in the second theorem may actually turn out to be everywhere dense, and may probably be of positive measure for arbitrarily small θ. This phenomenon is analogous to the "instability zones" discovered by Birkhoff when studying neighbourhoods of elliptic periodic trajectories [6].

3. As one of the special cases to which all that we have said earlier applies, we can indicate inertial motion on an analytic surface that is close to a triaxial ellipsoid.

§3. Are dynamical systems on compact manifolds "in general" transitive, and should we regard the continuous spectrum as the "general" case and the discrete spectrum as an "exceptional" case?

The hypothesis that the transitive case and the case of a continuous spectrum (mixing) are of predominant significance have more than once been stated in connection with the "ergodic" hypotheses in physics. In application to canonical systems, both hypotheses should naturally be attributed only to $(2s-1)$-dimensional invariant manifolds L_h^{2s-1} which are defined by the condition of constancy of energy

$$H = h,$$

and should relate only to the case of compact L_h^{2s-1}, since on non-compact L_h^{2s-1} even in the simplest problems there are "receding" trajectories (usually dominating in the sense of measure), which will be discussed later (in §4). When the first hypothesis is discarded, it is natural to relate the second one not to the entire manifold Ω^n (or L_h^{2s-1} in the case of canonical systems) but to the ergodic sets into which Ω^n is decomposed (on the understanding, of course, that we neglect ergodic sets whose union is of measure zero).

In application to analytic canonical systems, the answers to both questions are negative, since the theorem on the stability of the decomposition into tori which we stated for systems with two degrees of freedom remains valid for any number of degrees of freedom as well. If

$$H(q,p,\theta) = W(p) + \theta S(q,p,\theta)$$

in a $2s$-dimensional toroidal layer G in the phase space Ω^{2s}, then for $\theta = 0$ this layer decomposes in an obvious manner into invariant s-dimensional tori T_p^s on each of which the motion is conditionally periodic with s periods, and in the case

$$\left|\frac{\partial^2 W}{\partial p_\alpha \partial p_\beta}\right| \neq 0$$

the periods are independent on almost all tori T_p^s in the sense that

$$(n, \lambda) = \sum_\alpha n_\alpha \lambda_\alpha \neq 0$$

for any integers n_α (and therefore the trajectories wind around the torus everywhere densely), the s-dimensional Lebesgue measure on T^s is transitive, and the entire torus is a single ergodic set. Theorems 1 and 2 in my paper [22] assert that in the above-described situation the only change in the entire pattern for small θ is that some of the tori corresponding to systems of frequencies for which the expression (n, λ) decreases too rapidly with increasing

$$|n| = \sqrt{\sum n_\alpha^2}$$

may disappear while the majority of the tori T_p^s, retaining the character of motions occurring on them, are somewhat displaced in Ω^{2s}, and still fill for small θ the region G to within a set of small measure. Thus , under small variations of H the dynamical system remains non-transitive and the region G continues to be decomposable, to within a residual set of small measure, into ergodic sets with discrete spectra (of the indicated specific nature).

In this connection it is interesting to note that some physicists (see, for example, [7]) put forward the hypothesis that the "general case" of a canonical dynamical system without receding trajectories is this very decomposition of Ω^{2s} into s-dimensional tori T^s carrying conditionally periodic motions with s periods. This idea seems to be based only on the predominant attention given to linear systems and to a limited group of integrable classical problems, and it should in any case be noted that the methods for proving the above-mentioned theorem are essentially related to this specific decomposition of Ω^{2s} into tori T^s and are inapplicable to the decomposition into tori of any other dimension $r > s$ or $r < s$.

In its general form the hypothesis above can hardly be sustained, since it is very likely that for any s there exist examples of canonical systems with s degrees of freedom and stable transitivity and mixing on manifolds L_h^{2s-1}. I have in mind the motion along geodesics on compact manifolds V^s of constant negative curvature, that is, dynamical systems with

$$H(q,p) = \sum_{\alpha} g_{\alpha\beta}(q) p_\alpha p_\beta, \tag{1}$$

where q_α are coordinates on a compact manifold V^s of constant negative curvature and $g_{\alpha\beta}$ are the components of the metric tensor on V^s.

The stability of negative curvature with respect to small variations of the functions $g_{\alpha\beta}(q)$ does not require any elucidation. Here the only difficulties are that the variation of the functions $g_{\alpha\beta}(q)$ is not the only possible way of varying the function $H(q,p)$ and transitivity and mixing for $s > 2$ are proved only for the case of constant curvature, whereas under variation of $g_{\alpha\beta}$ the curvature ceases to be constant. In the case $s = 2$, for which transitivity has been proved for variable curvature as well, the latter difficulty disappears. As to the former, it disappears if we confine ourselves to functions $H(q,p)$ having the form

$$H(q,p) = U(q) + \sum_{\alpha\beta} g_{\alpha\beta}(q) p_\alpha p_\beta \tag{2}$$

(which, as a matter of fact, are dealt with in classical mechanics), since systems of the form (2) can be reduced to systems of the form (1) by transferring to a new metric.

If we recall what was said earlier about inertial motion on surfaces close to a triaxial ellipsoid, we arrive at the conclusion that even in the simplest problems of classical mechanics it is necessary to recognize as being stable, and hence equally deserving prime attention, at least the two above-mentioned cases, one related to transitivity on manifolds of constant energy and involving a continuous spectrum, and the other characterized by the absence of transitivity and involving a primarily discrete spectrum.

No similar results regarding the stability of a certain general type of behaviour of non-canonical dynamical systems with an integral invariant and a compact phase space Ω^n are known to me.

§4. Some remarks on the non-compact case

The distinctive property of the non-compact case is the possibility of the existence of trajectories receding from every compact region in Ω as $t \to \infty$ and $t \to -\infty$. Here we present some general concepts of ergodic theory suitable for arbitrary continuous flows S^t in locally compact spaces Ω. Since one-sided approach to infinity is possible only for trajectories constituting a set of measure zero, a receding point x is immediately defined by the requirement that for any compactum K there exist T such that all points $S^t x$ with $|t| > T$ lie outside K. Let Ω'' denote the set of all receding points. For a detailed analysis of specific classical dynamical systems it is advisable to construct an "individual ergodic theory" not in the purely metrical version presented in the book by Hopf [9], but following earlier works by Hopf [10] and Stepanov [11] and in certain respects following directly the presentation in the memoir by Krylov and Bogolyubov [12], although it deals with the compact case.

In this presentation, as in the compact case, the notion of a regular point remains of fundamental importance; by this I mean a point x such that there exists an invariant measure μ possessing the following properties:

1. $\mu(\Omega - I_x) = 0$, where I_x is the closure of the trajectory passing through x;

2. $\mu(V_y) > 0$ for any neighbourhood V_y of a point $y \in I_x$;

3. for any two continuous functions $f(x)$ and $g(x)$ with supports in compact sets the relation

$$\lim \frac{\int_a^T f(S^t x)dt}{\int_a^T g(S^t x)dt} = \frac{\int_\Omega f d\mu}{\int_\Omega g d\mu}$$

holds, provided that

$$\int_\Omega g d\mu \neq 0;$$

4. the measure μ is transitive.

Since no normalization condition is imposed, the measure μ is determined by the point x only to within a constant factor. Nevertheless, we will denote it by μ_x and call it the "individual measure" associated with the point x. Because of this, the definition of ergodic sets is slightly modified; namely, two points x and x' are said to belong to the same ergodic set if their individual measures are equal in the sense of equality to within a constant factor. Thus, the entire set Ω' of regular points is represented as a sum of ergodic sets:

$$\Omega' = \sum \epsilon.$$

Of course, the measures μ_ϵ are also determined by an ergodic set to within a constant factor.

The individual ergodic theorem asserts that

$$\Omega = \Omega' + \Omega'' + N, \text{ where } \lambda(N) = 0$$

for any invariant measure λ. For our purposes, however, all that is of primary interest is the fact that we always have

$$m(N) = 0.$$

Any transitive invariant measure μ is either a measure μ_ϵ of an ergodic set ϵ or has the form

$$\mu(A) = \nu_I \ (A \cap I),$$

where ν_I is the "temporal" measure on the receding trajectory I. In contrast to the latter, trivial case, it is natural to call measures of the former type ergodic, since they correspond to a set ϵ_μ such that

$$\mu_{\epsilon_\mu} = \mu.$$

The arguments which, in the case of a compact Ω, can be given in favour of the idea that a compact dynamical system of "general type" is transitive, when applied to non-compact dynamical systems, leads to the hypothesis that "in general" one of the following two cases holds: either the system is dissipative (that is, almost all its points recede) or the measure m is ergodic (obviously, in the latter case the receding points constitute only a set of measure zero).

This hypothesis is sometimes also applied to individual classical problems in the following form: if there are a certain number of first integrals for the given problem and there are no reasons to expect the discovery of new ones, then it is assumed that transitivity is likely to take place on the manifolds determined by specifying the values of these known first integrals. To justify this approach it can be noted that, according to the studies of Hedlund and Hopf, this alternative always takes place for motions along geodesics of constant negative curvature.

When it is known in advance that there is a set of positive measure consisting of receding points, then in accordance with what has been said, the conjecture arises that the system is dissipative. Probably Birkhoff's assumption that the three-body problem is dissipative is based on some consideration of this kind.

However, it is likely that the methods indicated in [22] can be applied to canonical systems to construct examples in which a dissipative part of positive measure and a positive region G filled out primarily by s-dimensional invariant tori are simultaneously present in Ω^{2s}.

We note that, among more elementary problems, particular problems dealing with receding trajectories of various specific types attract little attention of specialists in the qualitative theory of differential equations. A spectacular example of this is the fact that a disproof of Chazy's assertions that no "exchange" and "capture" are possible in the three-body problem [17, 18] was first carried out in a way which is cumbersome (and logically unconvincing without precise error estimates), using numerical integration (see Bekker [19] and Shmidt [20]), and only recently has Sitnikov [21] constructed an example of "capture" in a very simple manner and almost without calculations.

§5. Transitive measures, spectra, and eigenfunctions of analytic systems

We say that a measure μ in Ω^n is analytic if it can be represented in the form

$$\mu(A) = \int_{V^k \cap A} f(\xi) d\xi_1 \dots d\xi_k,$$

where V^k is a k-dimensional ($k \leq n$) manifold locally closed in Ω^n and the function f of the coordinates ξ_α on V^k (depending analytically on the coordinates x_α in Ω^n) is analytic.

The manifold V^k is uniquely determined by the measure μ (unless it is identically zero). Therefore, the number k can also be called the dimension of the measure μ.

We are particularly interested in transitive measures. In this case the manifold V^k must be invariant. Invariant manifolds of the same dimension are of no interest, and those of different dimensions can only be entirely contained one in another (a manifold of lower dimension is contained in one of higher dimension). Every invariant manifold carries at most one transitive measure. According to what has been said, a system with analytic transitive measure has a comparatively simple structure.

Up to now only analytic transitive measures have been known for analytic systems. Only recently has Grabar' [13] constructed an analytic analogue of Markov's example (that is, an analytic, irreducible, but not strictly ergodic dynamical system) and thus has given an example of a non-analytic transitive measure in an analytic system. However, it is not improbable that the union of all non-analytic ergodic sets is always negligible in the sense of the basic measure m.

Ergodic sets are uniquely determined by their measures μ_ϵ which by their very definition are transitive.

As to the ergodic sets corresponding to analytic transitive measures (not reducible to a measure μ_ϵ of a single trajectory), we only note that for an analytic measure μ_ϵ the ergodic set lies on the support V^r of the measure μ_ϵ and is everywhere dense in it, but even in some simple classical examples the difference $V^r - \epsilon$ may also be everywhere dense in V^r.

The spectral properties of transitive measures in analytic systems have not been studied enough.

Discrete spectra have so far been found only for a finite basis of independent frequencies

$$\lambda_1, \lambda_2, \ldots, \lambda_k,$$

and for analytic measures the number of independent frequencies coincides with the dimension in all known cases.

Quite recently, the continuous spectrum has been determined completely by Gel'fand and Fomin [14, 15] for some cases of motion along geodesics on surfaces of constant negative curvature. In these cases it turned out to be a countably-multiple Lebesgue spectrum.

It is not impossible that only these cases (a discrete spectrum with a finite number of independent frequencies and a countably-multiple Lebesgue spectrum) are admissible for analytic transitive measures or that, in a sense, only they alone are general typical cases.

For non-analytic transitive measures it is more likely that their structure can be quite arbitrary. This would be beyond doubt if an analytic analogue of Kakutani's theorem [16] on isometric embedding of an arbitrary flow in the flow of a continuous dynamical system were established.

With regard to eigenfunctions, we discuss only an example of an analytic dynamical system on a two-dimensional torus T^2 with a discrete spectrum and everywhere discontinuous eigenfunctions. However, in its origin, this example, which relates to the fact that the quotient $\gamma = \lambda_1/\lambda_2$ of average frequencies is approximated abnormally well by rational fractions r/s, indicates that we are rather dealing not with a typical but with an exceptional phenomenon.

To elucidate the problem in greater detail, let us consider again equations of motion on a two-dimensional torus and introduce a parameter θ varying within certain limits $[\theta_1;\theta_2]$:

$$dx_\alpha/dt = F_\alpha(x_1, x_2, \theta).$$

The functions $F_\alpha(x_1, x_2, \theta)$ are assumed to be analytic. It is obvious that in this case the quotient of the average frequencies $\gamma(\theta)$ also depends analytically on θ. If $\gamma(\theta)$ is not constant, then the set R of those values of θ for which the system can be transformed analytically to the form

$$d\xi_\alpha/dt = \lambda_\alpha$$

occupies almost the entire closed interval $[\theta_1, \theta_2]$. After the inverse transformation to the original coordinates x_1, x_2 is performed, the eigenfunctions

$$\phi_{mn} = e^{i(m\xi_1 + n\xi_2)}$$

become analytic functions of x_1 and x_2 for $\theta \in R$. However, generally speaking, even on the set R they are everywhere discontinuous functions of θ, and the discontinuity cannot be eliminated by deleting a set of measure zero from R. These factors are much more important than the fact that $\phi_{mn}(x_1, x_2, \theta)$ can be defined even at some points of the residual set $[\theta_1, \theta_2] - R$ of measure zero by admitting non-analyticity and discontinuity in x_1 and x_2.

Probably, the dependence of $\phi_{mn}(x_1, x_2, \theta)$ on the parameter θ in R is related to a class of functions of the type of Borel's monogenic functions [24] and, despite its everywhere discontinuous nature, admits of an investigation by some appropriate analytic means.

Conclusion

I will consider my objective accomplished if I have managed to convince the audience that, in spite of the great difficulties and the limited nature of the results already obtained, the problem I have posed of using general notions of modern ergodic theory for qualitatively analyzing motion in analytic and, particularly, canonical dynamical systems deserves great attention of scientists capable of comprehending the many-sided interrelations with the most varied branches of mathematics revealed here. In conclusion, I wish to express my gratitude to the Organizing Committee of the Congress for giving me the opportunity to read this paper and for their kind help in reproducing the abstract with formulas and bibliographic references, and to all those present for the attention shown to me on the final day of the Congress, when everybody is already saturated with the enormous amount of papers presented during the past few days.

References

1. J. von Neumann, *Mathematische Grundlagen der Quantenmechanik*, Berlin, 1932.
2. A.A. Markov, 'Almost periodicity and harmonizability', in *Proc. Second Math. Congress USSR, Vol. 2, USSR Acad. Sci.*, Leningrad-Moscow, 1935, pp. 227– 231 (in Russian).
3. A.N. Kolmogorov, 'On dynamical systems with an integral invariant on a torus', *Dokl. Akad. Nauk SSSR* **93**:5 (1953), 763–766 (in Russian). (See No. 51.)
4. H.J. Tallquist, 'Über die Bewegung eines Punktes, welcher von zwei festen Zentren nach dem Newtonschen Gesetze angezogen wird', *Acta Soc. Sci. Fenn. A, NS* **1**:5 (1927).
5. G.K. Badalyan, 'On the shape of the trajectory in the problem of two immobile centres', in *Proc. Second Math. Congress USSR, Vol. 2, USSR Acad. Sci.*, Moscow, 1935, 239–241 (in Russian).
6. G.D. Birkhoff, *Dynamical systems*, New York, 1927.
7. L. Landau and L. Pyatigorskii, *Mekhanika* (Mechanics), Gostekhizdat, Moscow, Leningrad, 1940 (in Russian).
8. E. Hopf, 'Statistik der geodätischen Linien in Mannigfaltigkeiten negativer Krümmung', *Ber. Verh. Sächs. Akad. Wiss. Leipzig* **91**:3 (1939), 261–304.
9. E. Hopf, *Ergodentheorie*, Springer, Berlin, 1937.
10. E. Hopf, 'Zwei Sätze über den wahrscheinlichen Verlauf der Bewegungen dynamischer Systeme', *Math. Ann.* **103** (1930), 710–719.
11. W. Stepanoff, 'Sur une extension du théorème ergodique', *Comp. Math.* **1** (1936), 239.
12. N. Krylov and N. Bogolyubov, 'La théorie générale de la mesure dans son application à l'étude des systèmes dynamiques de la mécanique non

linéaire', *Ann. Math.* **38** (1937), 65–113.

13. M.I. Grabar', 'On strongly ergodic dynamical systems, *Dokl. Akad. Nauk SSSR* **95**:1 (1954), 9–12 (in Russian).
14. I.M. Gel'fand and S.V. Fomin, 'Unitary representation of Lie groups and geodesic flows on surfaces of constant negative curvature', *Dokl. Akad. Nauk SSSR* **76** (1951), 771–774 (in Russian).
15. I.M. Gel'fand and S.V. Fomin, 'Geodesic flows on manifolds of constant negative curvature', *Uspekhi Mat. Nauk* **7**:1 (1952), 118–137 (in Russian).
16. S. Kakutani, 'Representation of measurable flow in Euclidean 3-space', *Proc. Nat. Acad. Sci. USA* **28**:1 (1942), 16–21.
17. I. Chazy, 'Sur l'allure finale du mouvement dans le problème des trois corps', *J. Math.* **8** (1929), 353–380.
18. I. Chazy, 'Sur l'allure finale du mouvement dans le problème des trois corps', *Bull. Astron.* **8** (1952), 403–436.
19. L. Bekker, 'On capture orbits', *Monthly Notices Roy. Astron. Soc.* **80**:6 (1920), 590–597.
20. O.Yu. Schmidt, 'On possible capture in celestial mechanics', *Dokl. Akad. Nauk SSSR* **58**:2 (1947), 213–216 (in Russian).
21. K.A. Sitnikov, 'On possible capture in the three body problem', *Mat. Sb.* **32**:3 (1953), 693–705 (in Russian).
22. A.N. Kolmogorov, 'On the preservation of conditionally periodic notions under small variation of the Hamilton function', *Dokl. Akad. Nauk SSSR* **98**:4 (1954), 527–530 (in Russian).
23. S. Kakutani, 'Ergodic theory' in: *Proc. Intern. Congress. Math.* Vol. 2, 1950, 128–142.
24. E. Borel, *Leçons sur les fonctions monogènes uniformes d'une variable complexe*, Paris, 1917.

54. SOME FUNDAMENTAL PROBLEMS IN THE APPROXIMATE AND EXACT REPRESENTATION OF FUNCTIONS OF ONE OR SEVERAL VARIABLES *

1. The difficulty of indicating to within ϵ a function f belonging to a class F can be considered from the standpoint of the "amount of information" contained in the indication. In this approach, a natural characteristic of the class F is the function

$$I_F^a(\epsilon) = \log N_F^a(\epsilon),$$

where $N_F^a(\epsilon)$ is the minimum number of points in an ϵ-net in F. This paper reviews earlier published and recently obtained estimates for the rate of growth of the function $I_F^a(\epsilon)$, as $\epsilon \to 0$, for some classes of analytic functions and functions possessing a given number of derivatives.

Further development of these studies seems to be important, in particular, for the correct estimation of the possibilities of various methods used in computational mathematics for approximately representing functions, their implementation on computers, and their storage in the computer memory.

In the estimation of the capacity of communication channels, transmitting signals in the form of continuous functions of time belonging to a class F, there naturally arises the problem of determining the maximum number $N_F^c(\epsilon)$ of "well distinguishable signals", that is, functions of class F lying at pairwise distances $> \epsilon$ in an appropriate metric. In these cases the asymptotic behaviour of the function $N_F^c(\epsilon)$ is quite similar to that of the function $N_F^a(\epsilon)$. A sufficiently thorough investigation of the asymptotic behaviour of

$$I_F^c(\epsilon) = \log N_F^c(\epsilon)$$

may probably facilitate the elucidation of some difficult problems of the theory of data transmission by communication channels (for example, the clarification of the actual meaning of the so-called Kotel'nikov theorem) even without resorting to notions of probability theory.

2. The second part of this paper is devoted to a number of special problems in the approximate and exact representation of functions of several variables by means of various formulas involving only arbitrary functions of a smaller

* In: *Proc.* III *Math. Congress USSR* Vol. 2, MGU Press, Moscow, 1956, 28–29.

number of variables. One of them is the problem of approximating a function of two variables by "nomographable" functions, that is, functions

$$z = \phi(x, y)$$

that can be defined by the relation

$$\begin{vmatrix} a(x) & b(x) & 1 \\ c(y) & d(y) & 1 \\ e(z) & f(z) & 1 \end{vmatrix} = 0.$$

The various possible approaches to problems of this kind are described and the results obtained by different authors are systematized. As an example of unexpected facts concealed here even in the case of the simplest problems we can mention two recent results by V.I. Arnol'd relating to the representation of functions of two variables by functions of the form

$$\chi(\phi(x) + \psi(y)) \tag{1}$$

with continuous χ, ϕ and ψ. Let

$$E(f) = \inf_{\chi, \phi, \psi} \sup_{x, y} |f(x, y) - \chi(\phi(x) + \psi(y))|.$$

V.I. Arnol'd proved that there exist: a) continuous functions with $E(f) > 0$; and b) continuous functions with $E(f) = 0$ which, however, cannot be represented in the form (1).

3. The special problems in §2 include Hilbert's problem on the existence of a function of several variables that is not representable by any finite superposition, however complicated, of continuous functions of a smaller number of variables. In application to functions of *four* variables, an unexpected result has been obtained; namely, any continuous function of four variables is representable in the form

$$f(x_1, x_2, x_3, x_4) = \sum_{r=1}^{4} \chi_r(x_4, \phi_r(x_1, x_2, x_3), \psi_r(x_1, x_2, x_3))$$

where the functions χ_r, ϕ_r and ψ_r are continuous. It is yet unknown whether an arbitrary function of three variables can be represented as a superposition of a finite number of continuous functions of two variables (if this were proved, a complete solution to the so-called 13th problem of Hilbert would be obtained). It has only been proved that any continuous function of three variables $f(x_1, x_2, x_3)$ can be approximated with arbitrary precision by a function of the form

$$\tilde{f}(x_1, x_2, x_3) = \sum_{r=1}^{2} \theta_r(x_3, \chi_r(\phi_r(x_3, x_1), \psi_r(x_2, x_3))),$$

where the functions θ_r, χ_r, ϕ_r and ψ_r are continuous.

The answer to the problem remains unknown when constraints involving the existence of a given number of derivatives are imposed on the functions participating in the superposition. A.G. Vitushkin only managed to prove the existence of functions such that in their representation using superpositions of functions of a smaller number of variables there inevitably occurs a loss in the degree of smoothness. The results by Vitushkin have found a clear explanation form the viewpoint of estimates for the rate of growth of the functions $N_F^a(\epsilon)$ dealt with in the first part of this paper.

55. ON THE REPRESENTATION OF CONTINUOUS FUNCTIONS OF SEVERAL VARIABLES AS SUPERPOSITIONS OF CONTINUOUS FUNCTIONS OF A SMALLER NUMBER OF VARIABLES *

Theorem 3, stated below, implies the following somewhat unexpected consequence: *any continuous function of an arbitrarily large number of variables is representable as a finite superposition of continuous functions of at most three variables.* For an arbitrary function of four variables the representation has the form

$$f(x_1, x_2, x_3, x_4) = \sum_{r=1}^{4} h^r[x_4, g_1^r(x_1, x_2, x_3), g_2^r(x_1, x_2, x_3)].$$

The question whether an arbitrary continuous function of *three* variables can be represented as a superposition of continuous functions of *two* variables remains open. The proof of the possibility of such a representation would give the complete solution to Hilbert's 13th problem [1], in the sense of a refutation of the conjecture put forward by Hilbert. Theorem 2 only shows that the representation of an arbitrary continuous function of three variables in the form of a superposition of continuous functions of two variables is possible if we admit as auxiliary variables some variables running over a *one-dimensional formation* somewhat more complicated than a closed interval on the number line, namely *a universal tree* (by a tree is meant a locally connected continuum not containing a homeomorphic image of a circle; as was shown by Menger [2], there exists a universal tree Ξ containing homeomorphic images of all the trees).

In what follows k, m, n and r are natural numbers; $a, b, c, C, d, M, R, x, y, u, v, f, F, g, h, \epsilon, \delta$ and ρ are real numbers; ξ, ϕ and ψ are tree elements; E^n is the n-dimensional unit cube; $0 \leq x_i \leq 1$; $i = 1, \ldots, n$.

Theorem 1. a) *For any $n \geq 2$ there are continuous functions*

$$\phi^1, \ldots, \phi^{n+1}$$

on E^n with values belonging to the universal tree Ξ such that any continuous real function f on E^n can be represented as

$$f(x_1, \ldots, x_n) = \sum_{r=1}^{n+1} h_f^s[\phi^r(x_1, \ldots, x_n)],$$

---

* *Dokl. Akad. Nauk SSSR* **108**:2 (1956), 179–182 (in Russian).

where $h_j^r(\xi)$ *are continuous real functions on* Ξ.

b) *The functions* $h_j^r(\xi)$ *can be chosen so that they depend continuously on* f *in the sense of the topology of uniform convergence in the spaces of continuous functions on* E^n *and* Ξ.

Theorem 1 implies almost immediately

Theorem 2. *For any* $n \geq 3$ *there are continuous functions*

$$\phi^1, \ldots, \phi^n$$

on E^n *with values belonging to* Ξ *such that any continuous function* f *on* E^n *can be represented in the form*

$$f(x_1, \ldots, x_n) = \sum_{r=1}^{n} h^r[x_n, \phi^r(x_1, \ldots, x_{n-1})],$$

where $h^r(x, \xi)$ *are continuous real functions on the product* $E^1 \times \Xi$.

The universal tree Ξ can be regarded (see [2]) as having a realization as a continuum in the unit square E^2. Denoting by g_1^r and g_2^r the coordinates of the point ϕ^r, we obtain as an immediate consequence of Theorem 2 the following proposition:

Theorem 3. *For any* $n \geq 3$ *there are continuous real functions*

$$g_1^1, \ldots, g_1^n;\ g_2^1, \ldots, g_2^n$$

on E^{n-1} *such that any continuous function* f *on* E^n *can be represented in the form*

$$f(x_1, \ldots, x_n) = \sum_{r=1}^{n} h^r[x_n, g_1^r(x_1, \ldots, x_{n-1}), g_2^r(x_1, \ldots, x_{n-1})],$$

where h^r *are continuous functions on* E^3.

Theorem 3 is trivial for $n = 3$; it is of actual interest only for $n \geq 4$.

It remains to indicate briefly a way of proving Theorem 1. The proof proceeds from the following lemma.

Main lemma. *For any* $n \geq 2$ *there is a system of functions*

$$u_{km}^r(x_1, \ldots, x_n)$$

on E^n, with indices r, k and m such that

$$1 \le r \le n+1, \quad 1 \le k < \infty, \quad 1 \le m \le m_k,$$

possessing the following properties:

1) $u^r_{km} \ge 0$;

2) $u^r_{km} \ne 0$ *only on a set* G^r_{km} *of a diameter* d_k*, where* $d_k \to 0$ *as* $k \to \infty$;

3) *two sets* G^r_{km} *and* $G^r_{km'}$ *with common indices* r *and* k *are disjoint for* $m' \ne m$;

4) *for any* k *at each point* $P \in E^n$,

$$c \le \sum_{r=1}^{n+1} \sum_{m=1}^{m_k} u^r_{km} \le C,$$

where c *and* C *do not depend on* k;

5) *the function* u^r_{km} *is constant on each set* $G^r_{k'm'}$ *with the same superscript* r *for* $k' > k$ *and arbitrary* m'.

The construction of the system of functions u^r_{km} cannot be presented within the framework of this paper. In what follows this system of functions is assumed to be given.

Lemma 1. a) *Any continuous function* f *on* E^n *can be represented as*

$$f(P) = \sum_{k=1}^{\infty} \sum_{r=1}^{n+1} \sum_{m=1}^{m_k} a^r_{km}(f) u^r_{km}(P), \tag{1}$$

where the coefficients $a^r_{km}(f)$ *do not depend on* P.

b) *The coefficients* $a^r_{km}(f)$ *can be chosen in the form of continuous functionals of* f *and so that*

$$|a^r_{km}(f)| \le a(\mathfrak{F}), \quad \sum_{k=1}^{\infty} a_k(\mathfrak{F}) < \infty$$

on each family $\mathfrak{F}$ *of uniformly bounded and equicontinuous functions* f.

The proof of Lemma 1 is based on properties 1), 2), and 4) of the system u^r_{km} and begins with estimation of the remainder R in the representation

$$f(P) = \sum_{r=1}^{n+1} \sum_{m=1}^{m_k} b^r_m u^r_{km}(P) + R,$$

where

$$b_m^r = \frac{1}{C} f(P_{km}^r)$$

and P_{km}^r are arbitrary points belonging to the corresponding sets G_{km}^r. It can easily be shown that for an appropriate choice of the coefficients b_m^r we have

$$|R| \leq (|1 - c/C| + \delta_k)M,$$

where

$$M = \sup_{P \in E^n} |f(P)|, \quad \delta_k = \sup_{\rho(P,P') \leq d_k} |f(P) - f(P')|.$$

The complete proof of Lemma 1 falls outside the framework of this paper.

We now write expansion (1) in the form

$$f(P) = \sum_{r=1}^{n+1} f^r(P), \quad f^r(P) = \sum_{k=1}^{\infty} \sum_{m=1}^{m_k} a_{km}^r u_{km}^r(P). \tag{2}$$

Properties 2), 3) and 5) of the system u_{km}^r readily imply the following property of the functions f^r.

Lemma 2. *The function $f^r(P)$ is constant on each component of any level set of the function*

$$F^r(P) = \sum_{k=1}^{\infty} \frac{1}{k^2} \sum_{m=1}^{m_k} u_{km}^r(P).$$

We now note that, as was shown by A.S. Kronrod [3], the components of the level sets of any continuous function form a tree in a certain natural topology. We denote the tree of the components of the level sets of the function F^r by Ξ^r and map the trees $\Xi^1, \dots, \Xi^{n+1}$ by means of the homeomorphisms

$$\psi_r(\Xi^r) = \Xi_0^r \subseteq \Xi$$

onto pairwise disjoint subsets of the universal tree Ξ. We put

$$\phi^r(P) = \psi_r(\xi^r)$$

if $P \in \xi^r \in \Xi^r$ and define continuous functions $h^r(\xi)$ on Ξ such that for $\xi \in \Xi^r$

$$h^r(\xi) = y, \text{ if } f^r(P) = y \text{ for } P \in \psi_r^{-1}(\xi).$$

It can easily be verified that

$$f^r(P) = h^r[\phi^r(P)]. \tag{3}$$

Formulas (2) and (3) lead to a proof of assertion a) of Theorem 1. Assertion b) of Theorem 1 is proved on the basis of assertion b) of Lemma 1.

In conclusion we also state without proof the following proposition.

Theorem 4. *Given any $n \geq 2$ and $\epsilon > 0$, for each continuous function f on E^n there exist polynomials*

$$b(u_1, \dots, u_{n-1}), \quad a_r(x), c_r(x); \quad r = 1, \dots, n+1,$$

such that

$$|f(P) - \tilde{f}(P)| < \epsilon$$

at all the points $P \in E^n$, where

$$\tilde{f}(x_1, \dots, x_n) = \sum_{r=1,2} a_r(x_n) b[c_r(x_n) + x_1, \dots, c_r(x_n) + x_{n-1}]. \tag{4}$$

For $n = 3$, by setting

$$d(u, v) = u + v, \quad g_r(x, y) = a_r(x)y, \quad h_r(x, x') = c_r(x) + x',$$

we obtain from (4)

$$\tilde{f}(x_1, x_2, x_3) = d(g_1\{x_3, b[h_1(x_3, x_1),\ h_1(x_3, x_2)]\}, \\ g_2\{x_3, b[h_2(x_3, x_1),\ h_2(x_3, x_2)]\}). \tag{5}$$

By virtue of Theorem 4, *any continuous function of three variables can be approximated arbitrarily accurately by an expression of the form* (5), *where d, g_r, b and h_r are polynomials in two variables.* This remark also illuminates from a new viewpoint the group of problems related to Hilbert's 13th problem.

5 May 1956

References

1. D. Hilbert, *Gesammelte Abhandlungen* Vol. 3, No. 17, Berlin, 1935.
2. K. Menger, *Kurventheorie*, Chapter 10, §6, Berlin, 1932.
3. A.S. Kronrod, *Uspekhi Mat. Nauk* **5**:1 (1950), 24–134 (in Russian).

56. ON THE REPRESENTATION OF CONTINUOUS FUNCTIONS OF SEVERAL VARIABLES AS SUPERPOSITIONS OF CONTINUOUS FUNCTIONS OF ONE VARIABLE AND ADDITION*

The aim of this paper is to present a brief proof of the following theorem:

Theorem. *For any integer $n \geq 2$ there are continuous real functions $\psi^{p,q}(x)$ on the closed unit interval $E^1 = [0;1]$ such that each continuous real function $f(x_1, \ldots, x_n)$ on the n-dimensional unit cube E^n is representable as*

$$f(x_1, \ldots, x_n) = \sum_{q=1}^{q=2n+1} \chi_q \Big[\sum_{p=1}^{n} \psi^{pq}(x_p) \Big], \tag{1}$$

where $\chi_q(y)$ are continuous real functions.

For $n = 3$, by setting

$$\phi_q(x_1, x_2) = \psi^{1q}(x_1) + \psi^{2q}(x_2), \quad h_q(y, x_3) = \chi_q[y + \psi^{3q}(x_3)],$$

we obtain from (1)

$$f(x_1, x_2, x_3) = \sum_{q=1}^{7} h_q[\phi_q(x_1, x_2), x_3], \tag{2}$$

which is a slight strengthening of a result by V.I. Arnol'd [2], who showed that any continuous function of three variables can be represented as a sum of *nine* summands of the same form as the *seven* summands involved in formula (2). The results of my paper [1] do not follow from the new theorem presented here in their exact statements, but their essence (in the sense of the possibility of representing functions of several variables by means of superpositions of functions of a smaller number of variables and their approximation by superpositions of a fixed form involving polynomials in one variable and addition) is obviously contained in the new theorem. The method for proving the new theorem is more elementary than that in [1, 2] and reduces to direct constructions and calculations. In particular, it is no longer necessary to use trees of components of level lines. However, the constructions used in this paper were in fact found by analyzing those employed in [1, 2] and discarding some of their details unnecessary for the derivation of the final result.

* *Dokl. Akad. Nauk SSSR* **114:5** (1957), 953–956 (in Russian).

§1. Construction of the functions ψ^{pq}

Everywhere in what follows, the indices p, q, k and i run over the integer values:

$$1 \leq p \leq n, \ 1 \leq q \leq 2n+1, \ k = 1, 2, \ldots, 1 \leq i \leq m_k = (9n)^k + 1.$$

When summing and multiplying within these limits we do not indicate the limits.

Consider the closed intervals

$$A^q_{k,i} = \left[\frac{1}{(9n)^k}\left(i - 1 - \frac{q}{3n}\right), \ \frac{1}{(9n)^k}\left(i - \frac{1}{3n} - \frac{q}{3n}\right)\right].$$

The intervals $A^q_{k,i}$ have lengths $\frac{1}{(9n)^k}\left(1 - \frac{1}{3n}\right)$ and are obtained from one another for fixed k and q by passing from i to $i' = i+1$ using a shift to the right over a distance $1/(9n)^k$, that is, not only do they not overlap, but they do not even overlap intervals of lengths $1/3n(9n)^k$ and to within the presence of these intervals, they cover the whole closed unit interval E^1. Accordingly, for fixed k and q the cubes

$$S^q_{k,i_1\ldots i_n} = \prod_n A^q_{k,i_p}$$

with edges of lengths $1/(9n)^k$ cover the unit cube E^n to within the separating slits of widths $1/3n(9n)^k$. It is easy to verify the following.

Lemma 1. *The system of all cubes $S^q_{k,i_1\ldots i_n}$ with constant k and variable q and $i_1, \ldots, i_n$ covers the unit cube E^n so that each point belonging to E^n is covered at least $n+1$ times.*

Using induction on k we can prove the following

Lemma 2. *The constants $\lambda^{pq}_{k,i}$ and ϵ_k can be chosen so that the following conditions hold:*

1) $\lambda^{pq}_{k,i} < \lambda^{pq}_{k,i+1} \leq \lambda^{pq}_{k,i} + 1/2^k$;

2) $\lambda^{pq}_{k,i} \leq \lambda^{pq}_{k+1,i'} \leq \lambda^{pq}_{k,i} + \epsilon_k - \epsilon_{k+1}$ *if the closed intervals $A^q_{k,i}$ and $A^q_{k+1,i'}$ do not intersect;*

3) *the closed intervals $\Delta^q_{k,i_1\ldots i_n} = \left[\sum_p \lambda^{pq}_{k,i_p}; \sum_p \lambda^{pq}_{k,i_p} + n\epsilon_k\right]$ are pairwise disjoint for fixed k and q.*

It is easy to note that 1) and 3) imply

4) $\epsilon_k \leq 1/2^k$.

On the basis of the above-indicated properties of the closed intervals $A^q_{k,i}$ and properties 1), 2) and 4) of the constants $\lambda^{pq}_{k,i}$ and ϵ_k, one can easily prove the following

Lemma 3. *For fixed p and q the conditions*

5) $\lambda^{pq}_{k,i} \leq \psi^{pq}(x) \leq \lambda^{pq}_{k,i} + \epsilon_k$ *for* $x \in A^q_{k,i}$ *uniquely determine a continuous function* ψ^{pq} *on* E^1.

Remark. It can easily be seen that, by construction, the functions ψ^{pq} are monotonically increasing. This property could have been included in the statement of the theorem.

From 5) and 3) it follows that

6) $\sum_p \psi^{pq}(x_p) \in \Delta^q_{k,i_1\dots i_n}$ *for* $(x_1, \dots, x_n) \in S^q_{k,i_1\dots i_n}$.

§2. Construction of the functions χ^q

On establishing the existence of the functions ψ^{pq} and the constants $\lambda^{pq}_{k,i}$ and ϵ_i possessing properties 1)–6) we proceed to the proof of the main theorem. The desired functions $\chi^q(y)$ will be constructed in the form

$$\chi^q = \lim_{r\to\infty} \chi^q_r,$$

where $\chi^q_0 \equiv 0$, while for $r > 0$ χ^q_r will be defined by induction on r simultaneously with the natural numbers k_r.

We will use the notation

$$f_r(x_1, \dots, x_n) = \sum_q \chi^q_r[\sum_p \psi^{pq}(x_p)], \tag{3}$$

$$M_r = \sup_{E^n} |f - f_r|. \tag{4}$$

It is obvious that

$$f_0 \equiv 0, \quad M_0 = \sup_{E^n} |f|.$$

Assume that the continuous functions χ^q_{r-1} and the number k_{r-1} have already been determined. In this way a continuous function f_{r-1} on E^n has also been determined. Since the diameters of the cubes $S^q_{k,i_1\dots i_n}$ tend to zero as

$k \to \infty$, we can choose k_r so large that the oscillation of the difference $f - f_{r-1}$ does not exceed $(1/(2n+2))M_{r-1}$ on any $S^q_{k_r,i_1\dots i_n}$.

Let $\xi^q_{k,i}$ be arbitrary points belonging to the corresponding closed intervals $A^q_{k,i}$. For the closed interval $\Delta^q_{k,i_1\dots i_n}$ we put

$$\chi^q_r(y) = \chi^q_{r-1}(y) + \frac{1}{n+1}[f(\xi^q_{k,i_1},\dots,\xi^q_{k,i_n}) - f_{r-1}(\xi^q_{k,i_1},\dots,\xi^q_{k,i_n})]. \tag{5}$$

Obviously, the values of the function χ^q_r fixed in this way satisfy the inequality

$$|\chi^q_r(y) - \chi^q_{r-1}(y)| \le \frac{1}{n+1}M_{r-1}. \tag{6}$$

Outside the closed intervals $\Delta^q_{k,i_1\dots i_n}$ the function χ^q_r is defined arbitrarily, with preservation of the same inequality (6) and continuity.

We now estimate $f - f_r$ at an arbitrary point $(x_1,\dots,x_n)$ belonging to E^n. It is obvious that

$$f(x_1,\dots,x_n) - f_r(x_1,\dots,x_n) = f(x_1,\dots,x_n) - f_{r-1}(x_1,\dots,x_n) - \\ -\sum_q\Big\{\chi^q_r\Big[\sum_p \psi^{pq}(x_p)\Big] - \chi^q_{r-1}\Big[\sum_p \psi^{pq}(x_p)\Big]\Big\}. \tag{7}$$

We represent the sum $\sum_q$ in (7) in the form $\sum' + \sum''$, where the sum $\sum'$ extends over certain $n+1$ values of q for which the point $(x_1,\dots,x_n)$ is contained in one of the cubes $S^q_{k,i_1\dots i_n}$ (by Lemma 1, such cubes exist) and the sum $\sum''$ extends over the remaining n values of q.

By virtue of (5), for each term in $\sum'$ we have

$$\chi^q_r\Big[\sum_p \psi^{pq}(x_p)\Big] - \chi^q_{r-1}\Big[\sum_p \psi^{pq}(x_p)\Big] = \\ = \frac{1}{n+1}[f(\xi^q_{k,i_1},\dots,\xi^q_{k,i_n}) - f_{r-1}(\xi^q_{k,i_1},\dots,\xi^q_{k,i_n})] = \\ = \frac{1}{n+1}[f(x_1,\dots,x_n) - f_{r-1}(x_1,\dots,x_n)] + \frac{\omega^q}{n+1}, \tag{8}$$

where

$$|\omega^q| \le \frac{1}{2n+2}M_{r-1}. \tag{9}$$

The terms in $\sum''$ are estimated using (6). Relation (5), together with (8), (9) and (6) implies

$$|f - f_r| = \Big|\frac{1}{n+1}\sum{}' \omega^q + \sum{}''(\chi^q_r - \chi^q_{r-1})\Big| \le \\ \le \frac{1}{2n+2}M_{r-1} + \frac{n}{n+1}M_{r-1} = \frac{2n+1}{2n+2}M_{r-1}. \tag{10}$$

Since inequality (10) holds at any point $(x_1, \ldots, x_n) \in E^n$, we have

$$M_r \leq \frac{2n+1}{2n+2} M_{r-1}, \quad M_r \leq \left(\frac{2n+1}{2n+2}\right)^r M_0. \tag{11}$$

From (6) and (11) it follows that the absolute values of the differences $\chi_r^q - \chi_{r-1}^q$ do not exceed the corresponding terms of the absolutely convergent series

$$\sum_r \frac{1}{n+1} M_{r-1}.$$

Therefore the functions χ_r^q converge uniformly to continuous limit functions χ^q for $r \to \infty$.

From relations (3) and (4) and estimate (11), passing to the limit for $r \to \infty$, we obtain relation (1), which completes the proof of the theorem.

20 June 1957

References

1. A.N. Kolmogorov, 'On the representation of continuous functions of several variables as superpositions of continuous functions of a smaller number of variables', *Dokl. Akad. Nauk SSSR* **108**:2 (1956), 179–182 (in Russian). (See No. 55.)
2. V.I. Arnol'd, 'On the representation of continuous functions of three variables as superpositions of continuous functions of two variables', *Dokl. Akad. Nauk SSSR* **114**:4 (1957), 679–681 (in Russian).

57. ON THE LINEAR DIMENSION OF TOPOLOGICAL VECTOR SPACES *

Two topological vector spaces E and E' are said to be isomorphic if it is possible to establish a one-to-one linear bicontinuous correspondence between them.

It is well known that all topological vector spaces of the same finite dimension n are isomorphic. In this trivial case the dimension of the space $d(E)$ satisfies the following conditions:

a) if E is isomorphic to a closed linear subspace of a space E', then $d(E) \leq d(E')$;

b) if there is a continuous linear mapping of E' onto E, then $d(E) \leq d(E')$.

In Chapter XII of his famous monograph [1], Banach suggested a generalization of the notion of dimension to the case of infinite-dimensional vector spaces, proceeding from the requirement that property a) be preserved. In §1 of this paper we introduce a linear dimension $\delta(E)$ satisfying both conditions a) and b). Here the classification of spaces according to dimension is of course somewhat less rich than Banach's classification but, in some respects, it is more natural. For the sake of simplicity and in order to maintain a direct relation with Banach's definition, we henceforth assume that all spaces under consideration are F-spaces (see [1, Chapter III]).

We assuime that the reader knows how an F-metric is introduced in the following spaces, which will be considered as examples in what follows:

$C_n^{(p)}$ is the space of real functions $f(x_1, \ldots, x_n)$ on the n-dimensional unit cube, having continuous partial derivatives up to order p inclusive, with the topology of uniform convergence of the functions and of their partial derivatives up to order p inclusive;

$B_n^{(\infty)}$ is the space of real periodic functions $f(x_1, \ldots, x_n)$ of n real variables $x_1, \ldots, x_n$ with period 2π with respect to each of the variables and having continuous partial derivatives of all orders, with the topology of uniform convergence of the functions themselves and of their partial derivatives of any order;

A_n^G is the space of functions $f(z_1, \ldots, z_n)$ of n complex variables $z_1, \ldots, z_n$, analytic in a bounded open region G of complex n-dimensional space, with the topology of uniform convergence on every compactum $K \subseteq G$.

* *Dokl. Akad. Nauk SSSR* **120** (1958), 239–241.

Tradition and experience in classical analysis make one expect that spaces of functions of a greater number of variables must be "richer" in elements than those of functions of a smaller number of variables in the sense that it is assumed that if the solution of a problem depends on an "arbitrary" function of one variable, the "arbitrariness" in the choice of the solution is narrower than in the case when the solution depends on an arbitrary function of two variables, etc. Below we will see that in the case of analytic functions this idea finds support in the corresponding inequalities for linear dimensions (see Theorems 4 and 10). It is this result which we believe to be the most interesting in this paper. By contrast, for a space of functions of restricted smoothness this expectation, based on experience in classical analysis, does not find confirmation in the properties of the linear dimension. For example, from the results presented in Chapter XII of Banach's monograph [1], it can readily be derived that, irrespective of the values of n and p, all the spaces $C_n^{(p)}$ have the same Banach dimension $\dim_l$. Since the relation $\dim_l(E) = \dim_l(E')$ always implies $\delta(E) = \delta(E')$, the same applies to the dimension $\delta(E)$ we introduce in this paper.

The consideration of spaces of infinitely differentiable functions does not change the situation. For instance, all the spaces $B_n^{(\infty)}$ have the same dimension $\dim_l$ and the same dimension δ since the following theorem holds.

Theorem 1. *All the spaces* $B_n^{(\infty)}$, $n = 1, 2, \ldots$, *are isomorphic* (see [5]).

§1. The linear dimension $\delta(E)$

Following Banach, we introduce the function δ by defining the meaning of the inequality

$$\delta(E) \leq \delta(E'). \tag{1}$$

This defines the function δ to within an order-preserving one-to-one mapping of the partially ordered set Δ of its values onto a new partially ordered set Δ'. By definition, relation (1) means that there is a closed linear subspace E'' of the space E' such that there exists a continuous linear mapping of E'' onto the space E. The transitivity of the relation (1) thus defined, which is needed for the correctness of the definition of the function δ, can easily be proved.

Here we note one of the properties of the dimension $\delta(E)$ which the Banach dimension $\dim_l(E)$ does not possess.

Theorem 2. *If E and E' are Banach spaces (B-spaces) and are reflexive, then inequality* (1) *is equivalent to the inequality*

$$\delta(\widetilde{E}) \leq \delta(\widetilde{E}') \tag{2}$$

for the dimensions of the dual spaces.

Theorem 2 and the results in Chapter XII of Banach's monograph [1] readily imply a number of results relating to the dimension $\delta(E)$ of Banach spaces, which we do not present here. Instead, here we state several theorems on the dimension δ of spaces of analytic functions.

Theorem 3. *If G and G' are bounded finitely connected domains in the complex plane, then*

$$\delta(A_1^G) = \delta(A_1^{G'}).$$

The generalization of Theorem 3 to functions of several variables has until now been proved only in the following form. Let $G_1, \ldots, G_n$ be bounded finitely connected domains in the complex plane and let $G = G_1 \times G_2 \times \ldots \times G_n$; then the following theorem holds.

Theorem 3a. *The dimension $\delta(A_n^G) = \alpha_n$ does not depend on the choice of the domains $G_1, G_2, \ldots, G_n$.*

For the dimension α_n introduced in Theorem 3a we have

Theorem 4. *If $n < n'$, then $\alpha_n < \alpha_{n'}$.*

The dimension $\delta(E)$ occupies an extreme position among all linear dimensions satisfying conditions a) and b).

Theorem 5. *Any function $d(E)$ satisfying conditions* a) *and* b) *can be represented in the form*

$$d(E) = f[\delta(E)],$$

and $\delta(E) \leq \delta(E')$ implies $d(E) \leq d(E')$.

Among these functions $d(E)$ that are subordinate to $\delta(E)$ and are poorer in the sense of the possibility of *distinguishing* between spaces according to dimension, we will consider only one, which we call "approximative dimension". We use this notion to prove Theorem 4 as an immediate consequence of Theorem 3a and Theorem 10 presented below.

§2. The approximate dimension $d_a(E)$

We associate with each F-space E a class $\Phi(E)$ of functions $\phi(\epsilon)$ defined for $\epsilon > 0$ by means of the condition that $\phi \in \Phi$ if for any compactum $K \subset E$ and any open neighbourhood U of the zero element θ in E there exists ϵ_0 such that for any $\epsilon < \epsilon_0$ there are $N \leq \phi(\epsilon)$ points $x_1, \ldots, x_n$ in the space E such that $K \subset \bigcup_{1 \leq m \leq N} (x_m + \epsilon U)$.

Two spaces E and E' are said to have the same approximate dimension $d_a(E) = d_a(E')$ if $\Phi(E) = \Phi(E')$.

By definition, the inequality $d_a(E) < d_a(E')$ holds if $\Phi(E) \supset \Phi(E')$.

The relations $>$, $\leq$, $\geq$ and $\|$ (incomparability) are defined in like manner.

In many cases the approximate dimension can be calculated by methods related to ϵ-entropy and ϵ-capacity of metric spaces (see [2–4]). We present some simple results along these lines.

Theorem 6. *For an n-dimensional space E^n with finite n the set Φ is determined by the condition that $\phi \in \Phi$ if*

$$\lim_{\epsilon \to \infty} (\epsilon^n \phi(\epsilon)) = \infty.$$

Theorem 7. *For an infinite-dimensional Banach space E the set Φ is empty.*

Thus, all infinite-dimensional Banach spaces have the same dimension $d_a(E) = z$, which is maximal among the dimensions $d_a(E)$. In our opinion, this result must not be interpreted as making the dimension d_a meaningless. It is a meaningful notion for spaces which, in a sense, are closer to finite-dimensional spaces, namely countably-normed spaces of the type $B_n^{(\infty)}$ and A_n^G which are becoming important in analysis.

Theorem 8. *The approximate dimension of the spaces $B_n^{(\infty)}$ does not depend on n and is determined by the condition that $\phi \in \Phi$ if there exists $q > 0$ such that*

$$\lim_{\epsilon \to \infty} (\epsilon^q \log \phi(\epsilon)) = 0.$$

Theorem 9. *The approximate dimension a_s of the spaces A_s^G (where G is an arbitrary bounded open region in the s-dimensional complex space) does not* *depend on the choice of the region G and is determined by the condition that* $\phi \in \Phi$ *if*

$$\lim_{\epsilon \to 0} \frac{\log \phi(\epsilon)}{(\log(1/\epsilon))^{s+1}} = \infty.$$

Theorem 9 immediately implies

Theorem 10. *If* $s < s'$, *then* $a_s < a_{s'}$.

18 February 1958

References

1. S. Banach, *Opérations linéaires*, Paris, 1933.
2. A.N. Kolmogorov, *Dokl. Akad. Nauk SSSR* **108**:3 (1956), 385–388 (in Russian).
3. V.M. Tikhomirov, *Dokl. Akad. Nauk SSSR* **117**:2 (1957), 191—194 (in Russian).
4. A.G. Vitushkin, *Dokl. Akad. Nauk SSSR* **117**:5 (1957), 745–747 (in Russian).
5. A. Grothendieck, *Mem. Amer. Math. Soc.* **16** (1955).

58. A REFINEMENT OF THE CONCEPT OF THE LOCAL STRUCTURE OF TURBULENCE IN AN INCOMPRESSIBLE VISCOUS FLUID AT LARGE REYNOLDS NUMBERS *

The programme of the symposium related to our colloquium includes a paper by A.M. Obukhov. [1] One of the sections of the paper is devoted to a refinement of the conception of the local structure of turbulence and is of particular interest to all specialists in the statistical theory of turbulence. At the request of some of the participants of the colloquium, I am going to discuss in more detail the origin of this refinement and present the latest results obtained by A.M. Obukhov in this area.

The concept of local structure of turbulence at large Reynolds numbers elaborated by me [1–3] and A.M. Obukhov [4, 5] in 1939–1941 was based on the pictorial idea of Richardson that in a turbulent flow there exist eddies of all possible scales $l < r < L$ between an "external scale" L and an "internal scale" l and a unified mechanism of energy transfer from large-scale eddies to small-scale ones.

This concept, which had also been put forward independently by some other authors, gained general recognition. However, very soon after its appearance, L.D. Landau remarked that it did not take into account a factor which follows directly from the assumption of an essentially random chaotic character of the mechanism of energy transfer from large-scale eddies to small-scale ones, namely, as the ratio L/l increases, the energy dissipation rate

$$\epsilon = \frac{\nu}{2} \sum_{\alpha} \sum_{\beta} \left(\frac{\partial U_\alpha}{\partial x_\beta} + \frac{\partial U_\beta}{\partial x_\alpha} \right)^2$$

must increase indefinitely. More precisely, it is natural to assume that the variance of the logarithm of ϵ has an asymptotic behaviour of the form

$$\sigma^2_{\log \epsilon} \sim A + k' \log(L/l) \tag{1}$$

for $L/l \gg 1$, where k' is a universal constant.

---

* In: *Mécanique de la turbulence: Colloq. Intern. CNRS, Marseille, août–sept. 1961*, Paris, 1962, pp. 447–458 (in French and Russian).

[1] 'Some specific features of atmospheric turbulence', *J. Fluid Mech.* **13**:1 (1962), 77–81.

Only recently has A.M. Obukhov found a way of refining the results in [1–5] taking into account the remark made by Landau. This way is based on consideration of the dissipation rate

$$\epsilon_r(\mathbf{x}, t) = \frac{3}{4\pi r^3} \int_{|\mathbf{h}| \leq r} \epsilon(\mathbf{x} + \mathbf{h}, t) d\mathbf{h}$$

averaged over a sphere of radius r and assuming that for a large ratio L/r the logarithm of $\epsilon_r(\mathbf{x}, t)$ has a normal distribution . It is natural to assume that the variance of the logarithm of $\epsilon_r(\mathbf{x}, t)$ has the form

$$\sigma_r^2(\mathbf{x}, t) = A(\mathbf{x}, t) + 9k \log(L/r), \tag{2}$$

where k is a universal constant and $A(\mathbf{x}, t)$ depends on the macrostructure of the flow.

My presentation of A.M. Obukhov's idea is related directly to [1] and is based on a corresponding modification of the two similarity hypotheses in [1]. As a third hypothesis, use will be made of the already stated hypothesis of normal distribution of the logarithm of ϵ_r, with formula (2) for the variance of the logarithm.

Under the new assumptions, the formula

$$B_{dd}(r) = C r^{2/3} \bar{\epsilon}^{-2/3}$$

from [1] is replaced by the formula

$$B_{dd}(r) = C(\mathbf{x}, t) r^{2/3} \bar{\epsilon}^{-2/3} (L/r)^{-k}, \tag{3}$$

where k is the constant involved in (2) and the factor $C(\mathbf{x}, t)$ depends on the macrostructure of the flow. Instead of the assumption of constancy of the skewness

$$S(r) = \frac{B_{ddd}(r)}{[B_{dd}(r)]^{3/2}}$$

for $l \ll r \ll L$ stated in [2], we now obtain

$$S(r) = S_0 (L/r)^{(3/2)k}, \tag{4}$$

where the coefficient S_0 also depends on the macrostructure of the flow.

We associate with the length scale r and the point $(\mathbf{x}, t)$ the time and velocity scales $T_r = r^{2/3}\epsilon_r^{-1/3}$, $U_r = r^{1/3}\epsilon_r^{1/3}$ and the "internal length scale" $l_r = \nu^{3/4}\epsilon_r^{-1/4}$.

It is clear that the Reynolds number formed from U_r and r can be expressed in terms of the quotient r/l_r by

$$\mathrm{Re} = U_r r/\nu = (r/l_r)^{4/3}. \tag{5}$$

The coordinates x'_α, t' of a point $(\mathbf{x}', t')$ belonging to a neighbourhood of the point $(\mathbf{x}, t)$ can be expressed in terms of dimensionless parameters ξ_α and τ: $x'_\alpha = x_\alpha + \xi_\alpha r$, $\quad t' = t + \tau T_r$.

We introduce the dimensionless relative velocities

$$V_\alpha(\xi, \tau) = \frac{U_\alpha(\mathbf{x} + \xi r, t + \tau T_r) - U_\alpha(\mathbf{x}, t)}{U_r}.$$

First similarity hypothesis. *If $r \ll L$, then for given values of $\xi_\alpha^{(k)}, \tau^{(k)}$, $\alpha = 1, 2, 3$; $k = 1, 2, \dots, n$, the conditional distribution of the $3n$ variables $V_\alpha(\xi^{(k)}, \tau^{(k)})$ for a fixed value of $\mathrm{Re}_r = \mathrm{Re}$ depends only on Re, and is the same for all turbulent flows.*

Second similarity hypothesis. *For $\mathrm{Re} \gg 1$ the distributions indicated in the first hypothesis do not depend on Re.*

Unconditional mathematical expectations (mean values) will be denoted by a bar above. Since $\bar{\epsilon}$ is almost constant in regions small relative to the external scale L, it can be assumed that for $r \ll L$ we have

$$\bar{\epsilon}_r = \epsilon. \tag{6}$$

Consider the differences of the longitudinal velocity components at two points at a distance r:

$$\Delta_{dd}(r) = U_1(x + r, x_2, x_3, t) - U_1(x_1, x_2, x_3, t).$$

It can easily be seen that

$$\Delta_{dd}(r) = V_1(1, 0, 0, 0) r^{1/3} \epsilon_r^{1/3}. \tag{7}$$

If

$$r \gg l,$$

where l is the supremum (except for negligibly improbable cases) of the internal scale l_r, then (6), (7) and hypotheses I and II imply that

$$\overline{|\Delta_{dd}(r))|^3} = Cr\bar{\epsilon}, \tag{8}$$

where C is an absolute constant.

We must now use a formula given by Obukhov, expressing the moments of variables having a logarithmic normal distribution. The formula is

$$\overline{\xi^p} = \exp(pm + p^2\sigma^2/2), \tag{9}$$

where m and σ^2 are the mean value and the variance of $\log \xi$. It follows from (2) and (7)–(9) that

$$\overline{[\Delta_{dd}(r)]^p} = C_p(\mathbf{x}, t)(r\bar{\epsilon})^{p/3}(L/r)^{kp(p-3)/2}; \tag{10}$$

in particular, for $p = 2$ we have

$$\overline{\Delta^2_{dd}(r)} = C_2(\mathbf{x}, t)(r\bar{\epsilon})^{2/3}(L/r)^{-k},$$

that is, formula (3).

Since the formula

$$B_{ddd}(r) = -(4/5)\bar{\epsilon}r$$

from [2] remains valid, (3) implies (4).

The special choice of the function $\epsilon_r(\mathbf{x}, t)$ underlying Obukhov's consideration can be excluded from our presentation. Then the first two similarity hypotheses are stated thus:

First similarity hypotheses. *If* $|\mathbf{x}^{(k)} - \mathbf{x}| \ll L$, $k = 0, 1, \ldots, n$, *then the conditional distribution of the* $3n$ *variables*

$$\frac{U_\alpha(\mathbf{x}^{(k)}) - U_\alpha(\mathbf{x})}{U_\alpha(\mathbf{x}^{(0)}) - U_\alpha(\mathbf{x})}, \quad \alpha = 1, 2, 3; \; k = 1, 2, \ldots, n, \tag{11}$$

depends only on the Reynolds number

$$\mathrm{Re} = \frac{|U(\mathbf{x}^{(0)}) - U(\mathbf{x})||\mathbf{x}^{(0)} - \mathbf{x}|}{\nu}.$$

Second similarity hypothesis. *For*

$$\mathrm{Re} \gg 1$$

the distributions indicated in the first hypothesis do not depend on Re.

The essence of the content of the additional assumptions made by Obukhov can be stated in the following way:

Third hypothesis. *The two groups of variables* (11) *are stochastically independent if*

$$|\mathbf{x}^{(k)} - \mathbf{x}| \geq r_1$$

in the first group,

$$|\mathbf{x}^{(k)} - \mathbf{x}| \leq r_2$$

in the second group, and $r_1 \gg r_2$.

Of course, in order to present a rigorous mathematical derivation of the desired logarithmic normal distribution of velocity differences and the formula for the variances of the logarithms of the differences analogous to (1) and (2) from the third hypothesis, one needs a more precise statement of the hypothesis.

1 September 1961

References

1. A.N. Kolmogorov, 'Local structure of turbulence in an incompressible viscous fluid at very large Reynolds numbers', *Dokl. Akad. Nauk SSSR* **30**:4 (1941), 299–303 (in Russian). (See No. 45.)
2. A.N. Kolmogorov, 'On the degeneration of isotropic turbulence in an incompressible viscous fluid', *Dokl. Akad. Nauk SSSR* **31**:6 (1941), 538–541 (in Russian). (See No. 46.)
3. A.N. Kolmogorov, 'Dissipation of energy in isotropic turbulence', *Dokl. Akad. Nauk SSSR* **32**:1 (1941), 19–21 (in Russian). (See No. 47.)
4. A.M. Obukhov, 'On the energy distribution in the spectrum of a turbulent flow', *Dokl. Akad. Nauk SSSR* **32**:1 (1941), 22–24 (in Russian).
5. A.M. Obukhov, 'On the energy distribution in the spectrum of a turbulent flow', *Izv. Akad. Nauk SSSR Ser. Geogr. i Geofiz.* **5**:4–5 (1941), 453–466 (in Russian).

59. P.S. ALEKSANDROV AND THE THEORY OF δs-OPERATIONS *

Sets of points on the number line admitting of a simple definition are either finite, countable, or have the power of the continuum. As is well known, the question whether this is true for any subset of the number line constitutes the *continuum problem*. At the beginning of the 20th century it seemed natural to seek the solution to the continuum problem by considering progressively more general classes of point sets.

In the sequence of classes of Borel sets

$$F, F_\sigma, F_{\sigma\delta}, F_{\sigma\delta\sigma}, \dots,$$

$$G, G_\delta, G_{\delta\sigma\delta}, \dots$$

the (positive) solution to the problem is obvious for the class F_σ (and, of course, for the classes F and G contained in it). In 1906 Young treated the class $G_{\delta\sigma}$ and in 1914 Hausdorff showed in the first edition of his "Mengenlehre" that the solution remains positive for sets of class $G_{\delta\sigma\delta\sigma}$. Naturally, further progress in this area is of strong interest for mathematicians concerned with set theory. The complete positive solution of the problem for all Borel sets was found in 1916 independently by P.S. Aleksandrov [1] and Hausdorff 4].

Both Aleksandrov and Hausdorff arrive at the solution to the continuum problem for Borel sets by proving that *an uncountable Borel set contains a perfect subset.* Both authors base their proof on the scheme of generation of an arbitrary Borel set:

$$\begin{aligned} E &= \bigcap_{j_1} \bigcup_{i_1} E_{j_1}^{i_1}, \\ E_{j_1}^{i_1} &= \bigcap_{j_2} \bigcup_{i_2} E_{j_1 j_2}^{i_1 i_2}, \\ &\dots\dots\dots\dots\dots \\ E_{j_1 \dots j_{\lambda-1}}^{i_1 \dots i_{\lambda-1}} &= \bigcap_{j_\lambda} \bigcup_{i_\lambda} E_{j_1 \dots j_{\lambda-1} j_\lambda}^{i_1 \dots i_{\lambda-1} i_\lambda}, \end{aligned} \tag{1}$$

where for a finite λ all the chains of sets

$$E_{j_1}^{i_1},\ E_{j_1 j_2}^{i_1 i_2},\ \dots,\ E_{j_1 j_2 \dots j_\lambda}^{i_1 i_2 \dots i_\lambda}$$

* *Uspekhi Mat. Nauk* **21**:4 (1966), 275–278 (in Russian)

terminate in a closed (Aleksandrov) or open (Hausdorff) set.

The total number of closed sets from which a Borel set E can thus be obtained is ***countable***. They can be indexed by the natural numbers:

$$F_1, F_2, \ldots, F_s, \ldots,$$

and the set E can be written as

$$E = \Phi(F_1, F_2, \ldots, F_s, \ldots),$$

where Φ symbolizes a certain operation on the sets F_s. It is easy to see that this operation is a) ***analytic*** and b) ***positive*** in the sense of the definitions introduced by Kantorovich and Livenson [5].

Definition 1. An operation

$$E = \Phi(E_1, \ldots, E_s, \ldots)$$

is said to be ***analytic*** if the equivalence of inclusion relations

$$x \in E_s \Leftrightarrow y \in E_s$$

(for any s) implies that of the relations

$$x \in E \Leftrightarrow y \in E,$$

that is, if the fact that a point x belongs to the set E is completely determined by the indices of the sets E_s containing the point x.

Definition 2. An operation

$$E = \Phi(E_1, \ldots, E_s, \ldots)$$

is said to be ***positive*** if the inclusion

$$E_s' \subset E_s$$

(for all s) implies

$$\Phi(E_1', \ldots, E_s', \ldots) \subset \Phi(E_1, \ldots, E_s, \ldots).$$

Any analytic operation can be replaced by an analytic positive operation on the sets E_s and their complements. As to the class of analytic positive operations, it coincides with the class of *δs-operations*, that is, operations that can be written in the form

$$\Phi(E_1, \ldots, E_s, \ldots) = \bigcup_{S \in \mathfrak{S}} \bigcap_{s \in S} E_s,$$

where $\mathfrak{S}$ is a set of subsets of the natural numbers.

In the space of subsets of the natural numbers there exists a metric topology. [1] In essence, the proof carried out by Aleksandrov in [1] comes down to the following.

1. *When the construction* (1) *is written as a δs-operation*

$$E = \Phi(F_1, \ldots, F_s, \ldots) = \bigcap_{s \in \mathfrak{S}} \bigcup_{s \in S} F_s, \tag{2}$$

the set $\mathfrak{S}$ turns out to be closed (in its most natural construction).

2. *If the set of sequences in formula* (2) *is closed, the operation Φ applied to closed sets F_s generates a set E which either is countable or contains a perfect set.*

There naturally arises the question as to which sets can be obtained by applying δs-operations with a closed set $\mathfrak{S}$ to: a) closed sets, b) open sets, and c) Borel sets.

The answer is that in all the cases *Suslin sets and only Suslin sets* are obtained. Here it is unnecessary to use all possible δs-operations involving closed sets $\mathfrak{S}$. There exists a standard δs-operation of this kind whose application to closed (open) sets generates all Suslin sets. It is well known that this operation can be written in a more explicit manner if the initial sets are indexed not by natural numbers but by "tuples" of natural numbers

$$s = \{s_1, \ldots, s_n\}.$$

The A-operation we are interested in is determined by the formula

$$A\{E_s\} = \bigcup E_{s_1} \bigcap E_{s_1 s_2} \bigcap E_{s_1 s_2 s_3}, \ldots,$$

---

[1] For instance, this topology is determined by the distance

$$\rho(E_1, E_2) = \sum_{n \in E_1 \Delta E_2} \frac{1}{n^2},$$

where $E_1 \Delta E_2$ denotes the symmetric difference of the sets E_1 and E_2.

where the union is taken over all sequences $s_1 s_2 s_3 \dots$. Suslin sets, proceeding from this operation, are also called A-sets.

The definition of an A-operation is in fact contained in the same paper by Aleksandrov [1]. An arbitrary Borel set E is obtained in this paper "to within a countable set" from "canonical sets" π_μ. Since these canonical sets are finite intersections of initial sets $E^{i_1 \dots i_\lambda}_{j_1 \dots j_\lambda}$ and are therefore closed, while neglecting a countable set is unimportant (and can easily be avoided), it follows that any Borel set results from an A-operation on closed sets. Unfortunately, Aleksandrov did not publish a more systematic presentation with an explicit definition of an A-operation and of the statement of the theorem asserting that each Borel set is an A-set. Naturally, there arose the question whether the class of A-sets is broader than that of Borel sets or is substantially broader. In that same year (1916) this problem was solved by M. Ya. Suslin, who showed that there are A-sets that are not Borel sets, which gave rise to the development of an independent theory of Suslin sets. In particular, Suslin proved that, in general, superposition of A-operations according to the scheme

$$E = A\{E_s\}, \; E_s = A\{E_s^r\}$$

results in sets E that can be obtained from the set E_s^r (taken with new indices and with repetitions) by means of a single A-operation. In the general theory of δs-operations such operations are called *normal.*

In his paper [2] Aleksandrov introduced one more δs-operation on sets, called a Γ-operation. To state its definition in a compact form it should be noted that associated with each tuple

$$s = (s_1, s_2, \dots, s_n)$$

is a set Δ_s of number sequences

$$s_1 s_2 \dots s_n s_{n+1} \dots$$

beginning with $s_1 s_2 \dots s_n$. A set $\mathfrak{S}$ of tuples is called a Γ-*chain* if the corresponding sets Δ_s cover the whole set of number sequences (which is called the *Baire space* in problems of this kind). The Γ-operation is determined by the formula

$$E = \bigcup_{\mathfrak{S} \in \Gamma} \bigcap_{s \in \mathfrak{S}} E_s,$$

where Γ is the set of all Γ-chains.

Aleksandrov establishes the formula

$$\overline{A\{E_s\}} = \Gamma\{\overline{E}_s\},$$

where $\overline{E}$ denotes the complement of the set E. In essence, Aleksandrov introduces for the special case of an A-operation a Γ-operation "complementary" to the former. The general definition of the δs-operation $\overline{\Phi}$ complementary to a given δs-operation Φ, and the formula

$$\overline{\Phi(E_1, \ldots, E_s, \ldots)} = \overline{\Phi}(\overline{E}_1, \ldots, \overline{E}_s, \ldots)$$

generalizing formula (3), lie at the foundation of my paper [6].

In [3] Aleksandrov uses the fact that the Γ-operation provides a "positive" definition of sets complementary to A-sets and proves the topological invariance of this class of sets.

The theory of δs-operations has been developed further by myself, Hausdorff (the second edition of "Mengenlehre"), Kantorovich, Livenson, Lyapunov and many other authors.

References

1. P.S. Alexandroff, 'Sur la puissance des ensembles mesurables', *C.R. Acad. Sci. Paris* **162** (1916), 323–325.
2. P.S. Alexandroff, 'Sur les ensembles complémentaires aux ensembles (A), *Fund. Math.* **5** (1924), 160–165.
3. P.S. Alexandroff, 'Sur l'invariance topologique des ensembles complémentaires aux ensembles (A)', *Mat. Sb.* **31** (1924), 310–318.
4. F. Hausdorff, 'Die Mächtigkeit der Borelschen Mengen', *Math. Ann.* **77** (1916), 430–437.
5. L. Kantorovich and E. Livenson, 'Memoir on the analytical operations and projective sets', *Fund. Math.* **18** (1932), 234–279.
6. A.N. Kolmogorov, 'On operations on sets', *Mat. Sb.* **35** (1928), 415–421 (in Russian). (See No.13.)

60. A QUALITATIVE STUDY OF MATHEMATICAL MODELS OF POPULATION DYNAMICS *

§1. Introductory notes

Looking over the literature on mathematical modelling of population dynamics published in recent years I have ascertained that, despite the extensive development of this area, in one respect the recent studies have remained at the level of the first famous work by Volterra, namely the whole investigation proves to depend on a special and inevitably arbitrary choice of mathematical expressions for the laws governing population dynamics. However, in many other fields of application of mathematics the possibility and importance of deriving qualitative results from qualitative prerequisites has long been understood. An attempt to revise Volterra's theory of struggle for existence between two species (predator and prey) was made in my paper [1], which appeared as early as 1936. Later I encountered references to this paper, which was published as a response to the paper by Volterra in the same little-known Italian journal, but the actual idea of drawing meaningful conclusions from purely qualitative prerequisites has remained unused. This induces me to give a new and revised presentation of the results obtained in [1].

It is well known that the idea itself of describing schematically the temporal evolution of the population sizes $N_i(t)$ $(i = \overline{1,s})$ of s interacting species with the aid of a system of differential equations

$$dN_i/dt = F_i(N_1, \dots, N_s), \ i = \overline{1,s}, \tag{1}$$

is highly imperfect. The application of the apparatus of differential equations of the type (1) to population dynamics is inadequate due, for instance, to the fact that usually substantial changes in the population sizes $N_i(t)$ occur during time intervals comparable with the lifetime of individuals, which makes it necessary to take into account the composition by age of the populations. For example, however great the improvement of Volterra's theory within the framework of the description of population dynamics by equations of the type (1) may be, its application in explaining the three-to-four-year oscillation cycles in the population of birds indicated in Severtsev's observational data is still

* *Probl. Kibernet.* **25**:2 (1972), 101–106.

not fully justified, owing to the above-mentioned fact. However, we must begin with the simplest models. If this paper facilitates the appearance of papers in which an analogous qualitative analysis is applied to more realistic schemes, its purpose will be accomplished.

§2. Volterra's equations and their generalization

To describe the prey and predator population sizes $N_1(t)$ and $N_2(t)$, Volterra suggested the equations

$$\frac{dN_1}{dt} = (\epsilon_1 - \gamma_1 N_2)N_1, \quad \frac{dN_2}{dt} = (-\epsilon_2 + \gamma_2 N_1)N_2. \tag{2}$$

The arbitrary character of the assumption made is seen even from the fact that the predator multiplication factor $K_2 = (-\epsilon_2 + \gamma_2 N_1)$ increases indefinitely with increasing prey population; for instance, in the presence of a sufficient number of hares the number of wolves will be doubled every day and even every hour. In essence, from the viewpoint of mathematics, equations (2) are merely the simplest example of equations of the form

$$\frac{dN_1}{dt} = K_1(N_1, N_2)N_1, \quad \frac{dN_2}{dt} = K_2(N_1, N_2)N_2, \tag{3}$$

where the prey multiplication factor K_1 decreases with increasing predator population and changes sign from plus to minus whereas the predator multiplication factor K_2 increases with the prey population and changes sign from minus to plus.

Unfortunately, the most interesting qualitative conclusion of Volterra's theory that, except for the stationary case $N_1 = \epsilon_2/\gamma_2$, $N_2 = \epsilon_1/\gamma_1$, any initial data $N_1(0), N_2(0)$ result in *periodic* oscillations of the numbers $N_1(t)$ and $N_2(t)$ with an amplitude depending on the initial data, is in fact a consequence of the special form of the equations (2) as chosen by Volterra. In my paper [1], I have made an attempt to investigate the behaviour of solutions of a system (3) under certain *qualitative* assumptions concerning the properties of the functions $K_1(N_1, N_2)$ and $K_2(N_1, N_2)$. In §3 we shall investigate in more detail the behaviour of solutions of the system

$$\frac{dN_1}{dt} = K_1(N_1)N_1 - L(N_1)N_2, \quad \frac{dN_2}{dt} = K_2(N_1)N_2, \tag{4}$$

which, in my opinion, is a sufficiently realistic model of the possible relationships between two species. The passage from the general equations (3) to equations (4) can be interpreted meaningfully in the following way.

1) It is assumed that the predators do not "interact", that is, the predator multiplication factor K_2 and the part L of the prey population devoured by a predator per unit time do not depend on N_2.

2) It is assumed that the increment of the prey population during short time intervals in the presence of predators is equal to its increment in the absence of predators minus the part of the prey population devoured by the predators. With regard to the functions $K_1(N_1), K_2(N_1)$, and $L(N_1)$, we shall impose only quite natural conditions on the qualitative nature of their dependence on N_1.

§3. Investigation of the behaviour of solutions of the system (4)

In order that the theory of differential equations be applicable without any complications, we shall assume that the functions $K_1(N_1), K_2(N_1)$ and $L(N_1)$ are continuous and continuously differentiable. It is natural to assume that they are defined in the positive quadrant $N_1 \geq 0$, $N_2 \geq 0$. However, it is necessary to make a stipulation. For $N_2 = 0$ the first of equations (4) goes into the natural equation $dN_1/dt = K_1(N_1)N_1$, describing the evolution of the prey population in the absence of predators. But at $N_1 = 0$ the first of the equations (4) would imply that in the case $L(0) > 0$ the prey population must become negative on passing through zero, a meaningless conclusion. It is advisable to assume that after the point of the quadrant lying on the N_2-axis has been attained, the first of the equations (4) no longer applies, the prey population size remains constant equal to zero, and the evolution of the number of predators still satisfies the second equation.

We now state three more constraints on the choice of the functions $K_1(N_1)$, $K_2(N_1)$, and $L(N_1)$.

3) $dK_1/dN_1 < 0$, $K_1(0) > 0 > K_1(\infty) > -\infty$. In essence, this constraint means that in the absence of predators the prey multiplication factor decreases monotonically with increasing prey population and passes from positive values to negative ones. This assumption, reflecting that the amount of food and other resources necessary for the existence of the prey are limited, seems to be quite natural.

4) $dK_2/dN_1 > 0, K_2(0) < 0 < K_1(\infty)$. The meaning of this constraint is in fact that with increasing prey population the predator multiplication factor increases, passing from negative values (in a situation when there is no food) to positive ones.

5) $L(N_1) > 0$ for $N_1 > 0$.

As to the *limiting* value of $L(N_1)$ at $N_1 = 0$, we prefer to retain both possibilities $L(0) = 0$ and $L(0) > 0$. The first of these reflects in an idealized form the situation when a small part of the prey population may take shelter in places where the predators cannot hunt it and thus survive the period of excessive multiplication of predators. We note that in Volterra's case we have $K_1(N_1) = \epsilon, K_2(N_1) = -\epsilon_2 + \gamma_2 N_1$ and $L(N_1) = \gamma_1 N_1$, that is, our requirement that $K_1(N_1)$ be negative for sufficiently large N_1 is violated.

Let us determine the *rest points* of system (4) in the positive quadrant. It can easily be seen that there are two or three such points: 1) the point $(0,0)$; 2) the point $(A,0)$, where A is found from the equation $K_1(A) = 0$; and 3) the point (B,C), where B and C are determined by the equations $K_2(B) = 0$, $K_1(B)B - L(B)C = 0$. The third point lies in the positive quadrant and is distinct from the second one only when $K_1(B) > 0$, that is, $B < A$.

We investigate the nature of these singular points in the usual manner by investigating the linearized equations for $\zeta = N_1 - N_1^0$ and $\eta = N_2 - N_2^0$ where N_1^0, N_2^0 are the coordinates of the singular point.

I) At the point $(0,0)$ we obtain

$$\frac{d\zeta}{dt} = K_1(0)\zeta - L(0)\eta, \quad \frac{d\eta}{dt} = K_2(0)\eta.$$

The roots of the characteristic equation are $\lambda_1 = K_1(0)$, $\lambda_2 = K_2(0)$; they are real and have different signs, so that this is a saddle point. The slopes of the separatrices are found from the equation $L(0)\kappa^2 - [K_1(0) - K_2(0)]\kappa = 0$. One of the separatrices (corresponding to $\kappa_1 = 0$) is obviously advancing: this is the N_1-axis, and the slope of the other is $\kappa_2 = [K_1(0) - K_2(0)]/L(0)$. If $L(0) = 0$, then the second separatrix coincides with the N_2-axis. The first separatrix tends to the saddle point and the other goes away from it.

II) At the point $(A,0)$ the linearized equations are written as

$$\frac{d\zeta}{dt} = K_1'(A)A\zeta - L(A)\eta, \quad \frac{d\eta}{dt} = K_2(A)\eta.$$

The roots of the characteristic equation are $\lambda_1 = K_1'(A)A$, $\lambda_2 = K_2(A)$.

IIa) If $B < A$, then $\lambda_1 < 0$ and $\lambda_2 > 0$ and we have a saddle point. The slopes of the separatrices are $\kappa_1 = 0$ and $\kappa_2 = [K_1'(A)A - K_2(A)]/L(A) < 0$. The separatrix having the slope κ_2 goes away from the saddle point and is directed inside the quadrant.

IIb) If $B > A$, then $\lambda_1 < 0$ and $\lambda_2 < 0$, and the singular point is a stable nodal point.

III) At the point (B, C) for $B < A$ we obtain the linearized equations

$$\frac{d\zeta}{dt} = -\sigma\zeta - L(B)\eta, \quad \frac{d\eta}{dt} = CK_2'(B)\zeta,$$

where $\sigma = -K_1(B) - K_1'(B)B + L'(B)C$. The determinant

$$\Delta = \begin{vmatrix} -\sigma & -L(B) \\ CK_2'(B) & 0 \end{vmatrix} = CK_2'(B)L(B)$$

is positive. Therefore the singular point is a focal or nodal point. Stability depends on the sign of σ, namely if $\sigma > 0$, then the rest point is stable, and if $\sigma < 0$, it is unstable. A focal point appears in the case of complex roots of the equation $\lambda^2 + \sigma\lambda + \Delta = 0$, and a nodal point corresponds to real roots.

Assuming that $B < A$, we shall also investigate the behaviour of the separatrix going away from the point $(A, 0)$ in the upward direction. It is easy to see that under the assumptions made, the trajectories *cannot recede to infinity*. Therefore, in essence, there are three possible cases here, namely the separatrix we are interested in either: α) intersects the N_2-axis; β) winds around a limit cycle; or γ) arrives at the point (B, C). In case α) we must also consider the origin of the separatrix tending to the point $(0, 0)$. It can either: $\alpha 1$) wind off a limit cycle; or $\alpha 2$) go away from the point (B, C). In cases β) and γ) the separatrix tending to $(0, 0)$ must come from infinity. Under the assumption that $L(0) = 0$, case α) is impossible. We thus obtain the following classification.

In cases 1), 2), 4), 7), and 8) our analysis provides a complete description of the qualitative nature of the solutions if cases 4), 7) and 8) are split into subcases 4a), 7a), and 8a) with a focal point at (B, C) and subcases 4b), 7b), and 8b) with a nodal point at (B, C). It can easily be seen that the point (B, C) is unstable in case 4) and stable in cases 7) and 8). Proceeding from our assumptions, cases 3), 5) and 6) are not subjected to an exhaustive

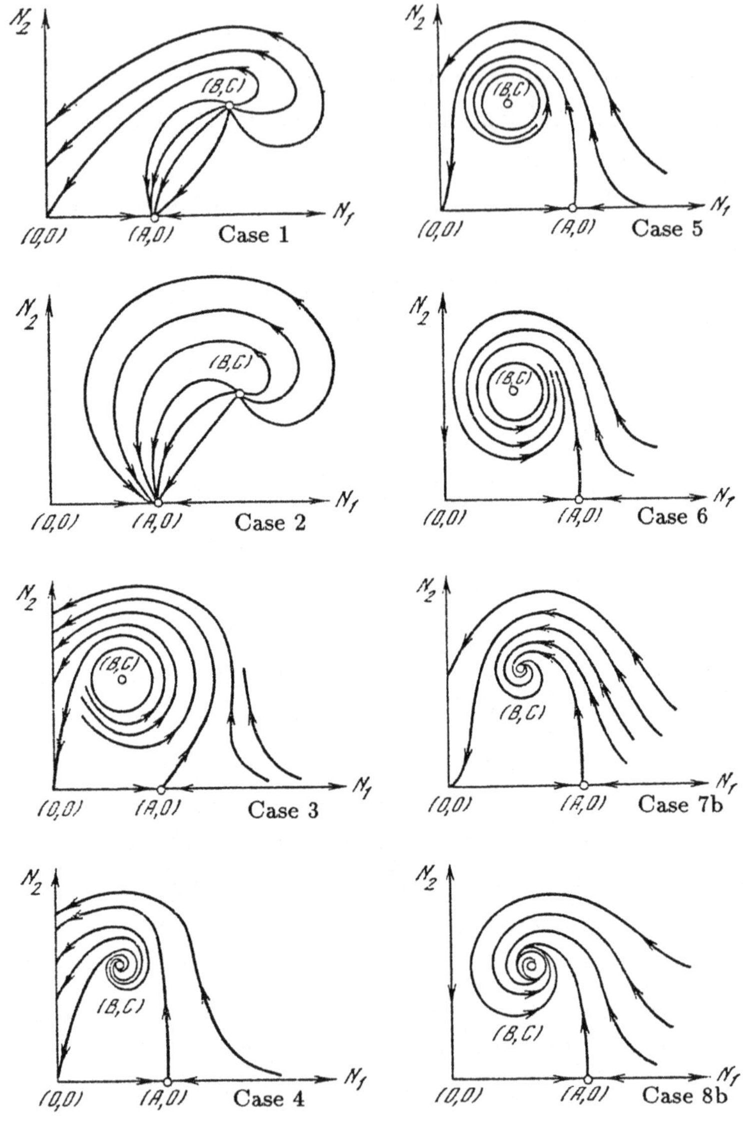

analysis, since the behaviour of trajectories inside the limit cycle may be rather complicated. It is however very likely that, in practice, it suffices to confine the consideration to the case when inside the limit cycle occurring in our analysis *there are no other limit cycles and, in general, closed trajectories.* In this case it remains to indicate whether the point (B,C) is a focal point (subcases 3a), 5a) and 6a)) or a nodal point (subcases 3b), 5b) and 6b)). The eight cases of different qualitative behaviour of the solutions are represented in the figures corresponding to the table below:

		$L(0) > 0$	$L(0) = 0$
$B > A$		1	2
$B < A$	$\alpha 1$	3	–
	$\alpha 2$	4	–
	β	5	6
	γ	7	8

Besides the above-mentioned cases, in which there can be additional closed trajectories inside the basic limit cycle, we have also omitted transient cases (that is, cases not "structurally stable" in the Andronov–Pontryagin sense), which can appear only under exceptional circumstances. For example, the separatrix going away from the point $(A, 0)$, when extended, may accidentally prove to be the separatrix tending to the point $(0, 0)$. When this is the case, the qualitative pattern continues to resemble cases 1) and 2).

15 April 1974

References

1. A.N. Kolmogoroff, 'Sulla teoria di Volterra della lotta per l'esistenza', *G. Ist. Ital. Attuar.* **7** (1936), 74–80.

Commentary

ON THE PAPERS ON THE THEORY OF FUNCTIONS AND SET THEORY

A.N. Kolmogorov

The most urgent and lively branches of the theory of functions and set theory requiring further development were separated by N.N. Luzin into two groups, which he called "metrical" and "descriptive". "Metrical" contained the theory of the integral, trigonometric series, boundary properties of analytic functions, etc. "Descriptive" embraced Baire's classification of functions, Borel classes of sets, analytic sets, etc., including the continuum problem. In relation to his pupils, Luzin had a definite idea of which of them was destined for working on "metrical" or "descriptive" problems. My predestination was to deal with "metrical", and in subsequent years I fulfilled this plan to some extent (see papers Nos. 1–8, 10–12, 14, 16, 21 etc. in the present book).

In 1921–1923 V.V. Stepanov conducted a seminar on the theory of trigonometric series, in which G.A. Seliverstov and I took part as the youngest participants. Naturally, the problems posed by Luzin attracted our particular attention. They included the problem of estimating the rate of decrease of the coefficients of a Fourier-Lebesgue series. The solution to the problem (see No. 2 in this book) turned out to be very simple for cosine series, so that I could not understand why nobody found it out before me. However, when Luzin learned about this result, he invited me rather solemnly to become one of his pupils and I started visiting him every week at a definite time fixed for one of his groups of pupils.

The whole of my work on trigonometric and orthogonal series stemmed from studies in V.V. Stepanov's seminar. Among the difficulties I had to overcome, the first seemed to be the paper in which an almost everywhere divergent Fourier-Lebesgue series was constructed (see No. 1). For rather a long time I was working in two directions, trying alternately to construct an example or to prove its impossibility. At the last stage I had been continuously lost in thought for a week, and then the desired construction suddenly appeared. A little later

I found rather easily an analytic version of the original purely geometric idea, which made it possible to strengthen the first result and to construct a series divergent everywhere.

The entire period of my research on trigonometric and orthogonal series awakens pleasant reminiscences of the harmonious work in the group headed by D.E. Men'shov (after Luzin ceased to be interested in this topic). I also often recall the collaboration with my friend G.A. Seliverstov, who died prematurely.

Simultaneously, in 1921–22 I had an extensive plan of investigations on descriptive set theory in a direction that was not at all envisaged by Luzin. The plan was partly realized in my paper "On operations on sets" completed at the beginning of 1922 and published in 1928 (see No. 13). The unpublished manuscript "On R-operations" [1] was closely related to this paper. The results contained in this manuscript received further development in the work by L.V. Kantorovich and E.M. Livenson and by A.A. Lyapunov.

Here I only mention the unfulfilled and perhaps unrealizable intention to extend the classification that begins with classes of B-sets and is continued by classes of C-sets, RC-sets, RRC-sets, etc. to classification over transfinite numbers of the second class.

TRIGONOMETRIC AND ORTHOGONAL SERIES

(P.L. Ul'yanov)

A.N. Kolmogorov published about a dozen papers on trigonometric and orthogonal series, each of which gave rise, in fact, to extensive studies which are continuing even now. Here we dwell only on results directly related to the papers by Kolmogorov, developing his ideas and characterizing this direction of investigation.

A. Divergence of trigonometric Fourier series

In 1922, at the age of nineteen, Kolmogorov obtained one of the most brilliant results in the theory of trigonometric series. Namely, in paper No. 1 he constructed an example of a 2π-periodic integrable function whose trigonometric Fourier series is divergent almost everywhere. The profound idea of the exam-

[1] The manuscript was found after this book was printed. It will be published in Volume 2 of the "Selected Works" of A.N. Kolmogorov.

ple in combination with geometric visuality and clarity is striking. In 1926 he also constructed (paper No. 11) an everywhere divergent Fourier series.

The constructions in these examples were a starting point for a large number of investigations of other authors. In 1936 Marcinkiewicz [1] constructed an almost everywhere divergent Fourier series whose partial sums are bounded at each point of a set $E \subset [0, 2\pi]$ of measure $mE = 2\pi$. We note that the case $E = [0, 2\pi]$ is impossible here.

In 1953 Sunouchi [2] indicated that there is an everywhere divergent Fourier series whose conjugate series is also a Fourier series.

As is well known, if for the integral modulus of continuity we have

$$\omega_1(\delta, f) = O\Big\{\frac{1}{(\log(1/\delta))^{1+\epsilon}}\Big\} \text{ as } \delta \to +0, \tag{1}$$

where $\epsilon > 0$, then the Fourier series of $f(t)$ is convergent almost everywhere. In this connection, Zygmund (see [4], Vol. 2) posed the question of the validity of the above-stated assertion for $\epsilon = 0$ as well. The answer to the question remains open.

Further, based on Kolmogorov's construction, Prokhorenko [5] (see also Tandori [6]) constructed an example of a function $f(t)$ whose Fourier series is divergent almost everywhere and whose modulus of continuity satisfies the condition

$$\omega_1(\delta, f) = O\Big\{\frac{1}{\log\log(1/\delta)}\Big\} \text{ as } \delta \to +0. \tag{2}$$

By an embedding theorem proved by me, relation (2) implies that

$$\int_0^{2\pi} |f(t)|\{\ln^+ \ln^+ |f(t)|\}^\alpha dt < \infty \text{ for any } \alpha \in (0,1). \tag{3}$$

Strangely enough, at present no essentially better results than (2) and (3) are known in relation to smoothness and class of integrability of functions with divergent Fourier series. The comparison of (1) and (2) demonstrates the "lacuna of uncertainty".

Another direction of investigation is related to the exact characterization of sets of points of divergence of Fourier series. It is known that the set of points of divergence of any series of continuous functions is a $G_{\delta\sigma}$-set (see, for example, [3], p. 433). Based on Kolmogorov's example of an everywhere unboundedly divergent Fourier series, Zeller [7] proved that each G_δ-set $E \subset [0, 2\pi]$ is the set of points of divergence of a Fourier series. This assertion is not valid for certain F_σ-sets E (see Körner [8]). Thus the problem of describing the sets of points of divergence of Fourier series of functions of a given class remains unsolved.

B. Divergence of orthogonal series

The main proposition in paper No. 12 is Theorem 1, asserting that there are an orthonormal (ON) system of functions $\{\phi_n(t)\}$ on the interval $[0,1]$ such that $|\phi_n(t)| \equiv 1$ and a series

$$\sum_{n=1}^{\infty} a_n \phi_n(t) \tag{4}$$

divergent everywhere on $[0,1]$, although $\{a_n\} \in l_2$.

Moreover, the numbers $\{a_n\}$ and the system $\{\phi_n\}$ can be chosen (see Theorem 2) so that for a given sequence $W(n) = o(\log n)$ the inequality

$$\sum_{n=1}^{\infty} a_n^2 W(n) < \infty$$

holds. These results were generalized in various directions. In particular, in 1938 Men'shov [12] proved that

1) for any number $K > 1$ there is an ON system $\{\phi_n\}$ for which $|\phi_n(t)| \leq K$ for $t \in [0,1]$ and $n = 1, 2, \ldots$;

2) the sequence $\{\log^2 n\}$ is an exact Weyl multiplier, that is fulfilment of the inequality (5) with $W(n) = \log^2 n$ implies the convergence of series (4) almost everywhere, whereas for any sequence $W(n) = o(\log^2 n)$ this is no longer true.

Men'shov [13] also constructed an ON system of algebraic polynomials which are bounded jointly and for which the sequence $\{\log^2 n\}$ is an exact Weyl multiplier.

What has been said means that the exact Weyl multiplier for the entire class of jointly bounded ON systems (and even for algebraic polynomials) is the same as that for the entire class of ON systems.

Further, in 1960 Ul'yanov [9] showed that in Theorem 1 one can take a rearranged Walsh system for $\{\phi_n(t)\}$. As to Theorem 2, its strongest generalization was established in 1975 by Kashin [10], who constructed an ON system $\{\phi_n(t)\}$ with $|\phi_n(t)| \equiv 1$ for which the sequence $\{\log^2 n\}$ is an exact Weyl multiplier. A simpler proof of this fact was suggested later by Tandori [11].

In the same paper No. 12, Kolmogorov's Theorem 3 is stated. It states that the trigonometric system can be renumbered in an appropriate manner as $\{\cos n_m t,\ \sin n_m t\}$ so that it becomes a system of divergence almost everywhere in the sense that there is a sequence $\{a_m, b_m\} \in l_2$ such that the series

$$\sum_{m=1}^{\infty} a_m \cos n_m t + b_m \sin n_m t \tag{6}$$

is divergent almost everywhere on $[0, 2\pi]$.

Later this assertion became a starting point for a large number of investigations on unconditional convergence almost everywhere (Zahorski, Ul'yanov, Olevskii, Tandori, Talalyan, Laindler, Arutyunyan, Garcia, Kashin, Poleshchuk, etc.). Zahorski [14] suggested in 1960 for the first time a brief scheme for constructing the corresponding example of a divergent series of the form of (6). A detailed and complete proof of this fact was given in [15]. In [15] Ul'yanov also proved an analogous assertion for series with respect to the Haar system. Based on this result, Ul'yanov [15] and Olevskii [16] showed that any complete ON system can be renumbered so that it becomes a system of divergence almost everywhere.

In §A we have stated Kolmogorov's result asserting that there exists a trigonometric Fourier series divergent on a set $E \subset [0, 2\pi]$ of measure $mE = 2\pi$. For ON systems jointly bounded on the interval $[0, 1]$ an analogue of this result was established by Bochkarev [17] for sets $E \subset [0, 1]$ of positive measure. However, there can be no exhaustive analogy here, since Kazaryan [18] constructed a complete jointly bounded ON system with respect to which any Fourier series converges on a set of positive measure.

C. Convergence of Fourier series

Consider a function $f \in L(0, 2\pi)$, and let $S_n(t, f)$ be the partial sums of the Fourier series of f.

In paper No. 3 Kolmogorov first proves Theorem 1, asserting that the sequence $\{S_{n_k}(t, f)\}_{k=1}^{\infty}$ converges almost everywhere if $f \in L^2(0, 2\pi)$ and the integers n_k satisfy the inequality $n_{k+1} > \lambda n_k$, where $\lambda > 1$ and $k = 1, 2, \ldots$, and then establishes Theorem 2, according to which every lacunary Fourier series is convergent almost everywhere.

In 1931 Littlewood and Paley [19] extended Theorem 1 to the case of functions $f \in L^p(0, 2\pi)$, $p > 1$. Gosselin [20] indicated that for $p = 1$ the analogous assertion is not true.

As to Theorem 2, it turned out to be a consequence of Theorem 1, since every lacunary Fourier series must be the Fourier series of a function $f \in L^p(0, 2\pi)$ (Zygmund [21]; see also [3], p. 684).

Theorems 1 and 2 gave an impetus to the active study of properties of lacunary series and the investigation of interrelations between convergence of subsequences of partial sums of a series and its summability by a certain method (see [3, 22] and also the review by Gaposhkin [23]).

In paper No. 4 it was proved, in particular, that for any $\epsilon > 0$ the sequence $\{(\log n)^{1+\epsilon}\}$ is a Weyl multiplier for convergence almost everywhere of trigonometric series. Using the methods of this paper, its authors Kolmogorov and Seliverstov (also see paper No. 10) and also Plessner [24] soon found that the above assertion remains valid for $\epsilon = 0$ as well. This Kolmogorov-Seliverstov and Plessner theorem had been for forty years the best result relating to convergence almost everywhere of trigonometric Fourier series of functions $f \in L^2$. Only in 1966 did Carleson (see [25, 26]) prove that the Fourier series of functions $f \in L^2(0, 2\pi)$ are convergent almost everywhere. Developing Carleson's technique, Hunt [27] showed that the Fourier series of functions belonging to any of the spaces $L^p(0, 2\pi)$, $p > 1$, are convergent almost everywhere. The strongest result of this kind was obtained by Sjölin [28], who proved that if

$$\int_0^{2\pi} |f(t)|\{\log^+ |f(t)|\}\{\log^+ \log^+ |f(t)|\}dt < \infty, \tag{7}$$

then the Fourier series of the function $f(t)$ is convergent almost everywhere. Note that, as yet, the distinction with respect to integrability of the function f in the "positive" and "negative" results (see (7) and (3)) lies in the multiplier $\log^+ |f(t)|$.

The proof of Carleson's theorem (see [25, 26]) is very complicated and therefore Fefferman [29] suggested a new one (its detailed presentation was given by Lukashenko [30]). However, Fefferman's proof is also rather complicated, and therefore it is of interest to find a simpler proof.

Finally, we note the great influence of paper No. 10, specifically the method of proof developed there, on studies of convergence of orthogonal series. In particular, Kaczmarz [31] (see also [22], Chapter 5) applied the method to prove that if for an ON system $\{\phi_n(t)\}$ the Lebesgue functions satisfy the relation

$$L_n(t) \le W(n) \uparrow \infty \text{ for } t \in E \text{ and } n = 1, 2, \ldots,$$

then the sequence $\{W(n)\}$ is a Weyl multiplier for convergence almost everywhere on a set E of Fourier series with respect to the system $\{\phi_n\}$. And, as was shown by Tandori [32], this result cannot be improved for the whole class of ON systems.

Finally we mention paper No. 25, in which Kolmogorov considered Fourier series with respect to an orthonormal system of polynomials $\{\omega_n(t)\}_{n=0}^{\infty}$ (with weight $p(t) \in L(a,b)$), where the degree of the polynomial $\omega_n(t)$ is equal to n and $p(t) > 0$ for $t \in (a,b)$. Kolmogorov proved that if for some $\epsilon > 0$ a function $f(t)$ satisfies the inequality

$$|f(t_1) - f(t_2)| \le \frac{C}{|\ln|t_1 - t_2||^{1+\epsilon}} \quad \text{for } a \le t_1 < t_2 \le b \ (C = \text{const}),$$

then its Fourier series

$$f(t) \sim \sum_{n=0}^{\infty} a_n(f)\omega_n(t), \quad a_n(f) = \int_a^b f(t)p(t)\omega_n(t)dt,$$

is convergent almost everywhere on (a,b).

This result became a cconceptually new fact, since earlier it was only known (see Natanson [52]) that if for some $\alpha > \frac{1}{2}$ we have

$$|f(t_1) - f(t_2)| < C|t_1 - t_2|^{\alpha} \quad \text{for } t_1 \ne t_2,$$

then the Fourier series of f with respect to the system $\{\omega_n\}$ is convergent almost everywhere.

The above theorem by Kolmogorov gave rise to the appearance of a number of papers on the convergence of Fourier series with respect to general orthonormal polynomials (Aleksich, Ul'yanov, Chen Kien Kwong, and others).

D. On Fourier-Lebesgue coefficients

We will assume that $a_n \to 0$ and that the conditions imposed on $\{a_n\}$ below guarantee the convergence of the series

$$\frac{a_0}{2} + \sum_{n=1}^{\infty} a_n \cos nt \tag{8}$$

everywhere on $(-\infty,\infty)$, except possibly at points $t\equiv 0\,(\mathrm{mod}\,2\pi)$, to a function $f(t)$. In paper No. 2 it was proved that the Fourier-Lebesgue coefficients can tend to zero arbitrarily slowly.[1] Incidentally, Kolmogorov showed that if the sequence $\{a_n\}$ is quasiconvex, that is,

$$\sum_{n=0}^{\infty}(n+1)|\Delta^2 a_n|<\infty, \tag{9}$$

where $\Delta^2 a_n=\Delta a_n-\Delta a_{n+1}$ and $\Delta a_n=a_n-a_{n+1}$, then $f(t)\in L(0,2\pi)$ and (8) is the Fourier series of f. Moreover, it was shown that if inequality (9) is fulfilled, then the condition

$$\lim_{n\to\infty} a_n\log n=0 \tag{10}$$

is necessary and sufficient for the series (8) to converge in the metric of the space $L(0,2\pi)$ to the function $f(t)$.

These results of Kolmogorov were generalized and extended to various cases (Moore [33, 34], Cesari [35], Sidon [36], Boas [37], Telyakovskii [38, 39], Fomin [40], and others). Here we state only the most general result known to us: Telyakovskii proved [38] that (8) is a Fourier-Lebesgue series if

$$\alpha\equiv\sum_{n=0}^{\infty}|\Delta a_n|<\infty \quad\text{and}\quad \beta\equiv\sum_{n=2}^{\infty}\left|\sum_{m=2}^{[n/2]}\frac{\Delta a_{n-m}-\Delta a_{n+m}}{m}\right|<\infty. \tag{11}$$

Moreover in this case there exists a constant C such that

$$\int_0^{\pi}|f(t)|dt\le C(a+\beta).$$

Conditions (11) are very cumbersome (because of the second inequality) but they imply rather general and easily verifiable conditions. For instance, if a sequence $\{a_n\}$ is such that

$$|\Delta a_n|\le A_n\downarrow \quad\text{and}\quad \sum_{n=1}^{\infty}A_n<\infty, \tag{S}$$

then (11) is also fulfilled. Conditions (S) are equivalent to Sidon's conditions (see [36, 39]), and when they are fulfilled, the series (8) converges to f in the metric of L if and only if (10) holds (see [39]).

[1] On the other hand, Littlewood showed (see [41]; 3, p. 226) that any condition imposed only on the $|a_n|$ and allowing that $\{a_n\}\in l_2$ cannot be sufficient for (8) to be a Fourier series.

We note that the question of integrability of a function $f(t)$ representable as a series (8), and also the estimation of the norm of the function using the sequence $\{a_n\}$, are closely related to the study of summability of Fourier series by linear methods (see, for example, [38]).

E. Conjugate functions and series

In paper No. 8 Kolmogorov first proved Theorem 1, asserting that for a function $f \in L(0, 2\pi)$ and its harmonic conjugate function $g(t)$ we have

$$mE\{t : t \in [0, 2\pi],\ |g(t)| > R\} \leq \frac{C}{R}\|f\|_1, \tag{12}$$

where $C = \text{const}$ and $R > 0$ is an arbitrary number.

From this relation one derives quite easily the inequality (Theorem 2)

$$\int_0^{2\pi} |g(t)|^{1-\epsilon} dt \leq \frac{C}{\epsilon}\Big\{\int_0^{2\pi} |f(t)| dt\Big\}^{1-\epsilon} \quad (0 < \epsilon < 1), \tag{13}$$

on the basis of which it is established (Theorem 3) that

$$\lim_{n \to \infty} \int_0^{2\pi} |f(t) - S_n(t, f)|^{1-\epsilon} dt = 0, \tag{14}$$

where the $S_n(t, f)$ are the partial sums of the trigonometric series $\sigma(f)$ of the function $f(t)$.

It is in this paper that for the first time the idea was suggested of representing the partial sums $S_n(t, f)$ in terms of certain conjugate functions such that the application of (13) to them immediately results in (14). Later the idea was used many times when studying questions of convergence of Fourier series and the conjugate series of functions belonging to various spaces.

For inequality (13) different versions for proving it were suggested, for instance, by Littlewood [45], Hardy [46], Titchmarsh [44], etc.

Inequalities of the type (12) became particularly important in calculus (and not only in the theory of trigonometric series). Direct proofs of (12) (that is, without using Privalov's theorem) were, in particular, given by Titchmarsh [44] and Loomis [47] (see also [3], Chapter 8, and [4], Chapter 7).

Further, attention should be drawn to the fact that inequality (12) probably gave impetus to the proof by Marcinkiewicz [48] of interpolation theorems for a class of operators of weak type (see also [4], Chapter 12). Zygmund mentions this in his article devoted to Marcinkiewicz (see [49], p. 19). Moreover, the very introduction of the notion of an operator of weak type seems to be suggested also by Kolmogorov's paper No. 8. Unfortunately, all this is not mentioned in the corresponding literature.

We also note that Kolmogorov derives the inequality (12) from the single assumption that the conjugate operator $g = Af$ exists, that is, the very existence of the operator A implies automatically the quantitative estimate (12). This indicates a definite relationship between paper No. 8 and later results of E. Stein on estimates of weak type for translation-invariant operators and their subsequent generalizations (Nikishin [50], and others).

Finally, we mention the book by Zhizhishvili [51], which is especially devoted to conjugate functions and series. Its presentation begins with Kolmogorov's inequality (13) and its generalizations.

We note one more theorem of Kolmogorov (see paper No. 14), establishing that for a function $f(t) \in L(0, 2\pi)$ the conjugate function[2] $g(t)$ is B-integrable on the interval $[0, 2\pi]$ and the conjugate series $\bar{\sigma}(f)$ of $\sigma(f)$ is the Fourier series of $g(t)$ in the sense of the B-integral, where B symbolizes the Denjoy-Boks integral (see also [3], Chapter 8 and [4], Chapter 7). This result immediately implies that

$$\bar{\sigma}(f) = \sigma(g) \tag{15}$$

provided that the functions f and g belong to the space $L(0, 2\pi)$.

The relation (15) was also obtained by Smirnov [43] and Titchmarsh [44].

Further, in his monograph [42] Kolmogorov introduced (see pages 73–75 there) the notion of generalized mathematical expectation, which led to the definition of A-integral.[3] Namely, a function $\phi(t)$ is said to be A-integrable on an interval $[a, b]$ ($\phi \in A(a, b)$) if

$$\operatorname{mes} E\{t : t \in [a, b],\ |\phi(t)| > n\} = o(1/n) \text{ as } n \to \infty$$

and the limit

$$\lim_{n\to\infty} \int_a^b [\phi(t)]_n \, dt = I$$

exists, where $[\phi(t)]_n = \phi(t)$ for $|\phi(t)| \leq n$ and $[\phi(t)]_n = 0$ for $|\phi(t)| > n$. By definition, the number I is called the A-integral of the function $\phi(t)$ over the

[2] In general, the function $g(t)$ is not Lebesgue integrable on $[0, 2\pi]$

[3] The term A-integral was also introduced by Kolmogorov.

interval $[a,b]$:

$$(A)\int_a^b \phi(t)dt = I.$$

The notion of A-integral was also applied and developed further by many authors (Ul'yanov, Tseretely, Ochan, Vinogradova, Khuskivadze, Bondi, Okano, Rubinshtein, Lukashenko, etc.) in their works relating to the study of various topics in the theory of functions (trigonometric series, boundary properties of analytic functions, integration theory, etc.). In particular, it was shown (see Titchmarsh [44] and Bari [3], Chapter 8) that for a function $f \in L(0,2\pi)$, its conjugate function g belongs to $A(0,2\pi)$ and the conjugate series (of the Fourier series of f) is the Fourier series of g in the sense of A-integration. Furthermore, Ul'yanov [53] showed that if a region G is bounded by a sufficiently smooth closed curve Γ, then every Cauchy-Lebesgue type integral

$$\Phi(z) = \frac{1}{2\pi i}(L)\int_\Gamma \frac{f(\zeta)}{\zeta - z}d\zeta \quad (z \in G,\ f(\zeta) \in L(\Gamma))$$

is a Cauchy A-integral, that is,

$$\Phi(z) = \frac{1}{2\pi i}(A)\int_\Gamma \frac{\Phi_i(\zeta)}{\zeta - z}d\zeta \quad \text{for } z \in G,$$

where the $\Phi_i(\zeta)$ are the limiting angular values of $\Phi(z)$ inside the region G as $z \to \zeta \in \Gamma$. On the other hand, Khuskivadze [54] proved that this result is no longer valid when the contour Γ is insufficiently smooth (for example, contains corner points).

References [4]

1. J. Marcinkiewicz, 'Sur les séries de Fourier', *Fund. Math.* **27** (1936), 38–69.
2. G. Sunouchi, 'Fourier series which belongs to the class H diverges almost everywhere', *Kodai Math. Sem. Rep.* **1** (1953), 27–28.
3. N.K. Bari, *Trigonometric series*, Fizmatgiz, Moscow, 1961 (in Russian).
4. A. Zygmund, *Trigonometric series*, Vols. 1, 2, Cambridge Univ. Press, 1959.

[4] The list of all works whose content is closely related to Kolmogorov's studies on the theory of trigonometric and orthogonal series is enormous. It is not even possible to indicate their approximate number. As to the papers and books mentioned here, they form a small part of this total list.

5. V.I. Prokhorenko, On divergent Fourier series', *Mat. Sb.* **75** (1968), 185–198 (in Russian).

6. K. Tandori, 'Ein Divergenz für Fourierreihen', *Acta Sci. Math.* **30** (1969), 43–48.

7. K. Zeller, 'Über Konvergenzmengen von Fourierreihen', *Arch. Math.* **6** (1955), 335–340.

8. T. Körner, 'Sets of divergence for Fourier series', *Bull. London Math. Soc.* **3** (1971), 152–154.

9. P.L. Ul'yanov, 'Divergent Fourier series of class L^p ($p \geq 2$)', *Dokl. Akad. Nauk SSSR* **137** (1961), 786–789 (in Russian).

10. B.S. Kašin, 'On Weyl's multipliers for almost everywhere convergence of orthogonal series', *Anal. Math.* **2** (1976), 249–266.

11. K. Tandori, 'Einfacher Beweis eines Satzes von B.S. Kashin', *Acta Sci. Math.* **39** (1977), 175–178.

12. D. Menchoff, 'Sur les séries de fonctions orthogonales bornées dans leur ensemble', *Mat. Sb.* **3** (1938), 103–120.

13. D. Menchoff, 'Sur les multiplicateurs de convergence pour les séries de polynomes orthogonaux', *Mat. Sb.* **1** (1936), 27–52.

14. Z. Zahorski, 'Une série de Fourier permutée d'une fonction de classe L^2 divergente presque partout', *C.R. Acad. Sci. Paris* **251** (1960), 501–503.

15. P.L. Ul'yanov, 'Divergent Fourier series', *Uspekhi Mat. Nauk* **16**:3 (1961), 61–142 (in Russian).

16. A.M. Olevskii, 'Divergent series in L^2 with respect to complete systems', *Dokl. Akad. Nauk SSSR* **138** (1961), 545–548 (in Russian).

17 S.V. Bochkarev, ' Fourier series with respect to an arbitrary bounded orthonormal system and divergent on a set of positive measure', *Mat. Sb.* **98** (1975), 436–449 (in Russian).

18. K.S. Kazaryan, 'On some questions in the theory of orthogonal series', *Mat. Sb.* **119**:2 (1982), 278–294 (in Russian).

19. J.E. Littlewood and R.E.A.C. Paley, 'Theorems on Fourier series and power series', *J. London Math. Soc.* **6** (1931), 230–233.

20. R.P. Gosselin, 'On the divergence of Fourier series', *Proc. Amer. Math. Soc.* **9** (1958), 278–282.

21. A. Zygmund, 'On the convergence of lacunary trigonometric series', *Fund. Math.* **16** (1930), 90–107.

22. S. Kaczmarz and H. Steinhaus, *Theorie der Orthogonalreihen*, Warszawa, Lvov: Z Subwencji Funduszu Kultury Narodwej, 1935.

23. V.F. Gaposhkin, 'Lacunary series and independent functions', *Uspekhi Mat. Nauk* **21**:6 (1966), 3–82 (in Russian).

24. A. Plessner, 'Über Konvergenz von trigonometrischen Reihen', *J. Reine Angew. Math.* **155** (1925), 15–25.

25. L. Carleson, 'On convergence and growth of partial sums of Fourier series', *Acta Math.* **116** (1966), 135–157.

26. L. Carleson, 'On convergence and growth of partial sums of Fourier series', *Matematika* **4** (1967), 113–132 (Russian transl. of 25).

27. R. Hunt, 'On the convergence of Fourier series', In: *Orthog. Expan. and their Contin. Anal. : Proc. Conf. Carbondale*: Illinois Press, 1968, pp. 235–255.

28. P. Sjölin, 'An inequality of Paley and convergence a.e. of Walch-Fourier series', *Ark. för Mat.* **7** (1969), 551–570.

29. C. Fefferman, 'Pointwise convergence of Fourier series', *Ann. of Math.* **98** (1973), 551–571.

30. T.P. Lukashenko, *Convergence almost everywhere of series of square integrable functions*, MGU Press, Moscow, 1978, pp. 3–109 (in Russian).

31. S. Kaczmarz, 'Sur la convergence et la sommabilité des développements orthogonaux', *Stud. Math.* **1** (1929), 87–121.

32. K. Tandori, 'Genaue Weylsche Multiplikatorfolgen', *Acta Sci. Math.* **20** (1959), 1–13.

33. C.H. Moore, 'On criteria for Fourier constants of L integrable functions', *Proc. Nat. Acad. Sci. USA* **19** (1933), 846–848.

34. C.H. Moore, 'On the use of Cesàro means in determining criteria for Fourier constants', *Bull. Amer. Math. Soc.* **39** (1933), 907–913.

35. L. Cesari, 'Sulle condizioni sufficenti per le successioni di Fourier', *Ann. Scuola Norm. Sup. Pisa* **3** (1934), 105–134.

36. S. Sidon, 'Hinreichende Bedingungen für den Fourier-Charakter einer trigonometrischen Reihe', *J. London Math. Soc.* **14** (1939), 158–160.

37. R.P. Boas, 'Absolute convergence and integrability of trigonometric series', *J. Rational. Mech. and Anal.* **5** (1956), 621–632.

38. S.A. Telyakovskii, 'Integrability conditions for trigonometric series and their application to the study of linear summability methods of Fourier

series', *Izv. Akad. Nauk SSSR Ser. Mat.* **28** (1964), 1209–1236 (in Russian).

39. S.A. Telyakovskii, 'On a sufficient condition of Sidon for integrability of trigonometric series', *Mat. Zametki* **14** (1973), 317–328 (in Russian).
40. G.A. Fomin, 'On a class of trigonometric series', *Mat. Zametki* **23** (1978), 213–222 (in Russian).
41. J.E. Littlewood, 'On the mean values of power series', *Proc. London Math. Soc.* **25** (1924), 328–337; *J. London Math. Soc.* **5** (1930), 179–182.
42. A.N. Kolmogorov, *Basic notions of probability theory*, ONTI, Moscow, Leningrad, 1936; 2nd ed. Nauka, Moscow, 1974 (in Russian).
43. V. Smirnoff, 'Sur les valeurs limites des fonctions analytiques', *C.R. Acad. Sci. Paris* **188** (1929), 131–133.
44. E.C. Titchmarsh, 'On conjugate functions', *Proc. London Math. Soc.* **29** (1929), 49–80.
45. J.E. Littlewood, 'On a theorem of Kolmogoroff', *J. London Math. Soc.* **1** (1926), 229–231.
46. G.H. Hardy, 'Remarks on three recent notes in the Journal', *J. London Math. Soc.* **3** (1928), 166–169.
47. L. Loomis, 'A note on Hilbert's transform', *Bull. Amer. Math. Soc.* **52** (1946), 1082–1084.
48. J. Marcinkiewicz, 'Sur l'interpolation d'opérations', *C. R. Acad. Sci. Paris* **208** (1939), 1272–1273.
49. J. Marcinkiewicz, *Collected papers*, Warszawa, 1964.
50. E.M. Nikishin, 'Resonance operators and superlinear operators', *Uspekhi Mat. Nauk* **25**:6 (1970), 120–191 (in Russian).
51. L.V. Zhizhiashvili, *Conjugate functions and trigonometric series*, Tbilisi Univ. Press, 1969 (in Russian).
52. I.P. Natanson, 'On the problem of expanding functions in orthonormal polynomials', *Izv. Akad. Nauk SSSR Ser. Fiz.-Mat.* 1 (1933), 85–88 (in Russian).
53. P.L. Ul'yanov, 'On Cauchy A-integrals for contours', *Dokl. Akad. Nauk SSSR* **112**:3 (1957), 383–385 (in Russian).
54. G.A. Khuskivadze, 'On conjugate functions and Cauchy singular integrals', *Dokl. Akad. Nauk SSSR* **140**:6 (1961), 1270–1273 (in Russian).

DESCRIPTIVE SET THEORY

(I.I. Parovichenko)

Paper No. 13 laid the foundations of the general theory of operations on sets. The history of the problem goes back to the French mathematical school at the beginning of the 20th century. In work by Borel, Baire, Lebesgue, and others, there arose the problem of describing procedures which, proceeding from an initial class of sets of a simple structure (first of all, sets on the number line), would lead to a meaningful description of more complicated sets.

Because of the work by Luzin, this problem attracted attention in the USSR, and soon a remarkable result was obtained. In connection with the solution of the problem of the cardinality of Borel sets or, as they are called, B-sets, P.S. Aleksandrov [10] introduced the notion of A-operation and Suslin [14] proved that application of the A-operation to closed sets on the line may result in sets whose complementary sets cannot be constructed in the same way. This implied the existence of A-sets that were not B-sets.

In paper No. 13 by Kolmogorov the notion of δs-operation is introduced. A δs-operation Φ is determined by a collection of subsets $\mathfrak{B}_0 = \{\mathfrak{z}\}$, $\mathfrak{z} = \{n_i\}_{i \in N}$ (number chains) of the natural sequence and a collection of sets $e = \{e(n)\}$, $n \in N$. There corresponds to each $\mathfrak{z} \in \mathfrak{B}_0$ a chain of sets $\{e(n_i)\}_{i \in N}$, $n_i \in \mathfrak{z}\}$ and the kernel $\bigcap_i e(n_i)$ of the chain. The union of the kernels of all chains corresponding to the given collection of number chains is the result of application of the operation Φ to the given collection of sets e. Thus,

$$\Phi(e) = \bigcup_{\mathfrak{z} \in \mathfrak{B}_0} \bigcap_{n \in \mathfrak{z}} e(n), \tag{1}$$

where $e : I \to \mathfrak{P}(Y)$, $\mathfrak{B}_0 \subset \mathfrak{P}(I)$ and I is a countable set. The author proves that there is a Φ-set, that is, a set obtained by means of a δs-operation from closed sets on the line, whose complementary set is not a Φ-set. (The conclusion of this theorem will be called the K-property of the corresponding space.)

A special recently published monograph [5] is devoted to the general theory of operations on sets, with the above paper by Kolmogorov as the starting point. Below we discuss some papers relating to this theory.

In [12] δs-operations were generalized to *set-theoretic operations* (STO), determined by Boolean polynomials in principal disjunctive normal form:

$$\Phi(e) = \bigcup_{\mathfrak{z} \in \mathfrak{B}} \Big(\bigcap_{n \in \mathfrak{z}} e(n) \cap \bigcap_{n \in I \setminus \mathfrak{z}} (Y \setminus e(n))\Big), \tag{1*}$$

where I is arbitrary. $\mathfrak{B}$ is called the *base* of the operation, and among the STO's the δs-operations are characterized by *base extensiveness* [12]:

$$\forall A \forall B,\ A \in \mathfrak{B}\ \&\ A \subseteq B \subseteq I \Rightarrow B \in \mathfrak{B}.$$

The system $\mathfrak{B}_0$ in (1) is called a *subbase* of $\mathfrak{B}$, since the smallest extensive system containing $\mathfrak{B}_0$ is none other than the base of $\mathfrak{B}$ for Φ in the sense of (1*). STO's are distinguished in the class of abstract functionals $e \to \Phi(e) \subseteq Y$, $e : I \to \mathfrak{P}(Y)$, by means of the following axiom: *for all* Y, Z *and* $f : Z \to Y$ *the formula* $\Phi(h \circ e) = h(\Phi(e))$ *holds*, where $h : A \to f^{-1}[A]$, and for these functionals the corresponding base of (1*) is given by the formula $\mathfrak{B} = \Phi(e_0)$ where $e_0 : I \to \mathfrak{P}(\mathfrak{P}(I))$ and $\forall n,\ e_0(n) = \{B | n \in B \subseteq I\}$. In particular, this provides a method for reducing any Boolean polynomial to principal disjunctive normal form (see [4, 5]).

The main theorem in the paper remains true without any changes for STO's as well. In this form it was generalized later. First of all, it was noted that for the K-property to exist in a space X it is sufficient that X topologically contain the Cantor discontinuum $D^{\aleph_0}$ (before the publication of [3] the unnecessary requirement that $D^{\aleph_0}$ be closed in X was added). Another generalization begins with replacing card $I = \aleph_0$ by card $I = \aleph_\alpha$; such an STO is called an $\aleph$-operation. In [9] Shneider proved that for an $\aleph_\alpha$-operation the K-property holds for the Tikhonov power $D^{\aleph_\alpha}$. Then in [3, 4], under the condition $\aleph_\alpha = \sum_{\beta<\alpha} 2^{\aleph_\beta}$, the K-property was proved for the $\aleph_\alpha$-operation in the *transfinite arithmetic continuum* C_α (see [2]). C_α is a transfinite generalization of the number line obtained by Dedekind's completion of a linearly ordered set of Hausdorff normal type η_α (see the first edition of Hausdorff's monograph [11]). Finally, in [8] Choban proved for the $\aleph_\alpha$-operation, that for the K-property in a space X to hold it is sufficient that X contains a regular subspace X_0 of weight $\aleph_\alpha$ that can be mapped continuously onto $D^{\aleph_\alpha}$ or $J^{\aleph_\alpha}$ ($J = [0,1]$); this theorem generalizes the foregoing ones and gives a new result, even in the classical case $\alpha = 0$.

In [6] Ponomarev proved Suslin's theorem for the first time, as well as the theorem on the non-emptiness of classes of Borel sets in a conceptually different situation, namely for perfect normal bicompacta. In [3] the theorem on the non-emptiness of classes was proved in the formulation of Kolmogorov's paper under discussion. Finally, in [7] this theorem was generalized to $\aleph_\alpha$-operations.

In conclusion we present the definition of an R-operation and a theorem on this operation stated and proved by Kolmogorov, which he did not publish but allowed Kantorovich and Livenson to include them in their paper [13]. Let $\Phi_{\mathfrak{B}_0}$ be a δs-operation with the natural sequence as index set and with subbase $\mathfrak{B}_0$. By an *R-operation* on $\Phi_{\mathfrak{B}_0}$, denoted by $R\Phi_{\mathfrak{B}_0}$ is meant a δs-operation whose index set is the set K of all ordered n-tuples of natural numbers and whose subbase is the system of all sets $M \subseteq K$ for which the following conditions hold for each $(s_1, s_2, \dots, s_n) \in K$:

$$(1) \quad \forall i \leq n, (s_1, s_2, \dots, s_i, \dots, s_n) \in M \Rightarrow (s_1, s_2, \dots, s_i) \in M,$$

$$(2) \quad \{s_1 | (s_1) \in M\} \in \mathfrak{B}_0 \text{ and } (s_1, s_2, \dots, s_n) \in M \Rightarrow$$
$$\Rightarrow \{s_{n+1} | (s_1, s_2, \dots, s_n, s_{n+1}) \in M\} \in \mathfrak{B}_0.$$

For example, an A-operation is an R-operation on the operation of countable summation.

Theorem. *If the power of the operation $\Phi_{\mathfrak{B}_0^+}$ is greater than that of the operation $\Phi_{\mathfrak{B}_0}$, then the power of $R\Phi_{\mathfrak{B}_0^+}$ is greater than that of $R\Phi_{\mathfrak{B}_0}$.*

R-operations were studied later (for example, see [1, 5]). In particular, the above theorem has recently been re-proved and generalized (see [5]).

Paper No. 59 was devoted to the 70th anniversary of P.S. Aleksandrov's birthday.

We now make some supplementary remarks to the definition of a Γ-operation. The base of the operation complementary to a given δs-operation consists of all sets intersecting each set in the base of the initial operation (cf. §2 in paper No. 13). This readily implies the characteristic property of Aleksandrov's Γ-chains presented in the paper under discussion in terms of Baire intervals Δ_s, since by definition, the complementary A-operation has as subbase a system of sets of the form $\{(s_1), (s_1, s_2), (s_1, s_2, s_3), \dots\}$ for all sequences $s_1, s_2, s_3, \dots$ of natural numbers ("regular chains of ordered n-tuples"). Baire intervals either do not intersect or are contained in one another, and $\Delta_{s^+} \subset \Delta_s$ if the n-tuple s^+ continues s. Therefore the maximal intervals of any covering of the space of sequences by Baire intervals corresponding to a Γ-chain $\mathfrak{S}$ form a disjunctive and therefore minimal subcovering to which a minimal Γ-chain $\mathfrak{S}_0 \subset \mathfrak{S}$ corresponds. Minimal Γ-chains form a subbase of the Γ-operation, whose formula can thus be written more economically, namely using only the Γ-chains $\mathfrak{S}_0$

corresponding to disjunctive coverings by Baire intervals. A subbase of a δs-operation is said to be *rigid* if its sets are not contained in one another. Along with the Γ-operation, the A-operation also possesses a rigid subbase; its role is played by the system of all regular chains of ordered n-tuples in the original definition, which is obvious in this case.

References

1. A.A. Lyapunov, 'R-sets', *Trudy Inst. Mat. Akad. Nauk SSSR* **40** (1953), 1–68 (in Russian).
2. I.I. Parovichenko, 'On the theory of sets not satisfying the countability axiom', *Uchen. Zap. Kishinev. Ser. Fiz.-Mat.* **29** (1957), 15–24 (in Russian).
3. I.I. Parovichenko, 'On descriptive set theory in topological spaces', *Dokl. Akad. Nauk SSSR* **196**:5 (1971), 1024–1027 (in Russian).
4. I.I. Parovichenko, 'On set-theoretic operations', *Dokl. Akad. Nauk SSSR* **201**:1 (1971), 40–43 (in Russian).
5. I.I. Parovichenko, *Theory of operations on sets*, Shtiintsa, Kishinev, 1981 (in Russian).
6. V.I. Ponomarev, 'On Borel sets in perfectly normal spaces', *Dokl. Akad. Nauk SSSR* **170**:3 (1966), 520–523 (in Russian).
7. M.M. Choban, 'Modifications of topology and non-empty classes', *Serdika (B"lg. Mat. Spisanie)* **1** (1973), 133–143 (in Russian).
8. M.M. Choban, 'On operations on sets', *Sibirsk. Mat. Zh.* **16**:6 (1975), 1332–1351 (in Russian).
9. V.E. Schneider, 'Descriptive set theory in topological spaces', *Uchen. Zap. MGU Ser. Mat.* **135**:2 (1949), 37–85 (in Russian).
10. P. Alexandroff, 'Sur la puissance des ensembles mesurables B', *C. R. Acad. Sci. Paris* **162** (1916), 323–325.
11. F. Hausdorff, *Grundzüge der Mengenlehre*, Leipzig, 1914.
12. L. Kantorovich and E. Livenson, 'Memoir on analytical operations and projective sets (1)', *Fund. Math.* **18** (1932), 234–279.
13. L. Kantorovich and E. Livenson, 'Memoir on analytical operations and projective sets (2)', *Fund. Math.* **20** (1933), 55–57.
14. M. Souslin, 'Sur une définition des ensembles mesurables B sans nombres transfinis', *C. R. Acad. Sci. Paris* **164** (1917), 88–91.

MEASURE THEORY AND THEORY OF THE INTEGRAL

(V.A. Skvortsov)

On the paper "Studies on the concept of integral"

The versions of the definition of integral introduced in paper No. 16 are known in the literature as "Kolmogorov's integral" for finite or countable partitions [1]. The construction of Kolmogorov's integral is presented in the textbook [2] and monographs [3, 4]. The significance of Kolmogorov's integral in the development of the theory of the integral is indicated in the monographs [5, 6] and the papers [7, 8]. The construction of Kolmogorov's integral was a result of synthesis and profound generalization of hitherto diverse investigations on the theory of Stieltjes-type integrals for functions on abstract sets and, in particular, of the ideas of Fréchet, Moore and Burkill.

Later Kolmogorov's integral and its generalizations were studied, for instance, in [3, 4, 9–23]. In particular, the interrelation between the two versions of Kolmogorov's integral (for finite and countable partitions) and the relation of these integrals to other types of integral were considered in [3, 13, 17, 23]. Kolmogorov's integral was successfully applied to finding new important properties of various types of integrals which are special cases of Kolmogorov's integral [12]. In [3] double and repeated Kolmogorov integrals are studied and the corresponding version of Fubini's theorem is established. In [3] a study is made of some applications of Kolmogorov's integral to functional analysis. It has been shown that for certain spaces, general linear operators and functionals admit of a representation in the form of the corresponding Kolmogorov integral; existence and uniqueness theorems have been proved for certain non-linear integral equations involving such integrals. Kolmogorov's integral also found application in mathematical physics [20, 24]. This integral was also considered for the case of functions with range in a topological group [22].

An important role in the further development of the theory of differentiation of additive set functions has been played by Appendix 2 to paper No. 16 (see [4]). In it, the possibility of differentiating an additive set function with respect to another set function is established, irrespective of any geometrical properties of the space. It was probably noted for the first time that it is possible to define differentiation via integration, that is, the definition of the derivative of a set function can be based on its integral representation. In modern literature the derivative in this sense is often called the Radon-Nikodym derivative [25].

The idea put forward in the introduction to paper No. 16 (see also paper No. 6), namely that attempts to combine within a unified scheme "absolute" integrals and integrals involving the order relation on the number line inevitably encounter conceptual difficulties, was confirmed and developed later in many works devoted to the interrelation between the various notions of the integral. In this connection we should mention numerous papers devoted to the A-integral, whose definition (in the form of a generalized mathematical expectation) was also stated by Kolmogorov [26]. The fact that the A-integral does not depend on a measure-preserving change of variable (although it is not an absolute integral) underlies the examples showing that it can contradict Denjoy integrals and other ordered integrals (see [27–29]). We note that in recent decades, the theory of ordered integration has seen a number of definitions of generalized integrals meant for solving various problems in function theory. These definitions do not admit of a unified scheme either, since they can result in different values of the integral on the intersection of the classes of functions for whose integration they are introduced [30–32].

On the paper "On the limits of generalization of the integral"

Paper No. 6 (this is its first publication) is not only devoted to proving the assertions stated without proof in paper No. 5, but also contains a number of results that were not published earlier. It would be interesting to compare the description of the class of generalizations of Denjoy's broad integral based on the introduction of a generalized notion of density of a set with the various generalizations of Denjoy's integral in modern literature (for example, see [33]). The remarks in the introduction to paper No. 6 on the importance of considering discontinuous primitive functions in some branches of mathematics (they were also made earlier by Luzin [34]), were confirmed, in particular, in many works on the application of the theory of integration to the theory of orthogonal series [32, 35–37]. However, no attempts have yet been made to construct an axiomatization of these integrals similar to the one realized in papers Nos. 5 and 6.

Paper No. 7 is also related to the ideas of the limits of possible generalizations of the notion of integral as developed in the introduction to paper No. 6.

On the paper "On the Denjoy integration process"

The result of this paper is presented in the monograph [38]. Similar questions were considered in [39–41]. See also [42–46] and the commentary "Trigonometric and orthogonal series" by Ul'yanov.

On the paper "On measure theory"

The significance of paper No. 21 for general measure theory was noted in many reviews and monographs (see [8, 47, 48]).

The peculiarity of this paper is that it considers problems of measure theory not only from the geometrical, but also from the set-theoretic viewpoint. Besides the standard conditions of semi-additivity and σ-additivity (conditions I and II) and the normalization condition (IV), Kolmogorov imposes on functions of sets lying in the Euclidean space R^n the following geometrical condition (III): under non-expanding mappings the measure of the set does not increase. This results in the existence of two measures for $k \leq n$, the upper and lower measures $\bar{\mu}^k$ and $\underline{\mu}^k$, such that any k-dimensional measure μ satisfies the inequality $\underline{\mu}^k(E) \leq \mu(E) \leq \bar{\mu}^k(E)$ for any A-set $E \subset R^n$, and it turns out that $\underline{\mu}^k(E) = \bar{\mu}^k(E)$ if E is the image of a set belonging to R^n under a non-expanding mapping.

The ideas and methods developed in this paper were widely applied in subsequent studies. A measure function satisfying conditions I to IV stated in the introduction to paper No. 21 received in the literature the name Kolmogorov measure, and the conditions themselves were called Kolmogorov axioms. In particular, these axioms are discussed in [48–50].

Besicovitch [51] investigated the nature of the class of sets whose Kolmogorov measure is single-valued in the case of a one-dimensional measure for sets in the plane. He also gave in [51] a negative answer to the hypothesis put forward at the end of §6 in paper No. 21, asserting that the Hausdorff length of a set and its Kolmogorov minimal linear measure coincide. This result was strengthened in the example (constructed by Vitushkin, Ivanov, and Mel'nikov [52]) of a set in the plane having finite positive Hausdorff length and zero Kolmogorov linear measure. Thus, these measures are not only different but even incommensurable.

The problem of uniqueness of a k-dimensional measure in an n-dimensional space posed in paper No. 21 was considered later in [53, 49, 55, 56].

The relationship between Kolmogorov measure and other measures was studied by Nöbeling [53, 54]. He proved that for surfaces of bounded extension many measures coincide with Kolmogorov measure although, in general, they do not satisfy Kolmogorov's axioms.

Some investigations devoted to k-dimensional measures in an n-dimensional space are related to the idea of introducing an axiomatization of measure function, making it possible to include measures related to projection and such that measure may increase under mappings not increasing the distances. Axiomatics of this kind were stated by Vitushkin [57].

On the paper "On the notion of the mean"

For the role of the generalized form of the mean found in paper No. 17 see the paper by Khinchin [47] (also see [59, 60]). Some results similar to the theorem in paper No. 17 were obtained by Nagumo [58]. For further extension of the axiomatics see the series of papers by Aczel [61, 62].

References

1. *Mathematical encyclopedia*, Vol. 2, Sovetskaya Entsiklopediya, Moscow, 1979 (in Russian).
2. V.I. Smirnov, *A course in higher mathematics*, Vol. 5, Gostekhizdat, Moscow-Leningrad, 1947 (in Russian).
3. V.I. Glivenko, *Stieltjes' integral*, ONTI, Moscow, Leningrad, 1936 (in Russian).
4. D.F. Goguadze, *On Kolmogorov integrals and some of their applications*, Metsniereba, Tbilisi, 1979, (in Russian).
5. F.A. Medvedev, *Development of the notion of integral*, Nauka, Moscow, 1974 (in Russian).
6. I.N. Pesin, *Development of the notion of integral*, Nauka, Moscow, 1966 (in Russian).
7. N.K. Bari, A.A. Lyapunov, D.E. Men'shov and G.I. Tolstov, 'Metrical theory of functions of a real variable', In: *Mathematics in the USSR during thirty years*, Gostekhizdat, Moscow, Leningrad, 1948 (in Russian).
8. P.S. Alexandrov and A.Yu. Khinchin, 'A.N. Kolmogorov (on his 50th birthday)', *Uspekhi Mat. Nauk* **8**:3 (1953), 194–200.

9. V.I. Glivenko, 'Attempt at a general definition of integral', *Dokl. Akad. Nauk SSSR* **14** (1937), 61–64 (in Russian).
10. L.Ya. Leifman, 'On conditions for the existence of the Kolmogorov integral and the notion of differential equivalence', *Uspekhi Mat. Nauk* **12**:3 (1957), 343–352 (in Russian).
11 L.Ya. Leifman, 'On passage to the limit under the Kolmogorov integral sign', *Izv. Vuzov. Mat.* **2** (1958), 182–196 (in Russian).
12. L.Ya. Leifman, 'Passage to the limit under the integral sign from the general point of view of the theory of the Kolmogorov integral', *Izv. Vuzov. Mat.* **1** (1960), 139–153 (in Russian).
13. M. Kotlar and Y. Frenkel, 'Sobre la intégral de Kolmogoroff', *Contrib. Cient. Ser. A Univ. Buenos Aires* **1**:3 (1950), 48–63.
14. B.S. Paliivets, 'On a generalized notion of integral', *Vestn. MGU Ser. Mat., Mekh.* **5** (1968), 31–40 (in Russian).
15. B.S. Paliivets, 'On the space of J-integrable functions', *Vestn. MGU Ser. Mat., Mekh.* **6** (1968), 28–35 (in Russian).
16. B.S. Paliivets, 'On the notion of diagonal equivalence of countably additive functions with a parameter', *Vestn. MGU Ser. Mat., Mekh.* **5** (1969), 37–43 (in Russian).
17. D.F. Protsenko, 'On the relation between two types of Kolmogorov integrals', *Uspekhi Mat. Nauk* **20**:5 (1965), 237–242 (in Russian).
18. B.C. Getchel, 'On the equivalence of two methods of defining Stieltjes integral', *Bull. Amer. Math. Soc.* **41** (1935), 413–418.
19. F. Maeda, 'Space of differential set functions', *J. Sci. Horoshima Univ. Ser. A* **6** (1936), 19–45.
20. F. Maeda, 'Representation of linear operators by differential set functions', *J. Sci. Hiroshima Univ. Ser. A* **6** (1936), 115–137.
21. T.H. Hildebrandt, 'Definitions of Stieltjes integrals of the Riemann type', *Amer. Math. Monthly* **49** (1938), 265–278.
22. I.Ya. Chernyavskii, 'A multiplicative integral in a topological group', *Izv. Vuzov. Mat.* **5** (1961), 102–111 (in Russian).
23. I.Yu. Chernyavskii, 'On conditions under which the Burkill and Kolmogorov integrals coincide', *Volzhsk. Mat. Sb.* **3** (1965), 339–342 (in Russian).
24. J.L. Destouches, *Corpuscules et systèmes de corpuscules*, Vol. 1, Paris, 1941.

25. G. de Barra, *Measure theory and integration*, London, 1981.

26. A.N. Kolmogorov, *Basic notions of probability theory*, ONTI, Moscow, Leningrad, 1936 (in Russian).

27. I.A. Vinogradova, 'On the indefinite A integral', *Izv. Akad. Nauk SSSR Ser. Mat.* **25**:1 (1961), 113–142 (in Russian).

28. I.A. Vinogradova, 'On representing a measurable function by an indefinite A integral', *Izv. Akad. Nauk SSSR Ser. Mat.* **26**:4 (1962), 581–604 (in Russian).

29. O.D. Tsereteli, 'An integral analogue of Riemann's theorem on convergence conditions for series', *Soobshch. Akad. Nauk Gruzin. SSR* **30**:4 (1963), 385–392 (in Russian).

30. V.A. Skvortsov, 'Du Bois-Reymond's theorem for generalized integrals and trigonometric series summable by the Abel-Poisson method', *Vestn. MGU Ser. Mat., Mekh.* **4** (1964), 16–20 (in Russian).

31. V.A. Sklyarenko, 'On Denjoy's integrable sums of everywhere convergent trigonometric series', *Dokl. Akad. Nauk SSSR* **210**:3 (1973), 533–536 (in Russian).

32. I.A. Vinogradova and V.A. Skvortsov, 'Generalized integrals and Fourier series', *Itogi Nauki Mat. Anal.* (1971), 65–107 (in Russian).

33. D.N. Sarkhel and A.K. De, 'The proximally continuous integrals', *J. Austral. Math. Soc. Ser. A* **31** (1981), 26–45.

34. N.N. Luzin, *The integral and trigonometric series*, Gostekhteorizdat, Moscow, Leningrad, 1951.

35. A. Denjoy, *Leçons sur le calcul des coefficients d'une série trigonométrique*, Paris, 1941–1949.

36. J. Marcinkiewicz and A. Zygmund, 'On the differentiability of functions and summability of trigonometrical series', *Fund. Math.* **26** (1936), 1–43.

37. J.C. Burkill, 'Integrals and trigonometrical series', *Proc. London Math. Soc.* **1**:1 (1951), 46–57.

38. A. Zygmund, *Trigonometric series*, Cambridge Univ. Press, 1959.

39. E.C. Titchmarsh, 'On conjugate functions', *Proc. London Math. Soc.* **29** (1929), 49–80.

40. V.I. Smirnov, 'Sur les valeurs limites des fonctions analytiques', *C. R. Acad. Sci. Paris* **188** (1929), 131–133.

41. V.I. Smirnov, 'Sur les valeurs limites des fonctions régulières à l'intérieur

d'un cercle', *Zh. Leningr. Fiz.-Mat. Obshch.* **2**:2 (1929), 22–37.

42. T.P. Lukashenko, 'Conjugate functions and the narrow Denjoy integral', *Mat. Sb.* **104**:1 (1977), 89–139 (in Russian).
43. T.P. Lukashenko, 'Denjoy integrals and conjugate functions' In: *Materials of the USSR School on function theory*, Baku, 1977, Azerb. Univ. Press, 1980, pp. 163–166 (in Russian).
44. T.P. Lukashenko, 'Conjugate functions and integrals', *Izv. Akad. Nauk SSSR Ser. Mat.* **36**:2 (1972), 435–449 (in Russian).
45. T.P. Lukashenko, 'Bok's integrable non-measurable functions', *Mat. Zametki* **17**:1 (1975), 49–56 (in Russian).
46. N.K. Bari, *Trigonometric series*, Fizmatgiz, Mosvow, 1961 (in Russian).
47. *Mathematics in the USSR during 15 years*, Gostekhteorizdat, Moscow, 1932 (in Russian).
48. H. Hadwiger, *Vorlesungen über Inhalt, Oberfläche und Isoperimetrie*, Springer-Verlag, Berlin-Göttingen-Heidelberg, 1957.
49. L.D. Ivanov, *Variations of sets and functions*, Nauka, Moscow, 1975 (in Russian).
50. A.G. Vitushkin, *On multidimensional variations*, Gostekhizdat, Moscow, 1955 (in Russian).
51. A.S. Besicovitch, 'On the Kolmogoroff maximum and minimum measures', *Math. Ann.* **113**:3 (1936), 416–423.
52. A.G. Vitushkin, L.D. Ivanov and M.S. Mel'nikov, 'Incommensurability of minimal linear measure and length of a set', *Dokl. Akad. Nauk SSSR* **151**:6 (1963), 1256–1259 (in Russian).
53. G. Nöbeling, 'Über den Flächeninhalt dehnungsbeschränkter Flächen', *Math. Z.* **48** (1942), 747.
54. G. Nöbeling, 'Über die Flächenmasse in Euklidischen Raum', *Math. Ann.* **118** (1943), 687–701.
55. H. Federer, 'The (ϕ, k) rectifiable subsets of n space', *Trans. Amer. Math. Soc.* **62**:1 (1945), 114–142.
56. V.A. Zhil'tsov, 'On uniqueness of hypermeasure in Euclidean spaces', *Izv. Akad. Nauk SSSR Ser. Mat.* **37**:6 (1973), 1428–1436 (in Russian).
57. A.G. Vitushkin, 'Proof of the upper semicontinuity of variations of sets', *Dokl. Akad. Nauk SSSR* **166**:5 (1966), 1022–1026 (in Russian).

58. M. Nagumo, 'Über eine Klasse der Mittelwerte', *Jap. J. Math.* **7** (1930), 71–79.

59. B. de Finetti, 'Sul concetto di media', *G. Ist. Ital. Attu. Ann.* **2** (1931), 369–396.

60. G.H. Hardy, J.E. Littlewood and G. Pólya, *Inequalities*, Cambridge Univ. Press, 1934.

61. J. Aczél, 'The notion of mean values', *Kgl. Norske Vid. Selsk. Forh.* **19**:23 (1946), 83–86.

62. J. Aczél, 'A generalization of the notion of convex functions', *Kgl. Norske Vid. Selsk. Forh.* **19**:24 (1946), 87–90.

POINTS OF DISCONTINUITY OF FUNCTIONS

(E.P. Dolzhenko)

In papers Nos. 22 and 24 some delicate metrical properties of derivatives of bounded discontinuous functions were discovered. Here for the first time the basic theorem on contingencies of plane sets was established, which reads: for an arbitrary set in the plane, the set of points at which the contingency of the given set is neither the plane nor a half-plane nor a straight line has zero length, and all points at which the contingency is not the plane are located on a countable number of rectifiable curves. The theorem is obtained as a simple consequence of the main result of these papers, asserting that the set of non-ordinary points of a function is located on a countable number of rectifiable curves and that the set of non-linear non-ordinary points has zero length on each of these curves. Simultaneously, (1934) Besicovitch published his result, according to which the points of a plane set at each of which the set has a tangent (that is, the contingency is a straight line or a ray) is the union of at most a countable number of sets of finite length.

The basic theorem on contingencies is a clear and very important means in the study of goemetrical properties of sets and in the theory of differentiation of functions. In particular, this theorem gives a very simple geometrical interpretation of Denjoy's theorem on derived numbers. A little later, Roger independently proved the basic theorem on contingencies for plane and spatial sets. Subsequently, the investigations of Verchenko and Kolmogorov were continued in the studies of Verchenko and Shmidov (1935). So-called metrical contingencies were considered by Shmidov (1943). Later the ideas used in the proof of contingency theorems were applied in geometrical measure theory, the theory of multidimensional variations, the theory of differentiation of functions, and the theory of monogenic functions and mappings.

THEORY OF APPROXIMATION

(S.A. Telyakovskii and V.M. Tikhomirov)

1. Paper No. 27 generated a great number of investigations, and gave rise to the creation of a new branch in the theory of approximation of functions.

The generalizations of Kolmogorov's results were carried out in different directions. Suprema of deviations were studied not only for partial sums of Fourier series but also for other types of means constructed using Fourier series or interpolation polynomials. The suprema of deviations were taken over other function classes, and deviations in other metrics were considered. This problem was also considered for approximation with algebraic polynomials, for functions of several variables, etc.

The first remarkable continuation of Kolmogorov's results was in a series of papers by Nikol'skii, published in 1940–46. Therefore, in relation to the study of the asymptotic behaviour of suprema of deviations,

$$C_n(\mathfrak{M}, u) = \sup_{f \in \mathfrak{M}} \|f(x) - u_n(f, x)\|, \tag{1}$$

one sometimes speaks of the Kolmogorov-Nikol'skii problem or of the Kolmogorov-Nikol'skii constants.

Here we mention only some results that are a direct continuation of Kolmogorov's work and are related to approximation using partial sums $s_n(f, x)$ of Fourier series.

Let W^p be the class of periodic functions f whose $(p-1)$th derivatives satisfy the Lipschitz condition

$$|f^{(p-1)}(x) - f^{(p-1)}(y)| \leq |x - y|, \quad p = 1, 2, \ldots .$$

Nikol'skii [1, 2] extended Kolmogorov's estimate to the class $\overline{W^p}$ of functions conjugate to those of class W^p.:

$$C_n(\overline{W^p}, s) = \frac{4}{\pi^2} \frac{\log n}{n^p} + O\Big(\frac{1}{n^p}\Big).$$

He also proved [3] that these two estimates are valid for approximations of the corresponding function classes in the metric of L.

Nikol'skii [4] also considered approximation of functions of class H_ω whose modulus of continuity $\omega(f,\delta)$ satisfies the condition $\omega(f,\delta) \le \omega(\delta)$, where $\omega(\delta)$ is a convex modulus of continuity, and proved the relation

$$C_n(H_\omega, s) = \frac{2\log n}{\pi^2}\int_0^{\pi/2} \omega\Big(\frac{2t}{n}\Big)\sin t\,dt + O\Big(\omega\Big(\frac{1}{n}\Big)\Big).$$

Efimov [5] showed that for many classes $\mathfrak{M}$ the determination of the principal asymptotic term of $C_n(\mathfrak{M}, s)$ can be reduced to finding the supremum of the nth Fourier coefficient for functions of a certain class.

The question whether the suprema of the deviations $C_n(\mathfrak{M}, s)$ can be attained on an individual function $f \in \mathfrak{M}$ has also been studied. For instance, Doronin [6] proved that in the class W^1 there is a function f such that

$$\varlimsup_{n\to\infty} \frac{\|f - s_n(f)\|}{C_n(W^1, s)} = 1,$$

and Oskolkov [7] showed that here limes superior cannot be replaced by limit. More precisely, he found that

$$\max_{f\in W^1} \varliminf_{n\to\infty} \frac{\|f - s_n(f)\|}{C_n(W^1, s)} = \tfrac{1}{2}.$$

Interest was shown in studying the dependence on the parameter p of the remainder in Kolmogorov's asymptotic formula. Sokolov [8] established the inequality

$$C_n(W^p, s) \le \frac{4}{\pi^2}\frac{\log n}{n^p} + \frac{A}{n^p},$$

where A is an absolute constant. Telyakovskii [9] proved that the asymptotic relation

$$C_n(W^p, s) = \frac{4}{\pi^2 n^p}\log\frac{n}{\min(p,n)} + O\Big(\frac{1}{n^p}\Big)$$

holds uniformly with respect to p and n. This estimate yields the principal asymptotic term for $C_n(W^p, s)$ under the condition that $p = o(n)$, and shows that when this condition is not fulfilled, then $C_n(W^p, s) = O(1/p^n)$ uniformly in n and p. The principal asymptotic term for this case was found by Stechkin [10].

Approximation of functions of several variables with rectangular partial sums of Fourier series was studied by Bugaets, Stepanets, and others.

2. Paper No. 28 determined a new direction of investigation in approximation theory. Here a new characteristic of approximative properties of function classes was introduced, later called Kolmogorov's n-width or simply Kolmogorov's width. Since the beginning of the sixties much attention has been drawn to widths of function classes.

Using geometrical language, we can say that in works by Chebyshev and his followers distances from specific functions to specific sets were studied (sets of algebraic or trigonometric polynomials and rational functions). In the works by de la Vallée-Poussin, Bernshtein, Jackson, Favard, and others, deviations of given function classes from specific sets were studied.

In paper No. 28 Kolmogorov suggested that, given a class of functions, those sets should be studied for which the corresponding deviations are least. More precisely, in this paper he posed and for two special cases solved the problem of finding best approximating subspaces of a prescribed dimension. Subsequently, similar characteristics (other types of widths) were studied. In this commentatry we confine ourselves to Kolmogorov widths.

In the paper under consideration, Kolmogorov investigated widths of the function classes now known as Sobolev classes. The Sobolev class $F_p^{q^*}$ ($q \geq 1$, $p > 0$) is the class of periodic functions with absolute value of the pth derivative (derivative understood in Weyl's sense) bounded by unity in the space L^q. The corresponding non-periodic class is denoted by F_p^q. In paper No. 28, the widths of the classes $F_p^{2^*}$ and F_p^2 for integral values of p were studied. It was found that in the periodic case the best approximating subspace of dimension $2n+1$ is the space of trigonometric polynomials of degree n, and that in the non-periodic case this is a special subspace of functions satisfying a certain differential equation of the $2p$th order.

The appearance of the trigonometric system for approximations in the periodic case in the space L^2 looked so natural, that in paper No. 28 Kolmogorov put forward the assumption that this classical approximation apparatus is unique (to within a linear transformation). However, it turned out that the situation is more complicated. Namely, it was found that in the problem of widths non-uniqueness of the solution is essential; this will be discussed below.

Following paper No. 28 and to the end of the fifties, many papers were devoted to Kolmogorov widths. Rudin [11] investigated the widths of the classes F_1^1 in the L^2 metric. In [12, 13] the width of an n-dimensional octahedron in a Hilbert space was calculated; the results of these papers were applied by Stechkin [14] to evaluate the order of the widths of the classes F_p^∞ and F_p^1 in the spaces C and L^2 respectively.

At the end of the fifties works appeared that aroused new interest in these problems.

In [15, 16] and some subsequent papers (some results of these studies are summarized in [17]) Babenko started to investigate widths of classes of functions of several variables and related these questions to problems of numerical mathematics.

In papers by Tikhomirov [18, 19] the exact values of Kolmogorov widths of the classes F_p^∞ and $F_p^{\infty *}$ in the space C were calculated using topological methods, which were for the first time applied to this group of problems.

It was found that in this case the non-uniqueness of the best approximation apparatus is essential and, in particular, it was shown that even in the periodic case, along with the classical approximation apparatus (the space of trigonometric polynomials), Sobolev classes of spline spaces yield approximations of the same accuracy.

Subsequently, the problem of the order of widths of Sobolev classes $F_p^{q^*}$ in the $L^{q'}$ metric was investigated completely. It was found that in this case, for $q \geq q'$ and for $q < q'$, $1 \leq q' \leq 2$, the solution to the problem is given by trigonometric polynomials (for these results see [20]). Ismagilov [21] proved for the first time that there exist pairs $F_p^{q^*}, L^{q'}$ for which trigonometric approximation is optimal as regards order. The complete solution to the problem of the order of widths of Sobolev classes was obtained by Kashin [22]. Later it was found that in many cases the best order is given not by trigonometric polynomials but by the subspace spanned on exponential functions with distinct harmonics. The exact solution was found by Micchelli, Pinkus, etc. (see [27]).

Korneichuk [23] solved exactly the fundamental problem on widths in the space C for the class of periodic functions with a given estimate for the modulus of continuity of the pth derivative.

The widths of certain classes of analytic functions were also found. In this case the statement of the problem of widths led to conceptually new methods (Erokhin [24]).

The description of all extremal subspaces of an ellipsoid (the classes studied in paper No. 28 are special cases of ellipsoids) was given by Karlovitz [25]. For a detailed bibliography see [27].

3. Paper No. 49 was devoted to the extension to complex functions of certain theorems in the theory of uniform approximation known for functions of a real variable. Here Kolmogorov studied polynomial approximation with respect to a system of functions $\phi_1(z), \dots, \phi_n(z)$ and a generalization of Chebyshev's theorem on the characteristic property of the polynomial of best approximation and Haar's uniqueness theorem.

The characteristic property of the polynomial of best approximation contained in Theorem 1 was known before (Tonelli [26]) for approximation with classical polynomials, that is, for $\phi_k(z) = z^{k-1}$.

After Kolmogorov's paper No. 49 there appeared investigations of characteristic properties of best approximation elements and uniqueness problems in more general situations. Functions with range in abstract spaces and approximation to elements of Banach spaces were considered.

For example, Nikol'skii [28] found a criterion for the best approximation element when approximating in a complex Banach space by elements of an arbitrary space. The criterion is stated in terms similar to those in Theorem 1 in paper No. 49, and instead of values of functions at points, it involves values of linear functionals on elements of the Banach space. The question of what set of functionals should be taken in this case was studied by Singer [29], Choquet [30] and Garkavi [31].

References

1. S.M. Nikol'skii, 'Asymptotic estimate of the remainder when approximating by Fourier sums', *Dokl. Akad. Nauk SSSR* **32** (1941), 386–389 (in Russian).
2. S.M. Nikol'skii, 'Approximation of periodic functions by trigonometric polynomials', *Trudy Mat. Inst. Akad. Nauk SSSR* **15** (1945) (in Russian).
3. S.M. Nikol'skii, 'Approximation in the mean of functions by trigonometric polynomials', *Izv. Akad. Nauk SSSR Ser. Mat.* **10** (1946), 207–256 (in Russian).
4. S.M. Nikol'skii, 'Fourier series of functions with given modulus of continuity', *Dokl. Akad. Nauk SSSR* **52** (1945), 191–194 (in Russian).
5. A.V. Efimov, 'Approximation of functions with given modulus of continuity by Fourier sums', *Izv. Akad. Nauk SSSR Ser. Mat.* **23** (1959), 115–134.

6. G.Ya. Doronin, 'Some inequalities for approximation by trigonometric polynomials', *Dokl. Akad. Nauk SSSR* **69** (1949), 487–490 (in Russian).
7. K.I. Sokolov, 'Estimates for approximation of continuous functions by subsequences of Fourier sums', *Trudy Mat. Inst. Akad. Nauk SSSR* **134** (1975), 240–253 (in Russian).
8. I.G. Oskolkov, 'Remainders of Fourier series of differentiable functions', *Dokl. Akad. Nauk SSSR* **103** (1955), 23–26 (in Russian).
9. S.A. Telyakovskii, 'Approximation of differentiable functions by partial sums of their Fourier series', *Mat. Zametki* **4** (1968), 291–300 (in Russian).
10. S. B. Stechkin, 'Remainder estimates for Fourier series of differentiable functions', *Trudy Mat. Inst. Akad. Nauk SSSR* **145** (1980), 126–151 (in Russian).
11. W. Rudin, 'L_2-approximation by partial sums of orthogonal developments', *Duke Math. J.* **19**:1 (1952), 1–4.
12. A.N. Kolmogorov, A.A. Petrov and Yu.M. Smirnov, 'A formula of Gauss from the theory of the least-squares method', *Izv. Akad. Nauk SSSR Ser. Mat.* **11** (1947), 561–566 (in Russian).
13. A.I. Mal'tsev, 'Remark to reference [12] above', *Izv. Akad. Nauk SSSR Ser. Mat.* **11** (1947), 587–588 (in Russian).
14. S.B. Stechkin, 'On best approximation of given classes by arbitrary polynomials', *Uspekhi Mat. Nauk* **9**:1 (1954), 133–134 (in Russian).
15. K.I. Babenko, 'On approximation of periodic functions of several variables by trigonometric polynomials', *Dokl. Akad. Nauk SSSR* **132** (1960), 247–250 (in Russian).
16. K.I. Babenko, 'On approximation of a class of periodic functions of several variables by trigonometric polynomials', *Dokl. Akad. Nauk SSSR* **132**:5 (1960), 982–985 (in Russian).
17. K.I. Babenko, ed. *Theoretical foundations and construction of numerical algorithms for problems of mathematical physics*, Nauka, Moscow, 1979 (in Russian).
18. V.M. Tikhomirov, 'Widths of sets in function spaces and the theory of best approximation', *Uspekhi Mat. Nauk* **15**:3 (1960), 81–120 (in Russian).
19. V.M. Tikhomirov, 'Best methods for approximation and interpolation in $C([-1,1])$', *Mat. Sb.* **80**:2 (1967), 290–304 (in Russian).

20. V.M. Tikhomirov, *Some questions in approximation theory*, Moscow University Press, 1976 (in Russian).
21. R.S. Ismagilov, 'Widths of sets in linear normed spaces and approximation of functions by trigonometric polynomials', *Uspekhi Mat. Nauk* **29**:3 (1974), 161–178 (in Russian).
22. B.S. Kashin, 'Widths of certain classes of finite-dimensional sets and of classes of smooth functions', *Izv. Akad. Nauk SSSR Ser. Mat.* **41** (1977), 334–351 (in Russian).
23. N.P. Korneichuk, 'Extremal values of functionals and best approximation on classes of periodic functions', *Izv. Akad. Nauk SSSR Ser. Mat.* **35**:1 (1971), 93–124 (in Russian).
24. V.D. Erokhin, 'On best linear approximation of functions that can be analytically continued from a given contour into a given domain', *Uspekhi Mat. Nauk* **23**:1 (1968), 91–119 (in Russian).
25. L. Karlovitz, 'On a class of Kolmogorov n-width problems', *Atti Accad. Naz. Lincei. Rend. Cl. Sci. Fis., Mat.* **53**:3–4 (1973), 241–245.
26. L. Tonelli, 'I polinomi d'approssimazione di Tchebychev', *Ann. of Math.* **15**:3 (1908), 47–119.
27. A. Pinkus, *n-widths in approximation theory*, Springer-Verlag, Berlin, 1985.
28. V.I. Nikol'skii, 'Best approximation by elements of convex sets in linear normed spaces', *Uchen. Zap. Kalinin. Gos. Ped. Inst.* **29** (1963), 85–119 (in Russian).
29. I. Singer, 'Choquet spaces and best approximation', *Math. Ann.* **148**:4 (1962), 330–340.
30. G. Choquet, 'Sur la meilleure approximation dans les espaces vectoriels normés', *Rev. Math. Pures et Appl.* **8**:4 (1963), 541–542.
31. A.L. Garkavi, 'On criteria for elements of best approximation', *Sibirsk. Mat. Zh.* **5**:2 (1964), 472–476 (in Russian).

INEQUALITIES FOR DERIVATIVES

(V.M. Tikhomirov and G.G. Magaril-Il'yaev)

The first results relating to the investigation of inequalities for derivatives of the form

$$\|x^{(k)}\|_{L_q(I)} \leq K\|x\|^{\alpha}_{L_p(I)}\|x^{(n)}\|^{\beta}_{L_r(I)} \tag{1}$$

(where $0 \leq k < n$ are integers, $1 \leq p, q, r \leq \infty$, $\alpha, \beta \geq 0$, and $I = R$ or R_+) were obtained by Landau [1] and Hadamard [2]. In these works the exact constants in (1) for $k = 1$, $n = 2$, $p = q = r = \infty$ were found in the cases $I = R_+$ (Landau) and $I = R$ (Hadamard).

At the end of the thirties, Kolmogorov suggested to his student Shilov (who later became a famous mathematician) that he should extend Hadamard's inequality to arbitrary k and n. Shilov obtained a number of particular results (which formed his first scientific publication) and stated a correct conjecture relating to the general case. The conjecture was that for $p = q = r = \infty$ and $I = R$, an extremal function in the inequality (1) (that is, a function for which the equality sign appears in (1)) is a periodic function whose nth derivative takes values of the same modulus and alternating sign at equal intervals. Such functions had been investigated by Euler, and not long before the above-mentioned studies of Shilov interest in these functions was renewed in the well-known works by Favard, Akhiezer, and Krein. Shilov failed to prove his hypothesis. Then Kolmogorov himself got interested in the problem and obtained its complete solution. Even now this remains the most remarkable result in the whole group of problems relating to inequalities of the form of (1).

In paper No. 40, which we consider here, a more general problem was stated, namely the problem on inequalities for arbitrary finite sets of $L_\infty(R)$ norms of consecutive derivatives. This formulation of the problem is possible not only for R and R_+, but also for certain other manifolds. Some special results in this area were obtained for $L_\infty(R)$ by Rodov [3] and Dzyadyk and Dubovik [4, 5]. In the paper by Din' Zung and Tikhomirov [6] the above problem was solved completely for functions belonging to $L_2(R^N)$ and $L_2(T^1)$ and for fractional derivatives. The most important special case when $I = R^N$ and $p = q = r = 2$ was studied earlier in the thesis of Subbotin [7]. The situation where $p = q = r = 2$ is probably the only case when an exhaustive solution can be found for the general statement of the problem.

In the course of proving the main theorem, Kolmogorov established a number of auxiliary assertions which were subsequently used many times in the proofs of results in approximation theory and in other questions (Korneichuk, Taikov, Gabushin, and others).

Stechkin [8] discovered a close relation between the problem of calculating the exact constant in (1) and the problem of best approximation of the operator of differentiation on the corresponding function class. These investigations aroused new interest in problems relating to inequalities (1).

Results whose degree of completeness is similar to that of Kolmogorov's theorem (where the exact constant in (1) was calculated for all k and n and for certain p, q, and r) were obtained only in three cases for a line [by Hardy, Littlewood and Pólya [9] ($p = q = r = 2$); Stein [10] ($p = q = r = 1$); Taikov [11] ($q = \infty$, $p = r = 2$)], and in two cases for a half-line [by Lyubich [12], Kuptsov [13] ($p = q = r = 2$), and Gabushin [14] ($q = \infty, p = r = 2$)]. Qualitative and quantitative aspects of the problem for $p = q = r = \infty$, $I = R_+$ were investigated by Schoenberg and Cavaretta and by Tikhomirov. Moreover, we know about 20 special cases. Arestov: $I = R$ or R_+, $k = 1$, $n = 2$, $p = \infty$, q, r arbitrary; $I = R$ or R_+, $k = 1, 2$, $n = 3$, $p = q = \infty$, r arbitrary, Berdyshev: $I = R_+$, $k = 1$, $n = 2$, $p = q = r = 1$. Gabushin: $I = R$, $k = 0, 1$, $n = 2$, p arbitrary, $q = r = \infty$. Magaril-Il'yaev: $I = R_+$, $k = 0, 1, n = 2, p$ arbitrary, $q = r = \infty$; $I = R$ or R_+, $k = 0, 1$, $n = 2$, $p = 1$, $q = \infty$, $r > 1$. Matorin: $I = R_+$, $k = 1, 2$, $n = 3$, $p = q = r = \infty$. Nagy: $I = R$ or R_+, $k = 0$, $n = 1$, p, q, r arbitrary. Solyar: $I = R$, $2k = n$, $q = 2$, p arbitrary, $r = p/(p-1)$. References to all these results can be found in [15]. In [16] a duality theorem was proved for (1) in the case $q = \infty$. This theorem implies the following exact results: $I = R$ or R_+, $k = 0$, $n = 2$, p arbitrary, $q = \infty$, $r = 1$; $I = R$ or R_+, $k = 1$, $n = 3$, p arbitrary, $q = \infty$, $r = 1$. We also mention the paper [17], in which a special case of inequality (1) was considered for a fractional value of k and integral n.

Buslaev obtained exact inequalities for $I = R$, $0 < k < n$, $n = 2, 3$, $n/q = (n-k)/p$, $r = \infty$ [18]. No existence theorem is proved here; however, the best constant in the inequality is attained on a sequence of functions convergent to a limit function which is an extremal in Kolmogorov's proper inequality ($I = R$, $p = q = r = \infty$, $0 < k < n$).

Exact inequalities of this kind are of great interest in the theory of extremal problems; they serve as an experimentation area. Almost all inequalities can be established by directly applying standard methods of the theory or slight modifications of them (see, in particular, Alekseev, Galeev and Tikhomirov [19], §12). In our opinion, the amount of exact results cannot be significantly increased in the future, since this is related to the necessity of solving complicated non-linear problems. Therefore, of interest are various qualitative problems which can lay foundations for a general theory of non-linear equations, similar to those arising as Euler equations in the corresponding extremal problems. Among the papers related to this topic we mention those by Arestov [20], Gabushin [21, 22] (in [21] an existence criterion was obtained for the inequality (1)), Buslaev, Magaril-Il'yaev and Tikhomirov [23] (here the existence theorem for extremal functions in the inequality (1) was proved) and Kwong and Zettl [24].

Multiplicative inequalities of the type (1) for functions of several variables play an important part in the theory of partial differential equations and in the embedding theory for classes of smooth functions. Existence problems for inequalities of this kind were considered, for instance, by Besov [25], Besov, Il'in and Nikol'skii [26] and Magaril-Il'yaev [27]. As to exact results, in addition to the above-mentioned [6, 7], we mention the papers by Konovalov [28] and Buslaev and Tikhomirov [29].

Inequalities (1) play an analogous role for periodic functions as well. The related existence problems were considered in [25, 26] and by Klots [30]. A number of exact inequalities were obtained by Ligun.

Kolmogorov's classical inequality served as a starting point for understanding the general functional and analytical nature of multiplicative inequalities. In particular, inequalities of this type for powers of operators on abstract Hilbert and Banach spaces have been studied. In this connection we mention the pioneering work by Lyubich [12].

References

1. E. Landau, 'Einige Ungleichunge für zweimal differentzierbare Funktionen', *Proc. London Math. Soc.* **2**:13 (1913), 43–49.
2. J. Hadamard, 'Sur le module maximum d'une fonction et de ses dérivées', *C. R. Séances Soc. Math.* **41** (1914), 68–72.
3. A.M. Rodov, 'Dependence between suprema of arbitrary functions of a real variable', *Izv. Akad. Nauk SSSR Ser. Mat.* **10** (1946), 257–270 (in Russian).
4. V.K. Dzyadyk and V.A. Dubovik, 'On Kolmogorov's problem of the dependence between suprema of arbitrary real functions on the real axis', *Ukrain. Mat. Zh.* **26**:3 (1974), 300–317 (in Russian).
5. V.K. Dzyadyk and V.A. Dubovik, 'On Kolmogorov's inequalities for the

dependence between suprema of real functions on the real axis', *Ukrain. Mat. Zh.* **27**:3 (1975), 291–299 (in Russian).

6. Din' Zung and V.M. Tikhomirov, 'On inequalities for derivatives in the L_2 metric', *Vestnik MGU Ser. Mat. Mekh.*, 1979, No. 2, 7–11 (in Russian).
7. Yu.N. Subbotin, *Extremal functional interpolation and spline approximation*, Doctoral Thesis, Sverdlovsk, 1973 (in Russian).
8. S.B. Stechkin, 'Best approximation of linear operators', *Mat. Zametki* **1**:2 (1967), 137–148 (in Russian).
9. G.H. Hardy, J.E. Littlewood and G. Pólya, *Inequalities*, Cambridge Univ. Press, 1934.
10. E.M. Stein, 'Functions of exponential type', *Ann. of Math.* **65**:3 (1957), 582–592.
11. L.V. Taikov, 'Inequalities of Kolmogorov type and best formula for numerical differentiation', *Mat. Zametki* **4**:2 (1968), 233–238 (in Russian).
12. Yu.P. Lyubich, 'On inequalities for powers of linear operators', *Izv. Akad. Nauk SSSR Ser. Mat.* **24** (1960), 825–864 (in Russian).
13. N.P. Kuptsov, 'Kolmogorov estimates for derivatives in $L_2[0,\infty)$', *Trudy Mat. Inst. Akad. Nauk SSSR* **138** (1975), 94–117 (in Russian).
14. V.N. Gabushin, 'On best approximation of the differentiation operator on the half-line', *Mat. Zametki* **6**:5 (1969), 573–582 (in Russian).
15. G.G. Magaril-Il'yaev, *Embeddings of generalized Sobolev classes and inequalities for derivatives*, Candidate Thesis, Moscow University Press, 1980 (in Russian).
16. G.G. Magaril-Il'yaev, 'Inequalities for derivatives and duality', *Trudy Mat. Inst. Akad. Nauk SSSR* **161** (1983), 183–194 (in Russian).
17. G.G. Magaril-Il'yaev and V.M. Tikhomirov, 'On Kolmogorov's inequality for fractional derivatives on the half-line', *Anal. Math.* **7**:1 (1981), 37–47 (in Russian).
18. A.P. Buslayev, 'On exact constants in inequalities for derivatives' in *School on the theory of operators in function spaces. Abstracts*, Minsk, 1982, p. 29 (in Russian).
19. V.M. Alekseev, E.M. Galeev and V.M. Tikhomirov, *Collection of optimization problems*, Nauka, Moscow, 1983 (in Russian).
20. V.V. Arestov, 'On best uniform approximation of differentiation operators', *Mat. Zametki* **5**:3 (1969), 273–284 (in Russian).

21. V.N. Gabushin, 'Inequalities for norms of functions and their derivatives in L_p metrics', *Mat. Zametki* **1**:3 (1967), 291–298 (in Russian).
22. V.N. Gabushin, 'Best approximation of functionals on certain sets', *Mat. Zametki* **8**:5 (1970), 551–562 (in Russian).
23. A.P. Buslaev, G.G. Magaril-Il'yaev and V.M. Tikhomirov, 'On the existence of extremal functions in inequalities for derivatives', *Mat. Zametki* **32**:6 (1982), 823–833 (in Russian).
24. M.K. Kwong and A. Zettl, 'Ramifications of Landau's inequality', *Proc. Roy. Soc. Edinburgh* **86A** (1980), 175–212.
25. O.V. Besov, 'Multiplicative inequalities for integral norms of differentiable functions of several variables', *Trudy Mat. Inst. Akad. Nauk SSSR* **131** (1974), 3–15 (in Russian).
26. O.V. Besov, V.P. Il'in and S.N. Nikol'skii, *Integral representation of functions and embedding theorems*, Nauka, Moscow, 1975 (in Russian).
27. G.G. Magaril-Il'yaev, 'Problems on intermediate derivatives', *Mat. Zametki* **25**:1 (1979), 81–96.
28. V.N. Konovalov, 'Exact inequalities for norms of functions, third derivatives, second mixed or skew derivatives', *Mat. Zametki* **23**:1 (1978), 67–78 (in Russian).
29. A.P. Buslaev and V.M. Tikhomirov, 'On inequalities for derivatives in the multidimensional case', *Mat. Zametki* **25**:1 (1979), 59–74 (in Russian).
30. B.E. Klots, 'Approximation of differentiable functions by smoother ones', *Mat. Zametki* **21**:1 (1977), 21–32 (in Russian).

RINGS OF CONTINUOUS FUNCTIONS

(E.A. Gorin)

Paper No. 41 is one of the first in which, in essence, the following thesis is discussed: "if X is a topological space with a sufficiently good topology and $C(X)$ is the set of all real (or complex) continuous functions on X, then the space $C(X)$ endowed with a sufficiently rich algebraic or topological structure determines X to within a homeomorphism". There are (Hausdorff regular) topological spaces on which, apart from the constants, there are no continuous real functions at all. In this connection, according to the considerations presented in the paper, it is reasonable to assume that X is a completely regular and, at least to start with, compact space (not necessarily metrizable).

Of course, the additional operations and structures on $C(X)$ are assumed to be compatible with the natural pointwise operations and structures. In view of this, assuming for simplicity that X is a compact space, we can endow $C(X)$ with the structure of an additive group, a vector space, a multiplicative semigroup, a lattice, a metric space with metric generated by the sup-norm, a Banach space, a ring, an algebra, a topological vector space, a topological (Banach) algebra, etc. Some of these structures are too weak to recover the compactum X. For example, the additive groups of all separable Banach spaces are isomorphic. $C(X)$, regarded as a vector space, makes it possible to distinguish between infinite metric compact spaces and finite ones and in the latter case to determine the number of elements. First examples of non-homeomorphic metric compact spaces X with topologically isomorphic vector spaces $C(X)$ were indicated by Banach and Borsuk as early as the beginning of the thirties and in 1951 Milyutin established that all spaces $C(X)$ for which X is an uncountable metric compact space are linearly topologically isomorphic. On the other hand, the isometry class of $C(X)$ determines X to within a homeomorphism (Banach, 1932; the assumption that X is metrizable in this theorem was later removed by Stone). $C(X)$ regarded as a vector lattice also determines X (Kakutani, 1941).

In the first part of paper No. 41 it is shown that algebraic isomorphism between $C(X)$ and $C(Y)$ implies homeomorphism of the compact spaces X and Y. In our opinion, the significance of this part of the paper lies in the method, based on the identification of maximal ideals with points of the compact space, rather than in the fact that the above result is a formal strengthening of the earlier proved Stone-Shilov theorem, in which the existence of a topological isomorphism between the algebras $C(X)$ and $C(Y)$ was assumed. The second part, dealing with algebras $C'(X)$ of bounded continuous functions on an arbitrary completely regular space, is essentially based on the application of Čech's extensions. And in our opinion, in this case as well, the considerations making it possible to describe Čech's extensions in algebraic terms are more important than the theorems proved. Subsequently these ideas were developed further; namely, by taking for the "space of maximal ideals" a subalgebra instead of $C'(X)$ one can obtain various useful compactifications of the original space X. Of great interest also are the schemes of the proofs, various modifications of which were applied later in the investigation of maximal ideals of other symmetric function algebras.

The theorem mentioned at the beginning of the foregoing paragraph reduces in fact to the consideration of topological isomorphism of algebras. Indeed, if, for instance, the basic field is that of the real numbers, then non-negativity of a function is equivalent to its being representable as a square, and this property is retained under isomorphism. Therefore the relations $f \geq g$ are also retained, whence continuity follows. The very fact of automatic continuity of an isomorphism is retained in the cases of algebraic isomorphism and complex algebras of all continuous functions. Moreover, in subsequent studies by Gel'fand it was revealed that every homomorphism from a Banach algebra into a semisimple commutative Banach algebra is automatically continuous. Of course, this implies automatic continuity of isomorphisms of semisimple commutative Banach algebras. It turned out (Johnson, 1967) that commutativity is also unimportant here, since each epimorphism from a Banach algebra onto a semisimple Banach algebra is continuous. On the other hand, under the continuum hypothesis, for each infinite compactum X there are discontinuous monomorphisms from $C(X)$ into a Banach algebra (so that the norm on $C(X)$ admits of a discontinuous strengthening; Dales and Esterle, 1977).

If a compactum X possesses an additional structure, for example, is a smooth analytic manifold with boundary, then instead of the entire space $C(X)$ it makes sense to consider its subspaces and subalgebras. For simplicity, let X be a convex compactum in complex Euclidean space $\mathbf{C}^n$ and let $A(x)$ consist of the continuous functions analytic inside X. $A(X)$ is a closed subalgebra in $C(X)$ relative to the natural operations and the norm of $A(X)$ and, as in the case of $C(X)$, the maximal ideals of $A(X)$ correspond to points. An algebraic isomorphism between two such algebras $A(X)$ and $A(Y)$ turns out to be topological and induces a homeomorphism of X and Y under which the internal analytic structure is preserved. This is in fact an obvious consequence of general theorems of the theory of Banach algebras. In particular, if X is a ball, Y is a polydisk, and $n \geq 2$, then, by virtue of Poincaré's theorem, $A(X)$ and $A(Y)$ are non-isomorphic. A more subtle result was obtained by Khenkin (1968), namely in the latter case $A(X)$ and $A(Y)$ are non-isomorphic even when regarded as Banach spaces.

In the concluding part of the paper, Kolmogorov studies the algebra $C(X)$ of all continuous functions on an arbitrary completely regular space X. In general, an attempt to endow such an algebra with a topology with good properties fails, and therefore the accepted approach related to the consideration of $C(X)$ as a purely algebraic structure looks particularly natural in this situation. The results of this part of the paper received a direct development in works by Hewitt, Nagata, Shirota, and many other scientists. If, for instance, arbitrary spaces X are replaced by Stein manifolds and the set $A(X)$ of holomorphic functions is considered, then there arise questions of the type discussed above, but lying outside the limits of Banach spaces.

Let us endow $A(X)$ with the topology of uniform convergence on compact sets. In general, topological isomorphism of two vector spaces $A(X)$ and $A(Y)$ does not lead to (biholomorphic) equivalence of the manifolds; for example, such spaces in a ball or a polydisk of the same dimension are isomorphic. However, for polydisks of different dimensions isomorphism of spaces no longer exists, and the first proofs were based on estimates of ϵ-entropy of compact sets in these spaces (Kolmogorov, 1958). We also note that an algebraic isomorphism of algebras $A(X)$ is continuous and generates a biholomorphic equivalence of the manifolds.

Thus, this small paper can be regarded as one of the sources of many versatile investigations. Along with the papers by Stone and Shilov, it was one of the first in which the general algebraic notion of a maximal ideal and the new concept of a space of maximal ideals were successfully applied to problems of set-theoretic topology. In application to problems of mathematical analysis, this concept manifests itself no less successfully in Gel'fand's theory (which appeared at the end of the thirties) of commutative Banach algebras over the field of complex numbers. (By the way, herein lies its distinction from the general theory of commutative rings, where the principal role is played not by maximal but by prime ideals; in contrast to prime ideals, maximal ideals of a Banach algebra are closed.) However, it may also be noted that the success of this theory is primarily due to the choice of the field of complex numbers as the underlying field. This made it possible not only to establish a natural bijective correspondence between maximal ideals and complex homomorphisms, but also to apply the apparatus of complex analysis for studying algebras that are quite distinct from $C(X)$, which often leads to profound results having no simple real analogues.

CURVES IN A HILBERT SPACE

(Yu.A. Rozanov)

Papers Nos. 42 and 43 deal with functions $t \to \xi(t)$ of a real variable t with values $\xi(t) \in H$ in a Hilbert space for which translations and similarity transformations in the space of the variable t induce analogous transformations of $\xi(t)$ in the space H. When H is realized as a Hilbert space (of the type L_2) of random variables, one is dealing with random processes with stationary increments and their subclasses (stationary processes, processes with stationary orthogonal increments, and Wiener processes), in view of their correlation and spectral structure. As an immediate continuation of papers Nos. 42 and 43 there have been many investigations by different authors devoted to random functions on groups and homogeneous spaces as well as to generalized random functions whose corresponding classes (random fields with homogeneous increments, etc.) were characterized from the standpoint of their correlation and spectral structure, which in particular, required a corresponding characterization of positive definite functions, generalizing the well-known Bochner-Khinchin theorem (for the basic results and bibliography see, for instance, [1]). It should be noted that classes of random functions $\xi(t)$ whose various probabilistic properties are order-invariant under similarity transformations of the variable t proved important in various applications and in particular, in statistical physics (in this connection, see [2]).

References

1. I.M. Gel'fand and N.Ya. Vilenkin, *Some applications of harmonic analysis*, Fizmatgiz, Moscow, 1961 (in Russian).
2. Ya.G. Sinai, *Theory of phase transition*, Nauka, Moscow, 1980 (in Russian).

ON THE PAPERS ON INTUITIONISTIC LOGIC

A.N. Kolmogorov

Paper No. 9, "On the tertium non datur principle" (TNDP), was intended as an introduction to the realization of a wider idea. The construction of models of various parts of classical mathematics within the framework of intuitionistic mathematics was meant as a justification of their consistency (here the consistency of intuitionistic mathematics is regarded as a consequence of its intuitive obviousness). In the justification of consistency of classical propositional logic this was, of course, unnecessary, but it was believed that the method might prove applicable to the justification of consistency of classical arithmetic (cf. the work by Gödel of 1933).

Paper No. 19, "On the interpretation of intuitionistic logic" (IIL) was written with the hope that the logic of solutions of problems would later become a regular part of courses on logic. It was intended to construct a unified logical apparatus dealing with objects of two types — propositions and problems.

INTUITIONISTIC LOGIC

(V.A. Uspenskii and V.E. Plisko)

The two papers by Kolmogorov on mathematical logic [9] "On the tertium non datur principle" and [10] "On the interpretation of intuitionistic logic", (papers Nos. 9 and 19 in the present book which will be referred to as TNDP and IIL respectively) are devoted to intuitionistic logic. They were written when the study of non-classical logical systems had just begun. Many questions that were for the first time considered in these papers were later investigated by other authors. Wang Hao, a well-known specialist in the area of mathematical logic, wrote in his foreword [42] to the English translation of TNDP in [30] that this paper anticipated to a great extent not only Heyting's formalization of intuitionistic logic, but also results on transformability of classical mathematics into intuitionistic mathematics and that it established an important relationship between intuitionism and other studies on the foundations of mathematics. In the foreword to [30], van Heijenoort noted that TNDP was the first systematic investigation of intuitionistic logic.

The aim of this commentary is to trace the development of the ideas put forward by Kolmogorov in his works on mathematical logic.

The text of TNDP is reprinted in this book without any significant changes (only certain misprints are corrected). The paper IIL is translated from German. In the translation the notation in logical formulas is somewhat changed; namely, instead of the now rarely used point system of notation (see [24], §11), accepted in the original, the bracket system is employed, which is more familiar to the modern reader. The same bracket system is used both in TNDP and in this commentary.

When writing propositional formulas we keep to the symbolism used in

IIL. It differs somewhat from the symbolism accepted in TNDP. Namely, we write $\supset$ instead of $\rightarrow$ to denote implication, and $\neg A$ instead of $\overline{A}$ to denote negation; as propositional variables we will use small italic letters (in TNDP use was made of capital letters). The universal quantifier is written in TNDP and IIL as (x) and the existential quantifier is written in TNDP in the form (Ex) (no existential quantifier is encountered in IIL); here we use the modern notation $\forall x$ and $\exists x$, respectively.

I

The intuitionistic approach in mathematics appeared at the beginning of the 20th century, due to work by Brouwer, Weyl, etc. Philosophical and methodological prerequisites for intuitionism are presented, for instance in Heyting's book [3] in which, in particular, it is said that intuitionists are those mathematicians who accept the following two principles:

1. Mathematics has not only a purely formal but also a meaningful significance.

2. Mathematical objects are perceived directly by a mental process and hence mathematical knowledge does not depend on experience.

The first principle places intuitionism in opposition to the formalistic viewpoint as accepted as a working methodological principle in Hilbert's school for the foundation of mathematics. As noted in TNDP, Chapter 1, the formalistic viewpoint asserts that mathematics is a collection of propositions in a certain formalized language deducible from a system of axioms, the choice of the axioms being quite arbitrary and subject only to certain more or less conventional reasons of practical convenience as well as the necessary requirement of consistency. The question of truth or falsity of mathematical propositions is meaningless from the formalistic viewpoint. One can only speak of their provability or refutability, based on the axioms.

Thus, the first of the above principles of intuitionism assumes the existence of an object of mathematical research, whereas the other principle understands mathematics purely as a product of thought and denies any objective character of mathematical notions independent of thought.

Brouwer denies any relation between the truth of mathematical propositions and experience, and proclaims intuition as the only truth criterion. Here, as was indicated by Heyting [3], Brouwerian intuition should not be interpreted in a "mystical" sense. It is only meant that, according to Brouwer, mathematical objects are created by human thought and therefore the truth of propositions about them is completely determined by the ideas (about these objects) of the mathematician whose thought produced them. Strictly speaking, from the viewpoint of intuitionism there are as many kinds of mathematics as there are mathematicians. However, due to some general properties of human thought it is possible that in the consciousness of different people similar mathematical notions are formed. For instance, such is the notion of natural number. Proceeding from this notion and based on intuitionistic ideas, a specific mathematical theory can be developed.

The philosophical prerequisites for intuitionism were found unacceptable by many mathematicians and were subjected to justifiable criticism. For example, in the foreword to the Russian translation of Heyting's book [3], A.N. Kolmogorov wrote: "We cannot agree that mathematical objects are simply a product of constructive mental activity. For us, mathematical objects are abstractions of actually existing forms of reality which are independent of thought." Nevertheless, the intuitionistic criticism of classical set-theoretic mathematics also proved very fruitful for mathematics as a whole, since it drew attention to problems on the constructive formation of abstract mathematical notions and to the question of the limits of applicability of classical logic.

From the viewpoint of intuitionism, both the construction of mathematical objects and their consideration must satisfy the criteria of intuitive clarity and convincingness. In discussing relationships between mathematics and logic, Heyting [3] indicates that in intuitionistic mathematics conclusions are not drawn according to pre-assigned rules (as is done in Hilbert's formalistic approach), that is, no *a priori* logical system is fixed. That each logical step is convincing must be verified directly in agreement with intuition. However, this does not exclude the existence of general rules according to which true mathematical propositions produce other true mathematical propositions in an intuitively obvious manner. Thus, it makes sense to speak of intuitionistic logic as a collection of intuitively acceptable methods of mathematical deduction. Brouwer, the founder of intuitionism, carried out an analysis of the principles of Aristotelean logic, and came to the conclusion that the applicability of the tertium non datur principle expressed by the logical formula $a \vee \neg a$ cannot be considered obvious in all cases.

II

In TNDP a first attempt was made to construct a formal logical system containing only intuitionistically acceptable laws of propositional logic. To this end the system of axioms of classical logic suggested by Hilbert was subjected to a critical analysis. As a result, the following system of axioms of intuitionistic logic (denoted $\mathfrak{B}$) was stated:

1. $a \supset (b \supset a)$;
2. $(a \supset (a \supset b)) \supset (a \supset b)$;
3. $(a \supset (b \supset c)) \supset (b \supset (a \supset c))$;
4. $(b \supset c) \supset ((a \supset b) \supset (a \supset c))$;
5. $(a \supset b) \supset ((a \supset \neg b) \supset \neg a)$.

The other laws of propositional calculus must be derived from these axioms using the rules of substitution and implication (that is, the "modus ponens" rule, allowing one to pass from a formula $A \supset B$ to the formula B).

The system $\mathfrak{B}$ (as well as Hilbert's system of axioms for classical propositional logic considered in TNDP) contains logical laws related to implication and negation, whereas, say, disjunction and conjunction are not considered. That is why the suggested system is somewhat limited. However, we note that, from the logical standpoint, the connectives $\supset$ and $\neg$ express the most important relation of implication and operation of negation, while the additional axioms explaining the meaning of conjunction and disjunction (for example, Ackermann's axioms for disjunction mentioned in TNDP) can literally be taken in intuitionistic logic from classical logic. Moreover, it is the limited character of the system $\mathfrak{B}$ that makes the results on the possibility of embedding classical logic in this system particularly strong.

The system $\mathfrak{B}$ completed with the natural logical laws for disjunction and conjunction turns into so-called minimal calculus. This calculus and the term "minimal" were introduced by Johansson [33]. Church [24], §26 calls this minimal calculus the "minimal propositional calculus of Kolmogorov and Johansson". This terminology is justified not only by the leading role of Kolmogorov in the formation of minimal calculus, but also by the following fact: each formula deducible in minimal calculus and involving only the symbols of implication and negation is deducible in Kolmogorov's system $\mathfrak{B}$ as well.

In §3 in Chapter 5 of TNDP Kolmogorov considered the intuitively obvious predicative axioms:

I. $\forall x(A(x) \supset B(x)) \supset (\forall x A(x) \supset \forall x B(x))$;

II. $\forall x(A \supset B(x)) \supset (A \supset \forall x B(x))$;

III. $\forall x(A(x) \supset C) \supset (\exists x A(x) \supset C)$;

IV. $A(x) \supset \exists x A(x)$,

and a derivation rule P making it possible to pass from a formula A to the formula $\forall x A$. After these axioms and the rule P are added to the system $\mathfrak{B}$, there arises a version of intuitionistic predicate calculus. In an analogous way minimal predicate calculus is obtained from minimal propositional calculus.

Thus, the system $\mathfrak{B}$ introduced in TNDP is the first axiomatization of intuitionistic propositional logic, and its predicative extension is the first axiomatization of intuitionistic predicate logic. Subsequently certain other systems of intuitionistic logic (with a broader set of introduced formulas) were suggested by Glivenko [28], Heyting [31], and Gentzen [27]. They all turned out to be equivalent in the sense that the same logical principles are deducible in them. A specific property of the system $\mathfrak{B}$ and the minimal calculus, in which they differ from Heyting's system and systems equivalent to it, is the rejection of the logical principle

$$\neg a \supset (a \supset b).$$

In Kolmogorov's opinion, this formula has no intuitive justification since it asserts something about consequences of what is impossible; namely, we must assume that B is true if a true proposition A is assumed to be false (TNDP, Chapter 2, §4).

It is impossible to prove the adequacy of a representation of intuitionistic logic by means of a system of axioms if the logic itself has no exact semantics. Nevertheless, irrespective of any semantic specifications, widest recognition was gained by the system of axioms suggested by Heyting (as in IIL, we preserve the numbering of the axioms as given by Heyting [31]):

2.1. $a \supset a \wedge a$;

2.11 $a \wedge b \supset b \wedge a$;

2.12 $(a \supset b) \supset (a \wedge c \supset b \wedge c)$;

2.13 $(a \supset b) \wedge (b \supset c) \supset (a \supset c)$;

2.14 $b \supset (a \supset b)$;

2.15 $a \wedge (a \supset b) \supset b$;

3.1 $a \supset a \vee b$;

3.11 $a \vee b \supset b \vee a$;

3.12 $(a \supset c) \wedge (b \supset c) \supset (a \vee b \supset c)$;

4.1 $\neg a \supset (a \supset b)$;

4.11 $(a \supset b) \wedge (a \supset \neg b) \supset \neg a$.

This system and the systems equvalent to it were later called intuitionistic propositional calculus. Decision procedures were found for this calculus, that is, algorithms making it possible to find whether an arbitrary formula is deducible in the calculus or not (Gentzen [27], Jas'kowski [32], Pil'chak [15, 16], Vorob'ev [2], and others). Algebraic and topological interpretations of intuitionistic propositional logic were discovered, which generalize in a natural way the interpretations of classical logic by means of Boolean algebras (Stone [40], Tarski [41], McKinsey and Tarski [36, 37], and Rasiowa [38]). The semantics of intuitionistic propositional calculus suggested by Kripke [35] turned out to be closest to the proper logical content of this calculus (and intuitionistic predicate calculus). For a detailed presentation of the various interpretations of intuitionistic logic see [5, 20].

III

As has been mentioned above when referring to Heyting [3], the fixation of intuitionistic logic as a system of axioms is of no conceptual importance to intuitionism. However, the construction of such a system allows one to make intuitionistic logic a subject of mathematical research no longer depending on the methodological principles of intuitionism. After intuitionistic propositional calculus was formulated explicitly, the term "intuitionistic" only indicates the history of appearance of the calculus, and should not mislead one when assessing its essence as an object of mathematical logic.

Hence, the system $\mathfrak{B}$ in TNDP can be characterized as a subsystem of intuitionistic propositional calculus and, simultaneously, as an implication-negation fragment of Johansson's minimal calculus. It is this context within which the basic results of TNDP are formulated in the book by Church [24]. Namely, the results of Chapter 3 in TNDP are stated in [24] (Exercise 26.20) thus: if each variable entering in a theorem of classical propositional calculus involving no connectives other than implication and negation is replaced by its double negation, then the resulting formula is a theorem of minimal calculus. The results of Chapter 4 are stated in [24] in a somewhat generalized form (Exercise 38.12), as follows. For each formula A of predicate calculus we define a formula A^* (the "translation" of the formula A) by assuming that if A is an elementary formula, then A^* is $\neg\neg A$; $(A \wedge B)^*$ is $\neg\neg(A^* \wedge B^*)$; $(A \vee B)^*$ is $\neg\neg(A^* \vee B^*)$; $(A \supset B)^*$ is $\neg\neg(A^* \supset B^*)$; $(\neg A)^*$ is $\neg A^*$; if A is $\forall xB$, then A^* is $\neg\neg\forall xB^*$; and if A is $\exists xB$, then A^* is $\neg\neg\exists xB^*$. A formula A is a theorem of classical predicate calculus if and only if A^* is a theorem of minimal predicate calculus.

The "translation" of classical theorems into the language of the intuitionistic system is in fact carried out in TNDP not only for logical formulas. In essence, this paper outlines a scheme for obtaining for a broad class of mathematical theories the following mathematical result: if a proposition is proved in a classical theory using the law of the excluded middle or equivalent principles, then a proposition equivalent (from the classical viewpoint) to the former can also be proved without applying the law of the excluded middle, namely within the framework of minimal logic. In particular, if a contradiction is obtained in a theory by means of the law of the excluded middle, then the contradictory proposition can also be proved without using this law. Thus, application of the law of the excluded middle cannot be regarded either as a genuine cause of contradictions in classical mathematics or, in particular, as a cause of set-theoretic antinomies.

The results of TNDP can also be summed up as follows. Let the axioms of a mathematical theory be such that the above-described "translations" of these axioms are true from the intuitionistic viewpoint. Then the translation of any theorem in this theory must also be true from the intuitionistic viewpoint. This conclusion is of conceptual importance. It shows that the part of classical mathematics dealing only with intuitionistically acceptable objects (for example, arithmetic) can be interpreted intuitionistically with complete preservation of its classical content. Therefore restriction to intuitionistic logic is inessential, and the distinction between intuitionistic and classical mathematics lies primarily in the manner in which abstract mathematical notions are formed.

Embedding classical logical and mathematic-logical theories in corresponding theories based on intuitionistic logic later became an important means for investigating intuitionistic theories, since this often makes it possible to translate into intuitionistic results, without unnecessary efforts, the results obtained in the investigation of classical theories. Shanin [25] carried out a detailed investigation of various embeddings of classical arithmetic in intuitionistic arithmetic. He introduced the term "embedding operation" (or interpretation operation) for an algorithm such that, given an arbitrary arithmetical proposition A, the algorithm constructs an arithmetical proposition A^* with the property that A is provable in the classical arithmetic if and only if A^* is provable in intuitionistic arithmetic.

Hence, historically, TNDP presents the first example of an embedding operation. A little later a similar embedding of classical logic in intuitionistic logic was suggested by Gödel [29]. It is proved in [25] that the Kolmogorov and Gödel embedding operations are equivalent in the sense that the results of their application to an arbitrary formula are intuitionistically equivalent.

IV

In TNDP an embedding operation is constructed which makes it possible to give an intuitionistic interpretation to the major part of classical mathematics, while, in a sense, the paper IIL is devoted to the solution of the inverse problem of interpreting intuitionistic logic within the framework of ordinary mathematical notions, irrespective of the philosophical and methodological principles of intuitionism. This can be done by interpreting logical formulas not as schemes of propositions but as schemes of types of problem. Here, general validity or "truth" of a logical formula is understood as the existence of a general method for solving all problems of a given type. For instance, in this way the law of the excluded middle is "refuted" in the sense that the general validity of the formula $a \vee \neg a$ would mean the existence of a general method allowing one to find a solution to any specific problem or to reduce to a contradiction the assumption that its solution exists. Since such a method can hardly be possible, the formula $a \vee \neg a$ cannot be regarded as being generally valid. At the same time, all formulas deducible in intuitionistic propositional calculus turn out to be generally valid. Thus, an interpretation of this calculus as a logic of problems is established.

The above interpretation of intuitionistic logic had an important methodological significance. Within the framework of the notions of "problem" and "solution to a problem", which are not quite exact but nevertheless understandable to a mathematician, the necessity of considering logical schemes not containing the law of the excluded middle was justified. P.S. Novikov wrote in [14], pp. 54–55: "In 1932 Kolmogorov [10] suggested an interesting interpretation of this (intuitionistic – V.A. Uspenskii and V.E. Plisko) calculus not related to any new specific principles in the foundations of mathematics. According to Kolmogorov, along with the traditional logic systematizing proofs of theoretical truths, there can exist a logic of systematization of schemes for solving problems (for example, geometrical construction problems). And here "problems" and "solutions to problems" are regarded as underlying notions."

The new interpretation of intuitionistic logic, free of philosophical concepts of intuitionism, made it meaningful to investigate the logic as a calculus of problems. A characteristic fact is that in the fifties the very term "calculus of problems" was used in the Soviet mathematical logic literature to denote intuitionistic propositional calculus (see, for example, [15, 16]).

Thus, IIL presents in a rather detailed form the idea of interpreting logical formulas by means of problems. This idea later received various specific realizations, characterized by the choice of an exactly defined class of problems. The first mathematical development of Kolmogorov's idea of interpreting intuitionistic logic as a calculus of problems was worked out by his student Medvedev [12] (also see [21]). Each of the problems considered by Medvedev is a problem of finding an everywhere defined function with prescribed properties. In [12] such problems are called "mass problems". A mass problem is said to be algorithmically solvable (or, simply, "solvable") if among its solutions there is a general recursive function. A central part in the theory developed is played by the notion of reducibility of one mass problem to another, which is of algorithmic character. It proves possible to define logical operations on mass problems reflecting the ideas put forward by Kolmogorov. All formulas deducible in intuitionistic propositional calculus turned out to be schemes of algorithmically solvable mass problems.

A little later Medvedev [13] suggested another interpretation of logical formulas by means of problems. Here, as a restriction and specification of the notion of a problem, use was made of the so-called finite problems. A finite problem is determined by a non-empty finite set of feasible solutions from which a definite (possibly empty) subset of actual solutions is separated out. Logical operations on finite problems are defined. For instance, $A \supset B$ is a finite problem whose feasible solutions are arbitrary mappings of the set of all feasible solutions to the problem A into the set of feasible solutions to the problem B and whose actual solutions are those mappings under which actual solutions to the problem A go into actual solutions to the problem B. Now let $A(p_1, \ldots, p_n)$ be a propositional formula with variables $p_1, \ldots, p_n$. We fix non-empty finite sets $X_1, \ldots, X_n$, replace the variables $p_1, \ldots, p_n$ in the formula A by all possible finite problems $F_1, \ldots, F_n$ whose sets of feasible solutions are the sets $X_1, \ldots, X_n$, and vary the sets of actual solutions. Formula A is said to be finitely generally valid if for any choice of the sets $X_1, \ldots, X_n$ there always exists a common true solution y to all problems $A(F_1, \ldots, F_n)$ obtained from A by substitutions of the indicated type. In other words, the propositional formula $A(p_1, \ldots, p_n)$ is finitely generally valid if it is possible to indicate a true solution to the problem $A(F_1, \ldots, F_n)$, knowing only the sets of feasible solutions of the problems $F_1, \ldots, F_n$.

The logic of finite problems turned out to be of a rather complicated structure. It does not coincide with intuitionistic propositional calculus. For example, the formula $(\neg a \supset b \vee c) \supset (\neg a \supset b) \vee (\neg a \supset c)$ is finitely generally valid but is not deducible in this calculus. It is unknown whether this logic is decidable. It has only been proved that it cannot be defined by means of a finite set of axioms (see [11]).

Skvortsov has recently considered some natural generalizations of the notion of a finite problem and the corresponding interpretations of logical formulas (see [22, 23]). One of the generalizations consists in the consideration of not only finite problems, but also problems with an arbitrary set of feasible and true solutions. It has been proved that the resulting logic does not coincide with intuitionistic propositional calculus, but can be defined by means of a recursive system of axioms.

Another generalization of the notion of a finite problem introduced by Skvortsov [22] is related to the consideration of problems whose feasible and true solutions are natural numbers and the logical operations are of algorithmic character. For instance, a true solution to the problem $A \supset B$ is any natural number serving as the Gödel number of a partial recursive function mapping feasible solutions of the problem A to feasible solutions of the problem B, the true solutions of the problem A being mapped to true solutions of the problem B. This interpretation, which is called recursive general validity, can be extended to predicate formulas as well.

The above interpretations of logical formulas appeared as a direct development of the ideas of Kolmogorov put forward in IIL. We note that the interpretation of formulas by means of problems should not be regarded directly as an explanation of the intuitionistic meaning of formulas. On the contrary, under discussion is an interpretation of logical formulas *independent of intuitionistic concepts* and leading to a logic without the law of the excluded middle. However, in §2 of IIL Kolmogorov showed that the intuitionistic meaning of an existential proposition can in a natural way be related to a problem. On the other hand, according to Heyting, a mathematical proposition asserts the fact that a mathematical construction has been performed [4], and the question of truth of a proposition lies in the existence of this construction, that is, it is of an existential character. Therefore the ideas of the calculus of problems prove applicable to the interpretation of the intuitionistic meaning of arbitrary mathematical propositions.

The idea of interpreting a mathematical proposition as the requirement of a certain construction, which goes back to Heyting, was realized in exact mathematical form by Kleene in relation to arithmetical propositions [34] (see also [8, 14]). Associated with each arithmetical proposition is a certain (possibly empty) set of natural numbers, called realizations of the proposition. Roughly speaking, a realization encodes information about the intuitionistic truth of the proposition. For example, an elementary proposition of the form $s = t$ possesses a realization only when it is true in the ordinary sense. A number e is a realization of a proposition of the form $A \supset B$ if the partial recursive function with Gödel number e is applicable to any realization of the formula A and transforms it into a realization of the formula B. A formula is said to be realizable if there is a number realizing it. Recursive realizability is usually regarded as an analogue of intuitionistic (more precisely, constructive) truth. For example, it is possible to construct a proposition of the form $\forall x(A(x) \vee \neg A(x))$ which is not true in this sense.

The relationship between recursive realizability and the calculus of problems consists in the fact that the establishment of the realizability of an arithmetic proposition can be regarded as a problem in Kolmogorov's sense. (The idea of associating a problem with each arithmetical proposition was realized more explicitly in an algorithm for the constructive deciphering of mathematical judgements, suggested by Shanin [26].) It is possible to develop the corresponding logic for solving such problems by regarding as generally valid or realizable those logical formulas that are schemes of realizable arithmetical propositions.

The resulting propositional calculus was first investigated by Rose [39]. He found that not all realizable propositional formulas are provable in intuitionistic propositional calculus. Subsequently, the logic of recursive realizability was investigated in [1, 6, 7], etc. However, no characterization of the class of realizable propositional formulas has been established up to now. In particular, it is not known whether this class is decidable, or at least enumerable.

Greatest progress was achieved in the investigation of predicate calculus based on the notion of realizability. It was found [19] that this logic cannot be defined as a calculus and even is not arithmetical, that is, the class of all realizable predicate formulas cannot be determined by an arithmetical formula. It was incidentally shown [17] that if instead of the language of formal arithmetic a richer language is taken and the notion of realizability is defined for it, then the corresponding class of realizable predicate formulas turns out to be substantially narrower than in the case of Kleene realizability. Hence, the semantics of predicate formulas turns out to depend on the original language of mathematical logic. Based on the ideas of recursive realizability, Plisko [18] suggested the notion of absolutely realizable predicate formula, not depending on any specific language. The notion of absolute realizability is close to the above-mentioned notion of recursive realizability.

V

In conclusion we consider the question (posed at the end of §2 in Chapter 2 of TNDP) whether the system of axioms 1–4 is complete for formulas without negation; namely, whether an arbitrary formula deducible in classical propositional calculus and containing only symbols of implication is deducible based on the axioms 1–4 of Hilbert's system. Wang Hao [42], p. 416, gave a negative answer to the question. Namely, the formula $((a \supset b) \supset a) \supset a$ is deducible in classical propositional calculus since it is tautological. At the same time, it is not deducible from axioms 1–4 since it is not deducible in intuitionistic propositional calculus. Indeed, substituting $\neg a$ for the variable b into the indicated formula we obtain a formula equivalent to the law of double negation.

References

1. F.L. Varpakhovskii, 'On non-realizability of disjunctions of non-realizable formulas', *Dokl. Akad. Nauk SSSR* **161**:6 (1965), 1257–1258 (in Russian).

2. N.N. Vorob'ev, 'A new algorithm for derivability in constructive propositional calculus', *Trduy Mat. Inst. Akad. Nauk SSSR* **52** (1958), 193–225 (in Russian).
3. A. Heyting, 'Abriss einiger Studier über die Grundlagen der Mathematik, Springer-Verlag, Berlin, 1933 *Ergebn. der Mathematik und ihre Grenzgebiete* **3**:4.
4. A. Heyting, *Intuitionism. An introduction*, North-Holland Publ. Co., Amsterdam, 1956.
5. A.G. Dragalin, *Mathematical intuitionism. Introduction to proof theory*, Nauka, Moscow, 1979 (in Russian).
6. M.M. Kipnis, 'On a property of propositional formulas', *Dokl. Akad. Nauk SSSR* **174**:2 (1967), 277–278 (in Russian).
7. M.M. Kipnis, 'On realizations of predicate formulas', *Zap. Nauchn. Sem. LOMI* **20** (1971), 40–48 (in Russian).
8. S.C. Keene, *Introduction to metamathematics*, D. van Nostrand, Groningen; Noordhoff, New York-Toronto, 1952.
9. A.N. Kolmogorov, 'On the tertium non datur principle', *Mat. Sb.* **32** (1925), 646–667. English translation in: J. van Heijenoort, *From Frege to Gödel. A source book in mathematical logic* 1879–1931, Cambridge, 1967, 416–437. (see No. 9.)
10. A. Kolmogoroff, 'Zur Deutung der intuitionistischen Logik', *Math. Z.* **35** (1932), 58–65. (See No. 19.)
11. L.L. Maksimova, D.P. Skvortsov and V.B. Shekhtman, 'The impossibility of finite axiomatization of Medvedev's logic of finite problems', *Dokl. Akad. Nauk SSSR* **245**:5 (1979), 1051–1054 (in Russian).
12. Yu.T. Medvedev,'Degrees of difficulty of mass problems' *Dokl. Akad. Nauk SSSR* **104**:4 (1955), 501–504 (in Russian).
13. Yu.T. Medvedev, 'Finite problems', *Dokl. Akad. Nauk SSSR* **142**:5 (1962), 1015–1018 (in Russian).
14. P.S. Novikov, *Constructive mathematical logic from the viewpoint of classical mathematical logic*, Nauka, Moscow, 1977 (in Russian).
15. B.Yu. Pil'chak, 'On the solvability problem for the problem calculus', *Dokl. Akad. Nauk SSSR* **75**:6 (1950), 773–776 (in Russian).
16. B.Yu. Pil'chak, 'On the problem calculus', *Ukrain. Mat. Zh.* **4**:2 (1952), 174–194 (in Russian).

17. V.E. Plisko, 'Recursive realizability and constructive predicate logic', *Dokl. Akad. Nauk SSSR* **214**:3 (1974), 520–523 (in Russian).

18. V.E. Plisko, 'Some variants of the notion of realizability for predicate formulas', *Dokl. Akad. Nauk SSSR* **226**:1 (1976), 61–64 (in Russian).

19. V.E. Plisko, 'Non-arithmeticity of the class of realizable predicate formulas', *Izv. Akad. Nauk SSSR Ser. Mat.* **41**:3 (1977), 483–502 (in Russian).

20. H. Rasiowa and R. Sikorski, *The mathematics of metamathematics*, Pańￂstowe Wydawnictwo Naukove, Warszawa, 1963.

21. H. Rogers, *Theory of recursive functions and effective computability*, McGraw-Hill, New York, 1967.

22. D.P. Skvortsov, 'Two generalizations of the notion of a finite problem', in *Studies on non-classical logic and set theory*, Nauka, Moscow, 1979, pp. 201–240 (in Russian).

23. D.P. Skvortsov, 'The logic of infinite problems and Kripke models on atomic semi-lattice sets', *Dokl. Akad. Nauk SSSR* **245**:4 (1979), 798–801 (in Russian).

24. A. Church, *Introduction to mathematical logic*, Princeton Univ. Press, Princeton, 1956.

25. N.A. Shanin, 'On certain logical problems in arithmetic', *Trudy Mat. Inst. Akad. Nauk SSSR* **43** (1955), 1–10 (in Russian).

26. N.A. Shanin, 'On the constructive understanding of mathematical assertions', *Trudy Mat. Inst. Akad. Nauk SSSR* **52** (1958), 226–311 (in Russian).

27. G. Gentzen, 'Untersuchungen über das logische Schliessen', I, II, *Math. Z.* **39** (1934–35), 176–210, 405–431.

28. V.I. Glivenko, 'Sur quelques points de la logique de M. Brouwer', *Bull. Cl. Sci. Akad. Roy. Belg. Sér.* 5 **15** (1929), 183–188.

29. K. Gödel, 'Zur intuitionistischen Arithmetik und Zahlentheorie', *Ergeb. Math. Kolloq.* **4** (1933), 34–38.

30. J. van Heijenoort, *From Frege to Gödel. A source book in mathematical logic*, 1879–1931, Cambridge, 1967.

31. A. Heyting, Die formalen Regeln der intuitionistischen Logik', *Sitzungsber. Preuss. Akad. Wiss. Phys.-Math. Kl.* (1930), 45–52.

32. S. Jas'kowski, 'Recherches sur le système de la logique intuitioniste', *Actes Congrès Intern. Philos. Sci.*, VI *Philos. Math., Actual Sci. Industr.* **393** (1936), 58–61.

33. I. Johansson, 'Der Minimalkalkül, ein reduzierter intuitionistischer Formalismus', *Compos. Math.* **4** (1936), 119–136.

34. S.C. Kleene, 'On the interpretation of intuitionistic number theory', *J. Symbol. Log.* **10**:1 (1945), 109–124.

35. S.A. Kripke, 'Semantical analysis of intuitionistic logic. I', In: *Formal systems and recursive functions*, Amsterdam, 1965, pp. 92–129.

36. J.C.C. McKinsey and A. Tarski, 'On closed elements in closure algebras', *Ann. of Math.* **47** (1946), 122–162.

37. J.C.C. McKinsey and A. Tarski, 'Some theorems about the sentential calculi of Lewis and Heyting', *J. Symbol Log.* **13**:1 (1948), 1–15.

38. H. Rasiowa, 'Algebraic treatment of the functional calculi of Heyting and Lewis', *Fund. Math.* **38** (1951), 99–126.

39. G.F. Rose, 'Propositional calculus and realizability', *Trans. Amer. Math. Soc.* **75** (1953), 1–19.

40. M.H. Stone, 'Topological representation of distributive lattices and Brouwerian logics', *Cas. Pešt. Mat. a Fys.* **65** (1937), 1–25.

41. A. Tarski, 'Der Aussagenkalkül und die Topologie', *Fund. Math.* **31** (1938), 103–134.

42. Wang Hao, [The introductory notes to the English translation of [9].] In: J. van Heijenoort, *From Frege to Gödel. A source book in mathematical logic*, 1879–1931, Cambridge, 1967, pp. 414–416.

ON THE PAPERS ON HOMOLOGY THEORY

A.N. Kolmogorov

An impetus to the creation of these papers was given by de Rham's memoir (Sur l'Analysis situs des variétés à n-dimensions: Thèse; published in *J. Math. Pures et Appl. Sér.* 9 **10** (1931), 115–200), in which the duality between the Betti groups of differentiable manifolds and the Betti groups generated by flows on these manifolds was established.

After the thirties I did not deal with this topic, but even now the idea (presented in four papers in *C.R. Acad. Sci. Paris*; Nos. 32–35 in this book) to assume as a basis the duality between the groups of skew-symmetric functions of n points and the groups of skew-symmetric additive functions of n sets seems to be at least very instructive.

HOMOLOGY THEORY

(G.S. Chogoshvili)

Kolmogorov's papers on algebraic topology answered many important questions arising in this field of mathematics, and opened up new prospects in its development. In these papers the notion of cohomology was introduced, which took a central place in the further development of topology.

Kolmogorov came to the notion of cohomology proceeding from some problems of calculus, in order to "develop a certain kind of finite-difference calculus which, on the one hand, can lead to differential operations on skew-symmetric tensors (multivectors) by passing to the limit and, on the other hand, is closely related to notions of combinatorial topology". In its initial version the theory was reported at the Conference on Tensor Calculus and is presented in paper No. 29.

The same ideas were simultaneously put forward by J.W. Alexander, an American topologist [1, 2]. Reports on the new topological notions were presented by the two authors at the Moscow Conference on Topology in 1935. We note that Alexander posed a narrower problem, namely, he wanted to construct in a direct manner (without using Pontryagin's character theory) a discrete group whose character group is the homology group of a compact metric space. In his paper No. 29 Kolmogorov discusses in parallel both homology and cohomology theories (the corresponding groups are called Betti u and o-groups). An analysis of n-dimensional manifolds, n-dual pairs of cellular decompositions, and topological (locally compact) coefficient groups leads to a proof of the Poincaré-Alexander duality theorems for homology and cohomology.

Both authors, Kolmogorov (see paper No. 31) and Alexander defined a multiplication operation on the cohomology group which played an extremely important role and thus turned this group into a ring.

The four papers (Nos. 32–35) published in C.R. Acad. Sci. Paris form a separate fragment in Kolmogorov's research on these problems. In these papers the author deals with the homology and cohomology theories of locally bicompact spaces and, in particular, considers duality laws. Kolmogorov's papers were preceded by Alexander's paper [1] in which the cohomology group was defined for metric compacta. In paper No. 32 homology and cohomology groups are defined for locally compact spaces. The definition of cohomology groups was later improved by many topologists, and took the form now called

Alexander-Spanier cohomology (see [3–7]). Alexander-Spanier cohomology is defined for arbitrary Hausdorff spaces. Balavadze [8] showed that Alexander-Spanier cohomology is isomorphic to a modified Kolmogorov cohomology (it is only necessary to exclude condition (d) from Kolmogorov's axiomatics in paper No. 32). Kline [9] studied the relationship between Kolmogorov and Alexander cohomology. Alexander-Spanier cohomology was further improved in Massey's book [7] (see also Keesee [10]).

In well-developed theories such as the singular theory and the Aleksandrov-Čech spectral theory, both homology and cohomology are defined, and they are constructed by a general method and related by duality. Moreover, these theories admit of a spectral representation, for example, based on mappings of polyhedra into the given space; the singular theory represents homology as the limit of straight lines and cohomology as the limit of inverse spectra whereas, conversely, the spectral theory, based on mappings of the given space into polyhedra, represents homology and cohomology as the limit of inverse and direct spectra, respectively. In accordance with the stated problem of constructing cohomology dual to the Vietoris homology, Alexander uses the same means as Vietoris in constructing his homology, namely, basic notions of combinatorial topology and limit relations (see [2, 11]). Using the notions that he introduced to this end, other authors later showed (see [12, 3]) that Vietoris homology and Alexander cohomology are the inverse and direct limits of the homology group and cohomology groups, respectively, of the same spectrum of complexes (these complexes are analogues of nerves in the Aleksandrov-Čech theory). Therefore these two objects constitute a unified (co)homology theory in the sense that they are constructed by similar means and are dual to each other. In its further development Alexander cohomology was defined in a more simple and direct way, that is, without using explicitly notions of combinatorial topology and limit relations (see [13, 3, 14]). However, in this new approach one can no longer construct in a similar way the associated homology. In this connection, Eilenberg and Steenrod [15] noted that this approach lacks a similar simple dual construction for homology chains and groups. In the very first paper No. 32, Kolmogorov defines, along with cohomology, dual homology, and this definition is simple and direct; namely, it uses basic concepts of combinatorial topology and limiting processes. In this sense this is a completely new presentation of homology theory. This definition is analogous to the definition of cohomology in the sense that it is based on the general mathematical notion of a function — a point function in the case of cohomology and a set function in the case of homology. Homology thus defined is dual to cohomology in the sense of character theory. In paper No. 33 it is proved that the homology and cohomology groups of the same space and in the same dimension (the former over a compact coefficient group and the latter over a discrete coefficient group) are dual if the coefficient groups are dual. This is a new type of duality law in algebraic topology interrelating homology and cohomology. (The first two laws of Poincaré and Alexander, stemming from set-theoretic topology (Jordan's theorem) are discussed in papers Nos. 34 and 35.) Following his analytical approach, Kolmogorov stresses that in this case multiplication is defined as an n-fold multiple Radon-Stieltjes integral of the product of the corresponding functions. The simplicity of his definition also manifests itself in that when presenting homology associated with his cohomology, Kolmogorov does not use the one in the definition of the other. This is what distinguishes his approach from those of many other authors, for example Massey, who believes that the introduction of homology via Alexander-Spanier cohomology by means of the hom functor is an important, long-desired, and definitive result (see [7, 16]). Recall that when introducing cohomology, Alexander [2] intended to construct the resulting discrete group in a way more direct than that in which the group is defined via the character group, which is also a hom functor.

Paper No. 34 is devoted to interrelations between the author's homology theory and other theories, and the Poincaré law. Of no less value are, however, the methods that Kolmogorov employs to study the problems posed. In the modern terminology these are known as homology and cohomology of the second kind. Decompositions lead (see [17, 18]) to a version of Aleksandrov-Čech homology theory which, as distinguished from Aleksandrov-Čech theory itself, is exact and is associated in a natural way with Kolmogorov's (co)homology theory. As will be indicated below, infinite cycles and their homology soon attracted attention and found versatile applications [21]. Finite cochains led to cohomology, which appeared, as was noted above, after the long development of the Alexander-Spanier theory (see [10, 7]).

Kolmogorov used decompositions for comparing his homology theory with the Vietoris theory, which was the most well developed theory by that time. Vietoris homology was then defined for compact metric spaces. Therefore it sufficed for Kolmogorov to consider only sequences of decompositions, although he defined his homology for the broader class of locally compact spaces. He proved that in the case of compact metric spaces and compact coefficient groups his homology groups are isomorphic to the Vietoris homology groups; moreover, the isomorphism takes place at the level of chain complexes. This was done by using the Aleksandrov projection cycles. Later, using arbitrary (quasi-ordered) spectra based on all (finite) decompositions, Chogoshvili showed that the Kolmogorov homology groups of any locally compact spaces with a compact coefficient group are isomorphic to the Aleksandrov homology groups taken with respect to the so-called special subcomplexes, and therefore to the Aleksandrov-Čech homology groups for compact spaces (see [17, 18]). Lefschetz showed that the Alexander-Kolmogorov homology theory can be represented as the projection theory of a single-valued lattice spectrum [19]. Dowker [20] proved that the Aleksandrov-Čech and Vietoris homology groups are isomorphic for arbitrary spaces. It follows from these isomorphisms that the Kolmogorov homology group is isomorphic to the Steenrod homology group [21] (and therefore to the Sitnikov group, which is isomorphic to the latter; see [22, 23]) in the case of compact metric spaces (for which the Steenrod group was constructed), since Steenrod himself showed that his group is isomorphic to the Vietoris group if the coefficient group is compact. Further, we note that the isomorphism between homology groups readily implies that the dual cohomology groups are also isomorphic (cf. [24]).

Kolmogorov's homology theory has a number of remarkable properties. We consider in more detail one of them, the property of exactness, which is a desirable relationship between sets of cycles and bounding cycles of certain kinds, the latter notion being introduced in the forties [15]. Unlike the singular theory, the Aleksandrov-Čech theory does not possess the exactness property. The often mentioned compactness of the coefficient group is required because it guarantees that both Aleksandrov-Čech and Vietoris homology for compact spaces possess the exactness property. For a long time mathematicians have tried to find improved versions of the Aleksandrov-Čech and Vietoris theories which under certain additional conditions possess properties similar to exactness. Aleksandrov true cycles, Lefschetz projection cycles, Pontryagin's compactification of the coefficient group, and Vietoris homology groups played a central part in these efforts. (Actually, exactness was usually achieved at the expense of losing a certain useful property, for example, continuity [15].) The objective was accomplished when in 1940 Steenrod constructed a homology theory for compact metric spaces substantially based on the theory of infinite cycles. This theory is exact for an arbitrary coefficient group. In 1951 Sitnikov constructed an isomorphic theory (see [22, 23]). Chogoshvili obtained homology groups isomorphic to the Kolmogorov groups, which gave a spectral representation for them [17]. However, in his axiomatic study of Kolmogorov (co)homology with a changed coefficient group (discrete homology and compact cohomology, in contrast to the case studied by Kolmogorov) Balavadze [25] introduced notions which had been lacking (such as induced homomorphisms, etc.) and, among other theorems (for example, theorems on duality between homology and cohomology), proved that Kolmogorov's theory satisfies all Steenrod-Eilenberg axioms and, in particular, the exactness axiom. Mdzinarishvili (see [26, 27]) showed that this theory satisfies the conditions of the Milnor uniqueness theorem and consequently is isomorphic to the Steenrod theory with not only compact but also discrete coefficient group. Thus, the desired exact theory turned out to be constructed as a Kolmogorov theory four years earlier than Steenrod's theory, and for a broader class of locally compact spaces. We note that the (co)homology theories for locally compact spaces including those with values in (co)sheaves gave rise to a large number of papers (see, for example, [28–30]). The American Mathematical Society Subject Index 1980 has a special division "Steenrod-Sitnikov Homology" (this is, in fact, Kolmogorov homology) demonstrating the significance of this topic in the eighties. The topic itself and the above-mentioned papers are closely related to Kolmogorov's theory.

Poincaré's theorem (No. 34) is proved for homology and cohomology groups with finite and infinite (co)chains of the cellular decomposition of an open n-dimensional manifold. The coefficient groups are discrete in the case of finite (co)-chains and are compact when the (co)chains are infinite. The relations are expressed in terms of isomorphism and duality. The form in which Poincaré duality is represented, simultaneously in terms of homology and cohomology, makes the presentation particularly clear, complete and natural.

In the last paper (No. 35) it is proved that the Kolmogorov cohomology and homology groups both satisfy the Alexander duality law. At that time the Alexander duality law was assumed to indicate the expedience and usefulness of the suggested homological notion, and in this sense the Alexander law played the role of the modern (co)homology axioms. This is quite understandable, since the (co)homology axioms form the basic points in the proof of duality laws. For example, the excision axiom, which is a characteristic property of the (co)homology functor, is given by Kolmogorov separately in its strong form. The very form of Kolmogorov's theorems relating (co)homologies of neighbouring dimensions, and not complementary dimensions, expresses the main part of the (co)homology sequence, the so-called connecting homomorphism.

In the framework of his theory of (co)homology with special subcomplexes P.S. Aleksandrov proved duality theorems in Kolmogorov's form and generalized them in various directions (see [31, 32]). Here infinite cycles were replaced by their finite analogues; namely, cycles relative to the subcomplex of a nerve of a finite closed covering the closures of whose vertices are not compact. The author calls them Kolmogorov theorems. These theorems were further generalized in terms of the same cycles by Bokshtein (see [33, 34]) and Chogoshvili (see [35–38]). Sklyarenko studied theorems in Kolmogorov's form using so-called canonical homology [39] for any coefficient group. Mdzinarishvili excluded the compactness condition on the coefficient group in Kolmogorov's theorems and showed that the Steenrod duality theorem (and, consequently, the Sitnikov duality theorem) are consequences of this generalized theorem in the case of compact metric spaces (see [40, 41]). Various problems of Kolmogorov's (co)homology theory arising in connection with Aleksandrov's (co)homology theory were investigated by Bokshtein [42]. Finally, even after the spectral (co)homology theory for arbitrary (non-compact) spaces appeared in the mid forties and the duality theorems were extended from closed subsets to arbitrary subsets, Kolmogorov devised a cohomology-based duality form which is expressed in terms of isomorphisms and does not require the application of direct spectra of compact and, generally, topologized groups [43].

Cohomology rings for locally bicompact spaces are introduced in paper No. 30, which is specially devoted to products of (co)chains and cohomology groups in complexes, manifolds, and locally bicompact spaces.

Recently much effort has gone into the construction of a homology theory which would be more useful from the teaching point of view (see, for example, the book by Massey [7] and Sklyarenko's comment to its Russian translation). From this point of view Kolmogorov's theory has a number of advantages over other theories. It has no known anomalies of the singular theory and, unlike the Aleksandrov-Čech theory and Aleksandrov's theory with special subcomplexes, this theory is exact; further, it has an associated cohomology theory, in contrast to the Steenrod-Sitnikov theory, and within this theory, it is possible to define homology in a direct way irrespective of the cochain complex, the latter being characteristic of Massey's theory; finally, it suggests a simpler way of choosing initial sets, stating the definition of induced homomorphism, etc. than in the Kurosh-Sklyarenko theory based on canonical coverings. These advantages can be supplemented by the visual character of Kolmogorov's theory, its applicability to various research areas, and its close relation to other mathematical theories.

References

1. J.W. Alexander, 'On the chains of a complex and their duals. On the ring of a compact metric space', *Proc. Nat. Acad. Sci. USA* **21** (1935), 509–512.
2. J.W. Alexander, 'On the connectivity ring of an abstract space', *Ann. of Math.* **37** (1936), 698–708.
3. E.H. Spanier, 'Cohomology theory for general spaces', *Ann. of Math.* **49** (1948), 407–427.
4. E.H. Spanier, 'Tautness for Alexander-Spanier cohomology', *Pacific J. Math.* **75** (1978), 561–563.
5. R. Godement, *Topologie algébrique et théorie des faisceaux*, Hermann, Paris, 1958.
6. G.E. Bredon, *Sheaf theory*, McGraw-Hill, New York, 1967.
7. W.S. Massey, *Homology and cohomology theory*, M. Dekker, New York, 1978.
8. M.B. Balavadze, 'On Kolmogorov's homology theory', *Trudy Tbiliss. Mat. Inst. Razmadze Akad. Nauk Gruzin. SSR* **41** (1972), 5–40 (in Russian).
9. M. Kline, 'Note on homology theory for locally bicompact spaces', *Fund. Math.* **32** (1939), 64–68.
10. J.W. Keesee, 'Finitely-valued cohomology groups', *Proc. Amer. Math. Soc.* **1** (1950), 418–422.
11. J.W. Alexander, 'A theory of connectivity in terms of gratings', *Ann. of Math.* **39** (1938), 883–912.
12. W. Hurewicz, J. Dugunji and C.H. Dowker, 'Continuous connectivity groups

in terms of limit groups', *Ann. of Math.* **49** (1948), 391–406.

13. J.W. Alexander, 'Gratings and homology theory', *Bull. Amer. Math. Soc.* **53** (1947), 201–233.
14. E.H. Spanier, *Algebraic topology,* McGraw-Hill, New York, 1966.
15. S. Eilenberg and N. Steenrod, *Foundations of algebraic topology,* Princeton Univ. Press, Princeton, 1952.
16. W.S. Massey, 'How to give an exposition of the Čech-Alexander-Spanier type homology theory', *Amer. Math. Monthly* **85** (1978), 75–83.
17. G.S. Chogoshvili, 'On the homology theory of topological spaces', *Soobshch. Akad. Nauk Gruzin. SSR* **1** (1940), 337–340.
18. G.S. Chogoshvili, 'On equivalence of functional and spectral homology theories', *Izv. Akad. Nauk SSSR Ser. Mat.* **15** (1951), 421–438 (in Russian).
19. S. Lefschetz, *Algebraic topology,* New York, 1942.
20. C.H. Dowker, 'Homology groups of relations', *Ann. of Math.* **56** (1952), 84–95.
21. N.E. Steenrod, 'Regular cycles of compact metric spaces', *Ann. of Math.* **41** (1940), 833–851.
22. K.A. Sitnikov, 'The duality law for non-closed sets', *Dokl. Akad. Nauk SSSR* **18**, (1951), 359–362 (in Russian).
23. K.A. Sitnikov, 'Combinatorial topology of non-closed sets I. First duality law; spectral duality', *Mat. Sb.* **34** (1954), 3–54 (in Russian).
24. J.L. Kelley, 'Descriptions of Čech cohomology', In: *General topology and its relation to modern analysis and algebra, Proc. Symp.*, Prague, 1961, pp. 235–237.
25. M.B. Balavadze, *On Kolmogorov's homology theory, Review of thesis,* Tbilisi, 1975 (in Russian).
26. L.D. Mdzinarishvili, 'On the relation between Kolmogorov's and Steenrod's homology theories', *Dokl. Akad. Nauk SSSR* **203**:3 (1972), 528–531 (in Russian).
27. L.D. Mdzinarishvili, 'On equivalence of Kolmogorov and Steenrod homology theory', *Trudy Tbiliss. Mat. Inst. Razmadze Akad. Nauk Gruzin. SSR* **41** (1972), 143–163 (in Russian).
28. A. Borel and J.C. Moore, 'Homology theory for locally compact spaces', *Mich. Math. J.* **7** (1960), 137–160.

29. A. Borel, *Cohomologie des espaces localement compacts*, Lecture Notes in Math., No. 2 (1964).
30. G.E. Bredon, 'Cosheaves and homology', *Pacific J. Math.* **25**:1 (1968), 1–32.
31. P.S. Aleksandrov, 'General homology theory', *Uchen. Zap. MGU* **45** (1940), 1–60 (in Russian).
32. P.S. Aleksandrov, 'Homological properties of decompositions of complexes and closed sets', *Izv. Akad. Nauk SSSR Ser. Mat.* **6** (1942), 227–282 (in Russian).
33. M.F. Bokshtein, 'On the Alexander-Kolmogorov duality theorem', *Dokl. Akad. Nauk SSSR* **59** (1948), 631–633 (in Russian).
34. M.F. Bokshtein, 'Duality theorem for locally bicompact spaces', *Uchen. Zap. MGU* **146** (1949), 131–164 (in Russian).
35. G.S. Chogoshvili, 'On duality relations in topological spaces', *Dokl. Akad. Nauk SSSR* **46**:4 (1945), 143–145 (in Russian).
36. G.S. Chogoshvili, 'On the duality law in normal spaces', *Dokl. Akad. Nauk SSSR* **48**:4 (1945), 249–252 (in Russian).
37. G.S. Chogoshvili, 'On basic duality homomorphisms', *Trudy Tbiliss. Mat. Inst. Razmadze Akad. Nauk Gruzin. SSR* **18** (1951), 1–52 (in Russian).
38. G.S. Chogoshvili, 'On homological approximations and duality laws for arbitrary sets', *Mat. Sb.* **28** (1951), 89–118 (in Russian).
39. E.G. Sklyarenko, 'Homology theory and the exactness axiom', *Uspekhi Mat. Nauk* **24**:5 (1969), 87–140 (in Russian).
40. L.D. Mdzinarishvili, 'On Kolmogorov's duality law', *Dokl. Akad. Nauk SSSR* **216** (1974), 502–504 (in Russian).
41. L.D. Mdzinarishvili, 'Functional homology', *Trudy Tbiliss. Mat. Inst. Razmadze Akad. Nauk Gruzin. SSR* **59** (1978), 98–118 (in Russian).
42. M.F. Bokshtein, 'The equivalence of certain homological definitions in topology', *Izv. Vuzov.* **3** (1960), 62–80 (in Russian).
43. P.S. Aleksandrov, 'Basic duality theorems for non-closed sets', *Mat. Sb.* **21** (1947), 161–232 (in Russian).

ON THE PAPER "ON OPEN MAPPINGS"

A.N. Kolmogorov

The possibility of an increase in dimension under open mappings (No. 36) interested P.S. Aleksandrov very much. For some time we together tried to prove that increase in dimension is impossible. In these attempts we gradually understood why we failed. An analysis of the failure led us to a counterexample.

TOPOLOGY

On the paper "Finite coverings of topological spaces"

(A.V. Arkhangel'skii)

The results in parts 3 and 4 of paper No. 31 established by A.N. Kolmogorov are of an exhaustive character. They imply that the definition of dimension in terms of the multiplicity of closed coverings (that is, Lebesgue's approach) and the definition using lengths of closed coverings, which was suggested in this paper, yield the same result for all normal spaces. Kolmogorov gives a very simple elementary construction enabling one to pass from length to multiplicity and back. Therefore, in the further development of dimension theory the notion of covering length was of minor significance, since the studies were based on the notion of multiplicity and the classical definition of dimension (dim) given by Lebesgue (see [2]).

On the paper "On open mappings"

(A.V. Arkhangel'skii)

1. The example of an open mapping of a one-dimensional continuum onto a two-dimensional continuum constructed in paper No. 36 was the first example of an open mapping of a compactum onto another compactum under which the dimension increased. Of special interest are some additional properties of the constructed mapping, namely, its zero-dimensionality (the pre-image of any point is a zero-dimensional space) and the fact that the spaces related by the mapping are compact.

In connection with the example constructed by Kolmogorov, the following simple facts are worth mentioning. It is not difficult to represent an arbitrary space as the image of a zero-dimensional space under a continuous mapping, since a discrete space can be mapped continuously onto any space. It is somewhat more difficult to prove that each Tikhonov space is the image of a zero-dimensional space under a continuous open mapping (see [2]). However, in the class of compacta the latter assertion is no longer true. Indeed, a continuous mapping of a compactum into a Hausdorff space is always closed, while the image of a zero-dimensional space under an open or closed continuous mapping is obviously a zero-dimensional space. In particular, dimension cannot increase under continuous open mappings of zero-dimensional compacta. Thus, the compactness of the spaces involved in Kolmogorov's example is a fundamental feature, and the dimension 1 is the least dimension that can increase under an open mapping of a compactum.

The problem of dimension increase under open mappings was by far not exhausted by Kolmogorov's example; on the contrary, this paper gave an impetus to some further studies. In particular, attention is drawn to the fact that in Kolmogorov's example the image compactum is of a rather special form. A new step was made by L.V. Keldysh when in 1959 she constructed an open zero-dimensional mapping from a one-dimensional compactum onto a square (see [1, 2]). Proceeding from the fact that such a mapping exists, Pasynkov established a final result in this area (see [3, 2]); he showed that each bicompactum of positive dimension is the image of a bicompactum of dimension 1 (in the sense of dim, that is, in the sense of the Lebesgue definition) under a continuous open zero-dimensional mapping.

L.V. Keldysh obtained some rather subtle results concerning the behaviour of dimension under monotone open mappings (a mapping is monotone if the pre-image of each point is connected; obviously, monotonicity is "opposite" to zero-dimensionality). She constructed a remarkable example of a monotone open mapping from a three-dimensional cube onto a cube of arbitrary dimension n greater than 3 (see [4, 5]). In this case there is an increase in dimension under the open continuous mapping although there are no exotic properties in both the space to be mapped and the image space.

The idea underlying Kolmogorov's example was vividly elucidated in a recent paper by Kozlovskii (see [6]). Noting that Kolmogorov's example is based on an open two-fold mapping of a two-dimensional torus onto the Möbius band, the mapping being inessential in a natural sense, he develops a general method using inverse spectra of open mappings, which are inessential in the Aleksandrov sense, allowing one to construct r-dimensional pre-images of any n-dimensional polyhedra ($n \geq r \geq 1$) under both monotone open mappings and zero-dimensional open mappings.

2. An increase in dimension under an open mapping indicates indirectly that the structure of the mapping is rather complicated. In particular, such a mapping can be expected to be fundamentally different (even locally) from a projection mapping of a product space onto a factor.

Homomorphisms of topological groups onto quotient groups of them are important cases of open mappings. Kolmogorov's approach was extended to the study of the behaviour of dimension under such homomorphisms. It became clear that if a group is locally bicompact, then the dimension of any of its quotient groups does not exceed the dimension of the group (see [7]). At the same time, every topological group can be represented as a quotient group of a zero-dimensional topological group; this result was recently obtained in [8].

3. The example construction is described in the paper clearly and comprehensively. However, the discussion of its properties contains an incorrect argument. Namely, it is erroneously asserted in part 1 that the limit of the countable inverse spectrum of open mappings $f_n : X_n \to Y_N$ is an open mapping of the limit space X onto the limit space Y. In the notation of Kolmogorov's paper, the desired assertion holds if, in addition,

$$f_n(\phi^{-1}(x_{n-1})) = \phi_{n-1}^{-1}(f_{n-1}(x_{n-1})).$$

In Pasynkov's terminology, this relationship shows that the mapping f_{n-1} is filled with f_n (see [2], p. 465). The compactness of the spaces involved in the spectrum is also important. It is easily seen that the indicated additional condition holds in the Kolmogorov example, which implies the openness of the constructed mapping.

References

1. L.V. Keldysh, 'Zero-dimensional open mappings', *Izv. Akad. Nauk SSSR Ser. Mat.* **23** (1959), 165–184.
2. P.S. Aleksandrov and B.A. Pasynkov, *Introduction to dimension theory*, Nauka , Moscow, 1973 (in Russian).
3. B.A. Pasynkov, 'Zero-dimensional open maps increasing dimension', *Uspekhi Mat. Nauk* **18**:5 (1963), 183–190 (in Russian).
4. L.V. Keldysh, 'Monotone maps from a cube onto a cube of higher dimension', *Mat. Sb.* **41**:2 (1957), 129–158 (in Russian).

5. L.V. Keldysh, 'Transformations of a monotone irreducible map to a monotone open irreducible map, and monotone open maps of a cube increasing dimension', *Mat. Sb.* **43**:2 (1957), 187–226.
6. I.M. Kozlovskii, 'Dimension-increasing open maps from compacta onto polyhedra as spectral limits of inessential maps of polyhedra', In: The Leningrad International Conference on Topology, Nauka, Leningrad, 1982 (in Russian).
7. A. Weil, *L'intégration dans les groupes topologiques et ses applications*, Publications de l'institut de mathématiques de Clermont-Ferrand, Paris, 1940.
8. A.V. Arkhangel'skii, 'Classes of topological groups', *Uspekhi Mat. Nauk* **36**:3 (1981), 127–146 (in Russian).

AXIOMATICS OF PROJECTIVE GEOMETRY

(A.V. Mikhalev)

Let K be a field, M an $(n+1)$-dimensional vector space, and $\mathfrak{B}(M) = (G_0, G_1, \ldots, G_n)$ the n-dimensional projective geometry over the field K, that is, the lattice of subspaces of M (here G_k is the set of all $(k+1)$-dimensional subspaces of M).

At the beginning of the 20th century Hilbert (see [4]) showed that all abstract finite-dimensional Desargues projective geometries $\mathcal{P}^n = (G_0, G_1, \ldots, \ldots, G_n)$ are exhausted by systems of linear subspaces of finite-dimensional vector spaces over fields, i.e., they are isomorphic to n-dimensional projective geometries $\mathfrak{B}(M)$. Later this branch of mathematics relating to the axiomatization of geometry or to the introduction of coordinates in an axiomatically defined geometry was actively developed (see [1–3, 6, 10]), which made it possible to relate projective geometry to many areas of algebra (such as group theory, lattice theory and ring theory). This gave rise to a new branch of algebra, called "projective algebra" due to Kurosh.

In the thirties, A.N. Kolmogorov was one of the first Soviet mathematicians to study topological algebras, that is, sets with a topology and algebraic operations, the latter being continuous. In 1932 he published the paper [5] (No. 20 in this book) devoted to the axiomatics of projective geometries. If K is a topological field, a vector space M is endowed with the topology of the product of $(n+1)$ copies of the topological abelian group K, and the sets G_k

are topologized by means of the natural procedure of "stirring" the basis of a k-dimensional subspace, then a plane $a \in G_k$ passing through linearly independent points $a_0, a_1, \ldots, a_k \in G_0$ depends continuously on these points. If the field K is continuous (that is, locally compact and non-discrete), then all spaces G_k are compact, while if the field K is totally disconnected, then all spaces G_k are totally disconnected. Further, Kolmogorov states that an abstract projective geometry $\mathcal{P}^n = \{G_0, G_1, \ldots G_n\}$ is continuous if the set G_0 of points and the set G_1 of lines are compact infinite topological spaces, and if a straight line is a continuous function of any pair of points it passes through. It was proved in [5] that every continuous projective geometry $\mathcal{P}^n$ is isomorphic to a projective geometry $\mathfrak{B}(M)$ over a continuous field K (and if the space of points is connected, the topological field K is also connected). In view of this, Kolmogorov posed the problem of describing connected locally compact fields, which was solved by L.S. Pontryagin (see [7, 8]); it turned out that there are only three such fields, namely, the field of real numbers, the field of complex numbers and the quaternion field.

This paper by Kolmogorov played an important role in the further development of topological algebra. In particular, the description of all locally compact connected Desargues planes based on the above-mentioned results of Kolmogorov and Pontryagin was a starting point for the theory of topological geometries (for example, see the review [9] on topological projective planes). Specifically, the author of this commentary noted in his thesis (1967) that under certain conditions the structure of subspaces of a topological vector space over a topological field with a topologized set of one-dimensional subspaces determines completely the topological vector space, and hence the topological field K and the topological space M in Kolmogorov's theorem are determined uniquely (to within a topological semilinear mapping).

References

1. E. Artin, *Geometric algebra*, Wiley (Interscience), New York, 1957.
2. G. Birkhoff, *Lattice theory*, Colloq. Publ. Amer. Math. Soc., 1940.
3. R. Baer, *Linear algebra and projective geometry*, New York, 1952.
4. D. Hilbert, *Grundlagen der Geometrie*, Teubner, Stuttgart, 1956.
5. A.N. Kolmogoroff, 'Zur Begründung der projektiven Geometrie', *Ann. of Math.* **33**:1 (1932), 175–176. (see No. 20.)
6. J. von Neumann, *Continuous geometry*, Univ. Press, Princeton, 1960.

7. L.S. Pontrjagin, 'Über stetige algebraische Körper', *Ann. of Math.* **33**:1 (1932), 163–174.
8. L.S. Pontryagin, *Continuous groups*, Nauka, Moscow, 1973 (in Russian).
9. H.R. Salzmann, 'Topological planes', *Adv. in Math.* **2**:1 (1967), 1–60.
10. L.A. Skornyakov, *Complemented modular lattices and regular rings*, Gostekhizdat, Moscow, 1961 (in Russian).

ON THE PAPER ON THE DIFFUSION EQUATION

A.N. Kolmogorov

The statement of the problem, as well as the intuitive prediction of the expected qualitative results is mine. Along with some other papers, which will be incorporated in the second volume of this edition, the present paper (No. 38) appeared as a consequence of my long contacts with A.S. Serebrovskii and his collaborators N.P. Dubinin, A.A. Malinovskii and D.D. Romashov. The most profound ideas in the mathematical solution of the problem are due to I.G. Petrovskii.

At that same time I set a programme of further study, of which I mention here only the following two points:

(1) the study of the "advance profile" of a recessive gene;

(2) the study of the behaviour of an "island" populated by carriers of a dominant gene (such an island will "dissolve" if it is too small, and will grow infinitely if it is sufficiently large).

The commentary by Barenblatt shows to what extent this programme is fulfilled at present and who contributed to it.

The paper did not receive much response in the biological literature. It turned out to be more significant in many areas of physics, which is discussed in detail in Barenblatt's commentary.

THE DIFFUSION EQUATION

(G.I. Barenblatt)

1. In short, the mathematical content of paper No. 38 can be summarized as follows. The diffusion equation with non-linear right-hand side,

$$\partial v/\partial t - k\partial^2 v/\partial x^2 = F(v), \tag{1}$$

was considered, where $F(v)$ is a sufficiently smooth function defined on the interval $0 \leq v \leq 1$ such that $F(0) = F(1) = 0$; $F(v) > 0$ for $0 < v < 1$; and

$F'(0) = \alpha > 0$, $F'(v) < \alpha$ for $v > 0$, $k = \text{const} > 0$. It was shown that this equation has invariant solutions of travelling-wave type,

$$v = V(x + \lambda t + c), \tag{2}$$

satisfying the conditions $v(\infty, t) = 1$, $v(-\infty, t) = 0$, with any propagation rate λ satisfying the inequality $\lambda \geq \lambda_0 = 2\sqrt{k\alpha}$. The constant c remains indeterminate if the solution (2) is constructed directly from invariance relative to a translation group. It was further established that for $t \to \infty$ the solution of the Cauchy problem for equation (1) with arbitrary initial data such that $v(x, 0) \equiv 0$ for $x < a$, $0 < v(x, 0) < 1$ for $a < x < b$, and $v(x, 0) \equiv 1$ for $x > b$, tends to a solution of the type (2) corresponding to $\lambda = \lambda_0$ and a unique value of c.

The scientific vitality of the paper No. 38 can be explained by the fact that for the first time ever it presented a mathematically complete formulation of and a method of solution for the problem of "intermediate asymptotics" in a non-linear problem of mathematical physics, that is, invariant asymptotic behaviour of solutions when time tends to infinity while the system has not yet come to an equilibrium state. Indeed, as the process continues at an intermediate asymptotic state, it becomes independent of detailed properties of the initial conditions (of the distribution of $v(x, 0)$) since the process is described by an invariant solution (2). At this time the system is still far from the equilibrium state $v(x, \infty) = 1$.

Following paper No. 38, a large number of intermediate asymptotic solutions appeared in various areas of mathematical physics. The immense literature relating to this is essentially based on the ideas first put forward in paper No. 38. Basically, the generality and interrelation of those invariant intermediate asymptotics have become clear in recent years.

2. The idea of intermediate asymptotic solutions of travelling-wave type turned out to be first of all very useful in the mathematical theory of burning. A year after paper No. 38 was published, the Zel'dovich and Frank-Kamenetskii paper [1] appeared (a more detailed presentation was given in [2]), in which the same idea was applied in order to develop a heat theory of flame propagation. Under very simplest conditions this problem also reduced to constructing solutions of travelling-wave type for equation (1) but in contrast to paper No. 38 the condition $F'(v) < F'(0)$ for $0 < v < 1$ no longer holds here. According to

the physics of the problem, it was assumed in [1, 2] that the function $F(v)$, which expresses the rate of the chemical reaction, sharply increases together with v until v reaches values close to 1 and then drops to zero for $v = 1$. To obtain solutions of travelling-wave type it was additionally assumed in [1], [2] that $F(v) \equiv 0$ for $0 \le v \le \beta < 1$. In contrast to paper No. 38, this assumption results in a unique value for the flame propagation rate, that is, the parameter λ_0. Thus, the problem of choosing the travelling-wave rate, which was essential in paper No. 38, did not arise here. That is why it was proved only much later, in [3], that under this condition on $F(v)$ the travelling-wave solution corresponding to $\lambda = \lambda_0$ is actually an asymptotic solution to the Cauchy problem as $t \to \infty$; the proof was substantially based on the ideas and methods developed in paper No. 38.

Such proofs are significant in the theory of burning itself; in some more complicated problems of this theory, λ_0 is not determined uniquely and the choice of the corresponding value is non-trivial. In the mathematical theory of burning it has become possible only quite recently [4, 5] to get rid of the strong assumption that $F(v)$ is identically equal to zero on an interval near $v = 0$. It was shown that in rejecting this assumption, the intermediate asymptotics of the solution to the Cauchy problem are described by a travelling-wave solution (2) in this case as well. Remarkably, although the condition $F'(v) < F'(0)$ does not hold in problems in the theory of burning, the basic formula of paper No. 38, $\lambda_0 = 2\sqrt{kF'(0)}$, remains valid. This fact was first discovered numerically and then confirmed analytically [5].

The modern literature on the mathematical theory of burning includes thousands of papers (an extensive but far from exhaustive list can be found in [6]). However, the idea of intermediate asymptotic travelling-wave type solutions whose propagation rate is not known in advance and is determined in the process of solution has also found application in many areas of mathematical physics besides the theory of burning. Among them we mention the well-known problem of excitation pulse propagation along a nerve [7] and, specifically, its electrochemical model [8]. The study of this model based directly on the ideas of paper No. 38 was presented in [9]. We also note that various processes have recently been studied, including effects of plasma front propagation through electromagnetic, electric, and light (laser) fields (so-called discharge propagation waves [10–13]). These processes also lead to travelling-wave type solutions whose propagation rate is not known in advance and is determined in the course of solution. The study of these processes is entirely based on the ideas first formulated in paper No. 38.

3. The fact that the intermediate asymptotics is free of parameters characterizing details in the initial conditions (such as linear dimensions and time) and the structure of the basic equation (1) guarantee the invariance with respect to a subgroup of the translation group, of the intermediate solution in the problem of dominant gene propagation considered in paper No. 38.

There is a broad class of problems in which the intermediate asymptotic solutions are invariant relative to a subgroup of the group of similarity transformations and can be represented as so-called self-similar solutions. The ideas of paper No. 38 turned out to play a fundamental role in the study of these problems as well.

Indeed, let us pass to new independent variables $\xi = e^x$ and $\tau = e^{-t}$ in the problem considered in paper No. 38 and put $A = e^{-c}$. Then the travelling-wave solution (2) is written in the self-similar form

$$v = V(\ln \xi - \lambda \ln \tau + c) = U(\xi / A\tau^{\lambda}), \tag{3}$$

where the exponent λ is not known in advance and must be found. When the self-similar solution is being constructed, the constant A is determined directly; it is found by sewing together a solution of the form (3) and the solution of the Cauchy problem whose asymptotics is the self-similar solution (3). After paper No. 38 had been published, self-similar solutions of this type appeared in many areas and were called self-similar solutions of the second kind. The first problem of this type was the gas dynamics problem on converging shock waves, considered by Guderley [14], Landau, and Stanyukovich (see [15]). The solution obtained in these papers is a self-similar asymptotic solution of the problem of gas motion inside a sphere caused by a strong explosion on the surface of this sphere. These asymptotics are applicable in a region $r_0(t) < r < r_1(t)$ near the strong shock wave $r = r_0(t)$ (r is the distance from the centre of the sphere). For the self-similar asymptotic solution the functions

$$P = \frac{p}{\rho_0 r^2/t^2}, \quad R = \frac{\rho}{\rho_0}, \quad V = \frac{v}{r/t} \tag{4}$$

(p is gas pressure, ρ is gas density, v is velocity and ρ_0 is the initial gas density) depend on the variable

$$\zeta = r/At^{\lambda}, \tag{5}$$

where the exponent λ is obtained from the condition of existence of a singular self-similar solution (4) of the gas dynamics equation in the large. The constant A remains indeterminate and can be found only by sewing together the self-similar intermediate asymptotics and the initial solution which is not self-similar; this situation is completely analogous to that in paper No. 38.

Note that the problem of a strong converging shock wave does not make it possible to determine the exponent λ from the energy conservation law. Indeed, the energy E contained in the region $r_0(t) < r < r_1(t)$ in which the self-similar asymptotics apply tends to zero together with the radius of the shock wave. However, E decreases more slowly than r_0, and therefore the energy concentration and also the gas pressure at the wave front tend to infinity. It is a specific feature of gas dynamics problems that the exponent λ is chosen from a condition on a characteristic, that is, on a line $dr/dt = v - c$ where c is the velocity of sound, arriving at the centre $r = 0$ simultaneously with the wave front. For the system of ordinary differential equations obtained from the gas dynamics equations for the functions $P(\zeta)$, $R(\zeta)$, and $V(\zeta)$ the characteristic corresponds to a singularity of saddle-point type through which an integral curve passes only if λ has the chosen value. It is remarkable that when the adiabatic exponent γ is greater than a critical value $\gamma_{\text{cr}} \sim 1.87$, as in the problem of paper No. 38, there appears a continuous spectrum of admissible values of λ, while for $\gamma < \gamma_{\text{cr}}$ the exponent λ is determined uniquely. Gel'fand (see [16]) put forward the hypothesis that in the case $\gamma > \gamma_{\text{cr}}$ the situation is the same as in paper No. 38, that is, the smallest value of λ is always realized; however this question still remains open.

Another gas dynamics problem leading also to self-similar solutions of the second kind and attracting much attention is the "short impact" problem [17, 18, 16, 19]. Here self-similar intermediate asymptotics were constructed near a shock wave with an exponent not known in advance and determined in the course of solution for the case of gas motion caused by an impact of a rigid wall against a half-space filled with gas and bordering vacuum.

Recently self-similar solutions of the second kind have also appeared in many other branches of mathematical physics (see the bibliography in [20]). The extensive development of the approach based on intermediate asymptotic invariant solutions and the permanent use in new problems of the ideas first suggested in paper No. 38 attach fundamental significance to this famous work.

References

1. Ya.B. Zel'dovich and D.A. Frank-Kamenetskii, 'On the theory of homogeneous flame propagation', *Dokl. Akad. Nauk SSSR* **19**:9 (1938), 693–697 (in Russian).
2. Ya.B. Zel'dovich, 'On the theory of flame propagation', *Zh. Fiz. Khimii* **22**:1 (1948), 27–48 (in Russian).
3. Ya.I. Kanel', 'On stabilization of the solutions of the Cauchy problem for equations occuring in burning theory', *Mat. Sb.* **59** (1962), 245–288 (in Russian).
4. Ya.B. Zel'dovich, *Flame propagation in a mixture reacting at initial temperature*, Preprint, Chernogolovka, OIHF Akad. Nauk SSSR, 1978 (in Russian).
5. A.P. Aldushin, Ya.B. Zel'dovich and S.I. Khudyaev, *Flame propagation in a mixture of reacting gases*, Preprint , Chernogolovka, OIHF Akad. Nauk SSSR, 1979 (in Russian).
6. Ya.B. Zel'dovich, G.I. Barenblatt, V.B. Librovich and G.M. Makhviladze, *Mathematical theory of burning and explosion*, Nauka, Moscow, 1980 (in Russian).
7. A.L. Hodgkin, *The conduction of the nervous impulse*, Liverpool Univ. Press, 1964.
8. K.F. Bonhoeffer, 'Modelle der Nervenerregung', *Naturwissenschaften* **40**:11 (1953), 301–311.
9. G.I. Barenblatt, V.M. Entov and R.L. Salganik, 'On the propagation of excitations in an electrochemical diffusion model of nerves', *Prikl. Mat. Mekh.* **29**:6 (1965), 977–992 (in Russian).
10. E.P. Velikhov and A.M. Dykhno, 'Waves of non-equilibrium ionization in a gas', In: *Proc. 7th Internat. Symp. on Ionization in Gases*, Belgrade, 1965, pp. 47–60 (in Russian).
11. F.V. Bunkin, V.I. Konov, A.M. Prokhorov and V.B. Fedorov, 'Laser spark in a situation of slow burning', *Zh. Exsper. Teoret. Fiz. (Pis'ma)* **9**:11 (1969), 609–612 (in Russian).
12. Yu.P. Raizer, 'High-frequency discharge of a high pressure in a gas flow as a process of slow burning', *Zh. Prikl. Mekh. i Tekhn. Fiz.* **3** (1968), 3–10 (in Russian).

13. Yu.P. Raizer, *Laser spark and discharge propagation*, Nauka, Moscow, 1974 (in Russian).
14. G. Guderley, 'Starke kugelige und zylindrische Verdichtungstösse in der Nähe des Kugelmittelpunktes bzw. der Zylinderrachse', *Luftfahrt-Forschungsber.* **19**:9 (1942), 302–312.
15. K.P. Stanyukovich, *Unsteady motion of a continuous medium*, Gostekhizdat, Moscow, 1955 (in Russian).
16. K.V. Brushlinskii and Ya.M. Kazhdan, 'On self-similar solutions of certain equations in gas dynamics', *Uspekhi Mat. Nauk* **18**:2 (1963), 3–24 (in Russian).
17. C.P. Weizsäcker, 'Genäherte Darstellung starker instationärer Stosswellen durh Homologie-Lösungen', *Z. Naturforsch.* **9a**:4 (1954), 269–275.
18. Ya.B. Zel'dovich, 'Gas motion under the influence of short-time pressure (shock)', *Akusticheskii Zh.* **2**:1 (1956), 28–38 (in Russian).
19. Ya.B. Zel'dovich and Yu.P. Raizer, *Physics of shock waves and high-temperature hydraulic phenomena*, Nauka, Moscow, 1966 (in Russian).
20. G.I. Barenblatt, *Similarity, self-similarity and intermediate asymptotics*, Gidrometeoizdat, Leningrad, 1978 (in Russian).

ON THE PAPERS ON TURBULENCE

A.N. Kolmogorov

I became interested in turbulent liquid and gas flows at the end of the thirties. From the very beginning it was clear that the theory of random functions of many variables (random fields), whose development only started at that time, must be the underlying mathematical technique. Moreover, I soon understood that there was little hope of developing a pure, closed theory, and because of the absence of such a theory the investigation must be based on hypotheses obtained in processing experimental data. It was also important to have collaborators capable of combining theoretical and experimental research work.

In this respect I was lucky: A.M. Obukhov transferred from the Saratov University to Moscow and became my student in 1939 and then my postgraduate student; about that time M.D. Millionshchikov started his research work under my guidance as a postgraduate student at the Moscow Aviation Institute. Later A.S. Monin and A.M. Yaglom also became my postgraduate students.

In 1946 O.Yu. Schmidt suggested that I should become head of the Laboratory of Turbulence at the Institute of Theoretical Geophysics of the USSR Academy of Sciences (in 1949 Obukhov became head of this laboratory). I did not conduct direct experimental work but spent a lot of energy on processing numerical and graphical data obtained by other research workers.

TURBULENCE
(A.M. Yaglom)

There are not many papers of Kolmogorov on the mechanics of turbulent liquid and gas flows, and none of them is large. However, these few small papers radically changed the modern theory of turbulence and strongly affected the whole further development of this theory as well as the experimental research work on broad classes of turbulent flows.

Turbulence is known to be a phenomenon observed in a great majority of liquid and gas flows both in nature and in technical devices or laboratory installations. This phenomenon consists in random fluctuations (that is, chaotic changes in time and space) of velocity u, pressure p, and other hydrodynamic characteristics of the flows under consideration, due to which the corresponding hydrodynamic fields $u(x,t)$, $p(x,t)$ etc. sharply change and become extremely irregular. For this reason the study of individual hydrodynamic fields of turbulent flows turns out to be practically impossible, and it is interesting only to describe these flows statistically by studying some of their smoothened characteristics, which vary more smoothly and regularly.

The necessity of passing to smoothened characteristics was already clear to Reynolds [1], the founder of the whole theory of turbulence, who suggested as early as the end of the last century that the velocity field $u(x,t)$ of a turbulent flow should be decomposed into the average velocity $\bar{u}(x,t)$ determined by averaging the values of $u(x,t)$ over a time interval and the fluctuation velocity $u' = u - \bar{u}$, and that the dynamic equations (now called the Reynolds equations) should then be considered for the average velocity $\bar{u}$. However, the simplest averaging over a given time interval (or a spatial region) turned out to be in fact not very convenient, since in order to obtain sufficiently simple equations for the average field $\bar{u}$ the averaging operation must satisfy some general conditions which hold only approximately (and not exactly) for both time and space averaging. Instead of averaging individual hydrodynamic fields, it proved much more convenient to consider the whole ensemble of all possible turbulent flows admissible under fixed external conditions and then to use probabilistic means of ensembles of similar flows, that is, to assume that the hydrodynamic fields of turbulent flow characteristics are random functions of three spatial variables and time in the sense of modern probability theory. This "statistical approach" to turbulence mechanics was first used by Kolmogorov and his students (see §1 in paper No. 45 and the earlier paper by Millionshchikov [1] quoted there) and is now generally accepted.

We now pass to the content of the very important papers Nos. 45 and 47, devoted to the formulation of the general laws determining the statistical regime of small-scale fluctuations of a developed turbulence with a sufficiently large Reynolds number $\mathrm{Re} = UL/\nu$ (where U and L are the typical velocity and length scales for the flow under consideration and ν is the fluid's kinematic viscosity). Before these papers appeared, nobody had guessed that random turbulent fluctuations obey some simple quantitative relationships of a quite universal character, that is, they remain valid for all flows sufficiently distinct from laminar flows (characterized by a value of Re much greater than the lowest value R_{cr} of the Reynolds number for which turbulence is still possible). However, it is significant that as early as the twenties Richardson (for example, see [3], p. 66) put forward a qualitative scheme for a developed turbulence as a hierarchy of vortices (that is, perturbations or irregularities) of different orders, where the largest "first-order vortices" obtain energy from the mean flow and then transfer it to "second-order vortices", which then transfer it to "third-order vortices", etc. down to the smallest vortices, whose energy is dissipated under the action of molecular viscosity (that is, transformed into thermal energy as a result of viscous friction). A very comprehensive description of the Richardson scheme is given in the first two paragraphs of the footnote 2 in paper No. 45; the next two paragraphs in the same footnote substantially extend this scheme, thus revealing the behaviour of a non-linear system with a large number of degrees of freedom, such as a turbulent flow, and enabling us to draw quite specific quantitative conclusions from this new understanding.

Among the specific results in turbulence mechanics presented in papers Nos. 45 and 47 the so-called "2/3 law" described by formulas (23) and (24) in paper No. 45 (also see formula (9) in No. 47) became most famous. In paper No. 47 Kolmogorov described the first attempt to compare this law with the available experimental data obtained from measurements of the lateral velocity correlation function $R_{nn}(r)$ for turbulence behind the wind tunnel grid. Later Townsend [4] repeated an analogous verification of the "2/3 law" using special measurements of turbulence behind the wind tunnel grid and concluded, like Kolmogorov, that the results were in more or less good agreement with the theoretical prediction. Still later it was found, however, that some of Townsend's conclusions were wrong and that the values of Re basically used in the measurements of turbulence behind the wind tunnel grid were too small for the existence of an appreciable interval of values of r in which the "2/3 law" can hold (see, for example, Monin and Yaglom [5], §16.6 and §23.3). It thus became clear that the "2/3 law" can most conveniently be tested in natural turbulent flows in the atmosphere, seas, and oceans, where turbulence is characterized by much greater Reynolds numbers than for flows observed in laboratory conditions. Indeed, the first sufficiently convincing justification of the "2/3 law" was obtained by Obukhov [6], after processing comparatively old but very thorough measurements by Gödecke [7] of the difference between the velocities at a number of pairs of points in a surface layer of the atmosphere (1 meter above the meadow surface). Later the functions $B_{dd}(r)$ and $B_{nn}(r)$ (called the *longitudinal* and *lateral velocity structure functions* and introduced by Kolmogorov) were repeatedly measured in the atmosphere and sea (for the data of such measurements, see, for example, [5], §23.3 and its extended English translation [8], the review in [9] and also the recent paper [10]). In full agreement with Kolmogorov's prediction, all cases showed that both functions $B_{dd}(r)$ and $B_{nn}(r)$ are proportional to $r^{2/3}$ for a large interval of values of r (see, in particular, the typical Figure 1 taken from the paper by Van Atta and Chen [11]). The relation $B_{nn}(r) = (4/3)B_{dd}(r)$ predicted in paper No. 45 also turned out to be in good agreement with the majority of measurement data.

It was a more complex matter to test relation (7) in paper No. 47 resulting from the dynamical equation (5) derived in the paper, which interrelates the longitudinal structure functions $B_{dd}(r)$ and $B_{ddd}(r)$ of the second and third orders. The point is that a reliable evaluation of the mean cube of the velocity difference $B_{ddd}(r)$ imposes more stringent conditions on the accuracy of measurement and requires more extensive averaging than the determination of the corresponding mean square B_{dd} with the same accuracy. Moreover, the simultaneous determination of $\bar{\epsilon}$ in (7) is possible only when there is equipment that can

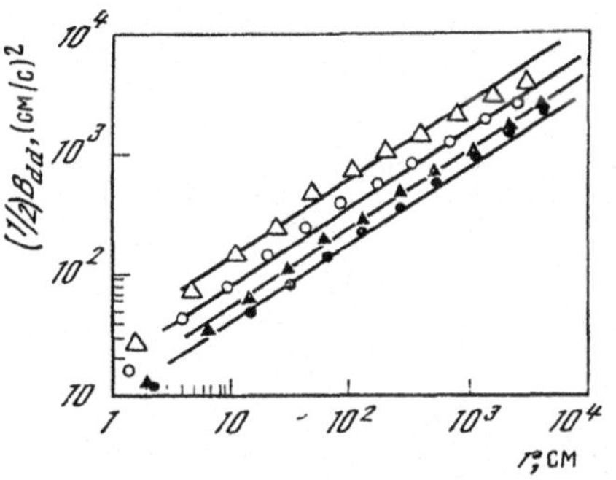

Fig. 1. Longitudinal wind velocity structure function $B_{dd}(r)$ according to data of measurements in the atmosphere above the sea surface.

Different symbols correspond to different series of measurements carried out at different altitudes (from 3 to 31 m) and for different values of the mean wind velocity. The straight lines correspond to the "two thirds law".

register without distortion the most fine-scale velocity fluctuations corresponding to the "Kolmogorov micro-scale length" given by formula (17) in paper No. 45. On the other hand, relation (7) is remarkable in that, by contrast with the above-mentioned "2/3 law", it does not involve any indeterminate numerical constants, so that its verification becomes particularly attractive. Therefore it is not surprising that recently a number of experimental groups have conducted measurements of the third-order structure function $B_{ddd}(r)$ along with $\bar{\epsilon}$ in both the atmosphere and in some laboratory turbulent flows for large values of Re. These experiments were aimed particularly at testing relation (7), and in all cases they were in very good agreement with Kolmogorov's prediction (see, for example, [10, 12] and the typical Figure 2 taken from [10]).

We also note that the basic equation (5) of paper No. 47 is derived from a modified Karman-Howarth equation (3), which follows directly from [13] only when the turbulent fluctuations of all scales are assumed to be homogeneous and isotropic. However, as follows from the conclusions of paper No. 45, the statistical regime of fine-scale fluctuations at large values of Re does not depend on the character of the large-scale fluctuations, and therefore the assumption in paper No. 47 that for $r \ll L$ equations (3) and (5) must remain valid for any (not necessarily isotropic) turbulence with sufficiently large value of Re seems to be quite legitimate (in this connection, see the derivation of equation (5) suggested by Monin [14] and also presented in [5], §22.1, in which it is

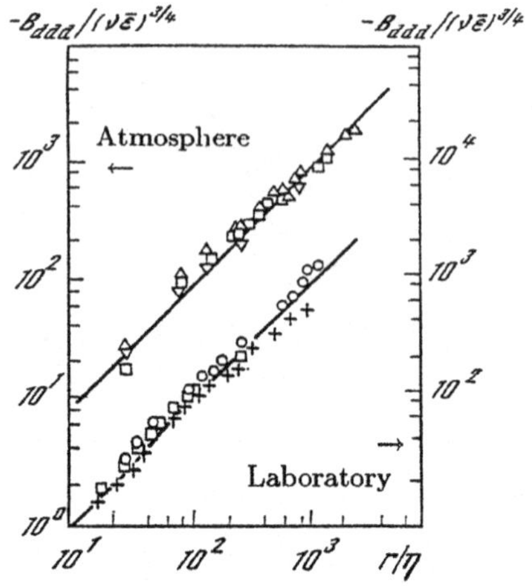

Fig. 2. Dependence of the normalized longitudinal third-order velocity structure functions $B_{ddd}(r)/(\bar{\varepsilon}\nu)^{3/4} = \beta_{ddd}(r)$ on the dimensionless distance $r/\eta = \rho$ according to data of measurements in the atmosphere and in several turbulent flows in a laboratory.

Different symbols correspond to different series of measurements; the straight lines correspond to Kolmogorov's prediction:

$$\beta_{ddd}(\rho) = -0.8\rho.$$

not assumed that the turbulence should be isotropic in the large). We note further that equation (6) in paper No. 45 (which is not related to the rest of the results), presented there without proof, is not quite obvious (in particular, this was noted by Batchelor in his paper [15], which played an important role in the popularization of Kolmogorov's results abroad); to prove (6) use should be made of spectral representations of locally homogeneous and locally isotropic random fields (see [16], pp. 334–335 or [5], §13.2).

Paper No. 46 deals with a more special problem in turbulence mechanics, concerning the degeneration (the decay with time) of totally homogeneous and isotropic turbulence in unbounded space. This paper is essentially based on Loitsyanskii's conclusion [17] that the parameter Λ in formula (11) remains constant for isotropic turbulence. Moreover, Karman's hypothesis of self-similarity of the correlation function is also used here (see formula (19) and the "2/3 law" (25)). The "2/3 law" is of secondary importance in paper No. 46, since all results can in fact be justified by using only Karman's self-similarity assumption and the constancy of Λ (see, for example, [5], §16.2). On the other hand, if the

"2/3 law" is assumed to be true (that is, if only very large values of Re are considered), then all the basic results of paper No. 46 can be obtained from the sole assumption that the "Loitsyanskii invariant" Λ is finite, and constant without any self-similarity requirements for the correlation functions (see [18]). It should be noted, however, that the basic assumption in paper No. 46 that Λ should be finite and constant in time is now considered far from being obviously correct (in this connection, see [19] or [5], §§15.5 and 15.6; see also [20, 21], in which the assumption that Λ is constant is replaced by a quite different hypothesis).

We now return to the general theory of locally isotropic turbulence as presented in papers Nos. 45 and 47. Results equivalent to the Kolmogorov basic "2/3 law" were established almost simultaneously by Obukhov [22, 23] in a completely different way. A little later, practically the same results were obtained independently by Onsager [24, 25], Weizsäcker [26], and Heisenberg [27] (this is briefly noted in paper No. 58). However, Kolmogorov's conclusions concerning the local structure of turbulence at very large values of Re are stated in papers Nos. 45 and 47 in the most general form. Here use is made only of certain clear and physically natural assumptions on the character of a developed turbulence in order to formulate two fundamental similarity hypotheses which hold for any developed turbulent flow and having very many different applications. Some of these applications, together with their experimental justification, are described in the extensive Chapter 8 in the books by Monin and Yaglom [5, 8] and in the recent review paper [9], which contains a large number of additional references. Very important applications of Kolmogorov's theory to light, sound and radio wave propagation in a turbulent medium can be found, for example, in [28]. In this commentary we consider only a few simple examples of the application of Kolmogorov's similarity hypotheses.

The second similarity hypothesis, when applied to higher moments of the difference between the longitudinal velocities $\Delta_{dd}(r) = u_d(x+r,t) - u_d(x,t)$ (where u_d is the projection of the vector u onto the direction of the vector r of length $r = |r|$), clearly yields the formula

$$\overline{|\Delta_{dd}(r)|^p} = C_p(\bar{\epsilon}r)^{p/3} \text{ for } L \gg r \gg \eta = (\nu^3/\bar{\epsilon})^{\frac{1}{4}}, \tag{1}$$

which turns into the "2/3 law" when $p = 2$ and is in agreement with formula (7) in paper No. 47 when $p = 3$. The two Kolmogorov hypotheses can be applied to the spectral density of kinetic energy of turbulence (or, in short, to the turbulence spectrum) $E(k,t)$ and the corresponding one-dimensional longitudinal and cross spectra $E_1(k,t)$ and $E_2(k,t)$, where k is the wave number. The spectrum $E(k,t)$ is determined by expanding the random velocity field $u(x,t) = u(x_1,x_2,x_3,t)$ into a triple Fourier integral, which is a generalization of the analogous expansion of random functions of one variable first suggested in Kolmogorov's paper No. 42, and the one-dimensional spectra $E_1(k,t)$ and $E_2(k,t)$ correspond to the expansion of the component $u_1(x_1,x_2,x_3,t)$ of the vector u into the one-fold Fourier integral with respect to the x_1-coordinate and x_2-coordinate (or the x_3-coordinate). There is a simple interrelation between these three spectra and the structure functions $B_{dd}(r,t)$ and $B_{nn}(r,t)$ (see, for example, [16] or [5], §13.3). By the first similarity hypothesis of Kolmogorov, the spectra E, E_1 and E_2 do not depend on t for $k \gg 1/L$ and are determined by the formulas

$$\begin{gathered} E(k) = (\nu^5\bar{\epsilon})^{\frac{1}{4}}\phi(\eta k), \\ E_1(k) = (\nu^5\bar{\epsilon})^{\frac{1}{4}}\phi_1(\eta k), \quad E_2(k) = (\nu^5\bar{\epsilon})^{\frac{1}{4}}\phi_2(\eta k), \end{gathered} \tag{2}$$

where ϕ, ϕ_1 and ϕ_2 are universal functions of one variable any two of which (and also the function β_{dd} in paper No. 45) can easily be expressed in terms of the third function. By analogy, the second similarity hypothesis implies the relations

$$E(k) = A\bar{\epsilon}^{2/3}k^{-5/3}, \quad E_1(k) = A_1\bar{\epsilon}^{2/3}k^{-5/3}, \quad E_2(k) = A_2\bar{\epsilon}^{2/3}k^{-5/3}, \tag{3}$$

which hold for $1/\eta \gg k \gg 1/L$; here A, A_1, and A_2 are universal constants, and it is not difficult to show that $A = [55/27\Gamma(1/3)]C \approx 0.76C$, $A_1 = (18/55)A \approx C/4$, $A_2 = (24/55)A \approx C/3$, where C is the coefficient in formula (23) of paper No. 45. The results (2) and (3) were first obtained (in application to the three-dimensional spectrum $E(k)$) by Obukhov [22, 23], who used his spectral equation of turbulence energy balance. Each of the relations in (2) is exactly equivalent to formula (20) of paper No. 45, and relations (3) (expressing the so-called "5/3 law") are equivalent to the "2/3 law" (23).

We also note that Obukhov [29] and Corrsin [30] extended Kolmogorov's results to the structure of the temperature field $T(x,t)$ (or the concentration $\theta(x,t)$ of an arbitrary passive impurity, that is, an impurity not affecting the dynamics) in a developed turbulent flow; it turned out that the "2/3 law" and the "5/3 law" are also valid for the scalar field T (or θ) (and, moreover, an equation similar to equation (5) in paper No. 47 holds for the structure functions; see [31]).

The spectral form (2) and (3) of Kolmogorov's fundamental laws of fine-scale turbulence is very convenient for experimental verification, since measurements of spectra of random fluctuations are widely used in modern science and technology, and the corresponding techniques are thoroughly developed. The simplest object for such measurements are the longitudinal (along the direction of the mean flow velocity) one-dimensional spectra, which can be expressed in terms of the frequency spectra of velocity fluctuations at a single fixed point.

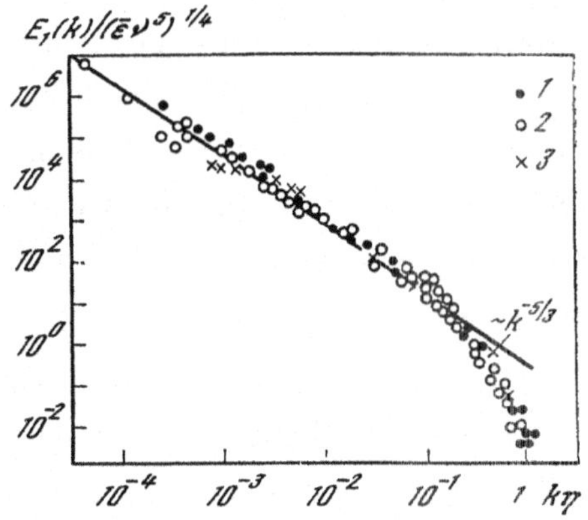

Fig. 3. Data of measurements of the one-dimensional longitudinal velocity spectrum for water flow in an oceanic stratum (1); for wind in the atmosphere near the Earth's surface (2); and for air flow in a boundary layer near the wall of a large wind tunnel (3).

A typical example shown in Figure 3 is taken from [5]. Here the one-dimensional longitudinal velocity spectra $E_1(k)$ measured for three different turbulent flows, water flow in an oceanic stratum, wind flow in the Earth's atmosphere near the surface, and flow in a large wind tunnel near its wall, are combined. It is seen that all three groups of measurement data, divided by $(\bar{\epsilon}\nu^5)^{\frac{1}{4}}$ and represented as functions of $k\eta$, can be sufficiently well approximated by a single curve, that is, are in good agreement with the theoretical formula (2). The "5/3 law" (3) is also in good agreement with these data.

The experimental results shown in Figure 3 can be used directly for determining the coefficient C in the "2/3 law" (which can simply be expressed in terms of the coefficients in (3)). The first attempt to find the value of this coefficient was made by Kolmogorov himself in paper No. 47, in which data of measurements behind the grid in a wind tunnel were used, and it was found that $C \approx 1.5$. However, as has already been mentioned, the Reynolds numbers for the experiments in paper No. 47 were insufficiently large to ensure the fulfilment of the "2/3 law" on a considerable interval of values of r. Therefore the first estimate of C given in paper No. 47 seems no longer reliable. Subsequently, a number of researchers conducted many experimental studies of the value of C using the structure functions $B_{dd}(r)$ and $B_{nn}(r)$ along with accurate measurements of $\bar{\epsilon}$, measurements of spectra and values of $\bar{\epsilon}$, and measurements of $B_{dd}(r)$ and $B_{ddd}(r)$, making it possible to employ formulas (10) and (8) of paper No. 47. The collection of data covering the results of 37 papers on the value of C is included in the review paper [9]. In addition, there are two more later papers [10, 32] devoted to the same topic. It is interesting that almost all measurements yielded results differing only slightly from Kolmogorov's first estimate of C; according to the new available data, $C \approx 2$ and the error of this estimate seems to be no greater than 10–15%.

The observed spread in the data on experimental values of C is to some extent due to measurement errors and can also be partly due to the influence of the variability of the energy dissipation rate $\epsilon(x,t)$, which was discussed in Kolmogorov's paper No. 58. This paper (based on the report at the International Colloquium on Turbulence Mechanics, Marseilles, 1961) is closely related to Obukhov's paper [33] and together with the latter marks a new stage in the development of the theory of locally isotropic turbulence based on the papers Nos. 45 and 47. In subsequent years a great number of theoretical and experimental investigations were conducted to refine the theory of the local structure of turbulence outlined in papers No. 58 and [33], and only part of these studies is discussed in §25 of [5] (see also [8]) devoted to these questions.

In paper No 58 it is suggested that the first and second similarity hypotheses in paper No. 45 should be replaced by two refined similarity hypotheses, relating to normalized differences of the velocities $V(\xi, \tau)$ and that a third hypothesis should be added. This third hypothesis postulates that the probability distribution of the energy dissipation rate ϵ_r averaged over a sphere of radius r is lognormal and that the variance of $\log \epsilon_r$ depends linearly on $\log(L/r)$. Hence, for the general velocity structure function $\overline{[\Delta_{dd}(r)]^p}$ of arbitrary order p one obtains formula (10) of paper No. 58 which generalizes the above formula (1) corresponding to $k = 0$ and shows that the function $\overline{[\Delta_{dd}(r)]^p}$ can be well approximated by a power function on a considerable interval of values of $r = |r|$ whose exponent, however, differs from $p/3$ and whose coefficient $C_p(x,t)L^{p(p-3)k/2}$ is not a universal constant and depends on the nature of the large-scale motion (and, in particular, on Re).

The short paper No. 58 offers no argument in favour of the third similarity hypothesis concerning the normality of the probability distribution of the parameter $\log \epsilon_r$ and its variance. Later it was shown, however, in [34] that a simple model of a cascade process of consecutive "splitting" of turbulent fluctuations with account of the self-similarity of the process provides a justification of this hypothesis (also see [35] and [5], §25.3). In [34] it was shown that the constant $\mu = 9k$, where $9k$ is the coefficient in formula (2) (see No. 58), also enters in the formula $E_{\epsilon\epsilon}(k) \sim k^{-1+\mu}$ describing the behaviour of the one-dimensional spectrum of fluctuations of the energy dissipation rate ϵ for a considerable range of values of k, after which the measurement data on the spectrum $E_{\epsilon\epsilon}(k)$ were used to obtain the first estimate $\mu \approx 0.4$ of the constant μ. Later many other authors made attempts to estimate μ experimentally; most of these attempts first yielded the same result as in [34], namely, $\mu \approx 0.4 - 0.5$ (see, for example, [11], [36], [37]), but recently a number of researchers have come to the conclusion that the earlier estimates were too high and that the estimate $\mu \approx 0.2$ is more plausible (see, for example, [38, 57]).

A number of experimental works (in particular, [11, 35, 36, 39–43]) dealt with measurements of the probability distribution of the parameter ϵ (more precisely, of the similarity parameters $\epsilon^{(1)} = 15\nu(\partial u_1/\partial x_1)^2$ and $\epsilon_2 = 7.5\nu\times (\partial u_2/\partial x_1)^2$ or the same parameters averaged over an area of size r in order to verify the hypothesis of paper No. 58 concerning the lognormality of these distributions. In all cases it was observed that the corresponding distribution function is in good agreement with the lognormal distribution function for a wide range of moderate values of the argument, but on the "tails" (that is, for very small or very large values of the argument) the empirical probability distribution deviates from the lognormal distribution. The deviations turned out to influence substantially the moments of higher orders of the probability distributions, since the higher order moments of ϵ_r can no longer be calculated accurately by means of formula (9) related to the lognormal distribution (see No. 58). In this connection Mandelbrot [44] noted that the fact that the probability distribution of ϵ_r tends asymptotically to the lognormal distribution as $\mathrm{Re} \to \infty$ by no means implies that the moments of ϵ_r must be close to the moments of this limit distribution for large values of Re. Later Novikov [45] showed that by postulating only the self-similarity of the process of cascade splitting of turbulent fluctuations one can obtain the formula $\overline{\epsilon_r^{p/3}}/\bar{\epsilon}_r^{p/3} \sim (L/r)^{\mu_p}$, where the μ_p are constants which, however, cannot increase more strongly than a linear function of p with increasing p (that is, they must increase more slowly than the exponent $\mu_p^{(0)} = p(p-3)\mu/18$ corresponding to the lognormal probability distribution of ϵ_r). Combining Novikov's formula with the first and second Kolmogorov hypotheses formulated in paper No. 58, we obtain the relation

$$\overline{[\Delta_{dd}(\mathbf{r})]^p} = C_p(\mathbf{x}, t)(r\bar{\epsilon}_r)^{p/3}(L/r)^{\mu_p}, \tag{4}$$

which generalizes formula (10) (see No. 58). Results of measurements of the velocity structure functions $\overline{[\Delta_{dd}(\mathbf{r})]^p}$ of various orders p (up to $p = 8, 9$ or even 12) in turbulent flows with large values of Re are described in [38, 46, 47]; according to all these papers, the functions $\overline{[\Delta_{dd}(\mathbf{r})]^p}$ are well approximated by power functions of the form $a_p r^{p/3-\mu_p}$ for a considerable range of values of $r = |\mathbf{r}|$, where $\mu_p > 0$ for all p (that is, $\overline{[\Delta_{dd}(\mathbf{r})]^p}$ increases with increasing r more slowly than is predicted by formula (1)), and μ_p also increases with increasing exponent p so that $\mu_p \approx \mu_p^{(0)} = (p(p-3)\mu)/18$ for comparatively small values of p, but as p increases further, the values of μ_p increase more slowly than those of $\mu_p^{(0)}$.

Kolmogorov's paper No. 48 occupies a special place among his other works on turbulence mechanics. This paper is devoted not only to the investigation of fine-scale turbulence, but also attempts to construct a complete (of course, approximate) system of equations describing turbulent motion which can be applied in order to calculate characteristics of actual flows that are very important in many areas of modern science and technology. The first very crude calculation methods were suggested between 1915 and 1935 by Prandtl, Taylor, and Karman; these methods were based on rigorous (although non-closed) Reynolds equations for the mean velocity, which then were closed artificially by means of a hypothesis on the turbulent viscosity coefficient and the notion of "path of mixing" used for computing this coefficient (see, for example, [48], §5.8). At the end of the thirties and the beginning of the forties more advanced models for closed equations of turbulence mechanics were suggested, which were based not only on Reynolds' equations but also on dynamic equations for certain second-order moments of velocity fluctuations and some other characteristics of turbulence. One of the earliest systems of such equations was introduced by Kolmogorov in paper No. 48. It included equations for the mean velocity $\bar{v}_i$, the turbulence intensity (that is, the fluctuation energy) b, and the "typical frequency" ω (or, equivalently, the turbulence scale $L = b^{\frac{1}{2}}/\omega$). The Kolmogorov equations proved sufficiently suitable (that is, in good agreement with measurement data) and convenient for practical computations; in this connection see the applied studies of Monin [49] and Barenblatt [50, 51], carried out under the guidance of Kolmogorov, in which these equations were used in a slightly modified form. Later a similar system of equations for turbulent motion was independently derived by Prandtl [52]; that is why in the literature the closed equations for the variables v_i, b and L are now called the Kolmogorov-Prandtl equations (see, for example, [53], Chapter 8). In subsequent years closed systems of equations of different degrees of complexity and describing approximately fluid and gas turbulent flows were extremely widely used, and the literature devoted to this area is very extensive, as can be seen, for example, in [54, 55] and also in [56], Chapter 5, and [53], Chapters 8 and 9.

References

1. O. Reynolds, 'On the dynamical theory of incompressible viscous fluids and the determination of the criterion', *Philos. Trans. Roy. Soc. London Ser.* **186** (1894), 123–161.
2. M.D. Millionshchikov, 'Degeneration of homogeneous isotropic turbulence in a viscous incompressible fluid', *Dokl. Akad. Nauk SSSR* **22**:5 (1939), 236–240 (in Russian).
3. L.F. Richardson, *Weather prediction by numerical process*, Cambridge Univ. Press, 1922.
4. A.A. Townsend, 'Experimental evidence for the theory of local isotropy', *Proc. Cambridge Philos. Soc.* **44**:4 (1948), 560–565.
5. A.S Monin and A.M. Yaglom, *Statistical fluid mechanics*, Vol. 2, Nauka, Moscow, 1967 (in Russian).
6. A.M. Obukhov, 'Local structure of atmospheric turbulence', *Dokl. Akad. Nauk SSSR* **67**:4 (1949), 643–646 (in Russian).

7. K. Gödecke, 'Messungen der atmosphärischen Turbulenz in Bodennähe mit einer Hitzdrahtmethode', *Ann. Hydrogr.* **10** (1935), 400–410.
8. A.S. Monin and A.M. Yaglom, *Statistical fluid mechanics*, Vol. 2, MIT Press, Cambridge (Mass.), 1975. (Translation of [5].)
9. A.M. Yaglom, 'Laws of small-scale turbulence in the atmosphere and oceans (on the occasion of 40 years of local istropic turbulence theory)', *Izv. Akad. Nauk SSSR Fiz. Atmosfery i Okeana* **17**:12 (1981), 1235–1257 (in Russian).
10. R.A. Antonia, A.J. Chambers and B.R. Satyaprakash, 'Kolmogorov constants for structure functions in turbulent shear flows', *Quart. J. Roy. Meteor. Soc.* **107** (1981), No. 453, 579–589.
11. C.W. van Atta and W.Y. Chen, 'Structure functions of turbulence in the atmospheric boundary layer over the ocean', *J. Fluid Mech.* **44**:1 (1970), 145–159.
12. J.T. Park and C.W. van Atta, 'Hot- and cold-wire sensitivity corrections for moments of fine scale turbulence in heated flows', *Phys. Fluids* **23**:4 (1980), 701–705.
13. Th. von Kármán, and L. Howarth, 'On the statistical theory of isotropic turbulence', *Proc. Roy. Soc. London* **A164** (1938), No. 917, 192–215.
14. A.S. Monin, 'On the theory of local isotropic turbulence', *Dokl. Akad. Nauk SSSR* **125**:3 (1959), 515–518 (in Russian).
15. G.K. Batchelor, 'Kolmogoroff's theory of locally isotropic turbulence', *Proc. Cambridge Philos. Soc.* **43**:4 (1947), 533–559.
16. A.M. Yaglom, 'Some classes of random fields in n-space, generated by stationary random processes', *Teor. Veroyatnost. i Primenen.* **2**:3 (1957), 292–337 (in Russian).
17. L.G. Loitsyanskii, 'Some basic laws of isotropic turbulent flow', *Trudy TsAGI* **440** (1939), 3–23 (in Russian).
18. G. Comte-Bellot and S. Corrsin, 'The use of a contraction to improve the isotropy of grid-generated turbulence', *J. Fluid Mech.* **25**:4 (1966), 657–682.
19. G.K. Batchelor and I. Proudman, 'The large-scale structure of homogeneous turbulence', *Philos. Trans. Roy. Soc. London Ser.* **A248** (1956), No. 949, 369–405.
20. P.G. Saffman, 'The large scale structure of homogeneous turbulence', *J. Fluid Mech.* **27**:3 (1967), 581–593.

21. P.G. Saffman, 'Note on decay of homogeneous turbulence', *Phys. Fluids* **10**:6 (1967), 1349.
22. A.M. Obukhov, 'On the energy distribution in the spectrum of a turbulent flow', *Dokl. Akad. Nauk SSSR* **32**:1 (1941), 22–24 (in Russian).
23. A.M. Obukhov, 'On the energy distribution in the spectrum of a turbulent flow', *Izv. Akad. Nauk SSSR Ser. Geografiya i Geofizika* **5**:4–5 (1941), 453–466 (in Russian).
24. L. Onsager, 'The distribution of energy in turbulence (abstr.), *Phys. Rev.* **68**:11–12 (1945), 286.
25. L. Onsager, 'Statistical hydrodynamics', *Nuovo Cimento* **6**:2 (1949), 279–287.
26. C.F. von Weizsäcker, 'Das Spektrum der Turbulenz bei grossen Reynoldsschen Zahlen', *Z. Phys.* **124**:7–12 (1948), 614–627.
27. W. Heisenberg, 'Zur statistischen Theorie der Turbulenz', *Z. Phys.* **124**:7–12 (1948), 628–657.
28. V.I. Tatarskii, *Wave propagation in turbulent atmosphere*, Nauka, Moscow, 1967 (in Russian).
29. A.M. Obukhov, 'Structure of the temperature field in a turbulent flow', *Izv. Akad. Nauk SSSR Ser. Geograf. Geofiz.* **13**:1 (1949), 58–69 (in Russian).
30. S. Corrsin, 'On the spectrum of isotropic temperature fluctuations in an isotropic turbulence', *J. Appl. Phys.* **22**:4 (1951), 469–473.
31. A.M. Yaglom, 'On the local structure of the temperature field in a turbulent flow', *Dokl. Akad. Nauk SSSR* **69**:6 (1949), 743–746 (in Russian).
32. A.J. Dyer and V.V. Hicks, 'Kolmogoroff constants at the 1976 ITCE', *Boundary-Layer Meteor.* **22**:2 (1982), 137–150.
33. A.M. Obukhov, 'Some specific features of atmospheric turbulence', *J. Geophys. Res.* **67**:8 (1962), 3011–3014; *J. Fluid Mech.* **13**:1 (1962), 71–81.
34. A.M. Yaglom, 'On the influence of fluctuations of energy dissipation on the shape of characteristics of turbulence in an inertial interval', *Dokl. Akad. Nauk SSSR* **166**:1 (1966), 49–52 (in Russian).
35. A.S. Gurvich and A.M. Yaglom, 'Breakdown on eddies and probability distributions for small-scale turbulence', In: *Phys. Fluids Suppl.*, Proc. Kyoto Sympos. on Boundary Layers and Turbulence, 1967, pp. 59–65.
36. C.H. Gibson, G.R. Stegen and S. McConnell, 'Measurements of the universal constant in Kolmogoroff's third hypothesis for high Reynolds number

turbulence', *Phys. Fluids* **13**:10 (1970), 2448–2451.

37. C.H. Gibson and P.J. Masiello, 'Observations of the variability of dissipation rates of the turbulent velocity and temperature fields', *Lecture Notes in Phys.*, Proc. Sympos., San Diego, July 1971, **12** (1972), 427–453.
38. R.A. Antonia, B.R. Satyaprakash and A.J. Chambers, 'Reynolds number dependence of velocity structure functions in turbulent shear flows', *Phys. Fluids* **25**:1 (1982), 29–37.
39. R.W Stewart, J.R. Wilson and R.W. Burling, 'Some statistical properties of small scale turbulence in an atmospheric boundary layer', *J. Fluid. Mech.* **41**:1 (1970), 141–152.
40. C.H. Gibson, G.R. Stegen and R.B. Williams, 'Statistics of a fine structure of turbulent velocity and temperature fields measured at high Reynolds number', *J. Fluid Mech.* **41**:1 (1970), 153–167.
41. A.Y-S. Kuo and S. Corrsin, 'Experiments of internal intermittency and fine structure distribution function in fully turbulent fluid', *J. Fluid Mech.* **50**:2 (1971), 285–320.
42. M.Z. Kholmyanskii, 'Measurements of micro-turbulent pulsations of the derived velocity of wind in an atmospheric layer near the Earth's surface', *Izv. Akad. Nauk SSSR Ser. Fizika Atmosfery i Okeana* **8**:8 (1972), 818–828 (in Russian).
43. F.N. Frenkiel and P.S. Klebanoff, 'On the lognormality of the small-scale structure of turbulence', *Boundary-Layer Meteor.* **8**:2 (1975), 173–200.
44. B. Mandelbrot, 'Possible refinement of the lognormal hypothesis concerning the distribution of energy dissipation in intermittent turbulence', *Lecture Notes in Phys.*, Proc. Sympos., San Diego, July 1971, **12** (1972), 333-351.
45. E.A. Novikov, 'Mixing and scale similarity in the structure of a turbulent flow', *Prikl. Mat. Mekh.* **35**:2 (1971), 266–277 (in Russian).
46. C.W. van Atta and J. Park, 'Statistical self-similarity and inertial subrange turbulence', *Lecture Notes in Phys.*, Proc. Sympos. San Diego, July 1971, **12** (1972), 402–426.
47. V.M. Vasilenko, M.M. Lyubimtsev and R.V. Ozmidov, 'On fluctuations of the velocity of dissipation of turbulent energy and on structure functions of higher order of velocity fields in the ocean', *Izv. Akad. Nauk SSSR Ser. Fiz. Atmosfery i Okeana* **11**:9 (1975), 926–932 (in Russian).

48. A.S. Monin and A.M. Yaglom, *Statistical fluid mechanics*, Vol. 1, Nauka, Moscow, 1965 (in Russian).
49. A.S. Monin, 'Dynamic turbulence in the atmosphere', *Izv. Akad. Nauk SSSR Ser. Geografiya i Geofizika* **14**:3 (1950), 232–254 (in Russian).
50. G.I. Barenblatt, 'On motion of suspended particles in a turbulent flow', *Prikl. Mat. Mekh.* **17**:3 (1953), 261–274 (in Russian).
51. G.I. Barenblatt, 'On motion of suspended particles in a trubulent flow occupying a half-space in a flat open canal of finite width', *Prikl. Mat. Mekh.* **19**:1 (1955), 61–88 (in Russian).
52. L. Prandtl, 'Über ein neues Formelsystem für die ausgebildete Turbulenz', *Nachr. Ges. Wiss. Göttingen. Math.-Phys. Kl.* (1945), 6–19; *Gesammelte Abhandlungen*, Springer-Verlag, Berlin, 1961, Part 2, pp. 874–887.
53. W. Frost and T.H. Moulden, eds., *Handbook of turbulence*, Plenum Press, New York, London, 1977.
54. S.J. Kline, M.V. Morkovin, G. Sovran and D.J. Cockrell, eds., *Computation of turbulent boundary layers*, Proc. 1968 AFOSR-IFP Stanford Conf. Vol. 1, University Press, Stanford, 1969.
55. B.E. Launder and D.E. Spalding, *Lectures on mathematical models of turbulence*, Academic Press, London, 1972.
56. P. Bredshaw, ed., *Turbulence*, Springer-Verlag, Berlin, Heidelberg, New York, 1978.
57. R.A. Antonia, B.R. Satyaprakash and A.K.M.F. Hussain, 'Statistics of fine-scale velocity in turbulent plane and circular jets', *J. Fluid Mech.* **119** (1982), 55–89.

ON THE PAPERS ON CLASSICAL MECHANICS

A.N. Kolmogorov

My papers on classical mechanics appeared under the influence of von Neumann's papers on the spectral theory of dynamical systems (see reference [1] in paper No. 53) and, particularly, under the influence of the Bogolyubov-Krylov paper of 1937 (see reference [12] in paper No. 53).

I became extremely interested in the question of what ergodic sets (in the sense of Bogolyubov-Krylov) can exist in the dynamical systems of classical mechanics and which of the types of these sets can be of positive measure (at present this question still remains open). To accumulate specific information we organized a seminar on the study of individual examples. My ideas concerning this topic and closely related problems aroused wide response among young mathematicians in Moscow.

This is discussed in greater detail in Arnol'd's commentary.

CLASSICAL MECHANICS

(V.I. Arnol'd)

Kolmogorov's classical papers Nos. 52 and 53 produced a very strong effect on the subsequent development of the theory of dynamical systems, and at present there are dozens of books developing or presenting the material of these papers. In this brief commentary it is impossible to embrace all applications of these results and we confine ourselves to some improvements introduced into the theory after 1954.

1. The basic problem of dynamics

The statement of the problem of the motion of systems that are close to the systems of classical mechanics, including the problem of orbit evolution in the three-body problem, dates back to Newton [1]. Laplace [2] stated explicitly the theorem on stability of the semimajor axes of Keplerian ellipses, which is a forerunner of Kolmogorov's theorem on preservation of tori, but proved it only in terms of approximate perturbation theory. On analyzing numerous attempts to justify and improve Laplace's argument, Poincaré [3] stated the problem in its modern form (to study the motion of a system whose Hamiltonian $W(p) + \theta S(q,p,\theta)$ is periodic in q) and called it the basic problem of dynamics (see [3], Chapter 1, §13). In the papers under consideration Kolmogorov solves this problem for the majority of initial conditions in the generic case ($\det \partial^2 W/\partial p^2 \neq 0$).

2. Rigid body and planetoid

Among the numerous problems of mechanics to which Kolmogorov's theorem on preservation of periodic motions can be applied we consider the following two classical problems:

(a) the problem of motion of an asymmetric heavy rigid body about a fixed point;

(b) the restricted circular three-body problem on the plane motion of a planetoid of negligibly small mass under the attraction of a heavy body S (Sun) and a body J (Jupiter) of small mass rotating around it.

In problem (a) Kolmogorov's theorem implies that the square of the total angular momentum changes little during infinite time, provided that at the initial instant the body rotates sufficiently fast (the kinetic energy is large as compared to the potential energy) (see [4]).

As to the planetoid, its instantaneous Keplerian elliptic orbit only rotates slowly in its plane under the influence of the "planet" J if at the initial instant the orbit does not intersect the circular orbit of the "planet" J and the ratio of the masses of the bodies J and S is sufficiently small.

It also follows directly from Kolmogorov's theorem that the majority of magnetic surfaces are stable relative to a sufficiently small variation of the magnetic field in toroidal systems of TOKAMAK type for non-zero "shear" (the derivative of the frequency ratio with respect to the torus number).

3. Limiting degeneration

Kolmogorov's theorem is not applicable to non-linear oscillations of a Hamiltonian system near equilibrium points or periodic motions (owing to degeneration, called limit degeneration by Born [5] when the tori contract to a point). However, a slight modification of this theorem leads to invariant tori even in the case of limiting degeneration. Therefore the Kolmogorov tori provide solutions, for example, to the Birkhoff problem of stability of general elliptic points of area-preserving mappings of a plane onto itself [6].

The most complete result concerning this problem is due to Moser [7]: the fixed point 0 of an area-preserving smooth mapping determined in polar coordinates by a formula

$$(r, \phi) \mapsto (r, \phi + \alpha + \beta r^2) + O(r^4)$$

is stable if $\alpha \neq 2\pi k/3,\ 2\pi k/4;\ \beta \neq 0$.

It follows from Kolmogorov's theorem, modified for the case of limiting degeneration, that an oscillating point always remains near an equilibrium state that is stable in the linear approximation, or near a periodic motion for the majority of initial conditions close to this motion, provided that the oscillatory Hamiltonian system under consideration satisfies some non-degeneracy conditions ("in general", these are satisfied).

In the case when the dimension of the phase space can be reduced to three, the two-dimensional tori whose existence is guaranteed by the generalized Kolmogorov theorem divide the phase space and therefore provide genuine Lyapunov stability of equilibrium states or periodic motions. Among the classical problems solved in this way we mention the following two:

(a) a proof of the stability of Lagrange periodic solutions of the restricted planar circular three-body problem [8, 9];

(b) a proof of the stability of an inverted friction-free non-linear pendulum under sufficiently fast vertical oscillation of the point of suspension.

4. Self-degeneration

There are many situations in which the unperturbed motion is described by a smaller number of frequencies than the number of the degrees of freedom (self-degeneration in the sense of Born; examples of this kind are Kepler's problem and the problem of charged particle motion in a strong magnetic field).

The extension of Kolmogorov's theory to this case [10] made it possible to prove, in particular, the perpetual adiabatic invariance of the action in non-linear systems with one degree of freedom for a slowly varying Hamilton function [11] as well as the perpetual motion of charged particles within axisymmetric magnetic traps [12].

A combination of these results with the study of oscillation near an equilibrium position (see §3) leads to a proof of the perpetual smallness of variations of the semimajor axes of planet orbits for the majority of initial conditions, such that the initial Keplerian ellipses differ slightly from non-intersecting circles lying in one plane and traversed in the same direction, provided that the masses of the planets are sufficiently small as compared to that of the central body [12].

The exceptional initial conditions leading to substantial changes in the semimajor axes of Keplerian ellipses presumably form a set of positive measure that is everywhere dense in the phase space.

5. Diffusion

If the invariant tori in Kolmogorov's theorem do not divide the phase space (reduced by taking into account the first integrals), then it is possible that after a long time a phase curve issuing from a slot between the tori goes far away from the invariant torus of the unperturbed system containing the initial point of this curve. In this case the values of the action variables (the integrals of the unperturbed system) diverge strongly from their initial values.

Examples of such divergence in a non-degenerate system were constructed in [13]. The average divergence rate in these examples is exponentially small (of the order of $e^{-1/\sqrt{\theta}}$ where θ is the magnitude of the disturbance). This exponentially slow divergence cannot be detected by any approximation of perturbation theory, where power series in θ are used and time intervals are of the order of negative powers of θ; such time intervals are insufficient for the action to go far away from its initial value. However, since the divergence develops in an unpredictable direction (strongly depending on the initial conditions) and has the character of a random walk between the Kolmogorov tori, physicists called this phenomenon diffusion [14].

Nekhoroshev [15–17] established an exponential upper estimate for the average rate of evolution of the action: the variation of the action remains small (of the order of θ^a, $0 < a < 1$) for an exponentially long time (of the order of $\exp(1/\theta^b)$).

The Nekhoroshev theorem explains why the evolution of the action variables cannot be revealed in any approximation of perturbation theory; in particular, in the theorems stated by Laplace and his successors.

Nekhoroshev estimated the parameters a and b in terms of some geometric characteristics of the curvature of the Hamiltonian level surfaces, the so-called steepness characteristics. At present these characteristics have been calculated for generic systems with at most three degrees of freedom [18].

6. Measure of the exceptional set

Kolmogorov's theorem asserts that (in a bounded region of the phase space) the measure of the complement of the invariant tori which are not destroyed under perturbation is small together with θ. A more thorough analysis makes it possible to obtain the upper bound $C\sqrt{\theta}$ for this measure (Lazutkin [19], Svanidze [20], Pöschel [21]). For an analytic system with two degrees of freedom and self-degeneration (one of the frequencies is proportional to θ), Neustadt established the exponential upper bound $e^{-C/\theta}$ for the measure of the complement of the invariant tori.

Estimates for the measure of the exceptional set play an important role in the application of Kolmogorov tori to the construction of quasiclassical asymptotic expansions for the majority of the eigenvalues of the Laplace operator in regions close to integrable billiards (for example, ellipsoids); see [22].

It is clear that the estimate $C\sqrt{\theta}$ is best possible, since $\sqrt{\theta}$ appears in all asymptotic expansions related to random degeneration (in the terminology of Born [5]); Poincaré [3] was already aware of the fact that in the unperturbed problem the destruction of a resonance torus results in a domain of influence of width of the order of $\sqrt{\theta}$ with respect to the action variable. This can easily be verified by averaging the disturbance over a torus of dimension smaller by 1 and integrating the resulting system with one degree of freedom (describing so-called phase oscillation in the resonance). We see that the amplitude of the phase oscillation is of the order of $\sqrt{\theta}$ and the period is of the order of $1/\sqrt{\theta}$.

However, in the indicated domain some phase curves again lie on tori of full dimension (so-called islands in the slots between tori close to unperturbed ones). The question whether the initial conditions corresponding to motions that are not conditionally periodic have positive measure remains open in the generic case (the answer is likely to be affirmative). The existence of such initial conditions was proved by Alekseev [23].

7. Smoothness

The theorem on the preservation of tori was proved by Kolmogorov for analytic systems and he did not assume that this result could be extended to at least infinitely differentiable systems. However, soon after Kolmogorov's method was first described in detail [24], Moser combined Kolmogorov's approach with the Nash smoothing method [25] and managed to prove the theorem on the preservation of tori for a system of finite smoothness [26, 27]. Initially, Moser used derivatives of the Hamilton function of a very high order (334 for $s = 2$), but later [28, 29] the order was reduced to $r > 2s$ where s is the number of degrees of freedom.

As was indicated by Kolmogorov, in the analytic case the dependence of the tori on the frequency of motion over them proved to be, in a sense, Borel monogenic (differentiable on a set that is a Cantor continuum extending to the real domain) [24]. For the smooth case Pöschel [21] proved that the dependence of the tori on the frequency is smooth in Whitney's sense on the Cantor set, provided that the Hamiltonian disturbance is r-smooth and $r > 3s - 1$.

On the other hand, Takens [30] has constructed an example of a 1-smooth mapping of an annulus that is area-preserving and close to rotation through a variable angle but has no invariant curves (4-smoothness would be sufficient for the existence of such curves). In none of the cases is it known exactly how many derivatives are needed for the existence of Kolmogorov tori.

Mather [31, 32] has interpolated the family of invariant curves of a mapping of an annulus to a one-parameter family in which all rotation numbers from the maximum possible value to the minimum possible value are realized (and not only the numbers belonging to the Cantor set as in Kolmogorov's theorem). Some of his "invariant curves" are in fact homeomorphic to the Cantor set and not to the circle. However all these curves are found by means of a unified variational principle introduced by Percival [33] for the approximate numerical calculation of Kolmogorov tori.

8. Non-Hamiltonian systems

The method used by Kolmogorov to construct his tori simultaneously proves many versions of the implicit function theorem in all areas of calculus where the solution of a linearized equation results in loss of smoothness (see [26, 34, 35]). Two theorems of this kind were proved as early as the forties, namely, Siegel's theorem (asserting that a holomorphic vector field is equivalent to the mapping given by its linear part at a singular point with normally incommensurable eigenvalues [36–38]) and Cartan's theorem [39].

The simplest non-Hamiltonian analogue of Kolmogorov's theorem is found in the theory of differential equations on a two-dimsensional torus (or the theory of diffeomorphic mappings of a circle onto itself). The most complete results in this area were established by Herman [40], who proved that for almost every rotation number an analytic diffeomorphism of a circle is analytically conjugate to a rotation.

In the multidimensional case it has only been proved that the majority of diffeomorphisms of tori that are close to translations are rectifiable [24].

Kolmogorov's method was successfully applied to the problem of finding invariant tori in non-Hamiltonian systems (Bogolyubov [41], Moser [42]). These results are particularly important in the theory of instability of self-excited oscillation since the bifurcation occurring in the generation of a torus from a cycle (and also of tori of higher dimension from tori of lower dimension) is one of the typical phenomena characteristic of the "stochastization" of motion of a dynamical system (for example, the passage from a laminar flow to turbulence in the theory of hydrodynamic stability). Applications of Kolmogorov's method to the theory of bifurcation of tori are discussed in the paper by Chenciner [43].

9. Numerical trials

After Kolmogorov established his theorem, an extensive series of numerical trials was conducted to verify the theorem. The trials were primarily aimed at studying the maximum magnitude of disturbances not destroying the Kolmogorov tori, at finding the measure of the complementary set of the Kolmogorov tori, and at analyzing the ergodic properties of the motion over this set [14, 44].

All the calculations confirm that: (1) the majority of tori are preserved under small disturbances; (2) the measure of the complementary set increases, beginning with the moment when the domains of influence of neighbouring resonances overlap (until this moment the measure of the complementary set is so small that the system can hardly be distinguished from an integrable system); (3) the motion in the complementary sets of the tori is stochastic in nature; namely, the neighbouring trajectories diverge exponentially.

Greene and Percival [45] have numerically examined the boundary of analyticity of Kolmogorov tori (for complex values of the angular variables) and have shown that this boundary is related to ergodic properties of the system in the complex domain and to the asymptotics of nearby periodic motions.

According to the experimental data, the magnitude of disturbances under which the Kolmogorov tori are preserved is not so small as predicted by rigorous proofs of estimates from above for this magnitude. Good agreement with experimental data is shown by the resonance overlap criterion [14, 44].

10. Exponential instability

Systems with stable transitivity and mixing on the energy level surfaces which Kolmogorov discusses at the end of §3 of the lecture at the Amsterdam Congress (paper No. 53) actually exist. Sinai and Anosov proved that geodesic flows on compact manifolds of negative curvature (along each two-dimensional direction) possess these properties (and also certain other stochastic properties, including the Ornstein-Weiss metric isomorphism to the "Bernoulli system" induced by coin tossing) [46–48]. Moreover, these properties are preserved under small perturbations not only in the class of Hamiltonian systems but also in the class of general dynamical systems. [1] The system also preserves its topological structure under perturbations, namely, in spite of the fact that the long-period trajectories form an everywhere dense set and that almost every trajectory is everywhere dense, a small perturbation of the vector field determining the system leaves it homeomorphic to itself (the Smale conjecture, proved by Anosov, [49]).

The discovery of such systems made the ergodic theory of classical dynamical systems extremely important not only for Hamiltonian mechanics but also for the study of phenomena in systems with energy dissipation, which also turned out to be capable of exhibiting stable stochastic behaviour. Incidentally, hydrodynamical systems are of this kind (for them, however, the situation is more complex due to an infinite-dimensional phase space). As early as 1958–59 Kolmogorov combined the ideas of the ergodic theory of classical dynamical systems and the theory of hydrodynamical instability in his seminar "Selected problems of calculus" [50].

The programme of this seminar displayed at the notice-board of the Mechanics-Mathematics Department of Moscow University announced the study of "models, albeit idealized, of hydrodynamic instability" from the standpoint of ergodic theory. Here Landau's model was meant primarily. This described the appearance of turbulence as consecutive bifurcations of a laminar flow resulting in periodic and conditionally periodic motions over tori whose dimensions increase together with the Reynolds number.

In Kolmogorov's opinion, the tori with conditionally periodic motions were preferred in the model to other dynamical systems because systems with mixing (such as geodesic flows on manifolds of negative curvature) were unknown to Landau (see §3 of the lecture in Amsterdam). When discussing Landau's model, Kolmogorov stressed that the steady-state motion appearing after stability loss can be generated "far from" the bifurcating "laminar" motion before it loses stability (so that the passage to complex "turbulent" motion can proceed in a jump-like manner as a result of "rigid" rather than "soft" stability loss). This opens up the possibility of appearance of motions unrelated to tori. At that

[1] With smooth invariant measure. A system with exponential instability need not have a smooth invariant measure, while, in general, there are a lot of ergodic non-smooth measures. In the case of discrete time the "Gibbs measure of maximum entropy" is singled out from these measures. For a review of the question of invariant measures see [61].

time Kolmogorov also predicted the existence of attracting sets ("attractors" in modern terminology) which are not manifolds but are pathologies at the set-theoretic level (the results of Lorenz' numerical experiments [51] appeared only several years later and, unfortunately, were not known to mathematicians for a long time).

However, along with finite-dimensional attractors, Kolmogorov suggested a model with an attractor in the form of an infinite-dimensional torus resulting from infinitely many bifurcations in the spirit of Landau's model, in the sense that in the experiment a spectrum with an infinite frequency basis is interpreted as a continuous spectrum and the motion is seen as being mixing and turbulent.

Smale's daring conjecture (1961) on the structural stability of dissipative systems with exponential divergence of trajectories was completely unexpected and its proof [49] in 1962 was solid evidence in favour of a model with finite-dimensional attracting set. A large number of papers on these problems have been published since that time (Yudovich, Ladyzhenskaya, Mallet-Paret, Foias, etc.). The finite-dimensionality of an attractor for any Reynolds numbers is now proved in two-dimensional hydrodynamics (first Il'yashenko estimated the dimension of an attractor from above in terms of the fourth degree of the Reynolds number for the Kolmogorov model (see [60]) in hydrodynamics on a two-dimensional torus and then Vishik and Babin estimated it in terms of the fourth degree of the Reynolds number for hydrodynamics on a two-dimensional compact manifold with boundary). As to the three-dimensional case, the question whether the attractor remains finite-dimensional for any Reynolds number is still open.

11. Exchange, capture, and oscillation

The problem of ultimate motions in the three-body problem discussed at the end of §4 of paper No. 53 was later thoroughly studied by Kolmogorov's students, Sitnikov and Alekseev.

Sitnikov [52] was the first to construct oscillatory motions, that is, to find initial conditions such that from time to time one body goes arbitrarily far away from the other two bodies, which remain at a finite distance from one another, and from time to time returns to them. In his example the oscillation occurs both as $t \to -\infty$ and as $t \to +\infty$.

Using his theory of quasi-random dynamical systems, Alekseev [23, 53] found all types of motions whose existence had remained doubtful after the papers of Shmidt and Sitnikov (and was denied by Chazi); namely, he showed that a body coming from afar can be captured by a two-body system so that a three-body system is formed (although the probability of this event is zero) or can start to oscillate, moving from time to time arbitrarily far away from the other two bodies. As $t \to +\infty$ oscillation can also appear in a system which remains bounded as $t \to -\infty$. The only problems remaining unsolved are those on the measure of the set of initial conditions leading to oscillation (which is likely to be positive).

12. Spectral properties

Kolmogorov's conjecture (§5 of paper No. 53) on the stability of a continuous (more precisely, countably-multiple Lebesgue) spectrum was proved by Sinai and Anosov [46, 47]. Thus far the conjecture that a discrete spectrum with a finite number of independent frequencies (not exceeding the phase space dimension) and a countably-multiple Lebesgue spectrum is typical has not been refuted for analytic systems. However, in the infinitely-differentiable case Anosov and Katok [54] constructed diffeomorphisms with a smooth invariant measure (for example, on a circle) which have: (a) a discrete spectrum with any (finite or infinite) prescribed number of basic frequencies; (b) a simple continuous singular spectrum in the absence of mixing; and (c) a combination of the former and latter. These examples are, however, exceptional since the diffeomorphisms constructed belong to a nowhere dense set in the function space of all diffeomorphisms endowed with the topology of C^∞ or C^n.

The assertion of paper No. 51 that an analytic flow on a torus can have a continuous spectrum is proved in [55]. In the case of continuous velocity the spectrum can be mixed [56]. A flow on a torus has no mixing in the rigorous sense and, consequently, no Lebesgue spectrum [57].

The conjecture in paper No. 51 that the divergence of the sum of squares of the Fourier coefficients of a formal solution to an equation with small divisors implies that there is no measurable solution and the spectrum is continuous proved false; Anosov [58] has constructed analytic systems with a measurable solution having neither a square summable solution nor a simply summable solution (the measurable solution is positive and therefore generalizations of the integral do not help here).

On the other hand, conditions of lacunarity type for Fourier series have been found which are sufficient for the non-existence of measurable solutions. It follows from these conditions that, for example, for any right-hand side of an equation with small divisors which is not a trigonometric polynomial there exists an irrational rotation number such that the equation has no measurable solutions [59].

References

1. I. Newton, *Philosophicae naturalis principia mathematica*, London, 1686.
2. P.S. de Laplace, *Traité de mécanique céleste*, Vol. 1, Paris, 1799.
3. H. Poincaré, *Les méthodes nouvelles de la mécanique céleste*, Vol. 1, Paris, 1892.
4. V.I. Arnol'd, 'Proof of a theorem of A.N. Kolmogorov on preservation of conditionally periodic motions under small change in the Hamilton function', *Uspekhi Mat. Nauk* **18**:5 (1963), 13–40 (in Russian).
5. M. Born, *Vorlesungen über Atommechanik*, Springer, Berlin, 1925.
6. V.I. Arnol'd, 'On stability of equilibrium positions of Hamiltonian systems of ordinary differential equations in the general elliptic case', *Dokl. Akad. Nauk SSSR* **137**:2 (1961), 255–257 (in Russian).
7. J. Moser, 'New results on the stability of periodic motions', *Proc. Internat. Congress Math.*, 1962, Uppsala, 1963.
8. A.M. Leontovich, 'On stability of Lagrange periodic solutions in the restricted 3-body problem', *Dokl. Akad. Nauk SSSR* **143**:3 (1962), 525–528 (in Russian).
9. A.P. Markeev, 'On stability of a triangle of points of libration in the circular restricted 3-body problem', *Prikl. Mat. Mekh.* **33**:1 (1969), 112–116 (in Russian).
10. V.I. Arnol'd, 'On generation of a conditionally periodic motion from a family of periodic motions', *Dokl. Akad. Nauk SSSR* **138**:1 (1961), 13–15 (in Russian).
11. V.I. Arnol'd, 'On the behaviour of the adiabatic invariant under slow periodic change in the Hamilton function', *Dokl. Akad. Nauk SSSR* **142**:4 (1962), 758–761 (in Russian).
12. V.I. Arnol'd, 'Small divisors and stability problems in classical and celestial mechanics', *Uspekhi Mat. Nauk* **18**:6 (1963), 92–192 (in Russian).

13. V.I. Arnol'd, 'On instability of dynamical systems with many degrees of freedom', *Dokl. Akad. Nauk SSSR* **156**:1 (1964), 9–12 (in Russian).
14. G.M. Zaslavskii and B.V. Chirikov, 'Stochastic instability of non-linear oscillations', *Uspekhi Fiz. Nauk* **105**:1 (1971), 3–40 (in Russian).
15. N.N. Nekhoroshev, 'On the behaviour of Hamiltonian systems close to integrable ones', *Funktsional. Anal. i Prilozhen.* **5**:4 (1971), 82–83 (in Russian).
16. N.N. Nekhoroshev, 'Exponential estimate of the stability time of Hamiltonian systems close to integrable ones I', *Uspekhi Mat. Nauk* **32**:6 (1977), 5–66 (in Russian).
17. N.N. Nekhoroshev, 'Exponential estimate of the stability time of Hamiltonian systems close to integrable ones II', *Trudy Sem. I.G. Petrovskogo* **5** (1979), 5–50 (in Russian).
18. E.E. Landis, 'Tangential singularities', *Funktsional. Anal. i Prilozhen.* **15**:2 (1981), 36–49 (in Russian).
19. V.F. Lazutkin, 'Existence of caustics for billiard problems in a convex region', *Izv. Akad. Nauk SSSR Ser. Mat.* **37**:1 (1973), 188–223 (in Russian).
20. N.V Svanidze, 'Small perturbations of integrable dynamical systems with integral invariant', *Trudy Mat. Inst. Akad. Nauk SSSR* **147** (1980), 124–146 (in Russian).
21. J. Pöschel, *Integrability of Hamiltonian systems on Cantor sets*, Zürich, 1981.
22. V.F. Lazutkin, *Convex billiard and eigenfunctions of the Laplace operator*, Leningrad Univ. Press, Leningrad, 1981 (in Russian).
23. V.M. Alekseev, 'Quasi-random dynamical systems I–III', *Mat. Sb.* **76**:1 (1968), 72–134; **77**:4 (1968), 545–601; **78**:1 (1969), 227–228 (in Russian).
24. V.I. Arnol'd, 'Small divisors I', *Izv. Akad. Nauk SSSR Ser. Mat.* **25**:1 (1961), 21–86 (in Russian).
25. J. Nash, 'The imbedding problem for Riemannian manifolds', *Ann. of Math.* **63** (1956), 20–63.
26. J. Moser, 'A new technique for the construction of solutions of non-linear differential equations', *Proc. Nat. Acad. Sci. USA* **47** (1961), 1824–1831.
27. J. Moser, 'On invariant curves of area preserving mappings of an annulus', *Nachr. Akad. Wiss. Göttingen. Math.-Phys. Kl.*, (1962), 1–20.
28. H. Rüssmann, 'Kleine Nenner I. Über invariante Kurven differenzierbaren

Abbildungen eines Kreisringes', *Nachr. Akad. Wiss. Göttingen. Math.-Phys. Kl.* No. 5, (1970),67–103.

29. J. Pöschel, *Über invariante Tori in differenzierbaren Hamiltonischen Systemen*, Bonn, 1980.
30. F.A. Takens, 'A C^1 counterexample to Moser's twist theorem', *Indag. Math.* **33**:4 (1971), 379–386.
31. J. Mather, *Existence of quasi-periodic orbits for twist homeomorphisms of the annulus*, Preprint, Princeton Univ., 1981.
32. J. Mather, 'Area preserving twist homeomorphism of the annulus', *Comment. Math. Helv.* **54** (1979), 397–404.
33. I.C. Percival, 'Variational principles for invariant tori and Cantori', In: *Symposium on non-linear dynamics and beam-beam interactions: Amer. Inst. Phys. Conf. Proc.* **37** (1980), 302–310.
34. V.I. Arnol'd, 'Singularities of smooth maps', *Uspekhi Mat. Nauk* **23**:1 (1968), 3–44.
35. E. Zehnder, 'Generalized implicit function theorems with applications to some small divisors problem. I', *Comm. Pure Appl. Math.* **28**:1 (1975), 91–140.
36. C.L. Siegel, 'Iteration of analytic functions', *Ann. of Math.* **43** (1942), 607–612.
37. C.L. Siegel, 'Über die Normalform analytischer Differentialgleichungen in der Nähe einer Gleichgewichtslösung', *Nachr. Akad. Wiss. Göttingen. Math.-Phys. Kl.* (1952), 21–30.
38. V.I. Arnol'd, *Supplementary chapters in the theory of ordinary differential equations*, Nauka, Moscow, 1978, pp. 194–199 (in Russian).
39. H.J. Cartan, 'Sur les matrices holomorphes de n variables complexes', *J. Math. Pures Appl.* **19** (1940), 1–26.
40. M.R. Herman, 'Sur la conjugaison differentiable des difféomorphismes du cercle à des rotations', *Publ. Math. IHES* **49** (1979), 5–234.
41. N.N. Bogolyubov, 'On quasiperiodic solutions of problems of non-linear mechanics', In: Proc. 1st Summer Math. School, Kiev, Naukova Dumka **1** (1964), 11–101 (in Russian).
42. J. Moser, 'A rapidly convergent iteration method and non-linear partial differential equations I, II', *Ann. Scuola Norm. Sup. Pisa* **20** (1966), 265–315, 499–535.

43. A. Chenciner, 'Courbes invariantes non normalement hyperboliques au voisinage d'une bifurcation de Hopf dégénérée de difféomorphismes de R^2', *C.R. Acad. Sci. Paris* **292**:10 (1981), 507–510.

44. G. Casati, B.V. Chirikov, J. Ford and F.M. Izrailev, *Stochastic behaviour in classical and quantum Hamiltonian systems*, Lecture Notes in Phys. **93**, 1979.

45. J.M. Greene and I.C. Percival, 'Hamiltonian maps in the complex plane', *Physica* **3D**:3 (1981), 530–548.

46. Ya.G. Sinai, 'Chemical dynamical systems with countable Lebesgue spectrum', *Izv. Akad. Nauk SSSR Ser. Mat.* **30**:1 (1966), 15–68 (in Russian).

47. D.V. Anosov, 'Geodesic flows on closed Riemannian manifolds of negative curvature', *Trudy Mat. Inst. Akad. Nauk SSSR* **90** (1967),

48. D. Ornstein, *Ergodic theory, randomness and dynamical systems*, Univ. Press, Newhaven, 1974.

49. D.V. Anosov, 'Structural stability of geodesic flows on compact Riemannian manifolds of negative curvature', *Dokl. Akad. Nauk SSSR* **145**:4 (1962), 707–709.

50. V.I. Arnol'd and L.D. Meshalkin, 'Kolmogorov's seminar on selected problems in analysis (1958–1959)', *Uspekhi Mat. Nauk* **15**:4 (1960), 247–250.

51. E. Lorenz, 'Deterministic non-periodic flow', *J. Atmos. Sci.* **20** (1963), 130–141.

52. K.A. Sitnikov, 'Existence of oscillating motions in the 3-body problem', *Dokl. Akad. Nauk SSSR* **133**:2 (1960), 303–306.

53. V.M. Alekseev, *Sur l'allure finale du mouvement dans le problème de trois corps* In: Act. Congrès Intern. Math. Nice, 1970, Paris, 1971.

54. D.V. Anosov and A.B. Kotak, 'New examples in smooth ergodic theory. Ergodic diffeomorphisms', *Trudy Moskov. Mat. Obshch.* **23** (1970), 3–36.

55. M.D. Shklover, 'On classical dynamical systems on a torus with continuous spectrum', *Izv. Vuzov. Mat.* **10** (1967), 113–124.

56. A.B. Krygin, 'Example of a continuous flow on a torus with mixed spectrum', *Mat. Zametki* **15**:2 (1974), 235–240.

57. A.V. Kochergin, 'On the absence of mixing for special flows over a twisted circle and for flows on the two-dimensional torus', *Dokl. Akad. Nauk SSSR* **205**:3 (1972), 515–518.

58. D.V. Anosov, 'On an additive functional homological equation related to ergodic twist of a circle', *Izv. Akad. Nauk SSSR Ser. Mat.* **37** (1973), 1259–1274.

59. A.Ya. Gordon, 'A sufficient condition for unsolvability of the additive homological equation related to an ergodic twist of a circle', *Funktsional. Anal. i Prilozhen.* **9**:4 (1975), 71–72.

60. L.D. Meshalkin and Ya.G. Sinai, 'Study of the stability of the stationary solution of a system of equations of planar motion of an incompressible viscous fluid', *Prikl. Mat. Mekh.* **6** (1961), 1140–1143.

61. Ya.G. Sinai, 'Gibbs measures in ergodicity theory', *Uspekhi Mat. Nauk* **27**:4 (1971), 22–64.

ON THE PAPERS ON SUPERPOSITIONS

A.N. Kolmogorov

Two basic papers of this series (Nos. 55 and 56) stemmed from the student seminar which I led in 1955. Among the participants of that seminar I remember the second year students V.I. Arnol'd and A.A. Kirilov and the third year student S.A. Smolyak. The seminar was devoted to the theory of approximate representation of functions of several variables including problems of approximate nomography. In my introductory lecture I formulated Hilbert's thirteenth problem as a long-term plan which almost surely would not be fulfilled. It is interesting to note that in connection with nomography problems, nomographable functions of two variables, that is, functions of the form $\chi(\phi(x) + \psi(y))$, attracted a certain amount of attention from the very beginning. Functions of this kind proved to be an essential element in the final representation of functions of several variables by functions of one variable and the operation of addition. However, Hilbert's problem was first solved on the basis of completely different ideas, using technique developed by A.S. Kronrod. In this way I proved the central theorem of these papers, which stated that any continuous function of $n \geq 4$ variables can be represented as a superposition of functions of three variables.

By improving the methods of those papers Arnol'd gave the final solution to Hilbert's problem in the form of a theorem asserting that any continuous function of $n \geq 3$ variables can be represented as a superposition of functions of two variables.

Only some months later, using completely different construction techniques, did I manage to prove the theorem of paper No. 56, which strengthens Arnol'd's theorem, according to which any continuous function of an arbitrary number of variables can be represented as a superposition of functions of one variable and addition.

SUPERPOSITIONS

(V.I. Arnol'd)

In Kolmogorov's opinion, the proof of the fact that every continuous function of n variables can be represented as a superposition

$$f(x_1,\dots,x_n)=\sum_{q=1}^{2n+1}\chi_q\Big[\sum_{p=1}^{n}\phi^{pq}(x_p)\Big] \tag{1}$$

of continuous functions of one variable and the operation of addition was his most technically difficult achievement. In this commentary basic improvements of the representation (1) are discussed, along with a very brief description of some other aspects of the problem of representability of functions by means of superpositions (cf. [1–3]).

In Kolmogorov's representation (1) the inner functions ϕ^{pq} are fixed and only the outer functions χ_q depend on the represented function f.

1. Representability theorems

Kolmogorov's paper No. 56 was preceded by paper No. 55, in which he proved that any continuous function of four variables can be represented as a superposition of continuous functions of three variables. The method developed in paper No. 55 reduces the representability of functions of three variables as superpositions of functions of two variables to a representability problem for functions defined on universal trees of three-dimensional space. The latter problem was solved by Arnol'd in [34], and thus continuous functions of three variables were for the first time expressed as superpositions of continuous functions of two variables. After that the representation (1) reducing all continuous functions of a finite number of variables to superpositions of continuous functions of one variable and addition was soon obtained by Kolmogorov. In this connection he noted that the constructions used in his paper No. 56 were found by analyzing those employed in No. 55 and [34] and discarding some details unnecessary for the derivation of the final result.

Thorough proofs of Kolmogorov's theorem and the lemmas in his paper of 1957 were published by Tikhomirov [4], Lorenz [2], Sprecher [5], Hedberg [43], and others. Lorenz [2] noted that the outer functions χ_q can be replaced by a single function χ. Sprecher [5] reduced all the inner functions to translations and extensions of a single function ψ with the property that there exist $\epsilon > 0$ and $\lambda > 0$ such that any continuous function of n variables can be represented as

$$f(x_1, \ldots, x_n) = \sum_{q=1}^{2n+1} \chi[\lambda^p \psi(x_p + \epsilon q) + q]. \tag{2}$$

Fridman [6] proved that the inner functions ϕ^{pq} in (1) can be chosen so that they satisfy a "Lipschitz" condition. Sprecher [7] extended this result to the representation (2) (the function ψ can be chosen to satisfy a Lipschitz condition). It follows from Kolmogorov's representation (1) and Bari's representation [8] of any continuous function of one variable as a sum of three superpositions of absolutely continuous functions $\sum f_k \circ g_k$ that all continuous functions of any number of variables can be represented by means of superpositions of absolutely continuous functions of one variable and the operation of addition.

Doss [11] developed the Kolmogorov representation and used Kahane's results [42, 44] to obtain a set of continuous functions f_k and g_k of one variable such that any function continuous on a square can be expanded in a series

$$\sum a_k \exp(i f_k(x) g_k(y)),$$

where $\sum |a_k| < \infty$. He showed that if the inner sum in (1) is replaced by a product, then it is possible to take "almost any" (in the sense of Baire category) functions ϕ, while as χ one can take a function independent of q and with an absolutely convergent Fourier transform.

Ostrand [39] extended the representation (1) to functions defined on an n-dimensional compactum, while Doss [12] extended (1) to functions defined on an unbounded set (in R^n).

2. Non-representability theorems

Bassalygo proved that for any *three* functions ψ_k continuous on a square there exists a continuous function f which cannot be represented as $\sum \chi_k \circ \psi_k$ for any continuous χ_k [9].

A similar expansion into a sum of *five* summands (1) is possible by virtue of Kolmogorov's theorem. Doss [10] showed that it is insufficient to take *four* summands instead of five to represent an arbitrary continuous function of two variables as a sum of the form (1) when the (fixed) inner functions ϕ^{pq} are monotone.

Vitushkin and Khenkin have proved that

(1) for any functions ϕ_k continuous on a square and any continuously differentiable functions ψ_k there exists a function analytic on the square that cannot be represented as a sum of products $\sum \phi_k \cdot (g_k \circ \psi_k)$ for any choice of continuous outer functions g_k of one variable; moreover,

(2) the set of such sums (with all possible continuous g_k) is closed and nowhere dense (in the uniform metric);

(3) there exists a polynomial $(x_1 + \nu x_2)^\mu$ not belonging to this set.

These results can be extended to functions defined on a cube of arbitrary dimension (see [13–15]).

Fridman proved that sets of functions of two variables representable in the form $\sum g_k \circ \psi_k$ with fixed continuously differentiable inner functions ψ_k are nowhere dense in L^2, as are sets of functions of three variables representable as $\sum g_k \circ \psi_k$ where the functions $\psi_k : R^3 \to R^2$ are twice continuously differentiable and the functions $g_k : R^2 \to R$ are continuous [35, 36].

3. Hilbert's thirteenth problem

The exact formulation of Hilbert's problem to which Kolmogorov referred in his papers reads (see [1]): "to show that the equation of the seventh degree

$$f^7 + xf^3 + yf^2 + zf + 1 = 0 \tag{3}$$

cannot be solved by means of any continuous functions depending only on two variables".

A great number of papers are devoted to the representability of functions as superpositions of functions depending on a smaller number of variables and satisfying certain additional conditions (algebraicity, analyticity, smoothness). Hilbert was, of course, aware of the fact that superpositions of discontinuous functions represent all functions of a larger number of variables.

Hilbert also knew about the existence of *analytic* functions of three variables that cannot be represented by any finite superpositions of *analytic* functions of two variables (he indicated this in [1]).

Ostrovski [16] proved that the function

$$h(x,y)=\sum x^k/y^k$$

cannot be represented as a finite superposition of inifinitely differentiable functions of one variable and algebraic functions of an arbitrary number of variables.

Vitushkin [17] proved that there exist p times continuously differentiable functions of n variables that cannot be expressed in terms of finite superpositions of $q\geq 1$ times continuously differentiable functions of $k<n$ variables, if $q/k>p/n$.

Kolmogorov [3] put forward the conjecture that there exist analytic functions of three variables that cannot be expressed as superpositions of continuously differentiable functions of two variables and that there are analytic functions of two variables that cannot be represented by superpositions of continuously differentiable functions of one variable and addition.

The qeustion of representability of a root of equation (3) in the form of a superposition of analytic or algebraic functions remains open.

4. The resolvent problem

In the statement of his problem, Hilbert proceeded from a result of Tschirnhausen [18], according to which a root of an algebraic equation of degree $n>5$, that is, a function $f(x_1,\ldots,x_n)$ determined by an equation

$$f^n+x_1f^{n-1}+\ldots+x_n=0, \tag{4}$$

can be expressed as a superposition of algebraic functions of $n-4$ variables. Hilbert assumed that the number $n-4$ cannot be reduced for $n=6,7,8$; he also proved that in order to solve an equation of degree $n=9$ it suffices to have functions of $n-5$ variables [19]. Wiman [20] extended the latter result to $n>9$, while Chebotarev [21] reduced the number of variables involved in the representation of functions to $n-6$ for $n\geq 21$ and to $n-7$ for $n\geq 121$.

Chebotarev was the first to attempt to find topological obstructions to the representability of algebraic functions as superpositions of algebraic functions, but his proofs were not convincing (see [22, 24]). Using topological notions (related to the behaviour of a many-valued algebraic function on and near a branching manifold) it is now possible to prove that algebraic functions cannot be represented by complete superpositions of integral algebraic functions (*completeness* means that the represented function must involve *all* the branches of the many-valued functions and not only one of them as, for example, in the formulas expressing solutions to equations of the 3rd and 4th degree).

Certain topological obstructions to the representation by a complete superposition of algebraic functions were constructed in this way (cohomology classes in the complement of the branching manifold which are invariant under the Tschirhausen transformation (see [25–29])). Lin [29] established the following, most complete, result: in any neighbourhood of the origin, for any $n \geq 3$, the root $f(x_1, \ldots, x_n)$ of equation (4) is not a complete superposition of entire algebroid functions of fewer than $n-1$ variables and single-valued holomorphic functions of an arbitrary number of variables.

Thus, from the standpoint of *complete* superpositions of *entire* algebraic functions, even fourth-degree equations cannot be solved without using functions of three variables.

The proof of Lin's theorem can be regarded as a modernization and a kind of rehabilitation of Chebotarev's considerations [23] and Morozov's attempt to correct them [24].

The use of non-entire functions reduces approximately by one the number of variables in the representation. For instance, the root of the equation $f^5 + xf + y = 0$ can be represented by superpositions of algebraic functions of one variable and arithmetic operations including division, and in any neighbourhood of zero it cannot be represented by a superposition of analytic functions of an arbitrary number of variables and algebroid functions of one variable (Khovanskii [45]).

It would be interesting to check whether Deligne's theorems on the behaviour of mixed Hodge structures under algebraic mappings give new obstructions to the representation of a function by a superposition of algebraic functions.

5. Approximations

Returning to Kolmogorov's papers on superpositions, we also mention the problems stated by him in relation to the closedness of the classes of functions representable by superpositions in the uniform metric. The theorem asserting that the function xy, which can be approximated by functions $(x + \epsilon)(y + \epsilon)$

representable in the form $\chi[\phi(x)+\psi(x)]$ cannot be represented in this form on the square $0 \leq x, y \leq 1$ was quoted in the lecture at the Third USSR Congress of Mathematicians and was proved in [30].

Kreines and Vainshtein [31] proved that the uniform limit of a sequence of functions representable in this form can be represented in the same form, if the limit is monotone in each variable. Mukhin showed [32] that the class of all functions nomographable by single-valued functions is non-closed and nowhere dense. Fridman [36, 37] discussed the closedness of certain linear superpositions on I^2. The best approximation of a continuous function on a rectangle by sums $\phi(x)+\psi(y)$ is always attainable (cf. [32]).

The question whether functions on plane curves are representable in this form was posed by Kolmogorov for trees, and the problem was solved in [34]; the same question for closed curves immediately leads to "small divisors" (see [35]).

The representability of functions on a curve $x_k = u_k(t)$ in the form

$$f(t)=\sum_{k=1}^{n} A_k\phi_k(u_k(t))$$

(with fixed coefficients A_k) is important in the theory of singularities of differentiable mappings. A typical example is the expansion of functions on the semicubical parabola:

$$f(t)=\phi(t^2)+\psi(t^2+t^3).$$

If a smooth function f admits of such a representation as a formal series, then there also exists a smooth representation (of class C^∞) (Dufour). However, if f is holomorphic, then as a rule, a holomorphic representation need not exist (Voronin [47]).

References

1. *Hilbert's problems*, Nauka, Moscow, 1969, pp.163–170 (in Russian).
2. G.G. Lorenz, 'Metric entropy, width, and superpositions of functions', *Amer. Math. Monthly* **69** (1962), 469–485.
3. D.A. Sprecher, 'A survey of solved and unsolved problems on superpositions of functions', *J. Approx. Theory* **6**:2 (1972), 123–134.
4. V.M. Tikhomirov, 'A.N. Kolmogorov's work on ϵ-entropy of function classes and superpositions of functions', *Uspekhi Mat. Nauk* **18**:5 (1963), 55-92 (in Russian).

5. D.A. Sprecher, 'On the structure of continuous functions of several variables', *Trans. Amer. Math. Soc.* **115** (1965), 340–355.
6. B.L. Fridman, 'Increasing smoothness of functions in Kolmogorov's theorem on superpositions', *Dokl. Akad. Nauk SSSR* **177** (1967), 1019–1022 (in Russian).
7. D.A. Sprecher, 'An improvement in the superposition theorem of Kolmogorov', *J. Math Anal. Appl.* **38** (1972), 208–213.
8. N.C. Bari, 'Mémoire sur la représentation finie des fonctions continues', *Math. Ann.* **103** (1930), 185–248, 598–653.
9. L.O. Bassalygo, 'On the representation of continuous functions of two variables by continuous functions of one variable', *Vestn. MGU Ser. Mat.-Mekh.* **21** (1966), 58–63 (in Russian).
10. R. Doss, 'On the representation of continuous functions of two variables by means of addition and continuous functions of one variable', *Colloq. Math.* **10** (1963), 219–259.
11. R. Doss, 'Representations of continuous functions of several variables', *Amer. J. Math.* **98**:2 (1976), 375–383.
12. R. Doss, 'A superposition theorem for unbounded continuous functions', *Trans. Amer. Math. Soc.* **233** (1977), 197–203.
13. A.G. Vitushkin, 'Proof of the existence of analytic functions of several variables that are not representable as linear superpositions of continuously differentiable functions of a smaller number of variables', *Dokl. Akad. Nauk SSSR* **156**:6 (1964), 1258–1261 (in Russian).
14. G.M. Khenkin, 'On linear superpositions of continuously differentiable functions', *Dokl. Akad. Nauk SSSR* **157**:2 (1964), 283–290 (in Russian).
15. A.G. Vitushkin and G.M. Khenkin, 'Linear superpositions of functions', *Uspekhi Mat. Nauk* **22**:1 (1967), 77–124 (in Russian).
16. A. Ostrovski, 'Über Dirichletsche Reihen und algebraische Differentialgleichungen', *Math. Z.* **8**:3–4 (1920), 241–298.
17. A.G. Vitushkin, *On multidimensional variations*, Gostekhteorizdat, Moscow, 1955 (in Russian).
18. W. Tschirnhausen, 'Methodus auferendi omnes terminos intermedios ex data equatione', *Acta Eruditorum*, 1683.
19. D. Hilbert, 'Über die Gleichung neunten Grades', *Math. Ann.* **97** (1927), 243–250.

20. A. Wiman, 'Über die Anwendung der Tschirnhausen Transformation auf die Reduktion algebraischer Gleichungen', *Nova Acta R. Soc. Sci. Uppsaliensis* vol. extraordin. editum, 1927, pp. 3–8.

21. N.G. Chebotarev, 'On the resolvent problem', *Uchen. Zap. Kazan. Univ.* **114**:2 (1954), 189–193 (in Russian).

22. N.G. Chebotarev, 'The resolvent problem and critical manifolds', *Izv. Akad. Nauk SSSR Ser. Mat.* **7** (1943), 123–146 (in Russian).

23. N.G. Chebotarev, 'The resolvent problem', In: *Collected Works*, Vol. 1, USSR Acad. Sci. Press, Moscow, 1949, pp. 327–340 (in Russian).

24. V.V. Morozov, 'On some questions in the resolvent problem', *Uchen Zap. Kazan. Univ.* **114**:2 (1954), 173–187 (in Russian).

25. V.I. Arnol'd, 'On cohomology classes of algebraic functions invariant under a Tschirnhausen transformation', *Funktsional. Anal. i Prilozhen.* **4**:1 (1970), 84–85 (in Russian).

26. V.I. Arnol'd, 'Topological invariants of algebraic functions II', *Funktsional. Anal. i Prilozhen.* **4**:2 (1970), 1–9 (in Russian).

27. V.Ya. Lin, 'On superpositions of algebraic functions', *Funktsional. Anal. i Prilozhen.* **6**:3 (1972), 77–78 (in Russian).

28. V.Ya. Lin, 'Algebraic functions with universal invariant manifold', *Funktsional. Anal. i Prilozhen.* **6**:1 (1972), 81–82 (in Russian).

29. V.Ya. Lin, 'Superpositions of algebraic functions', *Funktsional. Anal. i Prilozhen.* **10**:1 (1976), 37–45 (in Russian).

30. V.I. Arnol'd, 'On the representation of continuous functions of two variables in the form $\chi[\phi(x)+\psi(y)]$', *Uspekhi Mat. Nauk* **12**:2 (1957), 119–121 (in Russian).

31. I.A. Vainshtein and M.A. Kreines, 'Sequences of functions of the form $f(X(x)+Y(y))$', *Uspekhi Mat. Nauk* **15**:4 (1960), 123–128 (in Russian).

32. O.N. Mukhin, 'Certain properties of classes of surfaces nomographable by single-valued functions', *Vestn. MGU Ser. Mat.-Mekh.* **17**:1 (1962), 24–29 (in Russian).

33. Yu.P. Ofman, 'On best approximation of functions of two variables by functions of the form $\phi(x)+\psi(y)$', *Izv. Akad. Nauk SSSR Ser. Mat.* **25**:2 (1961), 239–252 (in Russian).

34. V.I. Arnol'd, 'On representation of continuous functions of three variables by superpositions of functions of two variables', *Mat. Sb.* **48** (1959), 3–74;

98 (1962), 392 (in Russian).

35. V.I. Arnol'd, 'Small divisors I. On maps of the circle onto itself', *Izv. Akad. Nauk SSSR Ser. Mat.* **25**:1 (1961), 21–86 (in Russian).
36. B.L. Fridman, 'Nowhere dense spaces of linear superpositions of functions of several variables', *Izv. Akad. Nauk SSSR Ser. Mat.* **36**:4 (1972), 814–846 (in Russian).
37. B.L. Fridman, 'Estimate of the dimension of the null space of linear superpositions', *Mat. Sb.* **82**:1 (1970), 111–126 (in Russian).
38. G.G. Lorenz, *Approximation of functions*, Holt, Rinehart and Winston, New York, 1966.
39. P.A. Ostrand, 'Dimension of metric spaces and Hilbert's problem 13', *Bull. Amer. Math. Soc.* **71** (1965), 619–622.
40. J. Sternfeld, 'Dimension theory and superposition of continuous functions', *Israel J. Math.* **20** (1975), 300–310.
41. R. Kaufman, 'Linear superposition of smooth functions', *Proc. Amer. Math. Soc.* **46** (1974), 360–362.
42. J.P. Kahane, 'Sur le réarrangements de fonctions de la classe *A*', *Studia Math.* **31** (1968), 287–293.
43. T. Hedberg, Appendix IIB, In: H.S. Shapiro, *Lecture Notes in Math.* **187** (1971).
44. J.P. Kahane, 'Sur le théorème de superposition de Kolmogorov', *J. Approx. Theory* **13** (1975), 229–234.
45. A.G. Khovanskii, 'On representability of algebroid functions by superpositions of algebraic functions and algebroid functions of one variable', *Funktsional. Anal. i Prilozhen.* **4**:2 (1970), 74–79 (in Russian).
46. D.E. Marshall and A.G. O'Farrell, *Approximation by a sum of two algebras: the lightning bolt principle*, Preprint 1–23, Maynooth College, Co. Kildare (Ireland), 1982.
47. S.M. Voronin, 'Analytic classification of pairs of involutions and its application', *Funktsional. Anal. i Prilozhen.* **16**:2 (1982), 7–15 (in Russian).

LIST OF WORKS BY A.N. KOLMOGOROV [1]

1923

* Une série de Fourier-Lebesgue divergente presque partout, *Fund. Math.* **4**, 324–328.

* Sur l'ordre de grandeur des coéfficients de la série de Fourier-Lebesgue, *Bull. Acad. Pol. Sci. A*, 83–86.

1924

* Une contribution à l'étude de la convergence des séries de Fourier, *Fund. Math.* **5**, 96–97.

* Sur la convergence des séries de Fourier, *C. R. Acad. Sci. Paris* **178**, 303–306 (in collaboration with G.A. Seliverstov).

1925

* La définition axiomatique de l'intégrale, *C. R. Acad. Sci. Paris* **180**, 110–111.

* Sur les bornes de la généralisation de l'intégrale. (Paper No. 6 in the present book. First publication.)

* Sur la possibilité de la définition générale de la dérivée, de l'intégrale et de la sommation des séries divergentes, *C. R. Acad. Sci. Paris* **180**, 362–364.

* Sur les fonctions harmoniques conjuguées et les séries de Fourier, *Fund. Math.* **7**, 24–29.

* On the tertium non datur principle, *Mat. Sb.* **32**:4, 646–667 (in Russian).

* Über Konvergenz von Reihen, deren Glieder durch den Zufall bestimmt werden, *Mat. Sb.* **32**:4, 668–677 (in collaboration with A.Ya. Khinchin).

1926

* Sur la convergence des séries de Fourier, *Atti Accad. Naz. Lincei. Rend.* **3**, 307–310 (in collaboration with G.A. Seliverstov).

* Une série de Fourier-Lebesgue divergente partout, *C. R. Acad. Sci. Paris* **183**, 1327–1329.

1927

Sur la loi des grands nombres, *C. R. Acad. Sci. Paris* **185 , 917–919.

---

[1] The works included in Volume 1 of the present edition are marked by an asterisk, those in Volume 2 are marked by two asterisks and those in Volume 3 are marked by three asterisks.

* Sur la convergence des séries de fonctions orthogonales, *Math. Z.* **26**:2/3, 432–441 (in collaboration with D.E. Men'shov).

1928

* On operations on sets, *Mat. Sb.* **35**:3/4, 414–422 (in Russian). [2]

Sur une formule limite de M.A. Khintchine, *C. R. Acad. Sci. Paris* **186, 824–825.

* Sur un procédé d'intégration de M. Denjoy, *Fund. Math.* **11**, 27–28.

Über die Summen durch den Zufall bestimmter unabhängiger Grössen, *Math. Ann.* **99, 309–319.

1929

Bemerkungen zu meiner Arbeit "Über die Summen zufälliger Grössen", *Math. Ann.* **102, 484–488.

**General measure theory and calculus of probabilities, In *Proc. Communist Academy Math. Sect.* Vol. I, pp. 8–21 (in Russian).

Contemporary disputes on the nature of mathematics, *Nauch. Slovo* No. 6, 41–54 (in Russian).

Über das Gesetz des iterierten Logarithmus, *Math. Ann.* **101, 126–135.

Sur la loi des grand nombres, *Atti Accad. Naz. Lincei. Rend.* **9:6, 470–474.

1930

Sur la loi forte des grands nombre, *C. R. Acad. Sci. Paris* **191, 910–912.

* Zur toplogisch-gruppentheoretischen Begründung der Geometrie, *Nachr. Ges. Wiss. Göttingen. Fachgr.* 1 (*Mathematik*) **8**, 208–210.

* Untersuchungen über den Integralbegriff, *Math. Ann.* **103**, 654–696.

* Sur la notion de la moyenne, *Atti Accad. Naz. Lincei. Rend.* **12**:9, 388–391.

1931

Über die analytischen Methoden in der Wahrscheinlichkeitsrechnung, *Math. Ann.* **104, 415–458.

Sur le problème d'attente, *Mat. Sb.* **38:1/2, 101–106.

The method of medians in the theory of errors, *Mat. Sb.* **38:3/4, 47–50 (in Russian).

**Eine Verallgemeinerung des Laplace-Liapounoffschen Satzes, *Izv. Akad. Nauk SSSR* OMEN, 959–962.

---

[2] The second part of this paper is published in Volume 2.

*Über Kompaktheit der Funktionenmengen bei der Konvergenz im Mittel, *Nachr. Ges. Wiss. Göttingen* **9**, 60–63.

1932

Theory of functions of a real variable, In: *Mathematics in the USSR during 15 years*, GTTI, Moscow, Leningrad, pp. 37–48 (in Russian).

Sulla forma generale di un processo stocastico omogeneo. (Un problema di Bruno de Finetti), *Atti Accad. Naz. Lincei. Rend.* **15, 805–808.

Ancora sull forma generale di un processo stocastico omogeneo, *Atti Accad. Naz. Lincei. Rend.* **15, 866–869.

* Zur Deutung der intuitionistischen Logik, *Math. Z.* **35**, 58–65.

* Zur Begründung der projektiven Geometrie, *Ann. of Math.* 33, 175–176.

Introduction to the theory of functions of a real variable, GTTI, Moscow-Leningrad, (in Russian) (in collaboration with P.S. Aleksandrov).

1933

Introduction to the theory of functions of a real variable (2nd ed.), GTTI, Moscow-Leningrad (in collaboration with P.S. Aleksandrov).

Grundbegriffe der Wahrscheinlichkeitsrechnung, Springer-Verlag, Berlin.

* Beiträge zur Masstheorie, *Math. Ann.* **107**, 351–366.

Zur Berechnung der mittleren Brounschen Fläche, *Phys. Z. Sow.* **4:1, 1–13 (in collaboration with M. Leontovich).

Sulla determinazione empirica di una legge di distrebuzione, *G. Ist. Ital. Attuar.* **4, 83–91.

**Über die Grenzwertsätze der Wahrscheinlichkeitsrechnung, *Izv. Akad. Nauk SSSR OMEN*, 366–372.

Zur Theorie der stetigen zufälligen Prozesse, *Math. Ann.* **108, 149–160.

Sur la détermination empirique d'une loi de distribution, *Uch. Zap. MGU* **1**, 9–10.

** On the question of the applicability of prognosis formulas found in a statistical manner, *Zh. Geofiz.* **3**:1, 78–82 (in Russian).

1934

* On points of discontinuity of functions of two variables, *Dokl. Akad. Nauk SSSR* 1:3, 105–106 (in collaboration with I.Ya. Verchenko).

* Zur Normierbarkeit eines allgemeinen topologischen linearen Raumes, *Stud. Math.* **5**, 29–33.

* Continuation of the study of points of discontinuity of functions of two variables, *Dokl. Akad. Nauk SSSR* **4**:7, 361–362 (in collaboration with I. Verchenko) (in Russian).

* On convergence of series in orthogonal polynomials, *Dokl. Akad. Nauk SSSR* **1**:6, 291–294 (in Russian).

Quelques remarques sur l'approximation des fonctions continues, *Mat. Sb.* **41**:1, 99–103.

On certain new trends in probability theory, In: Bull. USSR Second Congress of Mathematicians in Leningrad, June 24–30, 1934, USSR Academy of Sciences Press, Leningrad, 8 (in Russian).

Modern mathematics, *Front Nauki i Tekhniki*, No. 5/6, 25-28 (in Russian).

The Institute of Math. and Mech. at MGU, *Front Nauki i Tekhniki*, No. 5/6, 75–78 (in Russian).

Zufällige Bewegungen (Zur Theorie der Brownschen Bewegung), *Ann. of Math.* **35, 116–117.

1935

On certain modern trends in probability theory, In: Proc. USSR Second Congress of Mathematicians in Leningrad, June 24–30, 1934, Vol. 1, USSR Academy of Sciences Press, Leningrad, Moscow, 349–358 (in Russian).

Deviation from a formula of Hardy for partial isolation, *Dokl. Akad. Nauk SSSR* **3:3, 129–132 (in Russian).

* La transformation de Laplace dans les espaces linéaires, *C. R. Acad. Sci. Paris* **200**, 1717–1718.

* Zur Grössenordnung des restgliedes Fourierschen Reihen differenzierbarer Funktionen, *Ann. of Math.* **36**, 521–526.

1936

* Über die beste Annäherung von Funktionen einer gegebenen Funktionenklasse, *Ann. of Math.* **37**, 107–110.

Basic notions of probability theory, ONTI, Moscow, Leningrad (in Russian). Translated from the German: *Grundbegriffe der Wahrscheinlichkeitsrechnung*, Springer-Verlag, Berlin, 1933.

Equations, In: *BSE* **56**, 163–165 (in Russian).

Modern mathematics, In: Collection of papers on the philosophy of mathematics, ONTI, Moscow, pp. 7–13 (in Russian).

Theory and practice in mathematics, *Front Nauki i Tekhniki*, No. 5, 39–42 (in Russian).

*Über die Dualität im Aufbau der kombinatorischen Topologie, *Mat. Sb.* **1**, 97–102.

Anfangsgründe der Theorie der Markoffschen Ketten mit unendlich vielen möglichen Zustanden, *Mat. Sb.* **1**, 607–610.

* Homologierung des Komplexes und des lokalbikompakten Räumes, *Mat. Sb.* **1**, 701–706.

Zur Theorie der Markoffschen Ketten, *Math. Ann.* **112, 155–160.

On Plessner's condition for the law of large numbers, *Mat. Sb.* **1**, 847–849 (in Russian).

* Endliche Uberdeckungen topologischer Räume, *Fund. Math.* **26**, 267–271 (in collaboration with P.S. Aleksandrov).

* Les groupes de Betti des espaces localement bicompacts, *C. R. Acad. Sci. Paris* **202**, 1144–1147.

* Propriétés des groupes de Betti des espaces localement bicompacts, *C. R. Acad. Sci. Paris* **202**, 1325–1327.

* Les groupes de Betti des espaces métriques, *C. R. Acad. Sci. Paris* **202**, 1558–1560.

* Cycles relatifs. Théorème de dualité de M. Alexander, *C. R. Acad. Sci. Paris* **202**, 1641–1643.

Foreword to the Russian translation of the book by A. Heyting: Review of studies on foundations of mathematics. Introduction to proof theory, ONTI, Moscow, Leningrad (in Russian).

Sulla teoria di Volterra della lotta per l'esistenza, *G. Ist. Ital. Attuar.* **7**, 74–80.

1937

* Über offene Abbildungen, *Ann. of Math.* **38**, 36–38.

* Skew-symmetric quantities and topological invariants, In: Proc. Seminar Vector and Tensor Anal. and Appl. to Geometry, Mechanics, and Physics, **4**, GONTI, Moscow, Leningrad, pp. 342–347 (in Russian).

Markov chains with countable set of states, *Bul. MGU Mat. Mekh.* **1:3, 1–16 (in Russian).

* Studies on the diffusion equation, combined with increase in the amount of matter and its application to a problem in biology, *Bul. MGU Mat.*

Mekh. **1**:6, 1–26 (in Russian) (in collaboration with I.G. Petrovskii and N.S. Piskunov).

**On the satistical theory of crystallization of metals, *Izv. Akad. Nauk SSSR Ser. Mat.* No. 3, 355–360 (in Russian).

Ein vereinfachter Beweis der Birkhoff-Khintchineschen Ergodensatzes, *Mat. Sb.* **2**, 367–368.

Zur Umkehrbarkeit der statistischen Naturgesetze, *Math. Ann.* **113, 766–772.

Editing and supplementing the Russian translation of the book by F. Hausdorff: Set theory, GONTI, Moscow, Leningrad (in Russian) (in collaboration with P.S. Aleksandrov.

The continuum, In: *BSE* **34**, 139–140.

1938

Introduction to the theory of functions of a real variable (3rd ed. (revised)), GONTI, Moscow, Leningrad (in Russian) (in collaboration with P.S. Aleksandrov) (1st ed.: Moscow, Leningrad, 1932).

A.A. Markov, In: *BSE* **38**, 152–153 (in Russian).

Mathematics, In: *BSE* **38**, 359–402 (in Russian).

Mathematical induction, In: *BSE* **38**, 405–406 (in Russian).

Measure, In: *BSE* **38**, 831–832 (in Russian).

Multidimensional space, In: *BSE* **39**, 577–578 (in Russian).

Probability theory and its applications, In: *Mathematics and Natural Sciences in the USSR*, GONTI, Moscow, Leningrad, pp. 51–61 (in Russian).

On the section on information in the first issue of Uspekhi Mat. Nauk, *Uspekhi Mat. Nauk* No. 4, 326–327 (in Russian).

A remark on the foundations of geometry, *Uspekhi Mat. Nauk* , No. 4, 347–348 (in Russian).

From the board of editors (Series of articles on the theory of random processes), *Uspekhi Mat. Nauk* No. 5, 3–4 (in Russian).

Über die analytischen Methoden in der Wahrscheinlichkeitsrechnung, *Uspekhi Mat. Nauk* , No. 5, 5–41. Translated from the German, *Math. Ann.* **104, 415–458 (1931).

* Ein vereinfachter Beweis der Birkhoff-Khintchineschen Ergodensatzes, *Uspekhi Mat. Nauk* , No. 5, 52–56. Translated from the German, *Mat. Sb.* **2**, 367–368 (1937).

Some problems in the theory of functions of a real variable, *Uspekhi Mat. Nauk* No. 5, 232–234 (in Russian) (in collaboration with G.M. Fichtengol'ts and I.M. Gel'fand).

On the solution of a problem in biology, *Izv. NII Mat. i Mekh. Tomsk. Univ.* **2:1, 7–12 (in Russian).

Une généralisation de l'inégalité de M.J. Hadamard entre les bornes supérieures des dérivées successives d'une fonction, *C. R. Acad. Sci. Paris* **207**, 764–765.

Editing the Russian translation of the book by H. Lebesgue: On measurement of quantities, Uchpedgiz, Moscow (in Russian).

1939

Algebra, Chapter 1, Uchpedgiz, Moscow (in Russian) (in collaboration with P.S. Aleksandrov).

Orientation, In: *BSE* **43**, 342–344 (in Russian).

* On inequalities between least upper bounds of sequences of derivatives of an arbitrary function on an infinite interval, *Uch. Zap. MGU Matematika* **3**:30, 3–16 (in Russian).

* On rings of continuous functions on topological spaces, *Dokl. Akad. Nauk SSSR* **22**:1, 11–15 (in Russian) (in collaboration with I.M. Gel'fand).

Sur l'interpolation et extrapolation des suites stationnaires, *C. R. Acad. Sci. Paris* **208**, 2043–2045.

1940

Surfaces, In: *BSE* **45**, 746–748 (in Russian).

* Curves in a Hilbert space, invariant under a one-parameter group of motions, *Dokl. Akad. Nauk SSSR* **26**, 6–9 (in Russian).

* Wiener spirals and some other interesting curves in Hilbert space, *Dokl. Akad. Nauk SSSR* **26**, 115–118 (in Russian).

On the 60th birthday of S.N. Bernshtein, *Izv. Akad. Nauk SSSR Ser. Mat.* **4**, 249–260 (in Russian) (in collaboration with V.L. Goncharov).

V.I. Glivenko, *Uspekhi Mat. Nauk* , No. 8, 379–383 (in Russian).

Review of V.I. Romanovsku, Mathematical statistics, *Uspekhi Mat. Nauk* , No. 7, 327–329 (in Russian).

On a new argument for Mendel's laws, *Dokl. Akad. Nauk SSSR* **27, 38–42 (in Russian).

1941

Stationary sequences in Hilbert space, *Bul. MGU Mat.* **2:6, 1–10 (in Russian).

Interpolation and extrapolation of stationary random sequences, *Izv. Akad. Nauk SSSR Ser. Mat.* **5, 3–14 (in Russian).

* Points of local topological character of countably multiple open mappings of compacta, *Dokl. Akad. Nauk SSSR* **30**, 477–479 (in Russian).

* Local structure of turbulence in an incompressible viscous fluid for very large Reynolds numbers, *Dokl. Akad. Nauk SSSR* **30**, 299–303 (in Russian).

On the lognormal distribution for particle sizes under grinding, *Dokl. Akad. Nauk SSSR* **31, 99–101 (in Russian).

* On degeneration of isotropic turbulence in an incompressible viscous fluid, *Dokl. Akad. Nauk SSSR* **31**, 538–541 (in Russian).

* Energy scattering for local isotropic turbulence, *Dokl. Akad. Nauk SSSR* **32**:1, 19–21 (in Russian).

Properties of inequalities and the notion of approximate computations, *Mat. v Shkole*, No. 2, 1–12 (in Russian) (in collaboration with P.S. Aleksandrov).

Irrational numbers, *Mat. v Shkole*, No. 3, 1–15 (in Russian) (in collaboration with P.S. Aleksandrov).

Confidence limits for an unknown distribution function, *Ann. Math. Stat.* **12**:4, 461–463.

1942

Definition of the scatter centre and measures of exactness using a finite number of observations, *Izv. Akad. Nauk SSSR Ser. Mat.* **6**, 3–32 (in Russian).

* The equation of turbulent motion of an incompressible fluid, *Izv. Akad. Nauk SSSR Ser. Fiz.* **6**:1–2, 56–58 (in Russian).

1943

Nikolai Ivanovich Lobachevskii, 1793–1843, Gostekhizdat, Moscow, Leningrad (in Russian) (in collaboration with P.S. Aleksandrov).

A great Russian scientific innovator: on the 150th birthday of N.I. Lobachevskii, *Izvestiya* (2 September, 1943) (in Russian).

1945

The number of hits for several shots, and general principles for estimating the efficiency of shooting systems, *Trudy Mat. Inst. Akad. Nauk SSSR* **12**, 7–25 (in Russian).

Artificial scatters in the case of hitting in one shot and scatter in one measurement, *Trudy Mat. Inst. Akad. Nauk SSSR* **12**, 26–45 (in Russian).

1946

On the problem of the law of resistance in a turbulent flow in smooth tubes, *Dokl. Akad. Nauk SSSR* **52**, 669–671 (in Russian).

On the substantiation of the method of least squares, *Uspekhi Mat. Nauk* **1:1, 57–70 (in Russian).

On the foundations of the theory of real numbers, *Uspekhi Mat. Nauk* **1**:1, 217–219 (in Russian).

Newton and modern mathematical thought, In: *Moscow University – in commemoration of Isaac Newton*, 1643–1943, Moscow Univ. Press, Moscow, pp. 27–43 (in Russian).

Review of M.A. Velikanov's article 'Transfer of suspended pumps by a turbulent flow', *Izv. Akad. Nauk SSSR OTN* **5**, 781–784 (in Russian).

1947

Development of mathematics in the USSR, In: *BSE* Vol. "SSSR", 1318–1323 (in Russian).

Average quantities, In: *BSE* **52**, 508–509 (in Russian).

A formula of Gauss from the theory of the least squares method, *Izv. Akad. Nauk SSSR Ser. Mat.* **11, 561–566 (in Russian) (in collaboration with A.A. Petrov and Yu.M. Smirnov).

Branching random processes, *Dokl. Akad. Nauk SSSR* **56, 7–10 (in Russian) (in collaboration with N.A. Dmitriev).

Computation of final probabilities for branching random processes, *Dokl. Akad. Nauk SSSR* **56, 783–786 (in Russian) (in collaboration with B.A. Sevast'yanov).

Statistical theory of oscillation with continuous spectrum, In: *Collection of articles dedicated to the 30th anniversary of the Great October Socialist Revolution, Vol.* 1, USSR Acad. Sci. Press, Moscow, Leningrad, pp. 242–252 (in Russian).

The role of Russian science in the development of probability theory, In: *The role of Russian science in the development of world science and culture, Vol.* 1, Moscow Univ. Press, Moscow, pp. 53–64 (*Uchen. Zap. MGU*, No. 91).

1948

Statistical theory of oscillation with continuous spectrum, In: General meeting of the USSR Acad. Sci. dedicated to the 30th anniversary of the Great October Socialist Revolution, USSR Acad. Sci Press, Moscow, Leningrad, pp. 465–472 (in Russian).

Probability theory, In: Mathematics in the USSR during thirty years, 1917–1947, Gostekhizdat, Moscow, Leningrad, pp. 701–727 (in Russian) (in collaboration with B.V. Gnedenko).

On two theorems regarding probabilities: comments, In: P.L. Chebyshev, Complete collection of works, Vol. 3, Mathematical analysis, Gostekhizdat, Moscow-Leningrad, pp. 404–409 (in Russian).

Construction of complete metric Boolean algebras, *Uspekhi Mat. Nauk* **3**:1, 212 (in Russian).

* Notes regarding the Chebyshev polynomials of least deviation from a given function, *Uspekhi Mat. Nauk* **3**:1, 216–221 (in Russian).

E.E. Slutskii, *Uspekhi Mat. Nauk* **3**:4, 143–151 (in Russian).

Editing and writing the foreword to the Russian translation of the book by H. Cramer: Mathematical methods of statistics, Inost. Lit., Moscow (in Russian).

Algèbres de Boole métriques complètes, In: *VI Zjazd Mat. Polskich*, Warsaw, pp. 22–30.

1949

Limit distributions for sums of independent random variables, Gostekhizdat, Moscow-Leningrad (in Russian) (in collaboration with B.V. Gnedenko).

On the problem of 'geometrical selection' of crystals, *Dokl. Akad. Nauk SSSR* **65**, 681–684 (in Russian).

** Solution of a problem in mathematical statistics, related to the problem of the mechanism of fibre-formation, *Dokl. Akad. Nauk SSSR* **65**, 793–796 (in Russian).

On sums of a random number of random terms, *Uspekhi Mat. Nauk* **4:4, 168–172 (in Russian) (in collaboration with Yu.V. Prokhorov).

A local limit theorem for classical Markov chains, *Izv. Akad. Nauk SSSR Ser. Mat.* **13, 281–300 (in Russian).

Basic problems of theoretical statistics, In: Proc. 2nd USSR Conf. Math. Statistics, Tashkent, 27 Sept. –2 Oct., 1948, Uzbekgosizdat, Tashkent, pp. 216–

220 (in Russian).

The true meaning of results by dispersion analysis, In: Proc. 2nd USSR Conf. Math. Statistics, Tashkent, 27 Sept.–2 Oct., 1948, Uzbekgosizdat, Tashkent, pp. 240–268 (in Russian).

* On the breakage of drops in a turbulent flow, *Dokl. Akad. Nauk SSSR* **66**:5, 825–828 (in Russian).

Absolute quantities, In: *BSE*-2 **1**, 32 (in Russian).

J. Hadamard, In: *BSE*-2 **1**, 388 (in Russian).

Additive quantities, In: *BSE*-2 **1**, 394 (in Russian).

Axioms, In: *BSE*-2 **1**, 613–616 (in Russian).

Axonometry, In: *BSE*-2 **1**, 617 (in Russian).

1950

Unbiased estimators, *Izv. Akad. Nauk SSSR Ser. Mat.* **14, 303–326 (in Russian).

On the problem of determining the coefficient of temperature-conductivity of soil, *Izv. Akad. Nauk SSSR Ser. Geogr. i Geofiz* **2**, 97–98 (in Russian).

Algebra in secondary school, In: *BSE*-2 **2**, 61–62 (in Russian).

Algebraic expressions, In: *BSE*-2 **2**, 64 (in Russian).

Algorithms, In: *BSE*-2 **2**, 65 (in Russian).

Euclid's algorithm, In: *BSE*-2 **2**, 65–67 (in Russian).

A.D. Aleksandrov, In: *BSE*-2 **2**, 83 (in Russian).

P.S. Aleksandrov, In: *BSE*-2 **2**, 84 (in Russian).

Asymptote, In: *BSE*-2 **3**, 238–239 (in Russian).

Asymptotic expressions, In: *BSE*-2 **3**, 239 (in Russian).

N.I. Akhiezer, In: *BSE*-2 **3**, 565 (in Russian).

S. Banach, In: *BSE*-2 **4**, 183 (in Russian).

N.K. Bari, In: *BSE*-2 **4**, 245 (in Russian).

S.N. Bernshtein, In: *BSE*-2 **5**, 52 (in Russian).

Infinitely large, In: *BSE*-2 **5**, 66–67 (in Russian).

Infinitesimal, In: *BSE*-2 **5**, 67–71 (in Russian) (in collaboration with V.F. Kagan).

Infinitely distant elements In: *BSE*-2 **5**, 71–72 (in Russian) (in collaboration with B.V. Delone).

Infinity (in mathematics), In: *BSE*-2 **5**, 73–74 (in Russian).

Biharmonic function, In: *BSE*-2 **5**, 159 (in Russian).

Bilinear form In: *BSE*-2 **5**, 167 (in Russian).

Law of large numbers, In: *BSE*-2 **5**, 538–540 (in Russian).

1951

On the problem of differentiability of transition probabilities in time-homogeneous Markov processes with countable number of states, *Uch. Zap. MGU* **148, *Matematika* 4, 53–59 (in Russian).

** Generalizations of Poisson's formula to the case of selection from a finite sample, *Uspekhi Mat. Nauk* **6**:3, 133–134 (in Russian).

I.G. Petrovskii, *Uspekhi Mat. Nauk* **6**:3, 161–164 (in Russian).

Statistical quality control for admissible number of defective articles equal to zero, Leningrad, pp. 1–24 (in Russian).

L.E.J. Brouwer, In: *BSE*-2 **6**, 62 (in Russian) (in collaboration with S.A. Yanovskaya).

Variational series, In: *BSE*-2 **6**, 641 (in Russian).

H. Weyl, In: *BSE*-2 **7**, 106 (in Russian) (in collaboration with S.A. Yanovskaya).

Quantity, In: *BSE*-2 **7**, 340–341 (in Russian).

Probable deviation, In: *BSE*-2 **7**, 507 (in Russian).

Probability, In: *BSE*-2 **7**, 508–510 (in Russian).

Sampling method, In: *BSE*-2 **9**, 417–418 (in Russian) (in collaboration with T.I. Kozlov).

Introduction and commentary to the book: N.I. Lobachevskii, Complete collected works, Vol. 5, Gostekhizdat, Moscow, Leningrad, pp. 329–332; 342–348 (in Russian) (in collaboration with A.N. Khovanskii).

1952

On the problem of pressure in the velocity profile under turbulent flow in tubes, *Dokl. Akad. Nauk SSSR* **84**, 20–30 (in Russian).

Gaussian distribution, In: *BSE*-2 **10**, 275 (in Russian).

Geodesic curvature, In: *BSE*-2 **10**, 481 (in Russian).

D. Hilbert, In: *BSE*-2 **11**, 370–371 in Russian).

Histogram, In: *BSE*-2 **11**, 447 (in Russian).

B.V. Gnedenko, In: *BSE*-2 **11**, 545 (in Russian).

Homeomorphism, In: *BSE*-2 **12**, 21 (in Russian).

Homotopy, In: *BSE*-2 **12**, 35 (in Russian).

Motion (in geometry), In: *BSE*-2 **13**, 447 (in Russian).

Two-term, In: *BSE*-2 **13**, 518 (in Russian).

Real numbers, In: *BSE*-2 **13**, 570 (in Russian).

Division, In: *BSE*-2 **13**, 628 (in Russian).

Discreteness, In: *BSE*-2 **14**, 425 (in Russian).

Variance, In: *BSE*-2 **14**, 438 (in Russian).

Distributivity, In: *BSE*-2 **14**, 479 (in Russian).

Distributive operator, In: *BSE*-2 **14**, 479 (in Russian).

Differential, In: *BSE*-2 **14**, 498–499 (in Russian).

Differential equation, In: *BSE*-2 **14**, 520–526 (in Russian) (in collaboration with B.P. Demidovich and V.V. Nemytskii).

Confidence probability, In: *BSE*-2 **14**, 616 (in Russian).

Confidence bound, In: *BSE*-2 **14**, 617 (in Russian).

Mathematical symbols, In: *BSE*-2 **17**, 115–119 (in Russian) (in collaboration with I.G. Bashmakova and A.P. Yushkevich).

Significant digits, In: *BSE*-2 **17**, 135 (in Russian).

Isomorphism, In: *BSE*-2 **17**, 478–479 (in Russian) (in collaboration with V.I. Bityutskov).

Isotropic line, In: *BSE*-2 **17**, 509 (in Russian).

Concrete number, In: *BSE*-2 **17**, 557 (in Russian).

V.G. Imshenetskii, In: *BSE*-2 **17**, 607 (in Russian).

On the profession of a mathematician: A guide for those entering colleges, Sovetskaya Nauka, Moscow (in Russian).

1953

On the notion of algorithm, *Uspekhi Mat. Nauk* **8**:4, 175–176 (in Russian).

Some papers of the last decade on the limit theorems of probability theory, *Vestn. MGU* **10, 29–38 (in Russian).

* On dynamical systems with an integral invariant on a torus, *Dokl. Akad. Nauk SSSR* **93**, 736–766 (in Russian).

Mathematical induction, In: *BSE*-2 **18**, 146 (in Russian).

Integral, In: *BSE*-2 **18**, 250–253 (in Russian) (in collaboration with V.I. Glivenko).

Probability integral, In: *BSE*-2 **18**, 253 (in Russian).

Interpolation, In: *BSE*-2 **18**, 304–305 (in Russian).

Intuitionism, In: *BSE*-2 **18**, 319 (in Russian).

Elimination of variables, In: *BSE*-2 **18**, 483 (in Russian).

Testing, In: *BSE*-2 **18**, 604 (in Russian).

Method of exhaustion, In: *BSE*-2 **19**, 50–51 (in Russian).

Quadrant, In: *BSE*-2 **20**, 434 (in Russian).

Compactum, In: *BSE*-2 **22**, 282 (in Russian).

Constant, In: *BSE*-2 **22**, 416 (in Russian).

Continuum, In: *BSE*-2 **22**, 454-455 (in Russian).

Coordinates, In: *BSE*-2 **22**, 524–525 (in Russian).

Correlation, In: *BSE*-2 **23**, 55–58 (in Russian).

1954

* On the preservation of conditionally periodic motions under small variations of the Hamilton function, *Dokl. Akad. Nauk SSSR* **98**:4, 527–530 (in Russian).

* The general theory of dynamical systems and classical mechanics, In: Proc. Internat. Math. Congress, Amsterdam, 1954, Review lectures, USSR Acad. Sci. Press, Moscow, 1961, pp. 187–208 (in Russian).

Line (curve), In: *BSE*-2 **25**, 167–170 (in Russian).

Law of small numbers, In: *BSE*-2 **26**, 168 (in Russian).

A.A. Markov, In: *BSE*-2 **26**, 294 (in Russian).

Mathematics, In: *BSE*-2 **26**, 464–483 (in Russian).

Mathematical statistics, In: *BSE*-2 **28**, 485–490 (in Russian).

Mathematical physics, In: *BSE*-2 **26**, 490 (in Russian).

R. von Mises, In: *BSE*-2 **27**, 414 (in Russian).

Multidimensional space, In: *BSE*-2 **27**, 660 (in Russian).

Set theory, In: *BSE*-2 **28**, 14–17 (in Russian) (in collaboration with P.S. Aleksandrov).

Elements of theory of functions and functional analysis: a course of lectures. No. 1. *Metric and normed spaces*, Moscow Univ. Press, Moscow (in Russian) (in collaboration with S.V Fomin).

1955

Estimates of the minimal number of elements of ϵ-nets in various function classes and their application to the problem of representing functions of several variables by superpositions of functions of a smaller number of variables, *Uspekhi Mat. Nauk* **10**:1, 192–194 (in Russian).

Orientation, In: *BSE*-2 **31**, 188–189 (in Russian).

Foundations of geometry, In: *BSE*-2 **31**, 296 (in Russian).

Surface, In: *BSE*-2 **33**, 346–347 (in Russian) (in collaboration with L.A. Skvornyakov).

Ordinal number, In: *BSE*-2 **34**, 238 (in Russian).

Statistical quality control, In: *BSE*-2 **34**, 498–499 (in Russian).

Estimates of the minimal number of elements of ϵ-nets in various function classes and their application to the problem of representing functions of several variables by superpositions of functions of a smaller number of variables, *Dokl. Akad. Nauk SSSR* **101**, 192-194 (in Russian).

1956

On Skorokhod convergence, *Teor. Veroyatnost. i Primenen.* **1, 239–247 (in Russian).

Two uniform limit theorems for sums of independent terms, *Teor. Veroyatnost. i Primenen.* **1, 426–436 (in Russian).

**Zufällige Funktionen und Grenzverteilugssätze, In: *Bericht über die Tagung Wahrscheinlichkeitsrechnung und mathematische Statistik*, Berlin, pp. 113–126 (in collaboration with Yu.V. Prokhorov).

* Some principal problems in the approximate or exact representation of functions of one or several variables, In: Proc. 3rd USSR Math. Congress, Vol. 2, Moscow Univ. Press, Moscow, pp. 28–29 (in Russian).

On the Shannon theory of information transmission in the case of continuous signals, *IEEE Trans. Inform. Theory*, **IT-2**, 102–108.

* On the representation of continuous functions of several variables as superpositions of continuous functions of a smaller number of variables, *Dokl. Akad. Nauk SSSR* **108**:2, 179–182 (in Russian).

On some asymptotic characteristics of totally bounded metric spaces, *Dokl. Akad. Nauk SSSR* **108**, 385–388 (in Russian).

***On a general definition of amount of information, *Dokl. Akad. Nauk SSSR* **111**, 745–748 (in Russian) (in collaboration with I.M. Gel'fand and A.M. Yaglom).

Probability theory, In: Mathematics, its content, methods, and significance, Vol. 2, USSR Acad. Sci. Press, Moscow, pp. 252–284 (in Russian).

S.M. Nikol'skii: on his 50th birthday, *Uspekhi Mat. Nauk* **11**:2, 239–244 (in Russian) (in collaboration with S.B. Stechkin).

E.E. Slutskii, In: *BSE*-2 **39**, 378 (in Russian).

N.V. Smirnov, In: *BSE*-2 **39**, 406 (in Russian).

1957

***Theory of information transmission, In: USSR Acad. Sci. Session on Scientific Problems in Automatization of Industry, 15–20 Oct., 1956: Plenary meetings, USSR Acad. Sci. Press, Moscow, pp. 66–69 (in Russian).

* On the representation of continuous functions of several variables as superpositions of continuous functions of one variable and addition, *Dokl. Akad. Nauk SSSR* **114**:5, 953–956 (in Russian).

On the foundation of the theory of real numbers, *Mat. Prosveshchenie* **2**, 169–173 (in Russian).

1958

On the profession of a mathematician (2nd ed.), Moscow Univ. Press, Moscow (in Russian).

***The amount of information and entropy for continuous distributions, In: Proc. 3rd USSR Math. Congress, Vol. 3, USSR Acad. Sci. Press, Moscow, pp. 300–320 (in Russian) (in collaboration with I.M. Gel'fand and A.M. Yaglom).

Sufficient statistic, In: *BSE*-2 **51**, 106 (in Russian).

Information, In: *BSE*-2 **51**, 129–130 (in Russian).

Cybernetics, In: *BSE*-2 **51**, 149–151 (in Russian).

A new metric invariant of transitive dynamical systems and automorphisms of Lebesgue spaces, *Dokl. Akad. Nauk SSSR* **119, 861–864 (in Russian).

Sur les propriétés des fonctions de concentrations de M.P. Lévy, *Ann. Inst. Henri Poincaré* **16:1, 27–34.

* On the linear dimension of topological vector spaces, *Dokl. Akad. Nauk SSSR* **120**, 239–241 (in Russian).

On the definition of an algorithm, *Uspekhi Mat. Nauk* **13**:4, 3–28 (in Russian) (in collaboration with V.A. Uspenskii).

1959

On entropy in unit time as a metric invariant of automorphisms, *Dokl. Akad. Nauk SSSR* **124**, 754–755 (in Russian).

***ϵ-entropy and ϵ-capacity of sets in function spaces, *Uspekhi Mat. Nauk* **14**:2, 8–86 (in Russian) (in collaboration with V.M. Tikhomirov).

**Transition of branching processes into diffusion processes and related problems of genetics, *Teor. Veroyatnost. i Primenen.* 4, 233–236 (in Russian).

Notes on the works of R.A. Minlos and V.V. Sazonov, *Teor. Veroyatnost. i Primenen.* **4**, 237–239 (in Russian).

Probability theory, In: *Mathematics in the USSR during 40 years, Vol.* 1, Fizmatgiz, Moscow, pp. 781–795 (in Russian).

Foreword to the Russian translation of the book by W.R. Ashby: Introduction to cybernetics, Inost. Lit., Moscow (in Russian).

1960

On Bernshtein's work in probability theory, *Teor. Veroyatnost. i Primenen.* **5**, 215–221 (in Russian) (in collaboration with O.V. Sarmanov).

On the classes $\Phi^{(n)}$ of Ford and Blanc-Lapierre, *Teor. Veroyatnost. i Primenen* **5, 373 (in Russian).

Random functions of several variables, almost all realizations of which are periodic, *Teor. Veroyatnost. i Primenen.* **5, 374 (in Russian).

On N.V. Smirnov's work in probability theory: on his 60th birthday, *Teor. Veroyatnost. i Primenen.* **5**, 436–440 (in Russian) (in collaboration with B.V. Gnedenko, Yu.V. Prokhorov and O.V. Sarmanov).

On strong mixing conditions for Gaussian stationary processes, *Teor. Veroyatnost. i Primenen.* **5, 222–227 (in Russian) (in collaboration with Yu.A. Rozanov).

A.Ya. Khinchin, *Uspekhi Mat. Nauk* **15**:4, 97–110 (in Russian) (in collaboration with B.V. Gnedenko).

On the profession of a mathematician, 3rd ed., Moscow Univ. Press, Moscow (in Russian).

Elements of the theory of functions and functional analysis, No. 2. Measure, Lebesgue integral and Hilbert space, Moscow Univ. Press, Moscow (in Russian) (in collaboration with S.V. Fomin).

1961

Automata and life, *Mash. Perevod i Prikl. Lingvistika* **9**, 3–8 (in Russian).

Automata and life, *Tekhn. Molodezhi*, No. 10, 16–19; No. 11, 30–33 (in Russian).

Note on the paper by V.K. Lezerson, *Teor. Veroyatnost. i Primenen.* **6**, 367 (in Russian).

Properties of inequalities and the notions of approximate computations, In: *Questions of mathematics teaching in school*, Uchpedgiz, Moscow (in Russian) (in collaboration with P.S. Aleksandrov).

1962

On an estimate of parameters of a complex stationary Gaussian Markov process, *Dokl. Akad. Nauk SSSR* **146, 747–750 (in Russian) (in collaboration with M. Arato and Ya.G. Sinai).

The Rhythm of Mayakovskii's poem, *Vopros. Yazykoznaniya*, No. 3, 62–74 (in Russian) (in collaboration with A.M. Kondratov).

On a probabilistic problem in optimal control, *Dokl. Akad. Nauk SSSR* **145**, 993–995 (in Russian) (in collaboration with L.S. Pontryagin and E.F. Mishchenko).

On the work of B.V. Gnedenko in probability theory, *Teor. Veroyatnost. i Primenen.* **7**, 323–329 (in Russian).

A refinement of the conception of the local structure of turbulence in an incompressible viscous fluid at large Reynolds numbers, In: *Mécanique de la turbulence*: Colloq. Intern. CNRS, Marseille, Août–Sept, 1961, Paris, 447–458.

A refinement of the conception of the local structure of turbulence in an incompressible viscous fluid at large Reynolds numbers, *J. Fluid Mech.* **13**:1, 82–85.

1963

On approximating the distribution of a sum of independent terms of infinitely divisible distributions, *Trudy Moskov. Mat. Obshch.* **12, 437–451 (in Russian).

Discrete automata and finite algorithms, In: Proc. 4th USSR Math. Congress, Vol. 1, Leningrad Univ. Press, Leningrad, p. 120 (in Russian).

**Various approaches to estimating the difficulty of approximately specifying and computing functions, In: *Proc. Internat. Congress Math. Stockholm*, pp. 369–376 (in Russian).

On tables of random numbers, *Ind. J. Stat. Ser. A* **25:4, 369–376.

On the down-to-earthness of modern Russian poetry: a general characteristic, *Vopros. Yazykoznaniya*, No. 6, 84–95 (in Russian) (in collaboration with A.V. Prokhorov); On the study of Mayakovskii's rhythm, No. 4, 64–71 (in Russian).

Statistics and probability theory in the study of Russian poetry compositions, In: Abstracts and annotations symp. on complex investigation of artistic creative work, Nauka, Leningrad, p. 23 (in Russian) (in collaboration with A.V. Prokhorov).

How I became a mathematician, *Ogonek*, No. 48, 12–13 (in Russian).

Foreword to the Russian translation of the book by C.A. Shannon: *Works on information theory and cybernetics*, Inost. Lit., Moscow (in Russian).

1964

On the down-to-earthness of modern poetry: a statistical characteristic of this with Mayakovskii, Bagritskii, Akhmatova, *Vopros. Yazykoznaniya*, No. 1, 75–94 (in Russian) (in collaboration with A.V. Prokhorov).

On the metre in Pushkin's 'Songs for Eastern Farmers', *Rus. Lit.*, No. 1, 98–111 (in Russian).

Foreword to the Russian translation of the book by W. Feller: *Introduction to probability theory and its applications*, 2nd ed., Mir, Moscow (in Russian).

1965

***Three approaches to the definition of "amount of information", *Problemy Peredachi Informatsii* **1**:1, 3–11 (in Russian).

Geometric education in geometry courses at school, *Mat. v Shkole*, No. 2, 24–29 (in Russian).

On the contents of the school course of mathematics, *Mat. v Shkole*, No. 4, 53–62 (in Russian) (in collaboration with I.M. Yaglom).

Functions, graphs, continuous functions, *Mat. v Shkole*, No. 6, 12–21 (in Russian).

Notes on the analysis of the rhythm of Mayakovskii's "Verses on a Soviet passport", *Vopros. Yazykoznaniya*, No. 3, 70–75 (in Russian).

1966

Introduction to analysis, Moscow Univ. press, Moscow (in Russian).

* P.S. Aleksandrov and the theory of δs-operations, *Uspekhi Mat. Nauk* **21**:4, 275–278 (in Russian).

On the school definition of identity, *Mat. v Shkole*, No. 2, 33–35 (in Russian).

Geometry on the sphere and geology, *Nauka i Zhizn'*, No. 2, 32 (in Russian).

1967

***On the realization of sets in three-dimensional space, *Probl. Kibnernetiki* **19**, 261–268 (in Russian) (in collaboration with Ya.M. Bardzin').

New programmes and some basic questions on the advanced course in mathematics at secondary school, *Mat. v Shkole*, No. 2, 4–13 (in Russian).

1968

Some theorems on algorithmic entropy and algorithmic amount of information, *Uspekhi Mat. Nauk* **21**:2, 201 (in Russian).

On the study of the exponential function and logarithms in the 8-year school, *Mat. v Shkole*, No. 2, 23–25 (in Russian).

Generalizations of the notion of power and exponential functions, *Mat. v Shkole*, No. 1, 24–32 (in Russian).

Introduction to probability theory and combinatorics, *Mat. v Shkole*, No. 2, 63–72 (in Russian).

Elements of the theory of functions and functional analysis, (2nd ed. (revised and extended)), Fizmatgiz, Moscow (in Russian) (in collaboration with S.V. Fomin).

On the foundations of classical Russian metre, In: *Cooperation of sciences and mystery of creative work*, Iskusstvo, Moscow, pp. 397–432 (in Russian) (in collaboration with A.V. Prokhorov).

Example of the study of metre and its rhythmic variants, In: *Prosody*, Nauka, Leningrad, pp. 145–167 (in Russian).

1969

***On the logical foundations of information theory and probability theory, *Problemy Peredachi Informatsii* **5**:3, 3–7 (in Russian).

S.N. Bernshtein, *Uspekhi Mat. Nauk* **24**:3, 211–218 (in Russian).

Scientific foundation for a school course in mathematics, *Mat. v Shkole*, No. 3, 12–17; No. 5, 8–17 (in Russian).

New material in school mathematics, *Nauka i Zhizn'*, No. 3, 62–66 (in Russian).

1970

Z.V. Nikolaevich: to the memory of a mathematician who died in World War II, *Uspekhi Mat. Nauk* **25**:3, 243 (in Russian).

G.A. Seliverstov, *Uspekhi Mat. Nauk* **25**:3, 244–245 (in Russian).

Generalizations of the notion of number, a non-negative rational number, *Mat. v Shkole*, No. 4, 27–32 (in Russian).

What is a function?, *Kvant*, No. 1, 27–36 (in Russian).

What is the graph of a function?, *Kvant*, No. 2, 3–13 (in Russian).

Tilings from regular polyhedra, *Kvant*, No. 3, 24 (in Russian).

1971

Quantity, In: *BSE*-3 **4**, 456–457 (in Russian).

N. Wiener, In: *BSE*-3 **5**, 72 (in Russian).

D. Hilbert, In: *BSE*-3 **6**, 519 (in Russian).

On a system of basic notions and notation for a school course of mathematics, *Mat. v Shkole*, No. 2, 17–22 (in Russian).

Modern mathematics and mathematics at modern school, *Mat. v Shkole*, No. 6, 2–3 (in Russian).

A course of mathematics for physico-mathematical schools, Moscow Univ. Press, Moscow (in Russian) (in collaboration with V.A. Gusev, A.B. Sosinskii, and A.A. Shershevskii).

Summer School at Rubskoe Lake, Prosveshchenie, Moscow (in Russian).

1972

Complexity of specification and complexity of construction of mathematical objects, *Uspekhi Mat. Nauk* **27**:2, 159 (in Russian).

Integral, In: *BSE*-3 **10**, 300–302 (in Russian).

Method of Exhaustion, In: *BSE*-3 **10**, 586 (in Russian).

Elements of the theory of functions and functional analysis (3rd ed. (revised)), Fizmatgiz, Moscow (in Russian) (in collaboration with S.V. Fomin).

Scientific Handbook, In L. Neiman, *Delight of discovery*, Detskaya Literatura, Moscow, 160–164 (in Russian).

There is no substitute for teachers, *Koms. Pravda* (19 January, 1972) (in Russian).

A qualitative study of mathematical models of population dynamics, *Probl. Kibernetiki* **25**:2, 101–106 (in Russian).

1973

Continuum, In: *BSE*-3 **13**, 64 (in Russian).

Semilogarithmic and logarithmic sets, *Kvant*, No. 4, 2 (in Russian).

On the profession of the mathematician, *Kvant*, No. 4, 12 (in Russian).

1974

Basic notions of probability theory (2nd ed.), Nauka, Moscow (in Russian).

To the memory of I.G. Petrovskii (18 Jan 1901–15 Jan. 1974), *Trudy Moskov. Mat. Obshch.* **31**, 5–16 (in Russian) (in collaboration with P.S. Aleksandrov and O.A. Oleinik).

I.G. Petrovskii, *Uspekhi Mat. Nauk* **29**:2, 3–5 (in Russian).

A.A. Markov, In: *BSE*-3 **15**, 379 (in Russian).

Mathematics, In: *BSE*-3 **15**, 467–478 (in Russian).

Mathematical statistics, In: *BSE*-3 **15**, 480–484 (in Russian) (in collaboration with Yu.V. Prokhorov).

Multidimensional space, In: *BSE*-3 **16**, 372 (in Russian).

Orientation, In: *BSE*-3 **18**, 508–510 (in Russian).

Concern about adequate stock replacement, *Vestn. Vyssh. Shkoly*, No. 6, 26–33 (in Russian) (in collaboration with I.T. Tropin and K.V. Chernyshev).

New programmes: specialized schools, In: *Mathematics education today*, Prosveshchenie, Moscow, pp. 5–12 (in Russian).

The sieve of Eratosthenes, *Kvant*, No. 10, 2 (in Russian).

1975

Statistical quality control, In: *BSE*-3 **20** 572–573 (in Russian) (in collaboration with Yu.K. Belyaev).

Elements of Combinatorics, *Mat. v Shkole*, No. 2, 16–25 (in Russian).

Real numbers, infinite sequences and their limits, *Mat. v Shkole*, No. 2, 25–35 (in Russian) (in collaboration with O.S. Ivashev-Musatov).

1976

Elements of the theory of functions and functional analysis (4th ed. (revised)), Nauka, Moscow (in Russian) (in collaboration with S.V. Fomin).

Trigonometric functions, their graphs and derivatives, tenth grade school fashion, *Mat. v Shkole*, No. 1, 10–25 (in Russian) (in collaboration with S.I. Shvartsburd).

Moscow 38th Math. Olympiad, *Mat. v Shkole*, No. 4, 68–72 (in Russian) (in collaboration with G.A. Gal'perin).

Integral, tenth grade school fashion, *Mat. v Shkole*, No. 6, 15–17 (in Russian).

Transformation groups, *Kvant*, No. 10, 2–5 (in Russian).

1977

Infinity, In: *Math. Encyclopedia*, Vol. 1, pp. 455–458 (in Russian).

Quantity, In: *Math. Encyclopedia*, Vol. 1, pp. 651–653 (in Russian).

Probability, In: *Math. Encyclopedia*, Vol. 1, pp. 667–669 (in Russian).

The physics-mathematics school at MGU, *Kvant*, No. 1, 56–57 (in Russian) (in collaboration with V.V. Vavilov).

1978

On education on the levels of mathematics and physics of a dialectical-materialistic world view, *Mat. v Shkole*, No. 3, 6–9 (in Russian).

**Estimates of the spectral functions of random processes, In: 11th European Conf. Statistics, *Abstr. of Papers*. 14–18 August, 1978, Oslo, p. 36 (in Russian) (in collaboration with I.G. Zhurbenko).

1979

Linear sample estimates of sums, *Teor. Veroyatnost. i Primenen* **24**, 241–251 (in Russian) (in collaboration with A.V. Bulinskii).

Fiz.-Mat. Shkola at MGU: 15 years, *Kvant*, No. 1, 55–57 (in Russian) (in collaboration with V.V. Vavilov and I.T. Tropin).

1980

The dialectical-materialistic world view in school courses of mathematics and physics, *Kvant*, No. 4, 15–18 (in Russian).

1981

Elements of the theory of functions and functional analysis (5th ed.), Nauka, Moscow (in Russian) (in collaboration with S.V. Fomin).

Physico-mathematical school at Moscow University, Moscow Univ. Press, Moscow, (in Russian); (*Mat. Kibernetika*, No. 5) (in collaboration with V.V. Vavilov and I.T. Tropin).

Geometry for 6–8*th forms: Textbook* (3rd ed.), Prosveshchenie, Moscow (in Russian) (1st ed. — Moscow, 1979) (in collaboration with A.F. Semenov and P.S. Cherkasov).

On the notion of vector in courses of secondary school, *Mat. v Shkole*, No. 3, 7–8 (in Russian).

1982

Introduction to mathematical logic, Moscow Univ. Press, Moscow (in Russian) (in collaboration with A.G. Dragalin).

Introduction to probability theory, Nauka, Moscow (in Russian); (*Bibliotechka "Kvant"*, No. 23) (in collaboration with I.G. Zhurbenko and A.V. Prokhorov).

Mathematics, In: *Math. Encyclopedia*, Vol. 3, pp. 560–564 (in Russian).

Mathematical statistics, In: *Math. Encyclopedia*, Vol. 3, pp. 576–581 (in Russian) (in collaboration with Yu.V. Prokhorov).

Newton and modern mathematical thought, *Mat. v Shkole*, No. 6, 58 (in Russian).

1983

Algebra and elements of calculus: Textbook for the 9th and 10th forms (4th ed.), Prosveshchenie, Moscow (in Russian) (in collaboration with A.M. Abramov, B.E. Veits, O.S. Ivashev-Musatov and S.I. Shvartsburd) (1st ed. — 1980).

On the science textbook "Geometry 6–10", *Mat. v Shkole*, No. 2, 45 (in Russian).

Combinatorial foundations of information theory and computation of probabilities, *Uspekhi Mat. Nauk* **38**:4, 27–36 (in Russian).

Lightning Source UK Ltd.
Milton Keynes UK
UKOW05f2130230417
299725UK00001B/1/P

9 789401 053471